T0215697

More information about this series at http://www.springer.com/series/7409

Zhisheng Huang · Wouter Beek ·
Hua Wang · Rui Zhou ·
Yanchun Zhang (Eds.)

Web Information Systems Engineering – WISE 2020

21st International Conference
Amsterdam, The Netherlands, October 20–24, 2020
Proceedings, Part II

 Springer

Editors
Zhisheng Huang
VU Amsterdam
Amsterdam, The Netherlands

Wouter Beek ⓘ
VU Amsterdam
Amsterdam, The Netherlands

Hua Wang ⓘ
Victoria University
Melbourne, VIC, Australia

Rui Zhou ⓘ
Swinburne University of Technology
Hawthorn, VIC, Australia

Yanchun Zhang ⓘ
Victoria University
Melbourne, VIC, Australia

ISSN 0302-9743   ISSN 1611-3349   (electronic)
Lecture Notes in Computer Science
ISBN 978-3-030-62007-3   ISBN 978-3-030-62008-0   (eBook)
https://doi.org/10.1007/978-3-030-62008-0

LNCS Sublibrary: SL3 – Information Systems and Applications, incl. Internet/Web, and HCI

This Springer imprint is published by the registered company Springer Nature Switzerland AG
The registered company address is: Gewerbestrasse 11, 6330 Cham, Switzerland

Preface

Welcome to the proceedings of the 21st International Conference on Web Information Systems Engineering (WISE 2020), held in Amsterdam and Leiden, The Netherlands, during October 20–24, 2020. The series of WISE conferences aims to provide an international forum for researchers, professionals, and industrial practitioners to share their knowledge in the rapidly growing area of web technologies, methodologies, and applications. The first WISE event took place in Hong Kong, China (2000). Then the trip continued to Kyoto, Japan (2001); Singapore (2002); Rome, Italy (2003); Brisbane, Australia (2004); New York, USA (2005); Wuhan, China (2006); Nancy, France (2007); Auckland, New Zealand (2008); Poznan, Poland (2009); Hong Kong, China (2010); Sydney, Australia (2011); Paphos, Cyprus (2012); Nanjing, China (2013); Thessaloniki, Greece (2014); Miami, USA (2015); Shanghai, China (2016); Pushchino, Russia (2017); Dubai, UAE (2018); Hong Kong, China (2019); and this year, WISE 2020 was held in Amsterdam and Leiden, The Netherlands, supported by Vrije Universiteit Amsterdam.

A total of 190 research papers were submitted to the conference for consideration, and each paper was reviewed by at least three reviewers. Finally, 37 submissions were selected as regular papers (with an acceptance rate of 20% approximately), plus 44 as short papers. The research papers cover the areas of network embedding, graph neural network, social network, graph query, knowledge graph and entity linkage, spatial temporal data analysis, service computing, cloud computing, information extraction, text mining, security and privacy, recommender systems, database system, workflow, and data mining and applications.

In addition, special thanks are due to the members of the International Program Committee and the external reviewers for a rigorous and robust reviewing process. We are also grateful to Vrije Universiteit Amsterdam, Springer Nature, Atlantis Press, Ztone, Triply, and the International WISE Society for supporting this conference. The WISE Organizing Committee is also grateful to the workshop organizers for their great efforts to help promote web information system research to broader domains.

We expect that the ideas that have emerged in WISE 2020 will result in the development of further innovations for the benefit of scientific, industrial, and social communities.

October 2020

Zhisheng Huang
Wouter Beek
Hua Wang
Rui Zhou
Yanchun Zhang

Organization

General Co-chairs

Yanchun Zhang   Victoria University, Australia
Frank van Harmelen   Vrije Universiteit Amsterdam, The Netherlands
Marek Rusinkiewicz   New Jersey Institute of Technology, USA

Program Co-chairs

Zhisheng Huang   Vrije Universiteit Amsterdam, The Netherlands
Wouter Beek   Triply, The Netherlands
Hua Wang   Victoria University, Australia

Workshop Co-chairs

Rui Zhou   Swinburne University of Technology, Australia
Haiyuan Wang   Ztone Beijing, China

Tutorial and Panel Co-chairs

Kamal Karlapalem   International Institute of Information Technology, India
Yunjun Gao   Zhejiang University, China

Industry Chair

Wouter Beek   Triply, The Netherlands

Demo Chair

Yi Cai   South China University of Technology, China

Sponsor Chair

Ran Dang   Atlantis Press, France

Finance Chair

Qing Hu   Ztone International BV, The Netherlands

Local Arrangement Co-chairs

Ting Liu	Vrije Universiteit Amsterdam, The Netherlands
Xu Wang	Vrije Universiteit Amsterdam, The Netherlands
Wei Wei	Holiday Inn Leiden, The Netherlands

Publication Chair

Rui Zhou   Swinburne University of Technology, Australia

Publicity Co-chairs

Panagiotis Bouros	Johannes Gutenberg University of Mainz, Germany
An Liu	Soochow University, China
Wen Hua	The University of Queensland, Australia

Website Co-chairs

Haiyuan Wang	Ztone Beijing, China
Di Wang	Ztone Beijing, China

WISE Steering Committee Representative

Qing Li   The Hong Kong Polytechnic University, Hong Kong

Program Committee

Karl Aberer	EPFL, Switzerland
Marco Aiello	University of Stuttgart, Germany
Bernd Amann	LIP6, Sorbonne Université, France
Chutiporn Anutariya	Asian Institute of Technology, Thailand
Nikolaos Armenatzoglou	Amazon, USA
Wouter Beek	Vrije Universiteit Amsterdam, The Netherlands
Devis Bianchini	University of Brescia, Italy
Xin Cao	University of New South Wales, Australia
Jinli Cao	La Trobe University, Australia
Tsz Nam Chan	The Hong Kong Polytechnic University, Hong Kong
Jiefeng Cheng	The Chinese University of Hong Kong, Hong Kong
Dickson K. W. Chiu	The University of Hong Kong, Hong Kong
Dario Colazzo	LAMSADE, Université Paris-Dauphine, France
Alexandra Cristea	Durham University, UK
Valeria De Antonellis	University of Brescia, Italy
Anton Dignös	Free University of Bozen-Bolzano, Italy
Lei Duan	Sichuan University, China
Yixiang Fang	University of New South Wales, Australia
Yunjun Gao	Zhejiang University, China

Daniela Grigori	LAMSADE, Université Paris-Dauphine, France
Tobias Grubenmann	University of Bonn, Germany
Hakim Hacid	Zayed University, UAE
Jiafeng Hu	Google, USA
Haibo Hu	The Hong Kong Polytechnic University, Hong Kong
Zhisheng Huang	Vrije Universiteit Amsterdam, The Netherlands
Xin Huang	Hong Kong Baptist University, Hong Kong
Jyun-Yu Jiang	University of California, Los Angeles, USA
Panos Kalnis	King Abdullah University of Science and Technology, Saudi Arabia
Verena Kantere	University of Ottawa, Canada
Georgia Kapitsaki	University of Cyprus, Cyprus
Panagiotis Karras	Aarhus University, Denmark
Kyoung-Sook Kim	National Institute of Advanced Industrial Science and Technology (AIST), Japan
Hong Va Leong	The Hong Kong Polytechnic University, Hong Kong
Jianxin Li	Deakin University, Australia
Hui Li	Xiamen University, China
Kewen Liao	Australian Catholic University, Australia
An Liu	Soochow University, China
Guanfeng Liu	Macquarie University, Australia
Siqiang Luo	Harvard University, USA
Fenglong Ma	Penn State University, USA
Hui Ma	Victoria University of Wellington, New Zealand
Jiangang Ma	Federation University, Australia
Yun Ma	City University of Hong Kong, Hong Kong
Abyayananda Maiti	Indian Institute of Technology Patna, India
Silviu Maniu	Université Paris-Sud, France
Yannis Manolopoulos	Open University of Cyprus, Cyprus
George Papastefanatos	Information Management Systems Institute, Athena Research Center, Greece
Kostas Patroumpas	Information Management Systems Institute, Athena Research Center, Greece
Dimitris Plexousakis	Institute of Computer Science, FORTH, Greece
Nicoleta Preda	Université Paris-Saclay, France
Dimitris Sacharidis	Vienna University of Technology, Austria
Heiko Schuldt	University of Basel, Switzerland
Caihua Shan	The University of Hong Kong, Hong Kong
Jieming Shi	National University of Singapore, Singapore
Kostas Stefanidis	University of Tampere, Finland
Stefan Tai	TU Berlin, Germany
Bo Tang	Southern University of Science and Technology, China
Chaogang Tang	China University of Mining and Technology, China
Xiaohui Tao	University of Southern Queensland, Australia
Dimitri Theodoratos	New Jersey Institute of Technology, USA
Hua Wang	Victoria University, Australia

Jin Wang	University of California, Los Angeles, USA
Lizhen Wang	Yunnan University, China
Haiyuan Wang	Beijing University of Technology, China
Hongzhi Wang	Harbin Institute of Technology, China
Xin Wang	Tianjin University, China
Shiting Wen	Zhejiang University, China
Mingjun Xiao	University of Science and Technology of China, China
Carl Yang	University of Illinois at Urbana-Champaign, USA
Zhenguo Yang	City University of Hong Kong, Hong Kong
Hongzhi Yin	University of Queensland, Australia
Sira Yongchareon	Auckland University of Technology, New Zealand
Demetrios Zeinalipour-Yazti	University of Cyprus, Cyprus
Yanchun Zhang	Victoria University, Australia
Detian Zhang	Jiangnan University, China
Jilian Zhang	Jinan University, China
Yudian Zheng	Twitter, USA
Rui Zhou	Swinburne University of Technology, Australia
Lihua Zhou	Yunnan University, China
Feida Zhu	Singapore Management University, Singapore
Yi Zhuang	Zhejiang Gongshang University, China

Additional Reviewers

Olayinka Adeleye	Haridimos Kondylakis	Xiangchen Song
Ebaa Alnazer	Kyriakos Kritikos	Wenya Sun
Yixin Bao	Christos Laoudias	Xiangguo Sun
Dong Chen	Rémi Lebret	Pei-Wei Tsai
Hongxu Chen	Bing Li	Fucheng Wang
Tong Chen	Veronica Liesaputra	Jun Wang
Xuefeng Chen	Ting Liu	Qinyong Wang
Constantinos Costa	Tong Liu	Xu Wang
Alexia Dini Kounoudes	Xin Liu	Xin Xia
Thang Duong	Steven Lynden	Yuxin Xiao
Vasilis Efthymiou	Stavros Maroulis	Yanchao Tan
Negar Foroutan	Grace Ngai	Costas Zarifis
Eugene Fu	Rebekah Overdorf	Chrysostomos Zeginis
Ilche Georgievski	Xueli Pan	Shijie Zhang
Yu Hao	Jérémie Rappaz	Zichen Zhu
Wenjie Hu	Mohammadhadi Rouhani	Nikolaos Zygouras
Xiaojiao Hu	Brian Setz	
Firas Kassawat	Panayiotis Smeros	

Contents – Part II

Database System and Workflow

Data Mining and Applications

Contents – Part I

Social Network

Graph Query

Knowledge Graph and Entity Linkage

Service Computing and Cloud Computing

Information Extraction

TaskGenie: Crowd-Powered Task Generation for Struggling Search

Luyan Xu[1]([✉])(ID), Xuan Zhou[2](ID), and Ujwal Gadiraju[3](ID)

[1] Renmin University of China, Beijing 100872, China
xuluyan@ruc.edu.cn
[2] East China Normal University, Shanghai 200062, China
zhou.xuan@outlook.com
[3] Delft University of Technology, Delft, The Netherlands
u.k.gadiraju@tudelft.nl

Abstract. Search tasks provide a medium for the evaluation of system performance and the underlying analytical aspects of IR systems. Researchers have recently developed new interfaces or mechanisms to support vague information needs and struggling search. However, little attention has been paid to the generation of a unified task set for evaluation and comparison of search engine improvements for struggling search. Generation of such tasks is inherently difficult, as each task is supposed to trigger struggling and exploring user behavior rather than simple search behavior. Moreover, the everchanging landscape of information needs would render old task sets less ideal if not unusable for system evaluation. In this paper, we propose a task generation method and develop a crowd-powered platform called *TaskGenie* to generate struggling search tasks online. Our experiments and analysis show that the generated tasks are qualified to emulate struggling search behaviors consisting of 'repeated similar queries' and 'quick-back clicks', etc. – tasks of diverse topics, high quality and difficulty can be created using this framework. For the benefit of the community, we publicly released the platform, a task set containing 80 topically diverse struggling search tasks generated and examined in this work, and the corresponding anonymized user behavior logs.

Keywords: Web search · User interaction · SERPs · User behavior

1 Introduction

Modern search engines are adequately equipped to help users in locating accurate information for well-defined queries. Nevertheless, web searchers still experience difficulty in finding relevant information when their information need is ill-defined, complex or complicated. Therefore, recent work has paid more attention to understand and support struggling search with an aim to help searchers cope with the entailing search difficulty. The notion of *struggling sessions* was first

© Springer Nature Switzerland AG 2020
Z. Huang et al. (Eds.): WISE 2020, LNCS 12343, pp. 3–20, 2020.
https://doi.org/10.1007/978-3-030-62008-0_1

formally introduced by Hassan et al. as "those search sessions where users experience difficulty locating required information" [12]. Task difficulty is the main factor that leads to struggling search [2,6]. *Struggling search tasks* are defined by "topically coherent sub-sessions in which searchers cannot immediately find sought information" [22]. Researchers focusing on user behavior analysis found that whether a user is handling a struggling search task can be identified and predicted through features of his/her search activities such as queries and post-query clicks [2,10]. Building on these outcomes, previous works proposed different support mechanisms and systems to detect and ameliorate struggling search cases [12,13,21,29].

However, little attention has been paid to generating a unified set of tasks in this area. Akin to the role that TREC datasets play in typical information retrieval (IR) research, the generation of struggling search tasks is particularly important for further development and unified evaluation of new techniques in struggling search. Nevertheless, the ever-changing landscape of complex information needs would render old task sets less ideal if not unusable for system evaluation. Currently, for task generation related to struggling search, researchers tend to create struggling tasks manually by increasing task complexity, e.g. *"There are five countries whose names are also carried by chemical elements. France has two (Ga–Gallium and Fr–Francium), ... Please name the left country"* [25]. Others adhere to small-scale situated lab experiments, e.g. *"You once heard that the Dave Matthews Band owns a studio in Virginia but you don't know the name of it. The studio is located outside of Charlottesville and it's in the mountains. What is the name of the studio?"* [2].

These methods require extensive experience and fertile imagination of researchers and since there is no common pattern to follow, these may lead to only small-sized task sets. Though studies have shown that small task sets could work well in some experimental lab studies [2,25], they are not sufficient for large-scale and robust system evaluation. The potential effect of participant fatigue limits laboratory experiments to a small number of topics and similar situated tasks, making the evaluation inclined to side with a subjective or biased perspective [27]. This dictates the need for a robust and cost-efficient method to generate struggling search tasks (SSTs) for evaluation. Crowdsourcing has been shown to be a powerful means for recruiting low-cost participants who are readily available around the clock [8,9]. This provides us with an alternative source of acquiring reliable human input. We thereby propose the use of crowdsourcing to generate SSTs.

Original Contributions. In this paper, we focus on struggling search that manifests in fact finding or checking tasks. We propose a crowd-powered task generation framework and develop an online platform[1] that can be used to generate high-quality SSTs at scale. This method leverages paraphrased (redundant) information in wikis, and decompose SST task generation into several low-effort steps, suitable for crowd workflows to create questions that are difficult and can

---

[1] http://waps.io/study/?uid=123.

simulate struggling search. This method can easily be applied to topically dedicated wikis (e.g. `wikinews` for news, `wikivoyage` for travel, etc.)[2], while in this paper we take English Wikipedia as the resource to generate a topically diverse set of SSTs. Applying crowd-workers, we generated 80 SSTs across diverse topics. Getting insights from previous studies [12,13,22], we evaluate the quality of these tasks by carrying out rigorous user-centric experiments, analyzing the characteristics of user behaviors elicited by these tasks. Results confirm the quality of the generated SSTs. We consolidated the tasks and publicly released the task set[3] containing 80 SSTs with difficulty level and success rate, which can help in developing and evaluating support mechanisms for users in struggling search. Also, we released the anonymized user logs gathered during task evaluation.

2   Related Literature

We discuss related work in the following areas: struggling search and task design for struggling search.

Struggling Search. Struggling search describes a situation whereby a searcher experience difficulty in finding the information they seek [12]. Within a single search, struggling could lead to frustrating of difficulty and dissatisfying search experiences, even if searchers ultimately meet their search objectives [11]. Characteristics of user behaviors have been used to identify whether a user was dealing with struggling search tasks – searchers dealing with a struggling search tasks can experience difficulty in locating required information, tend to issue multiple similar queries and conduct quick-back clicks as they are cycling on finding useful information [2,11]. Struggling search has been studied using a variety of experimental methods, including log analysis [22], laboratory studies [2], and crowdsourced games [1]. Hassan et al., studied how to detect and support struggling search by extracting search sessions from real user logs [13,22]; Aula et al. evaluated the influence of task difficulty on struggling search behaviors by setting up a small-scale lab experiment and an IR-based online study [2]. We evaluate the quality of generated tasks by analyzing the user behaviors elicited by the tasks, based on the behavior features that have been shown to be useful for identifying struggling search [11,13].

Task Design for Struggling Search. Researchers in sense-making found that users will suffer difficulty when there is an information gap between what they know and what they want to know [23], as they can seldom describe their questions clearly or find a way to get close to the answer. This sheds light on task design for struggling searching tasks; key information or the task solving strategy should not be directly given by the task. Also, it has been found that task complexity can increase the task difficulty thus affect learner perceptions of struggling [24]. On the other hand, difficulty of tasks has been viewed from both objective and subjective perspectives [18]. From the subjective perspective, the

---

[2] Wikinews: https://en.wikinews.org/; wikivoyage: https://www.wikivoyage.org/.
[3] Anonymized URL– https://github.com/sst20190816/WISE2020.

same task could be difficult and complex to one without background knowledge while be easy for the other who is an expert in the related domain [2,5]. To some extent this indicates that task design for struggling search should either try to avoid the cases that are highly influenced by domain knowledge or try to cover as many topics as possible. From the objective perspective, task difficulty can be related to task characteristics and independent of task performers, which has been supported by other works [6] – task with unknown goals, unexplored information space, accompanied by uncertainty and ambiguity would consequently mean that it could lead to a high task complexity, in turn resulting in users struggling [18]. Getting inspiration from previous work [30], we propose an online task generation framework for generating struggling search tasks at scale, covering various knowledge domains and are objectively difficult.

3   Task Generation Framework

3.1   Intuition and Method

We focus on a particular type of search tasks that exhibit search behavior suggestive of struggling – fact finding/checking tasks ("Looking for specific facts or pieces of information" [14]). Struggling search tasks (SST) differ from typical information retrieval tasks in that the typical informational search tasks are more like information locating problems which are well-defined, systematic and routine [28]. For example, consider the following struggling search task— *"Dave Matthews Band owns a studio in Virginia, the studio is located outside of Charlottesville and it's in the mountains. What is the name of the studio?"* [2]. Consider that the answer to this question does exist in the document collection, but it cannot be simply matched to search queries or resolved using the state-of-the-art information retrieval techniques. Rather, it can only be described using fragmented pieces of information and obtained by searchers through navigating and comprehending content within the information space. A searcher needs to collect relevant information from the documents, comprehend it, reason about it, and very often repeat the process for several rounds, until he/she reaches a conclusion with a certain confidence. This process involves information-seeking behavior, including searching, browsing, berry-picking and sense-making [20].

How Can We Easily Find or Frame Questions with Implicit Answers at Scale? In this paper, we leverage paraphrased sentences, which are abundant in common writings. To create a clear and logical flow while writing, an author tends to perform reasoning narratively. This naturally results in redundancy [7]. For instance, a statement following a causative sentence connector (i.e. a conjunctive adverb) [17], such as *"in other words"* or *"that is to say"*, is likely to be a paraphrase which repeats the same meaning of the former sentence(s) in a more colloquial manner [4]. In theory, the information conveyed by the paraphrased sentences can be recovered by a searcher who has read through the preceding content. Thus, removing the paraphrased sentence will not cause information loss. The sentences following such connecting phrases are typically declarative statements. It is therefore straightforward to turn them into questions, with the statement containing the answer.

For example, in Fig. 1, we can hide the underlined sentence and turn it into a question – *"Does Polypteridae belong to the Actinopteri?"* – (since 'Polypteridae' and 'Actinopteri' appear elsewhere in the article in different forms). By hiding the specific sentence that contains the answer, the answer will not be directly identifiable through information locating. A searcher may identify text fragments like 'Polypteridae' and 'Actinopteri' as their starting points. However, to understand their relation and answer the question, the searcher may need to know more and therefore be forced to explore the Web or Wikipedia further.

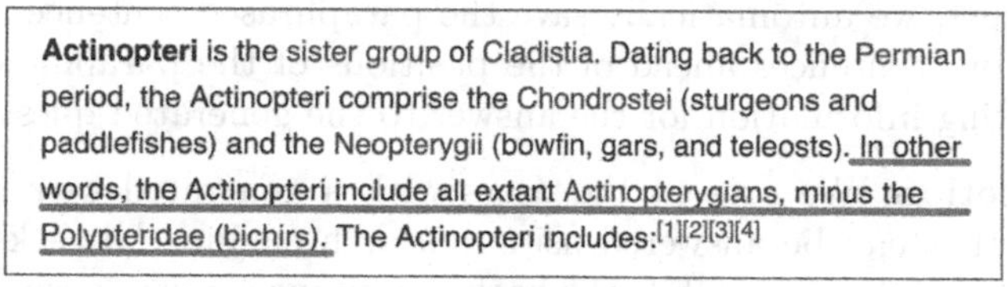

> **Actinopteri** is the sister group of Cladistia. Dating back to the Permian period, the Actinopteri comprise the Chondrostei (sturgeons and paddlefishes) and the Neopterygii (bowfin, gars, and teleosts). In other words, the Actinopteri include all extant Actinopterygians, minus the Polypteridae (bichirs). The Actinopteri includes:[1][2][3][4]

Fig. 1. Example of a paraphrased sentence in Wikipedia.

This inspires us to generate SSTs through the following steps:

1. Identify a paraphrased sentence;
2. Hide it from the document;
3. Create an informational question based on the given paraphrased sentence.

Since the answering sentence is hidden from the document, it is hard to obtain the answer through direct information locating; the paraphrased sentence usually lacks an accurate description or explanation of the entailing information points, a task generated based on it simulates a real-life situation where people have incomplete prior knowledge or means to meet their information need. This will elicit a searcher's struggling search behavior. The searcher may start from arbitrary documents that seem relevant, browse through parts or the whole collection, and reason about the possible answer. If the searcher is unfamiliar with the topic, he has to learn about it, since answering the question would require comprehension of related knowledge. Meanwhile, as the hidden sentence contains only redundant information, the searcher should be able to find the answer eventually.

3.2   TaskGenie – A Crowd-Powered Platform

Based on the task generation method, we built an online platform for task generation called *TaskGenie*, aiming to (i) generate struggling search tasks through crowdsourcing; (ii) study user behavior within the generated struggling search tasks. To this end, this platform serves in two phases: *Task Generation*, facilitating the creation of new struggling search tasks; and *Task Completion*, facilitating search experiment on solving the tasks.

Task Generation. For task generation, users are first guided to choose a conjunctive phrase from a drop-down list (e.g. *'in other words'*, *'that is to say'*). They are then presented with a filtered set of articles that contain (highlighted) statements with these conjunctive phrases. Users are asked to understand the highlighted sentence in the article context and grasp the information that the sentence contains. Finally, they are asked to create a question based on the paraphrased sentence, provide the answer and source page of the question. Assuming that a task generated from a paraphrased sentence is closely related to its surrounding context, we automatically save the paraphrased sentence and its context (i.e. the two sentences ahead of the positions of the paraphrase sentence) as the supporting information for the answer to the generated question.

Task Completion. We present the users with a generated task in the form of a question that can be answered using a search engine. All tasks are pulled randomly from our database while the background mechanism ensures each task is finally resolved equal times. Users can choose to change the task only once; if they do not like the task assigned to them. Users are tasked with finding the answer to the question by searching using our search engine. To ensure that the users are genuinely invested in reasoning, understanding and finding the correct answer and not merely guessing, we ask users to provide a justification in an open text field that supports their answer. Users are encouraged to copy-paste excerpts that provide evidence or justify their answers. Finally, we collect the users' opinions of the search task they completed from the following perspectives – (a) *Task Qualification* (whether or not the users found the question difficult in comparison to their usual experience of searching the Web or Wikipedia for answers); (b) *Task Difficulty Score* (how difficult/complex the users found the question to be). We divide the task difficulty scale into five equal parts using the following labels with corresponding score intervals on a sliding scale of 1 to 100 - *Easy (1–20)*, *Moderate (21–40)*, *Challenging (41–60)*, *Demanding (61–80)*, *Strenuous (81–100)*. Users could select the task difficulty level and indicate an exact score using the scrollbar. Next we asked the users to indicate the reasons due to which they found the question to be difficult, and provided options (using checkboxes) that were drawn from previous work analyzing struggling search [19]. To prevent forced choices in case users did not find the task to be difficult, they could select the checkbox with the label 'Not Difficult'.

3.3   System Implementation

Pluggable Web Search Engines. As a platform for task generation and evaluation, *TaskGenie* is designed to be compatible with main stream web search engines (e.g. Google, Bing) which provide a standardized search API. These search engines can be plugged into TaskGenie as a backend search system to support task generation and get evaluated in task completion. In this paper, Bing Web Search API is used in the experimental study.

Domain Controlling and User Activity Logging. *TaskGenie*, the search domain can easily be adjusted to support searching through different domains. For example, we set Wikipedia as the domain for task generation (i.e. get all the webpages containing paraphrased sentences from Wikipedia) and we set the entire web as the domain for task completion. During task generation and completion, we also logged worker activity on the platform including queries, clicks, key presses, etc. using PHP/Javascript and the jQuery library.

DOM Processing. During the task generation phase, it is useful to highlight paraphrased sentences to make it more convenient for searchers to locate a target sentence. During the task completion phase on the other hand, to emulate a struggling search situation, it is essential to hide the direct answers in the retrieved documents. So that in the two phases, we need to either highlight or hide the paraphrased sentences. Drawing inspiration from previous work[4], we implement this by filtering and manipulating DOM using Javascript. Given a retrieved webpage (DOM), we access all its child nodes recursively and match the regex of causative sentence connectors (*in other words* etc.) with the content of each node. The matched sentences are thereby either hidden or transformed into a different sentence according to their syntax.

4 Study Design

4.1 Task Generation

Wikipedia – Paraphrased Sentences. There are plenty of online archives or wikis. In this work, we choose Wikipedia as the source for our struggling search task generating framework, and *in other words* and *that is to say* as the conjunctive phrases to identify paraphrased sentences. Wikipedia is one of the richest sources of encyclopedic information on the Web, and generates a large amount of traffic. Prior work has highlighted the variety of factors that drive users to Wikipedia [26]. We explored the entire English Wikipedia (2018 version) and found 10,824 articles with on average one occurrence of the paraphrase *"in other words"*, and 2,195 articles with the paraphrase *"that is to say"*. Our findings suggest that Wikipedia is a good source for paraphrased sentences which can potentially serve in the creation of difficult search tasks across diverse topics.

Task Generation Experiment. We recruited 200 participants from Figure8[5], a premier crowdsourcing platform. At the onset, workers willing to participate were informed that the task entailed 'generating a task for others within the Wikipedia domain'. Workers were then redirected to the external platform, *TaskGenie*, where they completed the mission. We logged all worker activity

---

[4] https://j11y.io/snippets/.
[5] http://figure-eight.com/.

on the platform. During the *task generation* process, *TaskGenie* presents criteria to help a user control the quality of the generated question. We urge the users to ensure that (1) the selected sentence is a paraphrased sentence that contains enough information for creating a question; (2) they search for the answer on the Wikipedia to ensure that the generated question is challenging. This means that although the answer cannot be found easily, it can be eventually obtained through searching and exploring. We incentivize workers to strictly adhere to these criteria by rewarding workers with a post-hoc bonus payment if they successfully create a SST.

We restricted the participation to users from English-speaking countries to ensure that they understood the instructions adequately. On successfully creating a task, users received a mission completion code which they could then enter on the Figure8 platform to receive their monetary rewards. We compensated all users at an hourly rate of 7.5 USD ($\approx$1.5 USD and 12 mins per task).

Task Collection. To ensure the reliability of generated tasks, we filtered out workers in this phase using the following criteria:

i. Workers who did not follow the required syntax in creating a question in the *task generation*. Since the aim of this phase is to generate a readable question (we described the basic syntax of a question in our instructions), those who did not meet the criteria were discarded.
ii. Workers who create questions lacking a self-sufficient description in the way a question is phrased (for example, *"Reincarnation is possible?"*), and generate random questions ignoring the paraphrased sentence in the source page (for example, *"Is Wikipedia the best page to find anything?"*).

Using the aforementioned criteria, we filtered out 65 task generation cases, resulting 135 generated tasks. For the 135 generated tasks, we hired two students to search for the answer of each task on the web. We eliminated 55 tasks that either duplicated or for which the answer could be found within the two interactions with the search system. We finally got 80 tasks that qualified as struggling search tasks (SSTs).

4.2   Task Evaluation

To validate whether the generated tasks are struggling search tasks and are generally suitable for the study of struggling search, we conducted a web search experiment using the set of 80 generated tasks.

Through Figure8, we recruited 400 *Level-3 workers* (260 male and 140 female, with their age ranging from 18 to 57 years old). Workers willing to participate in the web-based task evaluation experiment were asked to "search for the answer of a given task" using our platform *Task-Genie: Task Completion*. For the web search experiment, we base the *TaskGenie* search system on top of the Bing Web Search API and extend the search domain to the entire web. We logged the user activity throughout the task completion. Using the task filtering criteria mentioned before, we filter out 31 spam workers who entered arbitrary strings for the answer or supporting information, and those who did not finish the experiment. Thus, the evaluation is based on the 369 valid search sessions.

5 Task Evaluation Results

In task evaluation, we validate the generated tasks are struggling search tasks by examining session-level features of search behaviors shown to be useful for identifying struggling search in previous work [12,22]: *topical characteristics, query characteristics, click characteristics* and *task difficulty.*

5.1 Topical Characteristics

We analyzed the topical distribution of the tasks and found that tasks in diverse topics could be generated through our task generation module. To categorized the generated tasks, we used the top-two-level categories of Curlie[6] (i.e. Open Directory Project; dmoz.org). Assuming that the topic of tasks is consistent with the topic of its wiki source pages, we categorize the generated tasks by analyzing the topics of their source wiki pages. To this end, we used an automatic url-based classifier [3] for topic categorization. We assigned the most frequently-occurring topic for the source web page as the topic of each generated task.

Figure 2 shows the prevalence of topics in the generated tasks. We note that the task generation domain we chose, Wikipedia, contains few articles which correspond to everyday activities. Thus, only a few generated tasks were about topics spanning our daily lives such as *Shopping, Entertainment*, etc. However, the generated tasks cover various topics.

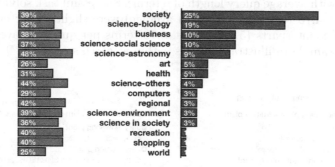

Fig. 2. Percentage of topics in generated tasks (gray color) and the corresponding success rate (green color) for each topic; category *'Science'* is further divided into second-level categories such as *'biology', 'astronomy'*, etc. (Color figure online)

Corresponding to each topic, we measured the success rate of tasks. For each generated task, we regard a corresponding answer is **successful** if: a searcher's

---

[6] http://curlie.org/.

answer is correct, and the searcher provides meaningful supporting information that corroborates the answer (i.e. the supporting information is semantically similar to the that given by the task creator). We evaluated the similarity between supporting information given by searchers and that given by the task creator using an automatic text-level similarity evaluation method [15]. Of all the search sessions across different topics in our set, around 37% correspond to successful cases, which is comparable lower to that observed from real user logs (i.e. 40% in [22]). As shown in Fig. 2, the success rate varied across the different topics, ranging from 25% in *world* to 48% in *science-astronomy*.

According to the type of answer that satisfies a given task, we further analyzed the generated tasks from two standpoints: *yes/no* tasks (37 in total, the answers to 19 of them are 'yes', the answers to 18 of them are 'no'), and *fact-finding* tasks (43 in total). Through a two-tailed T-test that compared the success rate across the two types of tasks, we did not find a significant difference. We also found no significant difference between tasks generated from "in other words" and those generated from "that is to say".

5.2   Query Characteristics

It has been found that searchers' struggling is reflected in their queries [2,11]. We examine the characteristics of queries elicited by the generated tasks focusing on the following features: **query features** (i.e. number of queries, query length), **query-transition features** (i.e. query similarity, query reformulation), which have been shown to be useful for determining struggling search sessions [12,13].

Query Features. Users in general issued more queries to handle a struggling search tasks [12,22]. On average, the generated tasks comprised 5 to 6 queries ($M = 5.55$) with average query length of 6 terms. Successful task solving sessions (5.48 queries, 4.76 terms per query on average) were slightly shorter than the unsuccessful counterparts (5.72 queries, 6.78 terms per query on average). We present an example to illustrate queries within a search session.

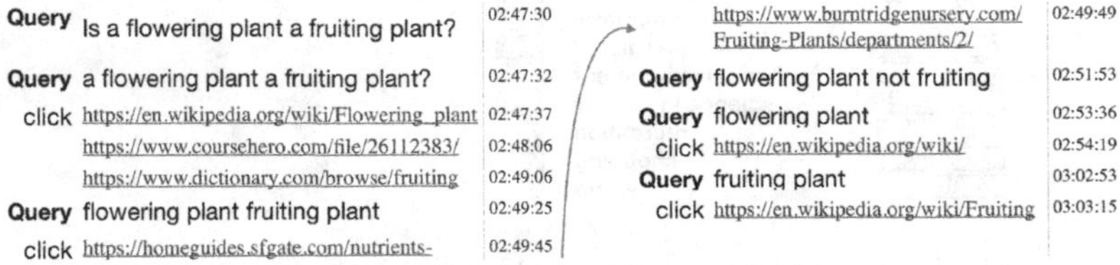

Query	Is a flowering plant a fruiting plant?	02:47:30		https://www.burntridgenursery.com/ Fruiting-Plants/departments/2/	02:49:49
Query	a flowering plant a fruiting plant?	02:47:32	Query	flowering plant not fruiting	02:51:53
click	https://en.wikipedia.org/wiki/Flowering_plant	02:47:37	Query	flowering plant	02:53:36
	https://www.coursehero.com/file/26112383/	02:48:06	click	https://en.wikipedia.org/wiki/	02:54:19
	https://www.dictionary.com/browse/fruiting	02:49:06	Query	fruiting plant	03:02:53
Query	flowering plant fruiting plant	02:49:25	click	https://en.wikipedia.org/wiki/Fruiting	03:03:15
click	https://homeguides.sfgate.com/nutrients-	02:49:45			

Fig. 3. Samples of search sessions in user logs

Figure 3 shows the sample process a searcher moved through a session to solve the task *"Is a flowering plant a fruiting plant?"*. We note that to solve a task generated in this work, a searcher generally issued even more queries with longer query length than the '3 to 4 queries averaging around 4 terms per query' observed from daily-life struggling search logs in previous work [22]. This difference may also be attributed to the difference in tasks that were studied. The information inquired by the generated informational tasks are more specific and difficult to resolve than the tasks studied in previous works (e.g. find a source-page of a video).

We observed that the first query in both successful and unsuccessful search sessions are typically the task description itself or an excerpt sentence extracted from the task description (8.93 terms on average) which are longer than the intermediate queries (5.81 terms on average) and the final queries (4.18 terms on average). Existing works show that there are generally two different cases that correspond to struggling with respect to the first query of a search session: (i) the query is too common as it is general and ambiguous, or (ii) the query is quite uncommon as it might be overly specified [22]. From this we note that the long over-specified first query does not lead searchers to a target page, and might elicit struggling search consequently. However, this struggling does not determine the final success or failure of the whole search session, which is consistent with the outcomes in prior work in [22].

Query Similarity. It has been shown that in a struggling search session the later queries can be quite similar to the initial query. Users experiencing the struggle tend to reformulate queries that closely resemble the initial search [12,22]. Based on prior works, we expect that in a struggling search task a user thinks of less diversified queries to explore alternatives. Thus in user logs, unique terms in the initial query persist through the future queries. To examine this, we measure the similarity between queries in the session. The similarity between any two queries Q_i and Q_j is computed using *Jaccard Index*:

$$\frac{|Q_i \cap Q_j|}{|Q_i| + |Q_j| - |Q_i \cap Q_j|} \tag{1}$$

where $|Q_i|$ is the number of unique terms in query Q_i, and $|Q_i \cap Q_j|$ is the number of matched terms in Q_i and Q_j.

Before measuring the similarity between queries in a session, we first normalize the queries; including lowercasing query text, deleting stop words, stemming, and unifying white space characters. For $|Q_i \cap Q_j|$, we consider two terms are matched if they are (i) **exact matched**: two queries match exactly; (ii) **approximate matched**: two queries match if the Jaro-Winkler distance (score) of them is larger than 0.6. In this work, we only consider the lexical-based query similarity. Assuming that for the concepts or information points in the generated tasks, users can seldom find alternative terms to search without learning through searching, we eliminate **semantic matched** cases (i.e. two queries match if semantic similarity of them over certain threshold [12]). Figure 4 shows the average similarity between queries to the first query. We found that in both successful and unsuccessful search sessions, searchers generally issue similar queries in the first three rounds. This is consistent with the outcomes in previous studies mentioned earlier [12,22]. We found that in successful sessions, queries gradually get less similar to the initial query as the searching progresses (though the difference was not found to be statistically significant using a two-tailed T-test at the 0.05 level). Prior work established that struggling searchers cycle through queries as they attempt to conceive a correct query to locate target information (i.e. the query similarity in struggling search sessions is generally greater than 0.4) [12]. Our findings corroborate that struggling search manifests during users' quest to satisfy the information need, even if they finally succeed in their search missions.

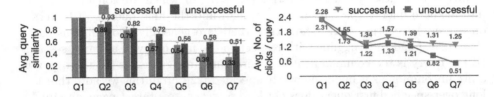

Fig. 4. Avg. query similarity in each step   **Fig. 5.** Avg no. of clicks per query

Query Reformulation. We delve into how users employ terms from one query to another in web search. We consider the three main query transition types which have been used in previous works [12]: **Term Addition:** ≥ 1 word added to the first query; **Term Removal:** ≥ 1 word removed from the first query; **Term Substitution:** ≥ 1 word substituted with other lexically matched terms. Term matching is done by using lexical matching described earlier.

We found that term removal is generally the most popular strategy; almost all the search sessions contain term removal cases. This can be explained by the task description that users consumed the information prior to beginning the search session. Due to the nature of Wikipedia, most generated tasks pertain to topics which people may not encounter in their daily life. Thus, we reason that most people struggled to come up with alternative terms to describe the vague information need in the tasks. In such cases, over 2 terms were removed on average in the last query ($M = 2.41, SD = 1.89$). The high standard deviation can be explained by differences between the generated tasks. For instance, a task with a long (short) information need description could elicit a long (short) initial query, finally converging to a few keywords. Term substitution occurs more frequently in successful sessions than in unsuccessful sessions (though not statistically significant, $p = 0.052$) which is consistent with previous work [12].

5.3   Clicks Characteristics

Prior works have shown that searchers experiencing 'struggle' tend to exhibit no click actions or *quick-back clicks* (i.e. result clicks with a dwell time less than 10 s [16]) after certain queries [2,12,22]. This has been attributed to the difficulty experienced in locating target information. We examine the characteristics of users' clicks on the SERPs in search sessions pertaining to the generated tasks.

On average, searchers exhibited 1.67 clicks after each query ($M = 1.67, SD = 1.49$), and over 62% of search sessions contain quick-back clicks. We further computed the average number of clicks for a sequence of queries in a session. Figure 5 shows the average change in the number of user clicks per query. We found that within the initial 4 queries there's no significant difference between successful and unsuccessful sessions in terms of the average number of clicks per query while the difference becomes more pronounced thereafter. Particularly, searchers in unsuccessful sessions issued less than 1 click on average after their last two queries. This is consistent with previous work, which also found that users in struggling search tasks tend to give up clicking on post-query results on the final query in an unsuccessful session [22]. From the click characteristics we find that solving the generated tasks, users are elicited with clicks in struggling, part of which could be the indicator of the eventual mission failure.

In contrast to our findings, Hassan et al. found that after several rounds of queries without locating any target information, struggling searchers tend to click on more results [12]. These contrasting findings can be explained by the difference of task types and difficulty levels. The generated tasks in our setup are generally fact-finding tasks with unambiguous final goals, while the tasks in previous works are more akin to open-ended exploratory tasks (e.g. 'software purchase advice', 'career development advice').

5.4   Task Difficulty Analysis

Corresponding to analysis of objective user behavior, we also investigate searchers' subjective perception of task difficulty. In general, participants scored the task difficulty as 57 on average ($M = 57, SD = 17$), which means tasks are in general *challenging* yet not *demanding*. We note that all participants agreed these tasks are much more difficult than the typical IR tasks. Among them 77% searchers thought the given tasks were more difficult in comparison to their general web search experience, rating task difficulty as 61 on average (i.e. *demanding*; $M = 61, SD = 13$).

Based on the reasons collected from previous work [19], we investigated the reasons why tasks made users perceive a 'struggling search' experience during web search through self-reports. Figure 6 illustrates the overall impact of different reasons that contribute to users experiencing a 'struggle' while completing the generated tasks across the entire web. We found that the top-3 main reasons cited for task difficulty were (1) *task complexity*, wherein workers believed that there were several components of the task that needed to be addressed; (2) *difficult to find useful pages*, wherein searchers met difficulties locating proper web pages to acquire target information; (3) *specific requirements*, wherein the struggle experience was due to the information need being so specific, consequently making it more difficult to find. While the reasons spread across various aspects including task features (40%), user aspects (26%), the interaction between user and system (24%), and the readability of documents (10%).

We found that within Wikipedia domain the paraphrased sentences are generally distributed across curated articles about history, literature, physics, biology,

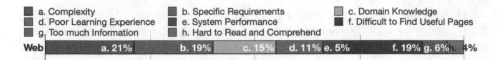

Fig. 6. Overview of the reasons why workers felt struggled in web search. Reasons are collected from 4 standpoints: task features (a, b), user aspects (c, d); user-system interaction (e, f); and document features (g, h).

etc., which people may not encounter in daily life. Thus, we observe the generated tasks correspond to subjective knowledge of users rather than more general scenarios that one may encounter in everyday life. This increases the task difficulty for most of the users; the information need of the generated tasks also requires users to process varied information from different perspectives. Moreover, self-reported difficulty reasons indicate that expanding the search domain increases the difficulty in locating useful pages to satisfy the information need (note that searchers were unaware of the fact that the source for all generated tasks was Wikipedia).

We also analyzed the influence of reasons on users' perception of struggling. Results of the generalized linear regression indicate that there was a collective significant effect between the reasons and users' perception of struggling in web search experiment ($\chi^2 = 83.1, p < .01$). The individual predictors were examined further and indicated that *complexity* ($t = 4.19, p < .001$), *specific requirements* ($t = 1.57, p < .05$), *domain knowledge* ($t = 2.03, p < .05$), *difficulty in finding useful pages* ($t = 6.88, p < .001$) and *too much information* ($t = 4.36, p < .001$) were significant predictors in the model, while searchers' *poor learning experience*, the *system performance*, and whether the *target document is hard to read* are not the key factors that influence users' struggling experience.

5.5   Publicly Released Task Set

For the benefit of the community, along with *TaskGenie* platform, we also publicly released the generated task set and user behavior logs (anonymized) gathered in our user study. We consolidated the 80 generated SSTs with different aspects including: question, answer, source page (i.e. suffixes of the sharing url "https://en.wikipedia.org/wiki/"), task type (i.e. "yes/no" or fact-finding), task topic (i.e. the ODP categories), task difficulty level (i.e. according to average difficulty score), and success rate. The complete task set is available online (the URL is provided in Introduction). This task set can be used to reliably simulate struggling search among users. For each task, we provide the basic success rate and task difficulty level that can be useful in the development and evaluation of methods to support users while they struggle in search tasks. Also, we provide the user behavior data collected in this work including queries, clicks, etc. Moreover, our proposed framework can be used to generate SSTs as per the topical/domain related needs at hand.

6   Discussion

Why We Need 'Humans'? Although paraphrased sentences are a good source to create difficult questions, framing these questions automatically is far more challenging due to the variety in paraphrased sentences and their context; existing methods cannot automatically generate SST tasks in this manner. Humans on the other hand, can easily identify those paraphrased sentences which are suitable for creation of SST tasks. *TaskGenie* allows us to collect and study user behavioral logs while they solve SST tasks, and also supports the generation of SST tasks. Note that *TaskGenie* can easily be customized to execute only a single phase (task completion or task generation) if desired.

Effects of the Document Collection. In this work, we chose Wikipedia as the domain for generating struggling search tasks. And for simplicity, we only considered paraphrased sentences using the conjunctions "in other words" and "that is to say" as the indicators for redundant information that is summarized. However, our framework can be easily customized to include other conjunctions concomitant with paraphrased sentences. We also showed that the generated tasks correspond to a variety of topics. Moreover, our framework can be readily used to generate SSTs for specific domains by depending on the corresponding wikis[7]. These include `WikiTravel` about traveling and places, `tvTrope` about television and movies, `WikiNews` about the news and events, etc. All these could be a potential source for paraphrased sentences. Thus, we argue that using this framework, a comprehensive SST task set that fits domain related requirements can be realized.

Effects of the Retrieval Model. In this work, generated tasks are not quantitatively balanced across topics. However, through a post study analysis we found that advanced searching grammar could help in balancing topics of generated tasks in a task set by locating paraphrased sentences pertaining to specific topics. For example, by issuing a call to the Bing API with an advanced option `''in other words'': recreation` targeting Wikipedia domain we could locate all the Wikipedia articles containing the phrase "in other words" and corresponding to the topic of "recreation". We observed that in the task generation phase, despite instructions that encourage workers to select articles with highlighted paraphrased sentences more arbitrarily and neglect the ranking order, some participants still selected the top-ranked results. As a consequence we found a few duplicates in the generated tasks. Nevertheless, we collected 80 distinct tasks generated by users within the task generation framework that adequately elicited struggling search behavior of users.

Task Pre-filtering Method. In this paper, authors manually filtered struggling search tasks from the generated set of tasks. A manual task filtering step guarantees the high quality of SSTs, but it gets progressively more expensive with the growing size of the task set. By analyzing the generated tasks, we note that when SSTs are expressed in natural language, they are potentially

---

[7] https://en.wikipedia.org/wiki/List_of_wikis.

more complex from a readability standpoint in comparison to typical IR tasks. Through K-means ($K = 2$; Euclidean distance) for task type clustering based on the two parameters of average word complexity and readability of the generated tasks, we found that the readability of tasks could be an indicator of SSTs. Such clustering resulted in identifying SSTs with an accuracy of 80%, providing a pre-filtering method for scalable filtering of the generated tasks that can be leveraged in the future.

We note that the reading comprehension ability of a worker plays an important role in the worker's understanding of the preceding context, and the accurate generation of a SST using a paraphrased sentence. In the current setup, we recruited *Level-3* workers from Figure8. However, we reason that to optimize the efficient generation of SSTs using our framework one can consider pre-screening crowd workers based on their proficiency in reading comprehension.

7   Conclusions and Future Work

By leveraging summarized (redundant) information in paraphrased sentences we proposed a task generation method and implemented it in an online crowd-powered framework. Through our task generation framework, we collected diverse questions from crowd workers with implicit task descriptions, and unambiguous answers that can be found by exploring the relevant information space. While this also results in some simple look-up tasks, these can be easily filtered out using existing criteria. We conducted a web search experiment to evaluate the task quality based on characteristics of elicited user behaviors. We showed that high quality struggling search tasks can be generated using our framework. We analyzed why searchers struggle in search sessions, and revealed insights into the independent impact of each of task characteristics that lead to users' struggle in a search session. We believe that our framework, the task set, together with our insights in this paper will help in advancing and developing methods to support users in struggling search. In the imminent future, we will test the SSTs in different search engines and explore a benchmark about how different search engines support such struggling fact finding or checking tasks.

References

1. Ageev, M., Guo, Q., Lagun, D., Agichtein, E.: Find it if you can: a game for modeling different types of web search success using interaction data. In: SIGIR, pp. 345–354 (2011)
2. Aula, A., Khan, R.M., Guan, Z.: How does search behavior change as search becomes more difficult? In: SIGCHI, pp. 35–44 (2010)
3. Baykan, E., Henzinger, M., Marian, L., Weber, I.: Purely URL-based topic classification. In: WWW, pp. 1109–1110 (2009)
4. Bhagat, R., Hovy, E.: What is a paraphrase? Comput. Linguist. **39**(3), 463–472 (2013)

5. Braarud, P.Ø., Kirwan, B.: Task complexity: what challenges the crew and how do they cope. In: Skjerve, A., Bye, A. (eds.) Simulator-based Human Factors Studies Across 25 Years, pp. 233–251. Springer, London (2010). https://doi.org/10.1007/978-0-85729-003-8_15

6. Capra, R., Arguello, J., O'Brien, H., Li, Y., Choi, B.: The effects of manipulating task determinability on search behaviors and outcomes. In: SIGIR, pp. 445–454 (2018)

7. De Beaugrande, R.A., Dressler, W.U.: Introduction to Text Linguistics, vol. 1. Longman, London (1981)

8. Gadiraju, U., et al.: Crowdsourcing versus the laboratory: towards human-centered experiments using the crowd. In: Archambault, D., Purchase, H., Hoßfeld, T. (eds.) Evaluation in the Crowd. Crowdsourcing and Human-Centered Experiments. LNCS, vol. 10264, pp. 6–26. Springer, Cham (2017). https://doi.org/10.1007/978-3-319-66435-4_2

9. Gadiraju, U., Yu, R., Dietze, S., Holtz, P.: Analyzing knowledge gain of users in informational search sessions on the web. In: CHIIR 2018 (2018)

10. Hassan, A.: A semi-supervised approach to modeling web search satisfaction. In: SIGIR, pp. 275–284 (2012)

11. Hassan, A., Jones, R., Klinkner, K.L.: Beyond DCG: user behavior as a predictor of a successful search. In: WSDM, pp. 221–230 (2010)

12. Hassan, A., White, R.W., Dumais, S.T., Wang, Y.M.: Struggling or exploring?: disambiguating long search sessions. In: WSDM, pp. 53–62 (2014)

13. Hassan Awadallah, A., White, R.W., Pantel, P., Dumais, S.T., Wang, Y.M.: Supporting complex search tasks. In: CIKM, pp. 829–838 (2014)

14. Kellar, M., Watters, C., Shepherd, M.: A goal-based classification of web information tasks. ASIST 43(1), 1–22 (2006)

15. Kenter, T., De Rijke, M.: Short text similarity with word embeddings. In: CIKM, pp. 1411–1420 (2015)

16. Kim, Y., Hassan, A., White, R.W., Zitouni, I.: Modeling dwell time to predict click-level satisfaction. In: WSDM, pp. 193–202 (2014)

17. Kolln, M., Funk, R.: Understanding English Grammar. Longman, London (1982)

18. Liu, C., Liu, J., Cole, M., Belkin, N.J., Zhang, X.: Task difficulty and domain knowledge effects on information search behaviors. ASIS&T 49(1), 1–10 (2012)

19. Liu, J., Kim, C.S., Creel, C.: Why do users feel search task difficult? In: The 76th ASIS&T. American Society for Information Science (2013)

20. Mai, J.E.: Looking for Information: A Survey of Research on Information Seeking, Needs, and Behavior. Emerald Group Publishing, Bingley (2016)

21. Mitra, B.: Exploring session context using distributed representations of queries and reformulations. In: SIGIR, pp. 3–12 (2015)

22. Odijk, D., White, R.W., Hassan Awadallah, A., Dumais, S.T.: Struggling and success in web search. In: CIKM, pp. 1551–1560 (2015)

23. Pirolli, P., Card, S.: The sensemaking process and leverage points for analyst technology as identified through cognitive task analysis. In: IA, vol. 5, pp. 2–4 (2005)

24. Robinson, P.: Task complexity, task difficulty, and task production: exploring interactions in a componential framework. Appl. Linguist. 22(1), 27–57 (2001)

25. Singer, G., Norbisrath, U., Lewandowski, D.: Ordinary search engine users carrying out complex search tasks. J. Inf. Sci. 39(3), 346–358 (2013)

26. Singer, P., et al.: Why we read Wikipedia. In: WWW, pp. 1591–1600 (2017)

27. White, R.W.: Interactions with Search Systems. Cambridge University Press, Cambridge (2016)

28. White, R.W., Roth, R.A.: Exploratory search: beyond the query-response paradigm. In: Synthesis Lectures on Information Concepts, Retrieval, and Services, vol. 1, no. 1, pp. 1–98 (2009)
29. Wilson, M.L., Kules, B., Shneiderman, B., et al.: From keyword search to exploration: designing future search interfaces for the web. Found. Trends® Web Sci. **2**(1), 1–97 (2010)
30. Xu, L., Zhou, X.: Generating tasks for study of struggling search. In: Proceedings of the 2019 Conference on Human Information Interaction and Retrieval, pp. 267–270 (2019)

Search Engine Similarity Analysis: A Combined Content and Rankings Approach

Konstantina Dritsa$^{(\boxtimes)}$, Thodoris Sotiropoulos$^{(\boxtimes)}$, Haris Skarpetis$^{(\boxtimes)}$, and Panos Louridas$^{(\boxtimes)}$

Athens University of Economics and Business, Athens, Greece
{dritsakon,theosotr,p3110180,louridas}@aueb.gr

Abstract. How different are search engines? The search engine wars are a favorite topic of on-line analysts, as two of the biggest companies in the world, Google and Microsoft, battle for prevalence of the web search space. Differences in search engine popularity can be explained by their effectiveness or other factors, such as familiarity with the most popular first engine, peer imitation, or force of habit. In this work we present a thorough analysis of the affinity of the two major search engines, Google and Bing, along with DuckDuckGo, which goes to great lengths to emphasize its privacy-friendly credentials. To do so, we collected search results using a comprehensive set of 300 unique queries for two time periods in 2016 and 2019, and developed a new similarity metric that leverages both the content and the ranking of search responses. We evaluated the characteristics of the metric against other metrics and approaches that have been proposed in the literature, and used it to (1) investigate the similarities of search engine results, (2) the evolution of their affinity over time, (3) what aspects of the results influence similarity, and (4) how the metric differs over different kinds of search services. We found that Google stands apart, but Bing and DuckDuckGo are largely indistinguishable from each other.

Keywords: Search engines · Distance metrics · Results ranking · Document similarity

1 Introduction

Search engine battles make headlines in the international media; changes in their algorithms have become topics of business analysts. Their rollout is eagerly followed across the globe, while their inner workings remain corporate secrets.

The battle for prevalence in the search engine market is an ongoing game. Recent developments, such as the advent of stricter data protection policies, have affected the dynamics of the market. The United States search engine market

---

K. Dritsa and T. Sotiropoulos—These authors contributed equally to this work.

© Springer Nature Switzerland AG 2020
Z. Huang et al. (Eds.): WISE 2020, LNCS 12343, pp. 21–37, 2020.
https://doi.org/10.1007/978-3-030-62008-0_2

developments over the last three years show an increase of Google's market share by 5.45%, a decrease of Bing's market share by 18.13%, while DuckDuckGo's market share rose almost by a factor of four [18].

Beyond the comparative evolution of search engines, the similarity between search engines' results has been a topic of interest as it widely affects users' exposure to diverse or similar views and perspectives, especially for informative search [2]. There are two different approaches for comparing search engines' results: (1) ranking-based approaches that consider only the ordering of web results, and (2) content-based approaches that exploit only the textual content (i.e., snippets) of web responses. However, search engines are evolving at a fast pace, returning far richer results than the "ten blue links" of the past [30] and their evolution has given prominence to new user interaction patterns [10]. As a result, while the existing approaches can still be used for search engine comparisons, they are essentially a first-order approximation of the problem that does not take into account the current heterogeneous user experience.

In this work we tackle the question of the similarity between Google, Bing, and DuckDuckGo, by investigating whether and how their search results are different. For our comparison, we propose a novel similarity metric that takes into account both the top k lists [12,17] of search results, and their semantic content, as shown by the titles and text snippets in their responses. We apply our metric to a comprehensive set of queries gathered from two time periods.

Contributions. Our work contributes to both search engine affinity analysis and the top k results literature.

- **A novel metric for search engine similarity:** We introduce a combined content and rankings approach that returns more expressive similarity scores and distinguishes important differences in search engine behavior that are not apparent using the existing metrics.
- **Search engine affinity:** We develop an experimental setting for assessing the affinity of search engines. By assembling a varied set of 300 unique queries and inspecting their top 10 results over two distinct periods, one in 2016 and one in 2019, we compare the behavior of different search engines across time.
- **Comparison findings:** While Google appears to be different than both Bing and DuckDuckGo, the last two are indistinguishable from each other.

The rest of the paper is organized as follows: Section 2 provides an overview of related work. We introduce our metric in Sect. 3 and its application on our data set in Sect. 4. Section 5 presents our conclusions and further discussion.

2   Background and Related Work

The issues of affinity, performance, and stability in search engines have attracted research attention since their early days in the 1990s. The oldest studies [11, 14,19] focused mainly on evaluating and comparing the performance of search engines, employing a few queries (2 to 20) and manually examining the relevance of the results with the queries. In 2004, Google and two defunct search engines were evaluated, with Google demonstrating the best performance [28].

On the affinity of search results, studies until the late 2000s indicated a low overlap with mostly unique results [5,7,14]. In a 2010 study, Zaragoza et al. [33] conducted an alternative approach with quantitative statements, on 1000 queries in Google, Microsoft Live Search, and Yahoo! Search. The three search engines gave satisfactory results for navigational queries (i.e., queries that referred to a particular web page or service) and for frequent non-navigational queries.

At the same time, Webber et al. [31] developed Rank-Biased Overlap, a similarity metric for ranked lists. The researchers created a set of 113 queries and inspected the top 100 URLs produced by 11 search engines. Google and Microsoft Live Search results were common by 25%. Moreover, when checking against the localized versions of the search engines (e.g., the .au domain), Google was found to use less localization than Yahoo and Microsoft Live.

In a subsequent work in 2011 investigating the ranking similarity between Bing and Google [9], Cardoso and Magalhães applied the Rank-Biased Overlap on the results of 40,000 queries, showing that the search engines differed considerably. Furthermore, they looked into the diversity of search results for a given query using the Jensen-Shannon divergence and came to the conclusion that Bing tended to interpret a given query more diversely than Google.

In 2014, Collier and Konagurthu [12] proposed a measure for the comparison of two ranking lists, based on the minimum length encoding framework developed by Wallace [29]. The investigators measured the similarity between Ask, Google, and Yahoo for up to the top 100 results of 250 queries. Their findings showed that the search engines results differed linearly on their ranks, or quadratically using the Spearman and Kendall distances. Agrawal et al. [1] proposed two methods, TensorCompare and CrossLearnCompare, to compare search engine affinity, and used them to compare Google and Bing. We will return to this study in Sect. 4.6.

Although semantic features are largely incorporated into the process of producing and ranking search results, they are not integrated in the commonly used rank-distance metrics [8]. In cases where the results of the search engines have very similar URLs and rankings but the snippets/titles differ, a rank-distance metric cannot reflect this dissimilarity. On the other hand, approaches that solely focus on the content of web results would not sufficiently represent reality, when comparing search engines with similar snippets/titles, but different rankings.

In addition, the trend towards aggregation of multiple information sources into search results has led to changes in the corresponding evaluation methodologies [3,30]. Studies highlight interesting user interaction patterns [10] where the ordering of search responses does not play the sole role in browsing result pages. Research has shown that snippets and titles *notably* affect the user's decision to click on a specific page [13,24,26].

Unlike previous work, we propose a metric tailored to the search engine similarity problem that leverages diverse criteria as to the rankings and the content of web results. Our combined metric aims to return more expressive, objective, and robust similarity scores, highlighting differences that are not apparent from the existing metrics. Our metric also views each search result as it is; a unified piece of information. Furthermore, to the best of our knowledge, none of the prior studies include privacy-friendly search engines.

3 The Metric

We introduce a new metric, which we call T, to study search engine similarity. In Sect. 3.1 we formulate the problem that the metric aims to resolve and the criteria that it should meet; in Sects. 3.2–3.5 we develop metric T step-by-step. Then, in Sect. 3.6 we compare it to other existing metrics.

3.1 Problem Formulation

In what follows, we assume that for two search engines A and B we have two lists $R_A = [a_1, a_2, a_3, \ldots, a_n]$ and $R_B = [b_1, b_2, b_3, \ldots, b_n]$ of the ranked top n results of search engine A and search engine B respectively. We denote the ith element of R_A with $R_A[i]$, and similarly for R_B.

Typically, search engine results consist of a URL, a result title, and a snippet describing the page content. Snippets and titles significantly affect the user's decision to click on a specific page [24,26]. To accurately appraise engine similarity, search engine comparisons should consider all these three aspects.

Motivating Example. To further highlight the importance of snippets and titles, consider Table 1 that shows the top result returned by Google and Bing for the query "Steven Wilson". Although search engines agree in the ordering of the same URL, they produce completely different snippets. The snippet produced by Bing focuses on artist's favorite film directors, while the snippet of Google gives emphasis on music news. Depending on user's search criteria, one snippet might be more effective on attracting user clicks than the other one.

Table 1. The top result retrieved for the query "Steven Wilson" on April 16, 2019.

	Bing	Google
Position	1	1
URL	http://stevenwilsonhq.com/sw/	http://stevenwilsonhq.com/sw/
Snippet	Steven is a film aficionado, and frequently cites cinema as one of the key inspirations for his music. Some of this favourite directors include Stanley Kubrick, David Lynch, Ben Wheatley, Jonathan Glazer, Shane Meadows and Christopher Nolan	The official website for songwriter/producer Steven Wilson. New live album/film 'Home Invasion: In Concert at the Royal Albert Hall' is out now!

Criteria. As the ranking of results does not fully capture their similarities, we need a comprehensive affinity metric that should meet the following criteria:

C1 The number of common elements (results). The more elements search engine
   A and B share in their top n results, the more similar they are.
C2 The distance of common elements. If an item appears in the results of both
   A and B, the affinity of A and B decreases as the distance of the element
   in the two result lists increases.
C3 The importance of agreement decreases as we go down in the results lists.
   For example, agreement at the top result is more important than that at
   the third or fourth result.
C4 If two search engines are similar, they produce similar titles and snippets,
   apart from returning similar results in a similar order.

3.2 Starting Point

As a starting point to define a metric for search engine affinity, we take the Jaro-Winkler distance, a variant of the Jaro distance [21], whose goal is to compute string similarity based on the common elements and the number of transpositions between them [32]. The Jaro distance of two strings S_1 and S_2 is given by:

$$d_j = \begin{cases} 0 & \text{if } m = 0 \\ \frac{1}{3} \left(\frac{m}{|S_1|} + \frac{m}{|S_2|} + \frac{m-t}{m} \right) & \text{otherwise} \end{cases} \tag{1}$$

In the above, m is the number of matching characters and t denotes the number of transpositions. Two characters are considered matching if they are the same and their positions do not differ by more than $(\max(|S_1|, |S_2|)/2) - 1$. The number of transpositions is defined as half the number of matching characters that are in different order in the two strings.

The Jaro-Winkler distance extends the Jaro distance by boosting it using a scaling factor p when the first l characters match exactly:

$$d_w = d_j + (l \times p \times (1 - d_j)) \tag{2}$$

In order to take into account the snippets and titles returned by the search engines, we adjust the Jaro-Winkler distance as follows:

$$S = \begin{cases} 0 & \text{if } m = 0 \\ \frac{1}{3n+1} (3m + 1 - a \cdot s - b \cdot h - c \cdot t) & \text{otherwise} \end{cases} \tag{3}$$

where n denotes the common length of the two result sets, $m = |R_A \cap R_B|$ is the number of common elements, t is the penalty from transpositions, s is the penalty from the differences between snippets, h is the penalty from the differences between titles, and $a, b, c \in [0, 1]$ are weights attached to the penalties accrued from snippets, titles, and transpositions respectively. To avoid division by zero, we add one to denominator.

Note that we compute the ratio of penalties $m - a \cdot s$ and $m - b \cdot h$ to the length n of results lists rather than the number of matching elements m, which is proposed by Jaro's metric. This gives us a more reliable estimation of the affinity between lists. For example, suppose we compare a pair of result rankings of length $n = 10$ and we get the number of matching elements as $m = 2$. According to Eq. 1, if $t = 0$ then the term $\frac{m-t}{m}$ is equal to 1 and it contributes $\frac{1}{3}$ to the overall similarity, which is a high number, considering the low number of matching items (two). Also, we use m/n instead of $m/|S_1| + m/|S_2|$, as R_A and R_B have a common length n.

3.3   Calculation of Penalties

Transpositions. To compute transpositions, we take the sum of the absolute differences of the positions of elements appearing in both lists. This is a variation of the deviation distance described by Ronald [25]. For lists R_A and R_B, the penalty is computed as follows, where $\sigma(R, e)$ is the position of e in list R:

$$t = \frac{\sum\limits_{e \in R_A \cap R_B} |\sigma(R_A, e) - \sigma(R_B, e)|}{t_{\max}}$$

This penalty is normalized on its upper bound. It can be proven that in the case of two lists of length n the upper bound for transpositions of $|R_A \cap R_B|$ is:

$$t_{\max} = \sum_{i=1}^{|R_A \cap R_B|} \phi(i, n)$$

where

$$\phi(i, n) = \begin{cases} n + 1 - i, & \text{if } i = 2k, k \in \mathbb{Z}^* \\ n - i, & \text{otherwise} \end{cases}$$

Snippets and Titles. The process of evaluating the penalties related to snippets and titles is common for both. We examine the sentences S_1, S_2 of snippets and titles that are produced by search engines A and B for a shared result. Then, we tokenize sentences S_1, S_2 and eliminate all stopwords as well as query terms. We get the union of all tokenized words that appeared in the two sentences and calculate the corresponding frequencies, forming two vectors V_1, V_2, that represent the actual snippets or titles. We then compute the cosine distance of the two vectors $d_s = 1 - \cos(V_1, V_2)$. The overall penalty is computed by iterating and repeating this process for all common results and summing all distances.

3.4   Similarity Boosting

The Jaro-Winkler metric treats all explicit matches at the first l characters of strings equally (recall Eq. 2). We, however, require a descending significance for agreement as we go down the list of results. To do that, we increase the value of S (Eq. 3) using weights w_i when there are common results in positions $1 \leq i \leq r \leq n$, with $w_1 > w_2 > \ldots > w_r$. This follows our third criterion, that exact or adjacent matches are more important at the beginning of results lists rather than the end. Moreover, in contrast to the Jaro-Winkler metric, the increase is not determined solely by the length of the matching prefix.

3.5 The Metric T

The final metric of similarity T combines the number of overlapping results as well as ordering, snippets, and titles of results, and it is given by:

$$T = S + \sum_{i=1}^{r} x_i w_i (1 - S) \qquad (4)$$

Table 2. Results for the comparisons described in Sect. 3.6.

	abcdef aghijk	abcdef abcghi	abcdef ghidef	abcdef defabc	abcdef abcdfe	abcdef abcfed
Spearman's footrule	1.0	1.0	1.0	0.0	0.89	0.78
Kendall's tau	1.0	1.0	1.0	0.85	0.98	0.95
G	0.29	0.71	0.57	0.29	0.95	0.90
M	0.48	0.82	0.36	0.18	0.98	0.97
Jaro-Winkler	0.44	0.77	0.67	0.0	0.96	0.96
T	$[0.24, 0.33]$	$[0.46, 0.68]$	$[0.25, 0.55]$	$[0.32, 0.48]$	$[0.59, 0.89]$	$[0.57, 0.88]$

where

$$x_i = \begin{cases} 0, & \text{if } R_A[i] \neq R_B[i] \\ 1, & \text{otherwise} \end{cases}$$

T meets all four criteria of Sect. 3.1. The calculation of overlapping items, m, fulfils C1. The computation of the penalty t fulfils C2, whereas boosting satisfies criterion C3. Finally, $a \cdot s$ and $b \cdot h$ cover C4.

3.6 Comparison with Other Metrics

In order to evaluate the behavior of our metric, we use a synthetic example and the criteria defined in Sect. 3.1 to contrast it with other metrics. Specifically, we compare it with Spearman's footrule and Kendall's tau (modified to measure similarity instead of distance) [17], the Jaro-Winkler metric, and the metrics G and M proposed by Bar-Ilan et al. [4,5].

Let $L_1 = [\text{abcdef}]$, a list that contains responses provided by one search engine. We compare L_1 with six other results lists $L_2...L_7$ using different metrics, as shown in Table 2. For the Jaro-Winkler metric we set $p = 0.1$, $l \leq 3$. In metric T we set $a = b = c = 1.0$ to penalize differences stemming from snippets, titles, and transpositions respectively, while we set $r = 3$, $w_1 = 0.15$, $w_2 = 0.1$, $w_3 = 0.05$ to reward matches at the first r elements.

Only metric T meets criterion C4 regarding snippets and titles. Thus, we present a lower and upper bound of our metric for every comparison. The lower bound corresponds to completely different snippets and titles among common results. The upper bound corresponds to snippets and titles that are identical.

In Table 2 we see that Spearman's footrule and Kendall's tau ignore mismatching elements, and compute similarity using only the common ones along with their distance, therefore, they do not meet criteria C1 and C3.

The Jaro-Winkler metric treats equally the transpositions of $d \leftrightarrow f$ and $e \leftrightarrow f$ in the comparisons (L_1, L_6) and (L_1, L_7) respectively, even though the former introduces a greater misplacement of elements. Thus, it violates criterion C2. Moreover, according to Eq. 2, it does not assign descending significance to agreements at the prefix of lists, which is required by criterion C3.

Both G and M metrics (the M metric to a greater extent) estimate the similarity of lists with emphasis to the ranking of items rather than the number of overlapping results. For example, we notice that even though L_1 and L_5 share all elements, the values of M and G show a decreasing importance to greater ranks, especially at the tail of lists. Also, a match in the first position, as in comparison (L_1, L_2), contributes 0.48 to the overall similarity according to the M metric, which is a great proportion relative to the number of matching items, i.e., *only* one out of total six. In essence, while M and G satisfy criteria C1–C3, they actually ignore matches or adjacent matches at the end of lists; in fact, the metric T can subsume G and M by using only the first q elements, for $q < n$.

Kumar and Vassilvitskii have proposed generalized versions of Spearman's footrule and Kendall's tau distances [22]; their versions take into account element weights, position weights, and element similarities in their calculations. It can be shown (omitted for reasons of space) that the generalizations overlook elements that appear only in a single list and thus miss criterion C1.

4   Evaluation

We compare Google, Bing, and DuckDuckGo (hereafter DDG), for numerous categories of queries, using our metric T. Google and Bing are the two dominant search engines, and have been the subject of comparative research. DDG adopts a different philosophy, placing a premium on user privacy. In our empirical evaluation, we try to answer the following research questions:[1]

RQ1 Do search engines produce similar web results? (Sect. 4.2)
RQ2 Is the similarity between search engines consistent over time? (Sect. 4.3)

---

[1] All data, results, and source code used on our experiments are available through https://doi.org/10.5281/zenodo.3980817.

RQ3 Which aspect of web results (i.e., rankings or content) influences the similarity of search engines the most? (Sect. 4.4)

RQ4 Do search engines produce similar results for different kinds of search services (i.e., news search service)? (Sect. 4.5)

RQ5 How do the results produced by the metric T correlate with the state-of-the-art? (Sect. 4.6)

4.1 Dataset

Table 3. Query categories

Books & Authors	Drinks & Food	Multinational companies	Music & Artists	Politicians
Regions	Software technologies	Sports	TV & Cinema	Universities

Our dataset consists of around 27600 top-10 result lists, spanning 10 categories of queries (Table 3). Each category contains around 30 queries; from these, 20 where taken from the U.S. version of Google Trends[2] in May 2016 and the rest were selected by us. Given that we cannot test all possible queries, we selected queries that affect a large number of users. For the data collection, we used the Bing Web Search API[3], the Google Custom Search API[4], and a web scraper that we developed for DDG. Our approach ensures that the search engines do not take user history into account, which would affect the final results [20]. We performed the queries daily, at the same time, using the American domain of each engine, for a period of one month (July–August) in 2016 and a period of 2 months (May–July) in 2019. We use both datasets to answer RQ2; for the rest of the research questions, both datasets gave consistent results, so, for brevity, we will focus on the 2019 dataset here.

Each result contains a URL, specifying its web location. Two identical URLs refer to the same result but a result could be pointed to by two different URLs [6]. To alleviate this issue, we applied standard normalization techniques [23] and resolved redirect HTTP responses to obtain the final target URL.

4.2 RQ1: Similarity of Search Engines

We estimate the similarity between Google, Bing, and DDG by employing metric T. For each search engine pair, we create a two-dimensional array D of the result similarity for every query and date. Each element D_{ij} represents the similarity between the two search engines in the day i for the query j.

---

[2] https://www.google.com/trends/topcharts.

[3] https://azure.microsoft.com/en-us/services/cognitive-services/bing-web-search-api/.

[4] https://developers.google.com/custom-search/.

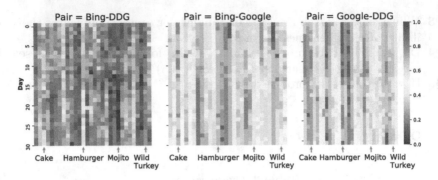

Fig. 1. Heatmaps of the DDG-Bing, Bing-Google and Google-DDG comparisons. The similarity between DDG and Bing is remarkable, while Google stands out. (Color figure online)

Recall from Eq. 4 that we need to define r and the weights $w_1, w_2, \ldots, w_r$ in order to reward matches at the first r elements of the ranking lists. In our experiments, we set $r = 5$ and $W = \{0.15, 0.1, 0.07, 0.03, 0.01\}$; we observed similar tendencies for different weight assignments. Regarding the importance of result factors, i.e., snippets, titles, and transpositions, we set $a = 0.8$, $b = 1$, $c = 0.8$. We use $b = 1$ as the weight for title penalties, because differences in titles are rare and in this way we could boost this factor (see Sect. 4.4).

Figure 1 presents the heatmaps of the similarity arrays D for the queries of the "Drinks & Food" category. These heatmaps are representative of all the other categories. Blue cells indicate cases where search engines are close to each other, while red cells reveal dissimilar web results.

We can see that Bing and DDG give very similar results for the vast majority of the queries. Despite its tiny market share, DDG still manages to offer a product comparable to that of the market leaders. The high Bing-DDG similarity could be explained by the fact that DDG -among other things- employs Bing to get its results [15]. Moving to Google-Bing and Google-DDG, the results of metric T indicate clear differences. However, there is still a number of queries where the search engines seem to have a high degree of resemblance, e.g., "Wild Turkey".

> **Finding #1.** Google stands apart from Bing and DDG for the majority of the queries, while the latter two are mostly identical to each other.

4.3   RQ2: Consistency of Search Engines

To estimate the consistency of search engine behavior over time, we calculate the pair-wise average similarity score of each day, as computed by metric T. Figure 2 presents the average similarity of every search engine pair in time. This figure clearly shows that the affinity of the search engines is almost constant

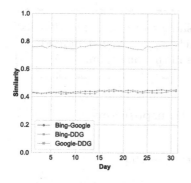 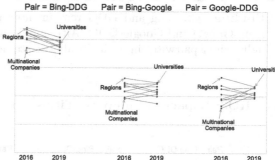

Fig. 2. Similarity evolution for a 31-day period in 2019. All pairs exhibit consistent behavior in the short-term.

Fig. 3. Comparison of search engines' similarity in 2016 and 2019. The similarity does not change considerably in the long-term. The DDG-Bing similarity is almost double than the others.

over time, with only a small number of trivial fluctuations. The findings from this experiment imply that either the search engines do not significantly change their behavior or that their behavior changes in the same way. In addition, the plots reveal that the similarity between Bing-DDG is almost double than that of Bing-Google and Google-DDG, strengthening our first finding.

> **Finding #2.** The behavior of all the search engines remains consistent in the short-term.

We also examined how the search engines' similarity changes in the long-term, by comparing the T similarities between the two time periods. Figure 3 shows the evolution of the pair-wise similarity for each query category from 2016 to 2019. Overall, we see that Bing and DDG moved from being very similar to slightly less so (their similarity decreases by 7.4%, on average). The affinity between of Bing-Google is almost stable (it drops by *only* 1.6%, on average), while DDG has come somewhat closer to Google, i.e., there is an increase in their similarity by 4.5%, on average. After inspecting the results, we found that these are due to changes in DDG's results within this time period.

Delving further into the data, we also examined how each search engine changed itself between these two points in time. We found that the average similarity between 2016 and 2019 is 0.37 for DDG, 0.43 for Bing, and 0.48 for Google; that is, Google's rankings and search algorithms changed the least and DDG has been updated to a greater extent, justifying its relative growth [18].

> **Finding #3.** Bing and DDG remain more similar to each other than Bing-Google and Google-DDG. Although search engines change individually, their pairwise similarity is almost stable in the long-term.

4.4   RQ3: Impact of Snippets, Titles, and Transpositions

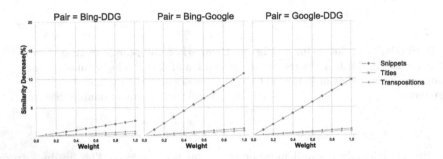

Fig. 4. The impact of snippets, titles and transpositions on each pair. The plots show the percentage decrease of similarity for various weight assignments of each factor. Google seems to construct different snippets compared to DDG and Bing.

Unlike existing approaches, metric T captures both the ordering (i.e., transpositions) and the content (i.e., snippets, titles) of results. Therefore, we can estimate how much each factor contributes to the differences of search engines. To do so, we instantiate the metric T with different weights for each factor (recall a, b, c from Eq. 3). We first consider the metric T_{base} as the baseline metric with weights $a = 0$, $b = 0$, $c = 0$. We compute the average similarity of every comparison pair for all the queries and days. Conceptually, T_{base} considers only the number of overlapping results and the agreements at the first $r = 5$ results. Then, we examine the effect of snippets by varying $a = 0.1, 0.2, \ldots, 1$ while keeping $b = c = 0$. Similarly, we examine the effect of titles and transpositions by varying b and c while keeping the other two weights pegged to zero.

In Fig. 4, each diagram shows the impact of every factor on the decrease of T_{base} for each search engine pair. It is clear that snippets have the biggest impact, while the difference in transpositions is much smaller, and in titles minimal. Google seems to construct different snippets compared to Bing and DDG, an observation that is consistent with our motivation example in Sect. 3.1.

> **Finding #4.** Snippets have the greatest impact on the differences among all the comparison pairs; Google yields more distinct ones though. All the search engines tend to place their common results in adjacent positions. Finally, all the search engines produce almost identical titles.

4.5 RQ4: Search Engine Similarity in Different Search Services

Apart from standard web search, search engines provide a list of additional search services. We investigated whether our findings apply to the news search tab. We created a set of 30 news queries; 20 of them were taken from the Google News trends of May 2019 and the remaining 10 were generic news topics, e.g., "flood".

The results show a low average similarity of 0.12, in contrast with the average 0.54 similarity of the results from the regular search. Furthermore, Bing-Google exhibits the highest similarity (0.15). This dissimilarity can be justified by the ephemeral nature of the news that requires quick evaluation, leading to daily ups and downs of topics and content. Also, the ranking algorithm of the news search results may be different than that of the regular search, certainly for Google [27].

> **Finding #5.** There is a considerable difference in the results produced by different search engines' services.

4.6 RQ5: Comparison with Other Approaches

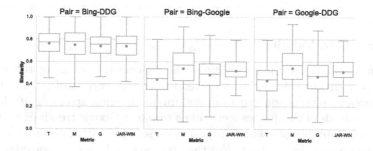

Fig. 5. Similarity of all search engine pairs using different metrics. Metric T exhibits lower box plots for the Bing-Google and Google-DDG comparisons, because it effectively captures the difference of their snippets and titles. (Color figure online)

Rankings-Based Approaches. We study how metric T correlates with three metrics that have been used in search engine comparisons (Sect. 3.6). Specifically, we use the metrics M, G, and Jaro-Winkler to compute the similarity between every search engine pair like we did in Sect. 4.2 using metric T.

Figure 5 shows the box plots of the search engine affinity for each metric. Every box plot contains the median similarity (horizontal line), the mean similarity (green circle), along with the maximum and minimum similarity values. The figure replicates our first finding (Sect. 4.2), that is, Google seems to produce more unique results when compared to Bing or DDG, as the corresponding box plots are lower than those of the Bing-DDG pair. Hence, the results of metric T are consistent with those of the three aforementioned metrics.

However, the box plots demonstrate that the metric T seems to distinguish from the others, especially in the Bing-Google, Google-DDG pairs. As shown by the average and the median similarity, metric T always produces lower values. This is explained by the fact that metric T *effectively* captures differences that stem from snippets and titles, which the other metrics ignore (Sect. 4.4).

Content-Based Approaches. Agrawal et al. [1] have proposed TensorCompare and CrossLearnCompare, two content-based methods that utilize tensor decomposition and supervised learning techniques. Both methods take into account the result snippets, but not their ordering. When applied on our data, the Tensor-Compare showed that the Bing-DDG pair is much more related than the rest, confirming our findings on their snippets similarity. The CrossLearnCompare, though, indicated an almost identical behavior for all search engines. An explanation for this could be that CrossLearnCompare actually predicts *queries* and not *search engines*, which may be distinguishable from each other. The Bing-DDG pair was more predictable than the rest, as we also find with metric T.

> **Finding #6.** Metric T, when compared to others, exhibits a consistent behavior. However, when the content similarity falls, the results of metric T differ from those of the other metrics.

4.7 Threats to Validity

The main threat to external validity is the representativeness of the selected queries. To mitigate this threat, we created a large corpus of $27,600$ lists of top-10 search results, assembled from 300 unique queries, spanning 10 different topics. Two-thirds of the queries were taken from the Google trends of 2016, that impact a large number of users. The rest were selected by us, aiming to include less popular queries that better reflect the average search use. We considered only the top n web results for every query, as previous studies of user behavior [16] have shown it is more likely for users to click on one of the first ten results.

The main threat to internal validity is the design of our metric T. To alleviate this threat, we meet four criteria (Sect. 3.1) that are considered very important in search engine comparisons. When compared with existing approaches, our metric T demonstrates consistency with both ranking-based and content-based metrics. Another threat comes from the methodology of web results collection. We used the REST APIs of Google and Bing that do not consider user history [20]. We queried all search engines at the same time every day, with the same parameters and standard URL normalization methods.

5 Conclusions and Discussion

In this work, we introduce a novel similarity metric for search engine comparison that combines the rankings of results and their semantic presentation. In contrast to the existing ranking-based or content-based approaches, our metric aims to be more expressive, robust and objective, following the aggregation of heterogeneous information into search results and the emergence of new user interaction patterns. Thus, it effectively captures differences that stem from snippets and titles, which the other metrics ignore.

By employing our metric, we were able to track engine similarity on both content and ranking across time, for a large and broad number of queries. Our results indicate that Google stands apart from Bing and DuckDuckGo, but these two are largely indistinguishable. The performance of DuckDuckGo may run counter to many expectations, taking into account the comparatively vast disparity of its resources. In our study we queried search engines without taking into account the user history. It is possible that when user history is employed, Bing would differ measurably from DuckDuckGo. Still, Google manages to differ from both Bing and DuckDuckGo even when it does not leverage personalized data.

Lately, search engines have started producing summaries, overviews, and compelling navigational aids, calling for more flexible comparison methodologies. Our approach consists a first step towards this direction, but the incorporation of semantically-rich features in search engine similarity measures seems a promising area for future research.

Acknowledgments. This work was supported by the European Union's Horizon 2020 research and innovation program "FASTEN" under grant agreement No. 825328.

References

1. Agrawal, R., Golshan, B., Papalexakis, E.: A study of distinctiveness in web results of two search engines. In: Proceedings of the 24th International Conference on World Wide Web (2015)
2. Agrawal, R., Golshan, B., Papalexakis, E.: Whither social networks for web search? In: Proceedings of the 21th ACM SIGKDD International Conference on Knowledge Discovery and Data Mining (2015)
3. Bailey, P., Craswell, N., White, R.W., Chen, L., Satyanarayana, A., Tahaghoghi, S.: Evaluating whole-page relevance. In: Proceedings of the 33rd International ACM SIGIR Conference on Research and Development in Information Retrieval (2010)
4. Bar-Ilan, J., Levene, M., Mat-Hassan, M.: Dynamics of search engine rankings–a case study. In: WebDyn@ WWW (2004)
5. Bar-Ilan, J., Mat-Hassan, M., Levene, M.: Methods for comparing rankings of search engine results. Comput. Netw. **50**(10), 1448–1463 (2006)
6. Bar-Yossef, Z., Keidar, I., Schonfeld, U.: Do not crawl in the DUST: different URLs with similar text. ACM Trans. Web **3**(1), 1–31 (2009)
7. Bharat, K., Broder, A.: A technique for measuring the relative size and overlap of public web search engines. Comput. Netw. ISDN Syst. **30**(1), 379–388 (1998)
8. Bian, J., Liu, T.Y., Qin, T., Zha, H.: Ranking with query-dependent loss for web search. In: Proceedings of the Third ACM International Conference on Web Search and Data Mining (2010)
9. Cardoso, B., Magalhães, J.: Google, Bing and a new perspective on ranking similarity. In: Proceedings of the 20th ACM International Conference on Information and Knowledge Management (2011)

10. Chen, D., Chen, W., Wang, H., Chen, Z., Yang, Q.: Beyond ten blue links: enabling user click modeling in federated web search. In: Proceedings of the Fifth ACM International Conference on Web Search and Data Mining (2012)
11. Chu, H., Rosenthal, M.: Search engines for the world wide web: a comparative study and evaluation methodology. In: Proceedings of the ASIS Annual Meeting, vol. 33 (1996)
12. Collier, J.H., Konagurthu, A.S.: An information measure for comparing top k lists. In: 2014 IEEE 10th International Conference on e-Science, vol. 1 (2014)
13. Cutrell, E., Guan, Z.: What are you looking for?: An eye-tracking study of information usage in web search. In: Proceedings of the SIGCHI Conference on Human Factors in Computing Systems (2007)
14. Ding, W., Marchionini, G.: A comparative study of web search service performance. Proc. ASIS Ann. Meet. **33**, 136–142 (1996)
15. DuckDuckGo: DuckDuckGo sources (2019). https://help.duckduckgo.com/results/sources/. Accessed 07 Aug 2019
16. Enge, E., Spencer, S., Fishkin, R., Stricchiola, J.: The Art of SEO. O'Reilly Media, Inc., Sebastopol (2012)
17. Fagin, R., Kumar, R., Sivakumar, D.: Comparing top k lists. SIAM J. Discrete Math. **17**(1), 134–160 (2003)
18. StatCounter GlobalStats: Statcounter globalstats (2019). http://gs.statcounter.com. Accessed 06 Aug 2019
19. Gordon, M., Pathak, P.: Finding information on the world wide web: the retrieval effectiveness of search engines. Inf. Process. Manag. **35**(2), 141–180 (1999)
20. Hannak, A., et al.: Measuring personalization of web search. In: Proceedings of the 22nd International Conference on World Wide Web. ACM (2013)
21. Jaro, M.A.: Advances in record-linkage methodology as applied to matching the 1985 census of Tampa, Florida. J. Am. Stat. Assoc. **84**(406), 414–420 (1989)
22. Kumar, R., Vassilvitskii, S.: Generalized distances between rankings. In: Proceedings of the 19th International Conference on World Wide Web. ACM (2010)
23. Lee, S.H., Kim, S.J., Hong, S.H.: On URL normalization. In: Gervasi, O., et al. (eds.) ICCSA 2005. LNCS, vol. 3481, pp. 1076–1085. Springer, Heidelberg (2005). https://doi.org/10.1007/11424826_115
24. Maxwell, D., Azzopardi, L., Moshfeghi, Y.: A study of snippet length and informativeness: behaviour, performance and user experience. In: Proceedings of the 40th International ACM SIGIR Conference on Research and Development in Information Retrieval (2017)
25. Ronald, S.: More distance functions for order-based encodings. In: 1998 IEEE International Conference on Evolutionary Computation Proceedings. IEEE World Congress on Computational Intelligence (Cat. No. 98TH8360), May 1998
26. Sachse, J.: The influence of snippet length on user behavior in mobile web search. Aslib J. Inf. Manag. **71**(3), 325–343 (2019)
27. The Economist: Seek and you shall find: Google rewards reputable reporting, not left-wing politics, June 2019. https://www.economist.com/graphic-detail/2019/06/08/google-rewards-reputable-reporting-not-left-wing-politics
28. Vaughan, L.: New measurements for search engine evaluation proposed and tested. Inf. Process. Manag. **40**(4), 677–691 (2004)
29. Wallace, C.S.: Statistical and Inductive Inference by Minimum Message Length. Springer, New York (2005). https://doi.org/10.1007/0-387-27656-4
30. Wang, Y., et al.: Optimizing whole-page presentation for web search. ACM Trans. Web **12**(3), 1–25 (2018)

31. Webber, W., Moffat, A., Zobel, J.: A similarity measure for indefinite rankings. ACM Trans. Inf. Syst. (TOIS) **28**(4), 1–38 (2010)
32. Winkler, W.E.: String comparator metrics and enhanced decision rules in the Fellegi-Sunter model of record linkage. In: Proceedings of the Section on Survey Research Methods (1990)
33. Zaragoza, H., Cambazoglu, B.B., Baeza-Yates, R.: Web search solved?: All result rankings the same? In: Proceedings of the 19th ACM International Conference on Information and Knowledge Management (2010)

Evaluating Similarity Measures
for Dataset Search

Xu Wang[(✉)], Zhisheng Huang, and Frank van Harmelen

Department of Computer Science, Vrije Universiteit Amsterdam,
Amsterdam, The Netherlands
{xu.wang,z.huang,Frank.van.Harmelen}@vu.nl

Abstract. Dataset search engines help scientists to find research datasets for scientific experiments. Current dataset search engines are query-driven, making them limited by the appropriate specification of search queries. An alternative would be to adopt a recommendation paradigm ("if you like this dataset, you'll also like..."). Such a recommendation service requires an appropriate similarity metric between datasets. Various similarity measures have been proposed in computational linguistics and informational retrieval. The goal of this paper is to determine which similarity measure is suitable for a dataset search engine. We will report our experiments on different similarity measures over datasets. We will evaluate these similarity measures against the gold standards which are developed for Elsevier DataSearch, a commercial dataset search engine. With the help of F-measure evaluation measure and nDCG evaluation measure, we find that Wu-Palmer Similarity, a similarity measure which is based on hierarchical terminologies, can score quite good in our benchmarks.

Keywords: Semantic similarity · Ontology-based similarity · Dataset search · Data science · Google Distance

1 Introduction

Sharing of datasets is becoming increasingly important in all branches of modern science [1,6,9]. Search engines dedicated to finding datasets that fill the needs of a scientist are now emerging rapidly, and similarity metrics for datasets are an important building block of such dataset search engines.

A scientific dataset is a set of data used by scientists or researchers for scientific experiments and scientific analysis. Usually, scientific datasets are categorized into three type: experimental datasets, computational datasets, and observational dataset [3].

Dataset search engines can help scientists to find such research datasets more efficiently. Dataset search engines are now emerging rapidly: DataSearch engine[1] (Elsevier), Dataset Search[2] (Google), Mendeley Data[3] (Mendeley) just to name

---

[1] https://datasearch.elsevier.com/.
[2] https://toolbox.google.com/datasetsearch.
[3] https://data.mendeley.com/.

© Springer Nature Switzerland AG 2020
Z. Huang et al. (Eds.): WISE 2020, LNCS 12343, pp. 38–51, 2020.
https://doi.org/10.1007/978-3-030-62008-0_3

a few. Elsevier's DataSearch engine is one of the most popular dataset search engines to date.

Although dataset search engine can be very helpful for scientists, the datasets returned by such search engines are strictly dependent on the appropriate specification of search queries. An alternative approach is the recommendation paradigm [2], where a search engine recommends datasets to a scientist based on similarity to datasets that are already known to be of interest to the researcher. Whereas the accuracy of queries is a limiting factor on dataset search, the quality of the similarity measure is crucial to dataset recommendation.

The goal of this paper is to answer which similarity measure is more suitable for a dataset search engine. In order to meet this goal, we propose a novel evaluation measure to evaluate the performance of a similarity measure for dataset search engines. The gold standard we used for evaluation is the gold standard ranking from a commercial dataset search engine (See footnote 1). However, this gives a gold standard for ranking, and not for similarity. To evaluate dataset *similarity* measures, we use of the similarity measures to reconstruct a ranking of datasets for a given query and then compare the reconstructed ranking to the gold standard ranking to get the accuracy of this reconstructed ranking. Usually, this accuracy is measured through the F-measure [4] and normalized Discounted cumulative gain (nDCG) measure [7]. We also propose a new F-measure to help us evaluate similarity measures because of the particularity of our gold standard ranking. In our experiments, we test our evaluation measures in Elsevier DataSearch engine with evaluating three measures (Wu-Palmer measures [12], Resnik measures [11] and Normalized Google Distance [5]). Then we using the evaluation measure to evaluate which similarity measure perform better in Elsevier DataSearch engine for these three similarity measures.

The main contributions of this paper are (1) to provide a new approach to evaluate similarity measures for dataset search engines, (2) to introduce two new kinds of F-measures (Brave and Cautious), and (3) to find out which similarity measure performs well on bio-medical datasets.

2 Preliminaries

In this section, we will introduce the similarity measures we used in this paper and the evaluation measures we used for evaluating the quality of our experiments results.

2.1 Similarity Measures

Various similarity measures have been proposed in computational linguistics and informational retrieval, such as topological similarity measures (for instance, Wu-Palmer Similarity measure [12] and Resnik Similarity measure [11]) and Statistical similarity measures (for instance Normalized Google Distance [5]). In NLP domain, word2vec is a popular measure to calculate the similarity between two terms.

Wu-Pamler Similarity. Wu-Palmer similarity measure [12] is a semantic similarity measure between two concepts based on the ontology structure. We use the Wu-Palmer Similarity measure in this paper because Wu-Palmer measure is an popular edge-based topological similarity measure. Wu-Palmer similarity between two concepts C_1 and C_2 is

$$Sim(C_1, C_2) = \frac{2 * N3}{N1 + N2 + 2 * N3} \tag{1}$$

where C_3 is the least common superconcept of C_1 and C_2, $N1$ is the number of nodes on the path from C_1 to C_3, $N2$ is the number of nodes on the path from C_2 to C_3 and $N3$ is the number of nodes on the path from C_3 to root.

Resnik Similarity. Resnik similarity measure [11] is a node-based topological similarity measure between two concepts based on the notion of information content, which combines the path based measure and the relative depth measure. We use the Resnik measure because most other node-based topological similarity measures are more or less based on the Resnik measure. In this measure, a function $p(c)$, which is the probability of encountering an instance of concept c, is introduced. $p(c)$ is computed as follows:

$$p(c) = \frac{\sum_{n \in words(c)} count(n)}{N} \tag{2}$$

where $words(c)$ is the set of concepts which are subsumed by concept c and N is the total number of nouns observed on the ontology structure.

Then the Resnik semantic similarity of two concepts C_1 and C_2 is defined as follows:

$$sim(C_1, C_2) = -\log p(C_3) \tag{3}$$

where C_3 is the least common super-concept of C_1 and C_2.

Normalized Google Distance. Normalized Google Distance (or Google Distance) [5] is a semantic similarity measure based on the number of hits from Google search engine. Different from Wu-Palmer and Resnik measures, Google Distance is a statistical similarity, and we use Google Distance measure as baseline to compare with two ontology-based similarity measures. For every two concepts x and y, the Google Distance between x and y is

$$NGD(x, y) = \frac{max\{logf(x), logf(y)\} - logf(x, y)}{logM - min\{logf(x), logf(y)\}} \tag{4}$$

where x and y are terms; $f(x)$ is the number of Google hits number of x; $f(x, y)$ is the number of Google hits for x and y; and M is the total number of web pages searched by Google multiplied by the average number of singleton search terms occurring on pages (estimated to be $25 * 10^9$).

Word2vec. Word2vec approach can produce word embedding and be used to calculate the similarity between two words. There are several popular NLP tools can implement word2vec algorithm. In this paper we use Gensim tools [10] for word2vec approach. The pretraining model we used for word embedding is the wiki-data[4]. Wikipedia can cover most concepts from every domain. So wiki-data is a suitable choose as the pretraining model for word2vec.

2.2 Evaluation Measures for Information Retrieval

Here, we will shortly introduce F-measure and nDCG measures, which we use in this paper.

F-measure. F-measure (also F-score or F1-score) is a measure of a test's accuracy [4]. F-measure considers two aspects: relevant and retrieved. Relevant always means document or dataset selected by given standard. Retrieved means the one selected by approach under evaluation. F-measure is defined as follow:

$$Precision = \frac{True\ pos}{True\ pos + False\ pos}, \quad Recall = \frac{True\ pos}{True\ pos + False\ neg} \quad (5)$$

$$F1 = 2 * \frac{Precision * Recall}{Precision + Recall} \quad (6)$$

where *True pos* is relevant and retrieved document/dataset; *False neg* is relevant and not-retrieved one; *False pos* is not-relevant and retrieved one; *True neg* is not-relevant and not-retrieved one.

nDCG. Discounted cumulative gain (DCG) is a measure of ranking quality [7]. Normalized Discounted cumulative gain (nDCG) is a normalized measure based on DCG measure [8]. nDCG through top rank position p is defined as follows:

$$DCG_p = \sum_{i=1}^{p} \frac{2^{rel_i} - 1}{\log_2(i+1)}, \quad IDCG_p = \sum_{i=1}^{|Rel_p|} \frac{2^{rel_i} - 1}{\log_2(i+1)}, \quad nDCG_p = \frac{DCG_p}{IDCG_p} \quad (7)$$

where DCG_p considers the list of documents (ordered by approach under evaluation); $IDCG_p$ considers the list of documents (ordered by given standard); rel_i is the relevant score in position i, which sometimes means the gold standard score of position i.

3   Similarity Measure and Evaluation Measure

3.1   Similarity Between Sets of Concepts

These similarity measures above calculate the similarity between two concepts (or between two terms representing those concepts). Then we also introduce the similarity measure to calculate the similarity between two *sets* of terms. The similarity between two sets A and B of terms is:

$$Sim(A, B) = \frac{sum\{Sim(a,b)|a \in A, b \in B\}}{(|C(A)| * |C(B)|)} \quad (8)$$

where $|C(A)|$ means the number of concepts in set A, and $|C(B)|$ means the number of concepts in set B.

---

[4] https://dumps.wikimedia.org/enwiki/.

3.2   Evaluation Measure

In this part, we will introduce the gold standard used for our experiments, the ranking reconstruction and the Caution/Brave F-measure for evaluating experiment's accuracy.

Gold Standard Ranking. For our evaluation, we have obtained a gold standard from the Elsevier DataSearch engine (See footnote 1). The gold standard consists of a set of queries together with the ranked results returned by the search engine for these queries. Expert scientist users had been invited to judge these results by giving a score to every search result. The range of judgement score is from −100 to 100, with the score 0 meaning that the experts cannot judge if this result is similar to the query. The gold standard aggregates these expert judgments into four levels: Likely satisfaction (which means the dataset is an excellent match for the query), possible satisfaction, possible dissatisfaction and likely dissatisfaction, according to the following score range:

- Likely dissatisfaction (level 3): from −100 to −51;
- Possible dissatisfaction (level 2): from −50 to −1;
- Possible satisfaction (level 1): from +1 to +50;
- Likely satisfaction (level 0): from +51 to +100.

Ranking Reconstruction. As described above, we have been given a gold standard for the *ranking* of datasets as query results, whereas we want to measure the *similarity* between datasets. In order to evaluate the similarity measures, we use each of the three similarity measure introduced above to "re-construct" a derived ranking. We can then compare these "derived rankings" with the given gold standard ranking, and find out which of our measures produces a better ranking.

Caution and Brave F-measure. As usual, we use the F-measure to evaluate our experiments of similarity metric. But because our gold standard gives us four categories of answer qualities (as described above), we redefine the original definition of the F-measure, into two more specific measure: the Cautious F-measure and the Brave F-measure.

For the Brave F-measure, we consider both dissatisfaction categories (possible dissatisfaction and likely dissatisfaction) as negative, and similarly we consider both satisfaction as positive.

For the Cautious F-measure, we consider only the stronger categories (*likely* (dis)satisfaction) as positive (resp. negative), while leaving the less pronounced *possible* (dis)satisfaction out of consideration.

In order to calculate these Cautious and Brave F-measures, precision and recall are defined in this paper as follows: For the cautious F-measure, the relevant number is the number of "likely satisfaction" results in the Gold Standard Ranking, denoted as $rel_{caution}$. For the brave F-measure, the relevant number is the number of "likely satisfaction" or "possible satisfaction" results in the Gold Standard Ranking, denoted as rel_{brave}. The retrieved number is the number of all results in the Reconstructed Ranking, donated as ret_num.

Then the brave/caution F-measure can be defined as follow:

$$precision_{caution} = \frac{\{rel_{caution}\} \cap \{ret\_num\}}{\{ret\_num\}}, \quad recall_{caution} = \frac{\{rel_{caution}\} \cap \{ret\_num\}}{\{rel_{caution}\}}.$$

(9)

$$precision_{brave} = \frac{\{rel_{brave}\} \cap \{ret\_num\}}{\{ret\_num\}}, \quad recall_{brave} = \frac{\{rel_{brave}\} \cap \{ret\_num\}}{\{rel_{brave}\}}. \quad (10)$$

$$F-measure_{caution} = \frac{2 * precision_{caution} * recall_{caution}}{precision_{caution} + recall_{caution}} \quad (11)$$

$$F-measure_{brave} = \frac{2 * precision_{brave} * recall_{brave}}{precision_{brave} + recall_{brave}} \quad (12)$$

nDCG with Gold Standard Ranking. In our gold standard introduced above, for every result in gold standard ranking list, the relevant score of it could be 0, 1, 2, 3 for level 0, level 1, level 2, level 3, respectively. So in this paper, we use nDCG with our gold standard ranking through all position in list.

4   Experiments and Results

In this section we introduce our experimental evaluation of the three similarity metrics.

4.1   Data Selection

For practical reasons, we restrict our experiment to the 3039 bio-medical datasets in Elsevier's DataSearch engine (See footnote 1). All the queries and ranked answers we used are given by the Gold Standard ranking obtained in Elsevier Data Search product testing.

Queries. In the Elsevier Gold Standard, 18 queries are listed as in the biomedical domain. These are listed in Table 1.

Ontology. In the biomedical domain, the MeSH terminology (Medical Subject Headings)[5] is an appropriate choice, since it is designed to capture biomedical terminology in the scientific domain.

Individual Datasets. Our datasets are characterized as a set of meta-data fields, and stored in JSON format (see Fig. 1). The restriction to meta-data seems appropriate, because in many real-life cases, the actual contents of the dataset might be entirely numerical, or encoded in some binary format, and hence not accessible for similarity measurements. To stay as close to this real-life situation as possible, we restricted ourselves to the meta-data fields only. Each dataset in the collection is data from an actual publication or scientific experiment.

---

[5] http://www.nlm.nih.gov/mesh.

Table 1. The corpus of 18 queries

Query	Content	Query	Content
E2	Protein Degradation mechanisms	E54	Glutamate alcohol interaction
E7	Oxidative stress ischemic stroke	E66	Calcium signalling in stem cells
E8	Middle cerebral artery occlusion mice	E67	Phylogeny cryptosporidium
E17	Risk factors for combat PTSD	E68	HPV vaccine efficacy and safety
E26	Mab melting temperature	E78	c elegans neuron degeneration
E28	Mutational analysis cervical cancer	E79	mri liver fibrosis
E31	Metformin pharmacokinetics	E80	Yersinia ruckeri enteric red mouth disease
E35	Prostate cancer DNA methylation	E89	Electrocardiogram variability OR ECG variability
E50	EZH2 in breast cancer	E94	Pinealectomy circadian rhythm

```
{
"id":"57525251:NEURO_ELECTRO",
"externalId":"3449",
"containerTitle":" Localization and function of the Kv3.1b ......",
"source":"NEURO_ELECTRO",
"containerDescription":"The voltage-gated potassium channel......",
"publicationDate":"2005",
"dateAvailable":"2005",
"containerURI":"http://neuroelectro.org/article/3449",
"firstImported":"2017-03-14T13:07:32.096Z",
"lastImported":"2017-03-14T13:07:32.096Z",
"containerKeywords":["Potassium Channels, Voltage-Gated"......],
"authors":["Mark L Dallas","David I Lewis","Susan A Deuchars"......],
"assets":......
}
```

Fig. 1. Meta-data fields in JSON

4.2 Extract Terms from Query or Dataset

As explained earlier, in order to calculate the similarity between a query and a dataset, we extract concepts and then consider the similarity between these two sets of concepts as the similarity between Query and Dataset.

MeSH has the following structure: a group of synonymous terms are grouped in a MeSH *concept*, and several MeSH concepts which are synonymous with each other are grouped in a **Descriptor**. A Descriptor is named by the preferred term of the preferred concept among all the concepts in this descriptor[6]. For example, for the MeSH descriptor Cardiomegaly:

---

[6] See also https://www.nlm.nih.gov/mesh/concept_structure.html.

```
Cardiomegaly                          [Descriptor]
    Cardiomegaly                      [Concept,  Preferred]
        Cardiomegaly                    [Term,  Preferred]
        Enlarged  Heart                 [Term]
        Heart  Enlargement              [Term]
    Cardiac  Hypertrophy              [Concept,  Narrower]
        Cardiac  Hypertrophy            [Term,  Preferred]
        Heart  Hypertrophy              [Term]
```

For each descriptor, we extract from the text-content of queries and datasets all the terms that occur with all concepts grouped under that descriptor. In the above example, extraction of any of the five terms Cardiomegaly, Enlarged Heart, Heart Enlargement, Cardiac Hypertrophy and Heart Hypertrophy would result in the annotation of the query or dataset with the descriptor Cardiomegaly. Since hierarchical relationships in MeSH are at the level of the descriptors, these are indeed suitable for calculating the ontology-based similarity measures.

4.3   Similarity Experiments

We calculate the similarity between a query and a dataset by using the Google distance measure, the Wu-Palmer measure, the Resnik measure and Word2vec. As described above, we will obtain these four similarity values, and use the ranking induced by these values with the ranking from the Elsevier gold standard, and thereby compare the performance of these metrics.

The sets of extracted MeSH descriptors are problematic in two distinct ways. Firstly, the number of MeSH descriptors varies widely between datasets, ranging from 1 to 110 per dataset, with an average of 15. Secondly, some of the extract MeSH descriptors are "noisy" and do not express the main meaning of the dataset. To correct both of these problems, we only consider the best 3, 4, 5 or 6 best scoring MeSH descriptor for each dataset, both balancing the number of descriptors across datasets, and avoiding the influence of poor descriptors (which are indicative of the dataset, and not of the quality of the similarity measure, which is what we are interested in).

4.4   Experimental Results

We will use the gold standard ranking from Elsevier Data Search engine to evaluate our reconstruction results.

First off, we reconstructed the similarity ranking as explained above. Two examples of such reconstructed rankings are shown in Fig. 2. Each column in these figures shows the datasets ranked according to their similarity under the indicated similarity metric, using the top 3, 4, 5 or 6 MeSH descriptors. We colored the result as green (resp. red) if this dataset is categorized as likely satisfaction (resp. dissatisfaction) in the gold standard ranking.

We can evaluate our similarity results by this visualization. From the right hand side table in Fig. 2, we can easily infer that the reconstructed ranking for query E80 is quite good, because most results from the top of the gold standard ranking (the green cells) are still located in top of the reconstructed ranking. This shows that the ranking based on the four similarity measures between query E80 and the datasets is similar to ranking given by the experts. The left hand side of

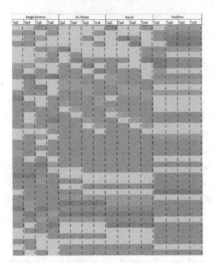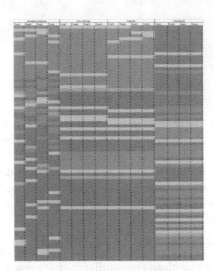

Fig. 2. Queries E79 (left) and E80 datasets ranking reconstructed with Google Distance, Wu-Palmer and Resnik similarity measures. (The tables are of different length because the gold standard contained more expert ratings for E80) (Color figure online)

Table 2. Cautious and Brave F-scores for queries E79 and E80 for the Google Distance, Wu-Palmer, Resnik and Word2vec

Google Distance				Wu-Palmer				Resnik				Word2vec			
Query	Top	Cautious	Brave	Query	Top	Cautious	Brave	Query	Top	Cautious	Brave	Query	Top	Cautious	Brave
E79	3	**0.55**	**0.84**	E79	3	**0.59**	**0.82**	E79	3	**0.59**	**0.82**	E79	3	0.40	0.76
E79	4	0.50	0.81	E79	4	**0.59**	**0.82**	E79	4	**0.59**	**0.82**	E79	4	0.40	0.76
E79	5	**0.55**	0.81	E79	5	**0.59**	**0.82**	E79	5	**0.59**	**0.82**	E79	5	0.40	0.76
E79	6	0.50	0.82	E79	6	**0.59**	**0.82**	E79	6	**0.59**	**0.82**	E79	6	0.40	0.76
E80	3	0.71	0.73	E80	3	**0.70**	**0.73**	E80	3	**0.70**	**0.73**	E80	3	0.07	0.07
E80	4	**0.78**	**0.77**	E80	4	**0.70**	**0.73**	E80	4	**0.70**	**0.73**	E80	4	0.07	0.07
E80	5	0.71	0.69	E80	5	**0.70**	**0.73**	E80	5	**0.70**	**0.73**	E80	5	0.07	0.07
E80	6	0.69	0.65	E80	6	**0.70**	**0.73**	E80	6	**0.70**	**0.73**	E80	6	0.07	0.07

Fig. 2 shows that for query E79, the similarity ranking by the metrics does not correspond to the ranking given by the experts, since for this query, most results in top of gold standard ranking are located in middle of reconstructed ranking.

The visual results from Fig. 2 are stated numerically in Tables 2 by computing the cautious and brave F-scores for all four similarity measures, again on queries E79 and E80 as examples. Again, we see that there is barely any difference between using 3, 4, 5 or 6 MeSH descriptors, with differences never bigger than ±0.01. However, these tables do reveal a difference in behaviour when graded on cautious or brave F-score. Using the cautious F-score, all three similarity metrics performed better on query E80, while for the brave F-Score, all three similarity metrics (except Word2vec) performed better on E79. Word2vec measure perform worse because it's hard for wiki-data to cover the bio-medical concepts as many as MeSH ontology and Google search engine do.

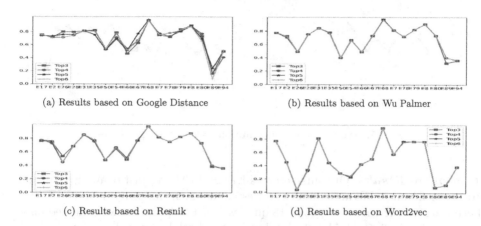

(a) Results based on Google Distance

(b) Results based on Wu Palmer

(c) Results based on Resnik

(d) Results based on Word2vec

Fig. 3. Brave F-scores of all queries based on Google Distance, Wu Palmer, Resnik and Word2vec

Table 3. Best F-score counts

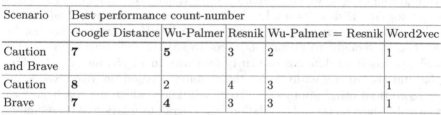

Scenario	Best performance count-number				
	Google Distance	Wu-Palmer	Resnik	Wu-Palmer = Resnik	Word2vec
Caution and Brave	7	5	3	2	1
Caution	8	2	4	3	1
Brave	7	4	3	3	1

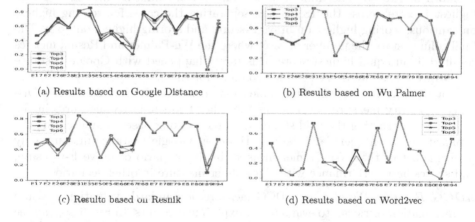

(a) Results based on Google Distance

(b) Results based on Wu Palmer

(c) Results based on Resnik

(d) Results based on Word2vec

Fig. 4. Caution F-scores of all queries based on Google Distance, Wu Palmer, Resnik and Word2vec

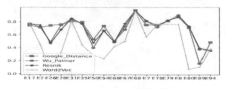 

(a) Average brave F-scores of all queries   (b) Average caution F-scores of all queries

Fig. 5. Average brave and caution F-scores of all queries

F-measure Results. Of course queries E79 and E80 are just examples to illustrate our findings. We are actually interested in which similarity measures scores better across our entire corpus of 18 queries. For this purpose, we collected data as shown in Fig. 3, 5a, 4, 5b, which tabulates the both brave and cautious F-score for all metrics on all queries when computing similarity based on the top 3, 4, 5, 6 MeSH descriptors, as well as the average one based on top 3, 4, 5, 6 Mesh descriptors.

In Fig. 3 and Fig. 4, we can roughly see the difference among every results of different top MeSH descriptors. In this two figures, we can easily find that the differences among every results of different top MeSH descriptors are not big. So we can go straight to average F-scores to see the differences among every approaches based on different similarity measures in Fig. 5a and Fig. 5b. From Fig. 5a and 5b, we can easily know that approach based on Word2vec have a lower score than other approaches. To clearly know which approach performs best, we also show an overall tabulation. The overall tabulation of these findings is shown in Table 3. The table lists how often each similarity measure has the highest F-score across the 18 queries, separating the cases for scoring highest on cautious, scoring highest on brave F-score, and scoring highest on both. The final column state the number of cases where the Wu-Palmer and Resnik metrics resulted in an equal highest score (this rarely happened with Google Distance and Word2vec, so we do not include columns for that).

For F1 score results, this final table shows conclusively that the Google Distance similarity measure outperforms the other two similarity measures in the task of reconstructing the gold standard search ranking based on measuring the similarity between query and dataset. However, Google Distance measure only performs better than Wu-Palmer measure in the scenario of brave F-measure, and shares best performance with Wu-Palmer measure in other scenarios.

nDCG Results. We also use nDCG measure, a standard evaluation measure for information retrieval, to evaluate our experiment results. In Fig. 6, we can see the nDCG scores of all reconstructing approaches based on Google Distance, Wu Palmer, Resnik and Word2vec of different top MeSH descriptors. We can easily find that there is no clear difference among different top MeSH descriptors for each approach. So we know that the difference on top MeSH descriptors would not impact the final results.

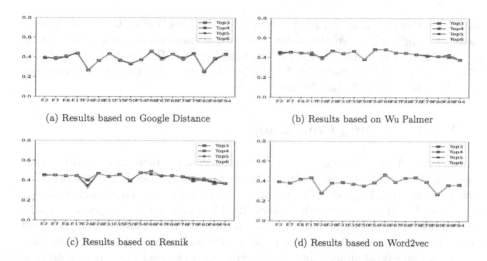

(a) Results based on Google Distance   (b) Results based on Wu Palmer

(c) Results based on Resnik   (d) Results based on Word2vec

Fig. 6. nDCG scores of all queries based on Google Distance, Wu Palmer, Resnik and Word2vec

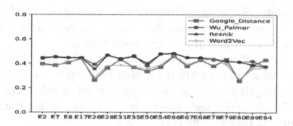

Fig. 7. Average nDCG scores of all queries

Table 4. Average nDCG score for each query.

Query	GoogleDistance	WuPalmer	Resnik	Word2Vec	Query	GoogleDistance	WuPalmer	Resnik	Word2Vec
E2	0.3969	0.4411	**0.4487**	0.3911	E54	0.3711	**0.48**	0.4762	0.3862
E7	0.3845	**0.4553**	0.4504	0.3781	E66	0.4609	**0.4835**	0.4833	0.466
E8	0.4066	**0.4476**	0.4451	0.4163	E67	0.3779	**0.4493**	0.4486	0.3908
E17	0.4429	0.4454	**0.448**	0.4336	E68	0.429	**0.4486**	0.447	0.4302
E26	0.2636	**0.3911**	0.354	0.2796	E78	0.379	0.4324	**0.4383**	0.4381
E28	0.3633	**0.4684**	0.4678	0.3807	E79	**0.4334**	0.4234	0.4172	0.3934
E31	0.433	**0.4394**	0.4383	0.3866	E80	0.2572	0.413	**0.4144**	0.2694
E35	0.3686	**0.4622**	0.456	0.3727	E89	0.3866	**0.4202**	0.3908	0.3605
E50	0.3345	0.3811	**0.4016**	0.3535	E94	**0.4305**	0.3772	0.373	0.3642

To find out which approach performs best in nDCG scores, we also collect the average nDCG scores for every approach, by taking average of all the scores of all different top MeSH descriptors for each approach. Average nDCG score results are shown in Fig. 7. According to this figure, we can intuitively know that approaches based on both Wu Palmer and Resnik outperform approaches based on Google Distance and Word2vec. To find out the winner of nDCG scores among these two approaches, we also collect the full average scores shown in Table 4. In Table 4, we can know that approach based on Wu Palmer has best average nDCG scores on 11 queries of all 18 queries. So we can say that Wu Palmer can score best in our nDCG benchmark.

5   Discussion and Conclusion

In modern science, sharing of datasets is becoming increasingly important. Search engines dedicated to finding datasets that fill the needs of a scientist are now emerging rapidly, and similarity metrics for datasets are an important building block of such dataset search engines. In this paper, we have reported experiments on four important similarity measures for datasets. Using a gold standard from a commercial search engine in experiments on biomedical datasets, we have found that the Wu-Palmer Similarity metric outperformed the other three candidates in nDCG benchmark, although it performed a bit worse than Google Distance and scores secondary in F-measure benchmark.

Future work would of course involve the extension of our results to other candidate similarity measures, including those based on embedding the biomedical domain vocabulary in a high-dimensional vector space.

Our test datasets are limited to the datasets in biomedical domain, because of the availability of a gold standard for this biomedical domain. Future work will have to show whether our conclusion can be extended to cover datasets in other domains. To this end, the development of similar gold standards is an important and urgent task for the community.

Acknowledgements. This work has been funded by the Netherlands Science Foundation NWO grant nr. 652.001.002, it is co-funded by Elsevier B.V., with funding for the first author by the China Scholarship Council (CSC) grant number 201807730060. We are grateful to our colleagues in Elsevier for sharing their dataset, and to all of our colleagues in the Data Search project for their valuable input.

References

1. Bauchner, H., Golub, R., Fontanarosa, P.: Data sharing: an ethical and scientific imperative. J. Am. Med. Assoc. **12**(315), 1238–1240 (2016)
2. Bobadilla, J., Ortega, F., Hernando, A., Gutiérrez, A.: Recommender systems survey. Knowl.-Based Syst. **46**, 109–132 (2013)
3. Borgman, C.L., Wallis, J.C., Mayernik, M.S.: Who's got the data? Interdependencies in science and technology collaborations. Comput. Supported Coop. Work (CSCW) **21**(6), 485–523 (2012). https://doi.org/10.1007/s10606-012-9169-z
4. Chinchor, N.: MUC-4 evaluation metrics. In: Proceedings of the 4th Conference on Message Understanding, MUC4 1992, pp. 22–29. Association for Computational Linguistics, New York (1992)
5. Cilibrasi, R.L., Vitanyi, P.M.: The google similarity distance. IEEE Trans. Knowl. Data Eng. **19**(3), 370–383 (2007)

6. Editorial: Benefits of sharing. Nature **530**(7589), 129 (2016). https://doi.org/10. 1038/530129a

7. Järvelin, K., Kekäläinen, J.: IR evaluation methods for retrieving highly relevant documents. In: Proceedings of the 23rd SIGIR Conference, SIGIR 2000, pp. 41–48. ACM, New York (2000)

8. Järvelin, K., Kekäläinen, J.: Cumulated gain-based evaluation of IR techniques. ACM Trans. Inf. Syst. **20**(4), 422–446 (2002). https://doi.org/10.1145/582415. 582418

9. McNutt, M.: Data sharing. Science **351**, 1007 (2016). https://doi.org/10.1126/ science.aaf4545

10. Řehůřek, R., Sojka, P.: Software framework for topic modelling with large corpora. In: Proceedings of the LREC 2010 Workshop on New Challenges for NLP Frameworks, pp. 45–50. ELRA (2010)

11. Resnik, P.: Using information content to evaluate semantic similarity in a taxonomy. CoRR abs/cmp-lg/9511007 (1995). http://arxiv.org/abs/cmp-lg/9511007

12. Wu, Z., Palmer, M.: Verbs semantics and lexical selection. In: Proceedings of the 32nd Annual Meeting on Association for Computational Linguistics, pp. 133–138. Association for Computational Linguistics (1994)

Towards Efficient Retrieval of Top-k Entities in Systems of Engagement

Anirban Mondal[1]([✉]), Nilesh Padhariya[2], and Mukesh Mohania[2]

[1] Ashoka University, Sonipat, India
anirban.mondal@ashoka.edu.in
[2] Indraprastha Institute of Information Technology Delhi, New Delhi, India
nilesh.iitd@gmail.com, mukesh@iiitd.ac.in

Abstract. Next-generation enterprise management systems are beginning to be developed based on the Systems of Engagement (SOE) model. We visualize an SOE as a set of entities. Each entity is modeled by a single parent document with *dynamic* embedded links (i.e., child documents) that contain multi-modal information about the entity from various networks. We address the problem of *efficiently* retrieving the top-k entities in an SOE for keyword-based queries. In particular, we propose an efficient bitmap-based approach for quickly identifying the candidate set of entities, whose parent documents contain all queried keywords. Moreover, we propose the two-tier HI-tree index, which uses both hashing and inverted indexes, for efficient document relevance score lookups. Our performance evaluation with both real and synthetic datasets demonstrates the overall effectiveness of our proposed schemes.

Keywords: Systems of engagement · Top-k entities · Keyword search · Indexing · Retrieval

1 Introduction

Enterprises have fully realized the value of data that they have about their customers in Customer Relationship Management (CRM) systems and transactional systems. To gain competitive advantage by knowing more about their customers, enterprises try to incorporate their customers' social data. This has led to the emergence of *next-generation* CRM systems that are built upon the paradigm of *"Systems of Engagement" (SOEs)*. Thus, we are witnessing a paradigm shift from the traditional "Systems of Records" (SORs) towards the SOEs [3].

While SORs typically allow only passive one-way transactional communication between the enterprise management system and the stakeholders, SOEs enable two-way communication, thereby allowing stakeholders to engage and collaborate with each other [2]. SOEs also incorporate social orientation by combining multi-modal information from different kinds of networks (e.g., social networks and business community networks), thereby establishing a better *context* for delivering greater agility and flexibility [1]. Interestingly, the shift towards a

© Springer Nature Switzerland AG 2020
Z. Huang et al. (Eds.): WISE 2020, LNCS 12343, pp. 52–67, 2020.
https://doi.org/10.1007/978-3-030-62008-0_4

model of engagement, as exemplified by SOEs, is also consistent with the increasingly important role that social networks, such as Facebook and Twitter, play in our lives today. Although we have motivated SOEs using CRM applications, SOEs also have significant commercial applications in important areas such as human resource management and supply chain management.

We visualize an SOE as comprising a set of entities e.g., customers. Each entity is modeled by a single parent document and possibly multiple child documents. Here, the links embedded in a given parent document constitute its child documents. These links refer to the interactions of the entity on various systems such as the email system, phone-call system, social network pages and product review pages. The parent document contains both structured data (e.g., customer name, date of birth) as well as unstructured data (e.g., hobbies) about the entity. SOE can encompass different types of data models (e.g., emails, phone-calls, social-pages etc.) for each entity, which can be integrated using data integration approaches [4]. We consider multi-modal data (e.g., speech data of phone-calls) that can be transcribed into text. Figure 1 depicts an example for parent and child documents for an SOE-oriented CRM application. Suppose Alice has recently purchased a Samsung Galaxy note 2 phone, and made comments on various systems about her phone charger problem. The company may wish to find its top-k customers who complained about phone charger problems. Here, top-k can be defined based on the frequency of occurrence of the keywords associated with the charger problem across various interaction networks of customers.

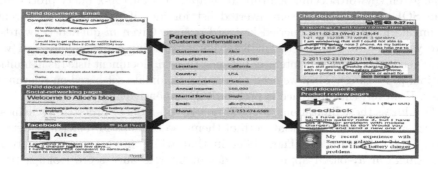

Fig. 1. Sample SOE for a CRM application

Incidentally, entities in an SOE are generally queried by means of *keywords*. Hence, we index both the structured and the unstructured data as a set of keywords. Our goal is to *efficiently* retrieve the top-k entities pertaining to a given keyword-based query [5,7,13,14,16,18] by considering the relevance scores of both their parent and child documents. Notably, the parent documents of a given entity remain relatively static, while the child documents may be *dynamic* (e.g., a new blog comment by the entity). Hence, we need to consider the dynamically changing relevance scores of the child documents as part of computing the top-k entities. For the sake of convenience, let us henceforth refer to the set of parent documents in which all the queried keywords occur together as the **candidate set**. (Thus, our query model uses 'AND' semantics.)

Related Work and Differentiation: Keyword search [5,7,13,14,16,18] has been studied for relational databases and graph-based approaches for keyword search over relational databases have also been proposed [5,7,14]. Works on keyword search in text cube models for multi-dimensional text database analysis have been proposed in [9,20,21]. Keyword search approaches using link analysis (e.g., the PageRank method [19] and the hubs & authority approach [15]) exploit the link structure of the document corpus for ranking the documents [8,11,17].

Our work fundamentally differs from existing works as follows. First, the *semantics* of the definition of an entity in existing works differ considerably from that of SOEs. Existing works consider every individual document in the corpus as a separate entity. In contrast, in an SOE, each document is *not* a separate entity since there is an *implicit semantic nested structure* in the corpus i.e., each entity is a parent document with embedded links (child documents). Thus, existing works fail to capture the semantic nested structure of SOE entities during keyword search. Second, for computing the top-k result, all keyword-search approaches compute a relevance score for each document in the corpus and then rank them. In contrast, we *efficiently prune* the search space by first determining the candidate set of parent documents and then compute their relevance scores, thereby considerably reducing computational costs.

Incidentally, inverted lists [6] focus on retrieving all the documents in which a *single* given keyword occurs. A brute-force approach for top-k query processing in an SOE could be to use an inverted list for identifying the list of parent documents in which each queried keyword occurs. Given m queried keywords, there would be m such document lists. For identifying the candidate set, we would need to exhaustively intersect between these m document lists (as in multi-way join processing). Then we need to sort the candidate set based on relevance scores. However, this approach is prohibitively expensive and non-scalable.

To address these limitations, we propose two top-k query processing schemes. The schemes are similar in that both of them use bitmaps for efficiently determining the candidate set. They differ in that while one scheme uses a two-dimensional array-based approach for document relevance score lookups (the two dimensions being document identifiers and keyword relevance scores), the other scheme uses our proposed index (designated as the HI-tree). Thus, we shall henceforth designate them as **Bitmap-based Scheme with Array (BSA)** and **Bitmap-based Scheme with HI-tree index (BSH)** respectively. BSA is a viable option for environments with large memory space (e.g., Cloud environments), while BSH is suitable for space-constrained environments. To reduce the memory consumption of BSA, we propose a variant, which keeps both the bitmap arrays (i.e., the candidate set identification bitmap and the document relevance score lookup array) of only the popular keywords in memory, while storing the bitmap arrays of the relatively less popular keywords in the disk.

We designate this variant as the **Bitmap-based Scheme with Array with reduction in Memory (BSAM)**. Our key contributions are three-fold:

1. We propose an efficient bitmap-based approach for quickly identifying the candidate set of entities, whose parent documents contain all queried keywords. A variant of this approach is also proposed to reduce memory consumption by exploiting skews in keyword popularity.
2. We propose the two-tier HI-tree index, which uses both hashing and inverted indexes, for efficient document relevance score lookups.
3. We perform detailed experiments with both real and synthetic datasets to show that our proposed schemes are indeed effective in providing good top-k result recall performance within acceptable query response times.

Ontologies and concept hierarchies [12] can also be used in conjunction with this work for query expansion purposes. The remainder of this paper is organized as follows. Section 2 presents the context of the problem, while Sect. 3 discusses the bitmap-based top-k query processing schemes. Section 4 presents the HI-tree index. Section 5 reports the performance study. We conclude in Sect. 6.

2   Context of the Problem

This section discusses the context of the problem. Recall that each entity has a single parent document with possibly multiple child documents. Let D denote the corpus of N parent documents $\{D_1, D_2, ..., D_N\}$ (one parent document per entity) such that each document D_i contains p_i embedded links $\{L_1, L_2, ..., L_{p_i}\}$. Each link points to a single document. Thus, each embedded link is a child document. Observe that the parent documents remain relatively static over time e.g., a customer's date of birth or email address generally remains the same over time. However, the child documents may be *dynamic* in that new text (e.g., a new review comment, a new blog or a new email) may be inserted. The number of child documents corresponding to different parent documents may vary. Furthermore, the number of child documents corresponding to a given parent document can vary over time due to the addition or deletion of links.

Top-k queries are of the form $\{Q_{id}, k, (a_1, a_2, ..., a_n)\}$, where Q_{id} is the unique identifier of a given query, k is the number of (top-k) entities that need to be retrieved and $\{a_1, a_2, ..., a_n\}$ are queried keywords. Given a top-k query Q, a parent document is considered to be in the candidate set if it contains at least one instance of each of the queried keywords in Q. (Note that if instances of any of the queried keywords occur in the child documents without occurring in the parent document, the parent document would not be in the candidate set.) Thus, the query model uses 'AND' semantics. However, the proposed schemes can also be cost-effectively extended to apply to queries with 'OR' semantics.

Relevance score $S_{Di}(a_j)$ of a given parent document D_i w.r.t. keyword a_j is computed as the weighted average of two components (a) frequency $f_{Di}(a_j)$ of a_j in D_i and (b) frequency $f_{Cw}(a_j)$ of a_j in D_i's child documents. $S_{Di}(a_j) = 0$ if a_j does not occur in the parent document (i.e., when $f_{Di}(a_j) = 0$), regardless of a_j's frequency in child documents. Thus, Eq. 1 below for computing $S_{Di}(a_j)$ is only applicable when a_j occurs at least once in the parent document.

$$S_{Di}(a_j) = [w_1 * f_{Di}(a_j)] + [w_2 * \sum_{w=1}^{v_i} f_{Cw}(a_j)] \qquad (1)$$

where w_1 and w_2 are weight coefficients such that $0 < w_1, w_2 < 1$ and $w_1 + w_2 = 1$. In this work, we set $w_1 = w_2 = 0.5$ (determined experimentally). The values of w_1 and w_2 are essentially application-dependent; we leave their determination to future work. Here, v_i is the number of child documents of D_i. The relevance score of D_i w.r.t. a set of queried keywords is computed as $\sum_{j=1}^{n} S_{Di}(a_j)$, where n is the number of queried keywords. Although we define document relevance scores based on keyword frequencies, other methods of determining document relevance scores [13, 16, 18] can be used in conjunction with our proposed schemes.

3   Top-K Query Processing Schemes

This section discusses the BSA and BSH top-k query processing schemes. Both of these schemes use bitmaps for efficiently determining the candidate set. For relevance score lookups, BSA uses an array-based approach, while BSH uses the HI-tree index, which we present in Sect. 4. Now let us discuss a variant of the brute-force approach to obtain some insights concerning top-k query processing.

Recall that the brute-force approach (discussed in Sect. 1) involves performing exhaustive multi-way join between m document lists, where each document list corresponds to one of the m queried keywords, for determining the candidate set. Consider a corpus of 1 million documents and a top-k query with four keywords. Suppose each queried keyword occurs in any 10% of the documents in the corpus, but all the queried keywords need not necessarily occur together in the same documents. Here, the brute-force approach would need to intersect four document lists each containing 100,000 documents for determining the candidate set, thereby making the processing prohibitively expensive.

To address the scalability issue, we adapt a variant of Fagin's algorithm [10] as follows: (a) Sort each keyword's document list in descending order of keyword frequencies. (b) Consider only upto a fixed percentage of each sorted document list for the multi-way join processing step, thereby reducing the processing cost. Since the number of documents would generally vary across the document lists, this becomes effectively equivalent to making a *zig-zag cut* across the document lists. Hence, we designate this variant approach as the **Zig-zag Cut Scheme (ZCS)**. In ZCS, the number k'_j of documents to be considered for the multi-way join processing step for the jth document list is computed as $P_D * |L_j|$, where P_D is the percentage factor and L_j is the document list corresponding to keyword a_j. We experimentally determine P_D in Sect. 5.

For keeping track of the relevance scores of documents w.r.t. a given keyword, ZCS uses an inverted list structure, which is essentially similar to that of Tier 2 of the HI-tree, which we shall describe in detail in Sect. 4. In particular, as we shall see later, this structure stores the keyword frequencies of the parent and child documents for recomputing the relevance scores when updates occur.

Although ZCS reduces processing costs, it may degrade the recall performance. For example, given queried keywords $\{a_1, a_2\}$, suppose the frequencies of a_1 and a_2 in document D1 are 1000 and 2 respectively. Here, ZCS may fail to identify D1 as being in the candidate set. This is because a_2's low frequency of occurrence in D1 may place D1 towards the end of the sorted document list of a_2, thereby in effect eliminating D1 from consideration for multi-way join processing. In essence, ZCS trades off recall performance for lower processing costs. Thus, there is a clear motivation for *efficiently* identifying the candidate set with high recall performance without incurring high processing cost. The big picture of our proposed three-step top-k query processing approach is presented in Fig. 2. Now we shall describe these steps.

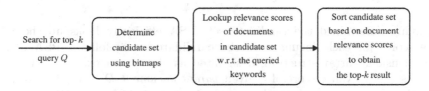

Fig. 2. Big picture of top-k query processing

Determining the Candidate Set Using Bitmaps: Each keyword is associated with a bitmap, where each bit corresponds to a *parent document* in the corpus. A bit is set to 1 if the keyword occurs in the parent document, otherwise it is set to 0. The length of a keyword's bitmap is equal to the total number of parent documents in the corpus. Hence, if there are one million parent documents in the corpus, each keyword would be associated with a bitmap comprising one million bits. By performing a bitwise-AND operation on the bitmaps of the queried keywords, we identify the candidate set as comprising those parent documents, whose corresponding positions in the result bitmap is 1.

Suppose there are N parent documents and M keywords in the system. Thus, the total memory Mem_{arr} consumed by the bitmap array equals $M * N$ bits. When $N = 1$ million and $M = 40{,}000$, $Mem_{arr} = 40$ Gbits or approximately 5 GB.

Figure 3 shows an example for determining the candidate set using bitmaps. Figure 3 shows a document corpus comprising parent documents D1 to D12 with five keywords a_1 to a_5. The bitmap for keyword a_1 is 001001111010, thereby indicating that a_1 occurs in documents $\{D3, D6, D7, D8, D9, D11\}$. Similarly, the bitmap for a_2 is 101011101011, which suggests that a_2 exists in documents $\{D1, D3, D5, D6, D7, D9, D11, D12\}$. Observe how the candidate set is *efficiently* computed for different sets of queried keywords using the bitwise-AND operation without incurring high costs. Notably, updates to the bitmaps occur infrequently since the bitmaps relate *only* to the parent documents, which are relatively *static*.

D1 to D12 : Document corpus							a_1 to a_5 : Keywords				
	D1	**D2**	**D3**	**D4**	**D5**	**D6**	**D7**	**D8**	**D9**	**D10**	**D11**	**D12**
a_1 →	0	0	1	0	0	1	1	1	1	0	1	0
a_2 →	1	0	1	0	1	1	1	0	1	0	1	1
a_3 →	1	1	1	1	0	1	0	0	1	0	1	0
a_4 →	0	0	1	0	0	1	1	1	1	1	1	0
a_5 →	0	0	1	1	1	1	1	1	1	1	1	1

Queried keywords	bitwise-AND result	Candidate set
a_2, a_3	101001001010	{ D1, D3, D6, D9, D11 }
a_4, a_5	001001111110	{ D3, D6, D7, D8, D9, D10, D11 }
a_1, a_2, a_3	001001001010	{ D3, D6, D9, D11 }
a_1, a_2, a_3, a_4, a_5	001001001010	{ D3, D6, D9, D11 }

Fig. 3. Computation of candidate set using bitwise-AND operation

Having determined the candidate set, BSA efficiently looks up the relevance scores for each document in the candidate set as follows. BSA uses a two-dimensional array structure, designated as Arr. An array entry $Arr(i,j)$ indicates the relevance score of a given *parent document* D_i w.r.t. keyword a_j. This relevance score is computed using Eq. 1 (see Sect. 2) by using the weighted keyword frequencies in both document D_i and its child documents. In case of updates to the child documents, keyword frequencies may change, thereby necessitating *periodic* recomputation of the relevance score entries in $Arr(i,j)$. Thus, the relevance scores are updated in Arr *periodically* (i.e., not in real-time when the updates actually occur). For recomputing a given relevance score, we store the keyword frequencies of the parent document and the *total* keyword frequencies of *all* its child documents in an auxiliary array data structure. Such relevance score computations are lightweight even if the child documents get updated frequently i.e., Arr is an update-efficient structure.

Observe how BSA enables efficient relevance score lookups in O(1) time for each document in the candidate set w.r.t. each queried keyword. Then for each candidate set document, it sums up the corresponding relevance scores for all the queried keywords to compute the document's relevance score. Finally, it sorts the documents in descending order of their relevance scores to obtain the top-k result. An illustrative example of BSA is depicted in Fig. 4 for the scenario of Fig. 3, where the candidate set was {D3, D6, D9, D11}. In Fig. 4, the relevance score of a given parent document w.r.t. a given keyword a_j is indicated as $S(a_j)$, and the queried keywords are a_1 to a_5.

Observe that although BSA enables quick keyword frequency lookups, it suffers from the drawback of consuming large amounts of memory space. In practice, a small percentage of popular keywords typically occur in a relatively large number of documents (i.e., high skew). Thus, the two-dimensional array used by BSA is likely to be sparse. Given N parent documents and M keywords, BSA consumes MN bits for storing the bitmap array. Additionally, it consumes $4MN$ bytes for storing the document score lookup array. Hence, its total memory

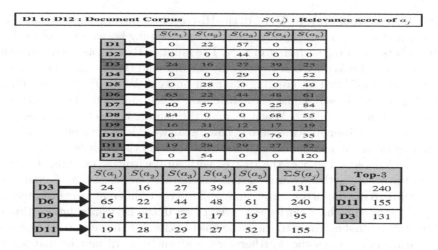

D1 to D12 : Document Corpus					$S(a_j)$: Relevance score of a_j

	$S(a_1)$	$S(a_2)$	$S(a_3)$	$S(a_4)$	$S(a_5)$
D1	0	22	57	0	0
D2	0	0	44	0	0
D3	24	16	27	39	25
D4	0	0	29	0	52
D5	0	28	0	0	49
D6	65	22	44	48	61
D7	40	57	0	25	84
D8	84	0	0	68	55
D9	16	31	12	17	19
D10	0	0	0	76	35
D11	19	28	29	27	52
D12	0	54	0	0	120

	$S(a_1)$	$S(a_2)$	$S(a_3)$	$S(a_4)$	$S(a_5)$	$\Sigma S(a_j)$		Top-3
D3	24	16	27	39	25	131	D6	240
D6	65	22	44	48	61	240	D11	155
D9	16	31	12	17	19	95	D3	131
D11	19	28	29	27	52	155		

Fig. 4. Illustrative example for BSA

consumption is $4.125MN$ (i.e., $(1/8)MN + 4MN$) bytes. For example, when $M = 1$ million and $N = 40000$, its total memory consumption equals 165 GB.

This motivates us to propose the **BSAM approach**, which is a memory-efficient variant of the BSA approach. BSAM keeps both the bitmap arrays (i.e., the candidate set identification bitmap and the document relevance score lookup array) of only the popular query keywords in memory, while storing both the bitmap arrays of the relatively less popular keywords in the disk. Since the vast majority of queries generally pertain to popular keywords, we do not expect BSAM to perform significantly worse than BSA. However, BSAM consumes much less memory than BSA. For example, given a dictionary of 40000 keywords, suppose there are 2000 popular keywords. Given 1 million parent documents, the memory consumption of BSAM would be (4.125 * 2000 * 106) = 8.25 GB. The next section presents another approach for memory-constrained environments.

4   The HI-TREE Index for Computing Relevance Scores

This section discusses the dynamic HI-tree index, which enables the efficient lookup of document relevance scores for facilitating top-k query processing. The HI-tree is a two-tier index, which uses a combination of hashing and inverted indexes. In Tier 1 of the HI-tree, dictionary keywords are mapped into hash buckets, and the hash buckets are organized in the form of a B+-tree to facilitate speedy access to any given hash bucket. Tier 2 comprises inverted indexes for facilitating keyword search within these hash buckets.

Notably, the HI-tree indexes only the *parent documents* because our top-k result set comprises only the parent documents. Moreover, it also keeps track of keyword frequencies in the child documents to enable it to efficiently recompute document relevance scores in case of updates to the child documents.

Tier 1 (Hash Buckets): In Tier 1 of the HI-tree, hashing is performed based on the notion of *pivots* on any given keyword a_j for mapping keywords a_1 to a_N into hash buckets B_1 to B_m such that $m \gg N$. The number P of pivots determines the number of hash buckets. When $P = 0$, only the first and the last characters in the keyword are considered for creating the hash buckets. When $P = 1$, the middle character is additionally considered. In effect, the middle character partitions the keyword into two parts. When $P = 3$, the middle characters of these two partitions are additionally considered. Thus, a keyword is recursively partitioned depending upon the number P of pivots to determine the hash buckets.

Figure 5 depicts Tier 1 of the HI-tree with an illustrative example. Observe that for the even-length keywords such as 'ameliorate', we select the left character among the two middle characters 'i' and 'o'. Moreover, different keywords may be mapped to the same hash bucket e.g., when $P = 0$, the keywords 'procrastinate' and 'penultimate' both map to the same hash bucket 'pe' because their first and last characters are the same. Furthermore, a given keyword may be mapped to the same hash bucket for different values of P e.g., 'net' gets mapped to the same hash bucket 'net' for $P = 1$ and $P = 3$ due to its short keyword-length.

Incidentally, the number N_B of hash buckets increases dramatically as the value of P increases. For example, when $P = 0$, $N_B = (26 * 26) = 676$. However, when $P = 1$, $N_B = (26 * 26 * 26) = 17576$. The trade-off here is that lower values of P result in fewer hash buckets albeit with possibly more keywords in each bucket, thereby increasing search time within the relevant bucket. Conversely, for higher values of P, finding the relevant hash bucket incurs more time since there are more buckets; however, search within the relevant bucket would be relatively fast because each bucket would have fewer keywords. Our preliminary experiments showed that $P = 1$ is a reasonable value to address this trade-off.

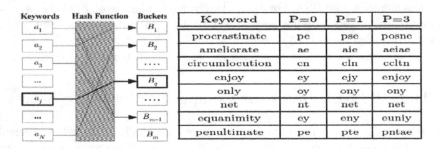

Keyword	P=0	P=1	P=3
procrastinate	pe	pse	posne
ameliorate	ae	aie	aeiae
circumlocution	cn	cln	ccltn
enjoy	ey	ejy	enjoy
only	oy	ony	ony
net	nt	net	net
equanimity	ey	eny	euniy
penultimate	pe	pte	pntae

Fig. 5. Mapping keywords into hash buckets

Figure 6 illustrates how search is conducted for finding a given bucket. For quickly locating the hash bucket, which is relevant to a given keyword-search query, the hash buckets are organized in the form of a B+-tree. The ordering of the hash buckets in the B+-tree is based on the dictionary order of the words, which we had assigned (as identifiers) to these buckets. For example, the bucket

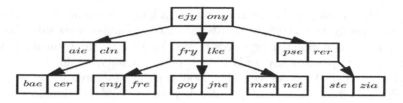

Fig. 6. Example: Hash tree to find relevant bucket

'aie' occurs to the left of the bucket 'eny' in the B+-tree because the word 'aie' appears before the word 'eny' in dictionary order.

Tier 2 (Searching for a Keyword within a Given Hash Bucket): In Tier 2 of the HI-tree, the keywords within a given hash bucket are organized in the form of a B+-tree structure, whose nodes are augmented with linked lists. The keyword ordering in the B+-tree is based on the dictionary word order. Each node of the Tier 2 index contains entries of the form $(kword, LL)$, where $kword$ is a given dictionary keyword and LL is a linked list. Each node of LL contains entries of the form (doc_id, S, f_D, f_c), where doc_id refers to the unique identifier of the document containing $kword$ and S is the relevance score of the document w.r.t. $kword$. Here, f_D and f_c represent the frequencies of $kword$ for the parent document and its child documents respectively.

The values of f_D and f_c are required for recomputing the value of the relevance score S from Eq. 1 (see Sect. 2), when updates to the child documents occur. We do not store the individual keyword frequencies for the child documents because when child documents get updated, we can incrementally add or subtract from the total keyword frequency for the child documents. Thus, the HI-tree is update-efficient as relevance score recomputations are lightweight even when the child documents get frequently updated.

Updates to the child documents are *periodically* incorporated into the HI-tree as follows. If new text is inserted into any child document of a given parent document D_i, the frequency of each keyword of the new text is first recorded. Then, for each keyword a_j, the HI-tree is searched. If a_j exists in the HI-tree, the search proceeds through the linked list attached to a_j. In case D_i occurs in this linked list, the value of f_C in that linked list node is updated by adding the frequency of a_j in the new text to the existing value of f_C. Otherwise, no action needs to be taken because the update did not impact any parent document.

5   Performance Evaluation

This section reports the performance evaluation using both real and synthetic datasets. The synthetic dataset had 1 million parent documents. For each parent document, the number of child documents was randomly selected to be between 8 and 12. The experiments use a set of 40,000 unique dictionary-keywords (maximum length of 25 characters). Each of parent and child documents have 1000 unique keywords, their frequencies randomly varying from 5 to 40.

In practice, a few popular keywords receive a large percentage of the queries. Hence, we classify the 40,000 keywords into 2,000 popular keywords, the rest being regarded as unpopular keywords. Thus, for BSAM, the bitmap arrays of only these 2,000 popular keywords were kept in memory, while the rest were stored on the disk. Here, we consider 2000 popular keywords as an example, although the actual number of popular keywords could be significantly less depending upon the application. Table 1 summarizes the performance study parameters. The experiments consider 10 different keyword classes. The number of keywords in each keyword class is determined based on a Zipf distribution with a zipf factor of 0.7 (i.e., high skew). The 40,000 unique dictionary-keywords are randomly assigned to any one of these keyword classes. Furthermore, 40,000 keywords are distributed across total number of documents with a *zipf* factor ZF_D of 0.7 (i.e., high skew). Here, we consider 10 different document buckets, in which the keywords are randomly assigned from any one of the 10 keyword classes. Moreover, the number of queries corresponding to each keyword class is determined based on a Zipf distribution with a *zipf* factor ZF_Q over 10 query buckets. Consistent with real-world scenarios, we set the value of ZF_Q to 0.7 (i.e., high skew) to ensure that a relatively small percentage of keywords receive a disproportionately large number of queries.

Table 1. Parameters for the performance study

Parameter	Default	Variations
Percentage of documents: P_D	60	20, 40, 80, 100
No. of documents: N_D (10^5)	10	2, 4, 6,8
No. of queried keywords: N_{QK}	5	2, 3, 4, 6
k (of top-k query)	100	
Zipf factor for Queries: ZF_Q	0.7	
Zipf factor for Data: ZF_D	0.7	0.1, 0.3, 0.5, 0.9

We could not obtain an SOE-related real dataset for our motivating CRM application scenario because such SOEs are still mostly beginning to be deployed. Hence, for our experiments, we used real data from the virtualtourist.com travel website for modeling parent and child documents. The real dataset used in the experiments comprised a total of 100,000 webpages (as the parent documents) from the virtualtourist.com travel website. Each of these web pages had 10 embedded links, each link pointing to the homepage of a specific hotel. (Each such homepage contains details such as hotel news, prices and user review comments.) Thus, the documents pointed to by each of these 10 links were downloaded to serve as the child documents for the real dataset.

The performance metric is **average response time (ART) of queries.** $ART = \frac{1}{N_Q} \sum_{q=1}^{N_Q} t_r$ where t_r is the query response time and N_Q is the total number of queries. Additionally, to quantify the top-k result accuracy, **query**

recall percentage (REC) is used as a performance metric. REC quantifies the percentage of correct top-k documents that a given scheme was able to retrieve. Notably, REC is computed as the average query recall percentage over all the queries. Thus, $REC = \frac{1}{N_Q} \sum_{q=1}^{N_Q} [|Tc \cap Tr|\,/\,|Tc|]$, where T_c represents the document set comprising the correct top-k query result, while T_r is the document set comprising the top-k query result retrieved by a given scheme. For example, if the correct top-5 result were D1 to D5, and the top-k result retrieved by a given scheme was {D1, D2, D3, D8, D10}, REC would be 60% because only three out of the five correct top-5 documents were retrieved.

We ran experiments on a machine with Intel Core-2Duo T6600 2.2 GHz processor with 64-bit OS and only 16 GB RAM. Recall that more than 160 GB of RAM is needed for running BSA completely in memory as BSA consumes huge amounts of memory due to the bitmap array for document relevance score lookups. Hence, we implemented BSA by keeping the relevance score lookup array in the disk and loaded it into the memory in 8 GB chunks, thereby increasing BSA's processing time. We ran each experiment 10 times and averaged the results. Confidence intervals for our experiments ranged from 92% to 95%.

As reference, we use the ZCS scheme (see Sect. 3). Recall that ZCS requires as input a percentage factor P_D, which addresses the trade-off between recall performance and query response times. Now, we find P_D experimentally.

(a) ART   (b) REC

Fig. 7. Determining percentage factor for ZCS

Determining the Percentage Factor for ZCS: Figure 7 depicts the results on varying P_D. As P_D is increased, a larger percentage of the documents in the corpus are considered for the multi-way join processing step, thereby improving REC albeit at the cost of increased ART. Observe that at PD = 60%, a reasonably good value of REC (i.e., 75%) can be obtained without incurring prohibitively high ART. Hence, we set PD = 60% for all the experiments involving ZCS.

Effect of Varying the Number of Documents: Figures 8a and 8b depict the effect of varying the number N_D of documents for the real dataset. As N_D increases, ART increases for all the schemes due to the increased computational costs of processing a larger number of documents. However, ART increases at a much slower rate for the proposed BSH, BSA and BSAM schemes than for ZCS.

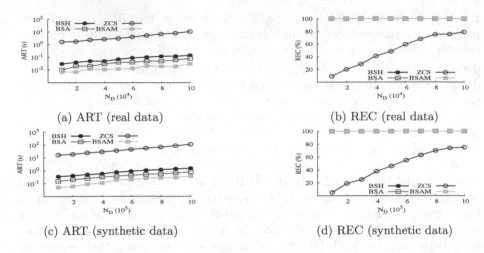

(a) ART (real data)   (b) REC (real data)

(c) ART (synthetic data)   (d) REC (synthetic data)

Fig. 8. Effect of varying the number of documents

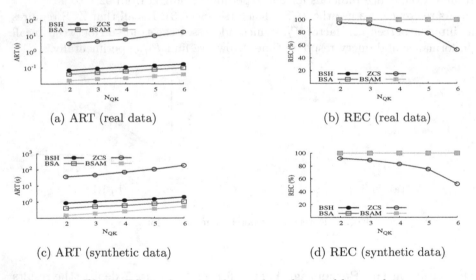

(a) ART (real data)   (b) REC (real data)

(c) ART (synthetic data)   (d) REC (synthetic data)

Fig. 9. Effect of varying the number of queried keywords

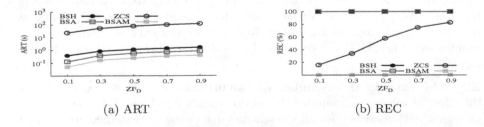

(a) ART   (b) REC

Fig. 10. Effect of skew in keyword distribution

This occurs because these schemes compute the candidate set efficiently by using the bitwise-AND operation, while ZCS uses costly multi-way join processing.

BSAM incurs lower ART than BSA because it reduces memory space consumption by exploiting skews in keyword popularity. Since most of the queries concern popular keywords, BSAM is able to avoid disk accesses for most queries, thereby saving on query processing times. BSA incurs significant processing times due to loading 8 GB chunks from disk into memory, thereby increasing ART. However, BSA incurs lower ART than BSH because its array-based structure allows it to look up document relevance scores quicker than BSH, which requires HI-tree index traversal for document relevance score lookups.

As N_D increases, REC remains constant at its highest possible value (i.e., 100%) for BSH, BSA and BSAM. This occurs because the bitwise-AND operation used by these schemes is able to efficiently generate the candidate set with 100% recall. However, for ZCS, REC degrades with increase in N_D because an increased number of documents in the document corpus do not get considered towards the top-k processing. The results indicate that both BSH and BSA indeed scale well w.r.t. the number of documents. Figures 8c and 8d depict the results for the synthetic dataset comprising 1 million documents. Observe that the results for real and synthetic datasets exhibit similar trend. ART varies across these results due to differences in dataset size.

Effect of Varying the Number of Queried Keywords: Figures 9a and 9b depict the effect of varying the number N_{QK} of queried keywords for the real dataset. As N_{QK} increases, ART increases for ZCS because multi-way join operations need to be performed across a larger number of document lists (one document list/keyword). ART also increases albeit less for BSH, BSA and BSAM because the bitwise-AND operation is performed on an increased number of bitmaps (one bitmap/keyword). REC remains constant at 100% for these schemes due to the recall accuracy of the bitwise-AND operation, as explained for the results in Fig. 8. As N_{QK} increases, ZCS exhibits lower values of REC partly because performing the multi-way join step across only 60% of the documents per document list degrades the recall performance and partly due to the reasons explained for the results in Figure 8. Figures 9c and 9d depict the results of this experiment for the synthetic dataset. Notably, the results for real and synthetic datasets exhibit similar trend.

Effect of Skew in Keyword Distribution: This experiment examines the effect of skew in the distribution of the keywords across the documents. Recall that higher values of ZF_D imply that a small percentage of the keywords (i.e., the 'popular' keywords) occur in a large percentage of the documents. Figure 10 depicts the results. For this experiment, queries were also highly skewed towards a small number of these 'popular' keywords (i.e., using the default value of ZF_Q = 0.7). Hence, as ZF_D increases, ART increases for all the schemes due to increase in the size of the document lists associated with each of the 'popular' queried keywords, thereby necessitating more documents to be processed. Moreover, as ZF_D increases, REC increases for ZCS primarily due to the involvement of a larger number of documents towards the top-k query processing.

6   Conclusion

This work has addressed the problem of *efficiently* retrieving the top-k entities in an SOE for keyword-based queries. We have proposed an efficient bitmap-based approach for quickly identifying the candidate set as well as a variant for reducing memory consumption by exploiting skews in keyword popularity. We have also proposed the two-tier HI-tree index, which uses both hashing and inverted indexes, for efficient document relevance score lookups. Our detailed evaluation with both real and synthetic datasets shows that our proposed schemes are indeed effective in providing good top-k result recall performance within acceptable query response times with good scalability. In the future, we plan to compare the performance of our proposed approach w.r.t. existing works.

References

1. http://blogs.hbr.org/cs/2011/10/moving_from_transaction_to_eng.html
2. http://www-01.ibm.com/software/ebusiness/jstart/systemsofengagement/
3. http://www.bersin.com/blog/post/Systems-of-Engagement-vs-Systems-of-Record--About-HR-software2c-design-and-Workday.aspx
4. Agrawal, R., Fuxman, A., Kannan, A., Shafer, J., Talukdar, P.P.: Associating structured records to text documents. In: WWW (2012)
5. Agrawal, S., Chaudhuri, S., Das, G.: DBXplorer: a system for keyword-based search over relational databases. In: ICDE (2002)
6. Baeza-Yates, R.A., Ribeiro-Neto, B.A.: Modern Information Retrieval. ACM Press (1999)
7. Bhalotia, G., Hulgeri, A., Nakhe, C., Chakrabarti, S., Sudarshan, S.: Keyword searching and browsing in databases using BANKS. In: ICDE (2002)
8. Chakrabarti, S., Dom, B., Indyk, P.: Enhanced hypertext categorization using hyperlinks. In: SIGMOD (1998)
9. Ding, B., Zhao, B., Lin, C., Han, J., Zhai, C.: TopCells: keyword-based search of top-k aggregated documents in text cube. In: ICDE (2010)
10. Fagin, R., Lotem, A., Naor, M.: Optimal aggregation algorithms for middleware. Comp. Sys. Sci. **66**(4) (2003)
11. Feldman, R.: Link analysis: current state of the art. In: KDD Tutorial (2002)
12. Han, J., Fu, Y.: Dynamic generation and refinement of concept hierarchies for knowledge discovery in databases. In: KDD Workshop (1994)
13. Hristidis, V., Gravano, L., Papakonstantinou, Y.: Efficient IR-style keyword search over relational databases. In: VLDB (2003)
14. Kimelfeld, B., Sagiv, Y.: Finding and approximating top-k answers in keyword proximity search. In: PODS (2006)
15. Kleinberg, J.M.: Authoritative sources in a hyperlinked environment. J. ACM **46**(5) (1999)
16. Liu, F., Yu, C., Meng, W., Chowdhury, A.: Effective keyword search in relational databases. In: SIGMOD (2006)
17. Lu, Q., Getoor, L.: Link-based classification. In: ICML (2003)
18. Luo, Y., Lin, X., Wang, W.: SPARK: Top-k keyword query in relational databases. In: SIGMOD (2007)
19. Page, L., Brin, S., Motwani, R., Winograd, T.: The PageRank citation ranking: bringing order to the web. Technical report, Stanford InfoLab (1999)

20. Wu, P., Sismanis, Y., Reinwald, B.: Towards keyword-driven analytical processing. In: SIGMOD (2007)
21. Xin, D., Han, J., Cheng, H., Li, X.: Answering top-k queries with multi-dimensional selections: the ranking cube approach. In: VLDB (2006)

A Compare-Aggregate Model with External Knowledge for Query-Focused Summarization

Jing Ya[1,2(✉)], Tingwen Liu[1,2], and Li Guo[1,2]

[1] Institute of Information Engineering, Chinese Academy of Sciences, Beijing, China
{yajing,liutingwen,guoli}@iie.ac.cn
[2] School of Cyber Security, University of Chinese Academy of Sciences, Beijing, China

Abstract. Query-focused extractive summarization aims to create a summary by selecting sentences from original document according to query relevance and redundancy. With recent advances of neural network models in natural language processing, attention mechanism is widely used to address text summarization task. However, existing methods are always based on a coarse-grained sentence-level attention, which likely to miss the intent of query and cause relatedness misalignment. To address the above problem, we introduce a fine-grained and interactive word-by-word attention to the query-focused extractive summarization system. In that way, we capture the real intent of query. We utilize a Compare-Aggregate model to implement the idea, and simulate the interactively attentive reading and thinking of human behavior. We also leverage external conceptual knowledge to enrich the model and fill the expression gap between query and document. In order to evaluate our method, we conduct experiments on DUC 2005–2007 query-focused summarization benchmark datasets. Experimental results demonstrate that our proposed approach achieves better performance than state-of-the-art.

Keywords: Query-focused summarization · Extractive summarization · Attention · External knowledge

1 Introduction

The purpose of query-focused summarization is to synthesize from a set of documents a brief, well-organized, fluent summary, which can meet the need for information related to a specific query [1]. Such task has various application scenarios, such as news services and search engines, where the results can be customized to specified requirements for information. This paper focuses on the task of query-focused extractive summarization. It directly selects sentences from original document to form a summary. The task of query-focused extractive summarization can be decomposed into two subtasks, *i.e.*, sentence scoring and sentence selection. The former scores a sentence to measure query relevance,

Z. Huang et al. (Eds.): WISE 2020, LNCS 12343, pp. 68–83, 2020.
https://doi.org/10.1007/978-3-030-62008-0_5

and the latter chooses sentences to create a summary by considering both query relevance and redundancy.

With advances of neural network models in natural language processing, deep learning has been used in text summarization task by generating better representations [2,3] than surface lexical features for sentences. There exist approaches utilizing attention mechanism to addressed the problem of query-focused summarization [4,5], which simulates human attentive reading behavior. Specifically, such approaches first encode sentences both in query and document into vectors with context information. Then, they leverage attention weight for each sentence to generate a global representation of the full document with considering query information. At last, they measure query relevance of each sentence in document with the generated representation.

However, such approaches are based on a coarse-grained sentence-level attention, which is likely to miss the real intent of query and causes relatedness misalignment. It must be emphasized that semantic similarity and semantic relevance are not same. The context level representation is always used in semantic similarity calculation, but fails in capturing the intent of query.

Take document cluster "d671g" in DUC 2005 as an example. The query is:

What are the benefits provided by the Salvation Army? To what extent do they support other organizations, both within and outside the US?

Here, "Salvation Army" and "organizations" are the key words in the query. And yet the main intent of the query is to look for "benefits" and "extent". When measuring relevance with a context level representation, sentences talking about the "Salvation Army" or "organizations within and outside the US" may have high scores. The reason is that the intent of query "benefits" has little proportion, it may be missed in the context information. Such words which represent intent of query are important for relevance measuring. Finding out such words and enhancing the weights of them can help to make a more effective summary related to the query.

To address the above problem caused by sentence-level attention, we introduce a fine-grained and interactive word-by-word attention to the query-focused extractive summarization system. In that way, we can capture the real intent of query. Specifically, we utilize a Compare-Aggregate framework to implement the idea. Furthermore, to fill the expression gap between query and document, we leverage conceptual knowledge to enrich the model.

To evaluate our proposed method, we conduct experiments on DUC 2005–2007 benchmark datasets. Experimental results show that the model achieves competitive performance and outperforms previous state-of-the-art models.

The main contributions of this paper are as follows:

- To address the problem of relatedness misalignment caused by the existing context level representation methods, we introduce the idea of a fine-grained and interactive word-by-word attention to a query-focused extractive summarization system.
- To implement the idea, we utilize a Compare-Aggregate model to simulate the interactively attentive reading and thinking of human behavior. We also

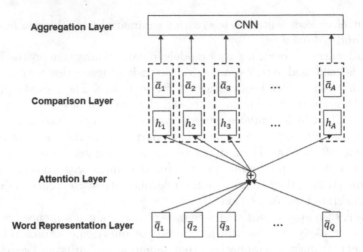

Fig. 1. Architecture of the standard compare-aggregate model

leverage external conceptual knowledge to fill the expression gap between query and document.

- We verify our proposed approach on the widely-used DUC 2005–2007 benchmark datasets. Experimental results demonstrate the superiority of our approach, which achieve state-of-the-art performance.

2   Background

2.1   Compare-Aggregate Model

Compare-Aggregate model is first proposed in studies on text sequence matching problem [6] In such a framework, comparison of two sequences is not done by comparing two vectors each representing an entire sequence. Instead,it first compares vector representations of smaller units such as words from sequences and then aggregates these comparison results to make the final decision. Studies have shown the effectiveness of such a "compare-aggregate" framework for sequence matching task, for example natural language inference [7] and answer selection [8,9].

Architecture of a standard Compare-Aggregate model is shown in Fig. 1. The model can be divided into the following four layers:

Word Representation Layer. The purpose of this layer is to obtain a new embedding vector for each word in each sequence(Q and A) that captures some contextual information in addition to the word itself, *i.e.*, $\bar{Q} = \{\bar{q}_1, \bar{q}_2, ..., \bar{q}_Q\}$, $\bar{A} = \{\bar{a}_1, \bar{a}_2, ..., \bar{a}_A\}$. $\bar{q}_i$ is the representation of i-th word q_i, and analogously for $\bar{a}_j$ and a_j.

Attention Layer. An attention mechanism is applied on $\bar{Q}$ and $\bar{A}$ to obtain attention weights over the column vectors in $\bar{Q}$ for each column vector in $\bar{A}$. With these attention weights, for each column vector $\bar{a}_j$ in $\bar{A}$, it obtains a corresponding vector h_j, which is an aligned sub-phrase(attention-weighted sum) of the column vectors of $\bar{Q}$.

Comparison Layer. A comparison function is used to combine each pair of $\bar{a}_j$ and h_j into a vector. This layer separately compares each aligned sub-phrase to produce a set of vectors.

Aggregation Layer. This layer aggregates the set of vectors from the comparison layer using a CNN model, and use the result to predict the label for the final text matching task.

Obviously, the Compare-Aggregate model can learn a fine-grained and interactive attention for smaller units from input sequence pairs. In this paper, we utilize the Compare-Aggregate model to calculate the relevance matching score between query and each sentence in document, with considering the interactive word-by-word attention.

2.2   External Knowledge

External conceptual knowledge refers to the knowledge of principles, generalizations, theories, models, or structures pertinent to a particular disciplinary area. It is useful to understand meanings of words with common sense knowledge in reality. Some previous work attempts to leverage external knowledge to deal with various tasks in natural language processing, for example, information retrieval [10], recommendation systems [11], question answering [12] and inference [13].

The prominent concern of choosing external knowledge is in terms of applicability to the task. For example, DBpedia is a generic knowledge base which extracts structured content from the information created in the Wikipedia infoboxes. By contrast, ConceptNet consists of common-sense knowledge acquired through crowdsourcing. WordNet is a similar knowledge base with ConceptNet, but is restricted to a small number of linguistic relationships among terms. DBpedia is usually used for entity-based tasks such as recommendation system and entity disambiguation. And, ConceptNet is more appropriate for tasks required common sense reasoning.

For the task of query-focused summarization, we need to understand the intent of query. While, words in query and document are often different in expression. For example, a query may contain conceptual words like "crime", "country"; while in documents, words which are relevant may be "smuggling" and "Mexico". Words in a query are always conceptual, and ones in document are often with detailed information. So, we need common sense reasoning to understand the real intent of query when measuring the relevance score of each sentence in document.

In this paper, we choose ConceptNet as external knowledge to enrich the Compare-Aggregate model. ConceptNet [14] consists of common-sense knowledge acquired through crowdsourcing, and is a freely-available semantic network, designed to help computers understand the meanings of words that people use. ConceptNet is optimized for making practical context-based inferences over real-world texts. In addition to its own knowledge about the English term astronomy, for example, ConceptNet contains links to URLs that define astronomy in WordNet, Wiktionary, OpenCyc, and DBPedia.

ConceptNet 5.7 provides a REST API[1] at `api.conceptnet.io` for users to acquire concepts from ConceptNet in JSON-LD format. User can easily map words or phrases to the corresponding concepts in ConceptNet, and obtain relationships between two concept nodes.

3   Approach

3.1   Problem Formulation

A classic framework for query-focused extractive summarization system is illustrated in Fig. 2. The framework takes query and document as raw input, and output a query-focused summary through the query-focused extractive summarization system. The system involves two main modules, namely Sentence Relevance Scorer and Sentence Selector, and a preprocess module named Sentence Tokenizer.

Sentence Tokenizer decomposes the input document into sentences. The tokenized sentences and query are sequentially processed to generate a query-focused summary finally.

Let Q denotes the input query, which contain the intent for information requirement. And, $S_1, S_2, ..., S_n$ are the sentences decomposed from input document.

Sentence Relevance Scorer quantifies the relevance matching between query and each sentence with a score $Score(Q, S_i)$. Sentence Selector selects sentences with high relevance scores and little redundancy to output a query-focused summary that meets the requirement.

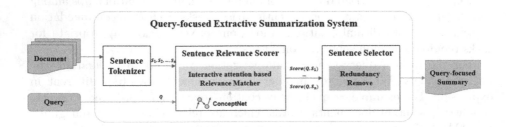

Fig. 2. Classic framework for query-focused extractive summarization

---

[1] https://github.com/commonsense/conceptnet5/wiki/API.

3.2 Sentence Relevance Scorer

The main contribution of this paper is to introduce a fine-grained and interactive word-by-word attention based relevance matching method for the Sentence Relevance Scorer.

An overview of the architecture of our proposed approach is shown in Fig. 3.

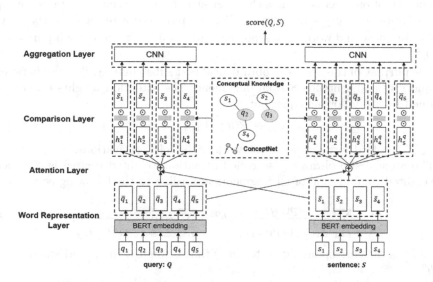

Fig. 3. Architecture of interactive attention based relevance matcher

Word Representation. This layer is to capture the contextual information contained in query and document by embedding each word into a vector, which is fed to subsequent steps of the model. As shown in Fig. 3, $\bar{q}_i$ is the representation vector of the i-th word q_i in query Q, and $\bar{s}_j$ is the representation vector of the j-th word s_j in the sentence S decomposed from input document.

Recent studies have shown that a pre-trained language model can help to capture the contextual meaning of the words in the sentence, and improve the performance of various tasks in natural language processing. We select a pre-trained BERT model [15] to generate the embedding of each word in input sentences, which is published by Google. Specifically, BERT model uses word-piece tokenization, and provides contextual embedding for the word-pieces. It is based on the methodology of transformers [16] and uses attention mechanism, is one of the most powerful context and word representations.

In this paper, we use a token level embedding library to obtain the embedding from BERT's pre-trained model[2]. Instead of building and fine-tuning for an end-to-end language model, we build our model by utilizing the token embedding.

---

[2] https://pypi.org/project/bert-embedding/.

Attention. We apply a standard attention mechanism on the query Q and sentence S to obtain attention weights over the column vectors in Q for each column vector in S. With these attention weights, for each column vector $\bar{s}_j$ in S, we obtain a corresponding vector h_j^Q, which is an attention-weighted sum of the column vectors of Q. In that way, we can fully learn the correlation between the query and each sentence.

The attention layer computes the inter-attention between the representation of each word in query and sentence. We compute the attention matrix e_{ij} of word q_i in query and word s_j in sentence, and obtain soft aligned sub-phrase h_j^Q and h_i^S.

The purpose of this step is to obtain local text substructure relevance between the query Q and sentence S from document. The attention weights e_{ij} can be obtained by the following equation:

$$e_{ij} = F(\bar{q}_i)^T F(\bar{s}_j). \tag{1}$$

where F is a feed-forward neural network with ReLU activations.

The aligned sub-phrases (also called attention-weighted vector) is obtained by weighted combination of all fragments according to the following equation:

$$h_j^s = \sum_{i=1}^{l_Q} \frac{exp(e_{ij})}{\sum_{k=1}^{l_Q} exp(e_{kj})} \bar{q}_i, h_i^q = \sum_{j=1}^{l_S} \frac{exp(e_{ij})}{\sum_{k=1}^{l_S} exp(e_{ik})} \bar{s}_j. \tag{2}$$

where h_j^s is the corresponding vector in Q that is aligned to s_j, and analogously for h_i^q.

It can be used to learn a word-by-word level and interactive attention between query and sentence. So, we can obtain the intent in query, and measure the relevance with a fine-grained model. Next we will combine h_j^s and $\bar{s}_j$ using a comparison layer.

Comparison. The goal of the comparison layer is to match each $\bar{s}_j$ (the j-th word in sentence S and its context in document) with h_j^s (a weighted version of query Q that best matches $\bar{s}_j$). A comparison function is used to match each word in the query and sentence to a corresponding attention-applied vector representation:

$$C^s = \bar{S} \odot h^s, C^q = \bar{Q} \odot h^q. \tag{3}$$

where $\odot$ denotes element-wise multiplication.

The comparison layer get the comparative information between query and sentence. While, in reality, query and document usually have discrepancy in expression of words. For example, a query may contain words like "crime", "country"; while in documents, words which are relevant may be "drug" and "Mexico". Words in a query are always conceptual, while the ones in documents are often with detailed information.

To address the discrepancy in expression of query and documents, we leverage external conceptual knowledge to enrich the model. Specifically, we use Concept-Net [14] to handle the expression gap.

External Conceptual Knowledge. In this paper, we construct a subgraph which is derived from ConceptNet based on the concepts mentioned in the input query and document by mapping words to concepts in the external knowledge graph. This subgraph includes connections between the concepts mentioned in the query and each sentence in document, via concepts available in ConceptNet. Such subgraph enhances the concepts mentioned in the query and document with additional information and can improve the performance of relevance matching in our task.

For each word in query and document, we use conceptual mapping with the knowledge graph from ConceptNet to map individual words to concepts. Those concepts make up the vertices of the Concepts. All edges connecting those vertices are also included in the subgraph. Then we use a feed-forward neural network to calculate the conceptual weight for each pair of words. Specifically, the concepts that act as the input for the neural network model is initialized with the corresponding concept embedding. The concept embedding vectors are pretrained by knowledge graph embedding [17], which are openly available for ConceptNet. These embeddings benefit from the fact that they have semi-structured, common sense knowledge from ConceptNet, giving them a way to learn about words that isn't just observing them in context.

For two words q_i and s_j, which have a edge between the corresponding concepts in ConceptNet, we calculate a conceptual attention weight c_{ij}.

Let $\bar{q}_i^c$, $\bar{s}_j^c$ denote the concept embedding vectors for words q_i and s_j respectively, the conceptual attention weight c_{ij} can be calculated by the following equation:

$$c_{ij} = G(\bar{q}_i^c)^T G(\bar{s}_j^c). \tag{4}$$

where G is a feed-forward neural network with ReLU activation.

Then, we alter the comparison function as:

$$C_a^s = \bar{S} \odot h^s \odot \sum_{i=1}^{l_Q} c_{ij}, C_a^q = \bar{Q} \odot h^q \odot \sum_{j=1}^{l_S} c_{ij}. \tag{5}$$

In that way, we enhance the affection of word pairs which have conceptual relationships. The performance of relevance matching can be improved by addressing the expression gap.

Aggregation. In the aggregation layer, we apply a comparison function to each pair of $\bar{s}_j$ and h_j^s to obtain a series of vectors. Finally, we aggregate these vectors using a one-layer CNN [18] with n-types of filters and calculate the relevance score for each sentence S. The output layer is used to score sentences for query-focused extractive summarization. In our approach, we try to predict the relevance matching score and develop a list-wise approach.

3.3   Sentence Selection

After obtaining the relevance score of each sentence, Sentence Selection selects sentences to generate a summary.

A summary is expected to provide both relevant and non-redeemable information. Our proposed approach focuses on sentence scoring. We employ a simple greedy algorithm used in previous work [4] to select sentences which can be chosen as summary. We sort the sentences in descending order according to the derived relevance matching scores. And, we iteratively dequeue the top-ranked sentence, and append it to the current summary if it is non-redundant. A sentence is considered non-redundant if it contains significantly new bi-grams compared with the current summary content. We set the threshold of the new bi-gram ratio to 0.5.

4   Experiments and Results

4.1   Experimental Setup

Datasets. We implement our experiment to evaluate the proposed approach on DUC 2005–2007 query-focused summarization benchmark datasets. A brief description of statistics for the three datasets are shown in Table 1.

DUC 2005–2007 datasets contain real-life complex questions, particularly used for query-based text summarization purpose. The task is to synthesize a fluent, well-organized 250-word summary from a set of 25–30 documents that answers the question in the query statement. All the documents are from news websites and grouped into various clusters. There are several reference summaries created by NIST assessors.

We use a Nltk package to decompose sentences from documents, and we discard sentences less than eight words. Our summarization model compiles the documents in a cluster into a single document. From Table 1, we can find that the data sizes of DUC are quite different. The number of sentence in DUC 2007 is only about a half of DUC 2005.

Implementation Details. As our approach is supervised and under a extractive framework, we need to label sentences in document. We annotate the dataset as a binary classification problem. Specifically, we calculate the ROUGE-1 scores of all the sentences in the training dataset. Those sentences with high ROUGE-1 scores are regarded as positive samples, and the rest as negative samples. We set the cut-off of score is 0.5.

In the training process, we apply the list-wise ranking strategy to tune model parameters. The initial learning rate is 0.1 and batch size is 32. And, we conduct a 3-fold cross-validation on the datasets, with two years of data as the training set and one year of data as the test set.

Table 1. Statistics of the DUC datasets

Dataset	Query	Sentences	Source
DUC2005	50	41516	TREC
DUC2006	50	28743	AQUAINT
DUC2007	45	19705	AQUAINT

Evaluation Metric. We evaluate our approach with the metric ROUGE. ROUGE is a standard and widely-used intrinsic-based metric by comparing candidate and reference summary. It measures quality of summary in terms of overlapping units such as the n-grams, word sequences and word pairs between the peer summary and reference summaries. In this paper, we use official metrics of ROUGE-1 (unigram-based), ROUGE-2 (bigram-based) on DUC 2005–2007 datasets for our experiment purpose.

4.2 Baselines

To evaluate the summarization performance, we compare our model with several other state-of-the-art summarization methods.

Firstly, as our method is data-driven, we compare it with a unsupervised method CES [19] which requires no label data. CES provides a solution to generate a summary by extracting a subset of sentences using the Cross-Entropy (CE) method based on a combination of several query-dependent and query-independent features.

Secondly, we compare our method with deep learning based supervised methods. AttSum [4] is a sentence-level global attention based methods. It is a neural attention summarization system that tackles query relevance ranking and sentence salience ranking jointly. It generates distributed representations for sentences as well as the document cluster. Meanwhile, it applies the attention mechanism that tries to simulate human attentive reading behavior when a query is given. We also compare our method with a BERT-only model, which is based on sentence-level pre-trained BERT model to extract summaries from the documents.

At last, we implement our method in two ways, one is the standard Compare-Aggregate model without external knowledge which called without_EK and the other is with external knowledge enriching the model called with_EK. Through this comparison, we want to see the affecting of the external knowledge.

4.3 Evaluation Results

The evaluation results are in presented in Table 2, including ROUGE-1 and ROUGE-2 scores of the different query-focused extractive summarization methods. From the overall results, our method outperforms better than other baseline method.

Table 2. ROUGE scores (%) of different models

Dataset	Model	Rouge-1	Rouge-2
DUC 2005	CES	37.57	7.39
	AttSum	37.01	6.99
	BERT	38.77	8.31
	without_EK	39.27	8.78
	with_EK	39.61	9.01
DUC 2006	CES	40.44	9.16
	AttSum	40.90	9.4
	BERT	41.65	10.24
	without_EK	41.93	10.60
	with_EK	42.48	10.90
DUC 2007	CES	42.41	11.34
	AttSum	43.92	11.55
	BERT	44.31	11.85
	without_EK	45.31	12.16
	with_EK	45.72	12.27

Firstly, although our method is totally data-driven, its performance is better than the unsupervised method CES. It is noted that CES heavily depends on hand-crafted features, and representations based on deep learning work better than such surface features.

Secondly, performance of our method is better than other deep learning methods. The reason is that we consider the fine-grained interactive word-by-word attention between query and sentence.

From the comparison between without_EK and with_EK, we can see that the external knowledge can improve the performance of summarization.

From the results shown in Table 2, we also observe that the performance on DUC 2005 is lower than the other two. As our approach is a supervised and data-driven method, and the performance of trained model is relying on the size of training data. Data size of DUC 2005 is highly larger than the other two, as demonstrated in Table 1. When using the three-fold cross-validation, the number of training data for DUC 2005 is the smallest among the three years. The reason may be the data size of training data. The lack of training data impedes the learning of sentence.

5   Related Work

In this section, we describe the related work. A wide variety of different approaches to summarization have been proposed. Nowadays, most summarization systems are under the extractive framework which directly selects existing sentences to form the summary. Approaches on extractive summarization can be divided into three categories, namely unsupervised methods, supervised methods and semi-supervised methods.

5.1 Unsupervised Methods

Early studies on the task of query-focused summarization mainly use unsupervised approaches, which do not need labeled data.

Query-focused summarization was first mentioned in the study of Maximum Marginal Relevance(MMR) [20]. Authors proposed MMR algorithm to address the redundancy in summarization. It used a greedy approach to select sentences and considered the trade-off between saliency and redundancy.

Graph-based models played a leading role in the extractive summarization area, due to its ability to reflect various sentence relationships. For example, Wan et al. [21] adopted manifold ranking to make use of the within-document sentence relationships, the cross-document sentence relationships and the sentence-to-query relationships. In a graph, nodes are sentences and the edge scores reflect the similarity between sentences, each node is given a relevance weight based on its relevance to the query.

Feigenblat et al. [19] proposed a unsupervised approach to generate a summary by extracting a subset of sentences using the Cross-Entropy (CE) method. Authors also suggest a instantiation based on a simple combination of several query-dependent and query-independent features.

There exited eep learning based methods recently. Zhong et al. [22] proposed a unsupervised method based on deep learning model, which was the first attempt to utilize deep learning in tasks of query-oriented multi-document summarization. Singh et al. [23] applied long-span n-gram-based and neural language models to the task of query-focused summarization by capturing larger context. It was the first attempt to apply long-span models to a query-focused summarization task in an unsupervised setting.

5.2 Supervised Methods

Following unsupervised methods, supervised machine learning approaches are applied to solve the task. Supervised methods can be divided into two categories: surface feature based methods and deep learning based methods.

Surface Feature Based Methods. Surface feature based methods use machine learning models with hand-crafted features. Different classifiers have been explored. Michel Galley [24] proposed a skip-chain Conditional Random Fields (CRF) to model non-local pragmatic dependencies between paired utterances that typically appear together in summaries. Ouyang et al. [25] applied regression models to query-focused multi-document summarization. They chosen to use Support Vector Regression (SVR) to estimate the importance of a sentence in a document set to be summarized through a set of pre-defined features. Li *et al.* [26] proposed a bigram based supervised method for extractive document summarization in the integer linear programming (ILP) framework. For each bigram, a regression model is used to estimate its frequency in the reference summary. Shen *et al.* [27] explored how to use ranking SVM to train the feature weights for query-focused multi-document summarization.

Many query-focused summarization systems are heuristic extensions of generic summarization methods by incorporating the information of the given query. A variety of query-dependent features were defined to measure the relevance, including TF-IDF cosine similarity, WordNet similarity, and word co-occurrence. However, these features usually reward sentences similar to the query, which fail to capture the real intent of query.

Deep Learning Based Methods. Following the popularization of deep learning, many summarization systems have also employed deep learning techniques. Cheng and Lapatta [28] treated single document summarization as a sequence labelling task by utilizing CNN as sentence encoder and LSTM (Long Short-Term Memory) as sentence extractor. Ren [29] proposed Query Sentence Relation (QSR) which also used CNN with attention mechanism. Kobayashi *et al.* [30] simply used the sum of trained word embeddings as sentence or document representation. AttSum [4] is a summarization system which joints query relevance ranking and sentence saliency ranking with a neural attention model.

The attention mechanism has been successfully applied to learn alignment between various modalities. This paper is based on a modification of the attention mechanism, and implement a supervised method to address the query-focused extractive summarization problem.

5.3   Semi-supervised Methods

Semi-supervised methods are proposed to address the problem of dataset lacking for supervised methods. Li *et al.* [31] extended the standard graph ranking algorithm by proposing a two-layer (*i.e.* sentence layer and topic layer) graph-based semi-supervised learning approach based on topic modeling techniques. Distant supervision framework [32] has been proposed that trained a question summarizer without annotation costs or question-title pairs, where sentences are automatically annotated by means of heuristic rules.

6   Conclusion and Future Work

In this paper, we introduce the idea of a fine-grained and interactive word-by-word attention to a query-focused extractive summarization system, and propose a novel method to measure the query relevance of sentences. We proved it to address the problem of relatedness misalignment caused by the existing context level representation methods. To implement the idea, we utilize a Compare-Aggregate framework to simulate the interactively attentive reading and thinking of human behavior. We also leverage external conceptual knowledge to fill the expression gap between query and document. To evaluate our approach, we conduct experiments on the widely-used DUC 2005–2007 query-focused summarization benchmark datasets. Experimental results demonstrate the superiority of our approach, which achieve state-of-the-art performance.

This paper is the first one to introduce the idea of interactive attention to a query-focused extractive summarization system. We believe that the interaction based methods can improve the performance of query-focused extractive summarization system by capturing the relation between query and document. In our future work, we would like to study the other kinds of interactive information which can improve the relevance of summary.

Acknowledgement. This work is supported in part by the Project, Grant No. BMKY2019B04-1 and the Strategic Priority Research Program of Chinese Academy of Sciences, Grant No. XDC02040400.

References

1. Dang, H.T.: Overview of Duc 2005. In: Proceedings of DUC, pp. 1–12 (2005)
2. Wenpeng, Y., Yulong, P.: Optimizing sentence modeling and selection for document summarization. In Proceedings of IJCAI, pp. 1383–1389 (2015)
3. Ziqiang, C., Furu, W., Sujian, L., Wenjie, L., Ming, Z., Houfeng, W.: Learning summary prior representation for extractive summarization. In: Proceedings of IJCAI, Short Paper, pp. 829–833 (2015)
4. Ziqiang, C., Wenjie, L., Sujian, L., Furu, W., Yanran, L.:. AttSum: joint learning of focusing and summarization with neural attention. In: Proceedings of COLING, pp. 547–556 (2016)
5. Preksha, N., Khapra, M.M., Anirban, L., Ravindran, B.: Diversity driven attention model for query-based abstractive summarization. In: Proceedings of ACL, pp. 1063–1072 (2017)
6. Shuohang, W., Jing, J.: A compare-aggregate model for matching text sequences. In: Proceedings of ICLR (2017)
7. Parikh Ankur, P., Oscar, T., Dipanjan, D., Jakob, U.: A decomposable attention model for natural language inference. In: Proceedings of EMNLP, pp. 2249–2255 (2016)
8. Weijie, B., Si, L., Zhao, Y., Guang, C., Zhiqing, L.: A compare-aggregate model with dynamic-clip attention for answer selection. In: Proceedings of CIKM, Short Paper, pages pp. 1987–1990 (2017)
9. Seunghyun, Y., Franck, D., Doo, K., Soon, B.T., Kyomin, J.: A compare-aggregate model with latent clustering for answer selection. In: Proceedings of CIKM, Short Paper, pp. 2093–2096 (2019)
10. Arbi, B., Xiaohua, L., Jian-Yun, N.: Integrating multiple resources for diversified query expansion. In: Proceedings of ECIR, pp. 437–442 (2014)
11. Sarasi, L., Sujan, P., Pavan, K., Amit, S.: Domain-specific hierarchical subgraph extraction: a recommendation use case. In: Proceedings of Big Data, pp. 666–675 (2017)

12. Sarasi, L., Sujan, P., Pavan, K., Amit, S.: Domain-specific hierarchical subgraph extraction: a recommendation use case. In: Proceedings of Big Data, pp. 666–675 (2017)
13. Qian, C., Xiaodan, Z., Zhen-Hua, L., Diana, I., Si, W.: Neural natural language inference models enhanced with external knowledge. In: Proceedings of ACL, pp. 2406–2417 (2018)
14. Robyn, S., Joshua, C., Catherine, H.: ConceptNet 5.5: an open multilingual graph of general knowledge. In: Proceedings of AAAI, pp. 4444–4451 (2017)
15. Jacob, D., Ming-Wei, C., Kenton, L., Kristina, T.: BERT: pre-training of deep bidirectional transformers for language understanding. In: Proceedings of NAACL-HLT, pp. 4171–4186 (2019)
16. Ashish, V., et al.: Attention is all you need. In: Proceedings of NIPS, pp. 5998–6008 (2017)
17. Robyn, S., Joanna, L.-D.: ConceptNet at SemEval-2017 Task 2: extending word embeddings with multilingual relational knowledge. In: Proceedings of SemEval workshop at ACL 2017, pp. 85–89 (2017)
18. Yoon, K.: Convolutional neural networks for sentence classification. In: Proceedings of EMNLP, pp. 1746–1751 (2014)
19. Guy, F., Haggai, R., Odellia, B., David, K.: Unsupervised query-focused multi-document summarization using the cross entropy method. In: Proceedings of SIGIR, Short Paper, pp. 961–964 (2017)
20. Jaime, C., Jade, G.: The use of MMR, diversity-based reranking for reordering documents and producing summaries. In: Proceedings of SIGIR, Short Paper, pp. 335–336 (1998)
21. Xiaojun, W., Jianguo, X.: Graph-based multi-modality learning for topic-focused multi-document summarization. In: Proceedings of IJCAI, pp. 1586–1591 (2009)
22. Sheng-hua, Z., Yan, L., Bin, L., Jing, L.: Query-oriented unsupervised multi-document summarization via deeplearning model. Expert Syst. Appl. **42**(21), 8146–8155 (2015)
23. Mittul, S., Arunav, M.: Long-span language models for query-focused unsupervised extractive text summarization. In: Proceedings of ECIR, pp. 657–664 (2018)
24. Michel, G.: A skip-chain conditional random field for ranking meeting utterances by importance. In: Proceedings of EMNLP, pp. 364–372 (2006)
25. You, O., Wenjie, L., Sujian, L., Qin, L.: Applying regression models to query-focused multidocument summarization. Inf. Process. Manage. **47**(2), 227–237 (2011)
26. Chen, L., Xian, Q., Yang, L.: Using supervised bigram-based ILP for extractive summarization. In: Proceedings of ACL, pp. 1004–1013 (2013)
27. Chao, S., Tao,L.: Learning to rank for query-focused multi-document summarization. In: Proceedings of ICDM, pp. 626–634 (2011)
28. Jianpeng, C., Lapata, M.: Neural summarization by extracting sentences and words. In: Proceedings of ACL, pp. 484–494 (2016)
29. Pengjie, R., Zhumin, C.: Sentence relation for extractive summarization with deep neural network. TOIS **36**(4), 1–32 (2018)
30. Kobayashi Hayato, M.N., Yatsuka, T.: Summarization based on embedding distributions. In: Proceedings of EMNLP, pp. 1984–1989 (2015)

31. Yanran, L., Li, S.: Query-focused multi-document summarization: combining a topic model with graph-based semi-supervised learning. In: Proceedings of COLING, pp. 1197–1207 (2014)
32. Tatsuya, I., Kazuya, M., Hayato, K., Hiroya, T., Manabu, O.: Distant supervision for extractive question summarization. In: Proceedings of ECIR, pp. 182–189 (2020)

An Active Learning Based Hybrid Neural Network for Joint Information Extraction

Yan Zhuang[1(✉)], Guoliang Li[2(✉)], Wanguo Xue[1], and Fu Zhu[2]

[1] PLA General Hospital, Beijing, China
joyear2008@163.com, xuewanguo@sina.com
[2] Department of Computer Science, Tsinghua University, Beijing, China
liguoliang@tsinghua.edu.cn, zhuf18@mails.tsinghua.edu.cn

Abstract. Joint information extraction with high quality and low annotation costs plays an important role in many natural language processing (NLP) scenarios. To tackle this challenging problem, we firstly propose a joint machine extraction method based on a hybrid neural network which takes three common NLP tasks—named entity recognition (NER), relation extraction (RE) and event extraction (EE) into consideration. Then, based on the joint model, we propose an efficient active learning algorithm to select the most beneficial sentences to be annotated for further improving the model quality in a batch mode. Experimental results show that the proposed joint framework achieves better performance than state-of-the-art information extraction approaches on standard datasets, and our active algorithm surpasses all baseline methods with just 25% of the original training data and saves more than 70% annotation costs in testing data.

Keywords: Information extraction · Neural network · Active learning

1 Introduction

Named entity recognition, relation extraction and event extraction are three fundamental yet challenging tasks for natural language understanding. These tasks play an important role in many NLP scenarios such as knowledge base population [12], event logic graph construction [5], and heterogeneous data integration [20]. Given a corpus, NER is to detect entity mentions, RE is to recognize their semantic relationships, and EE need to recognize event triggers with their specific types and their corresponding arguments with the roles. In many real-world applications, the three tasks need perform simultaneously to extract enough structured information from given sentences. For example, considering a sentence from a patient's electronic medical record,

The [patient]$_{\text{PER}}$ was [hospitalized]$_{\text{TRI}}$ on [December 21]$_{\text{DAT}}$ due to [pneumonia]$_{\text{DIS}}$ with [vomiting]$_{\text{SYM}}$.
[patient] is a person entity (**PER**), [hospitalized] is a trigger (**TRI**) of an event with a predefined event type like "Admission", [December 21] is a date value (**DAT**),

Z. Huang et al. (Eds.): WISE 2020, LNCS 12343, pp. 84–100, 2020.
https://doi.org/10.1007/978-3-030-62008-0_6

[pneumonia] is a disease entity (DIS), and [vomiting] is a symptom entity (SYM). Entity pneumonia and vomiting have a DIS-SYM relation. An admission event triggered by the word "hospitalized" has four arguments: patient (PER-Arg), December 21 (DAT-Arg), pneumonia (DIS-Arg), vomiting (SYM-Arg). All these informations are helpful for building reasoning models to assist diagnosis. The goal is to recognize all structured informations from the clinical free-text sentences to build more intelligent medical applications.

However, existing information extraction (IE) methods are limited by their quality and high annotation cost, and can hardly acquire satisfactory application results in many domains [8,31]. For example, biomedical annotation always need higher labeling quality, and requires a lot of effort from medical experts, which is costly. As such, the existing IE methods are unsuitable for large-scale medical applications. Traditional work treats these tasks as a pipeline of separated tasks which provides a simple and flexible model, but it is inefficient for multi-round annotations and does not take full advantage of the relevance among the sub-tasks. The emerging joint extraction models pay attention to the correlation and present a neural network-based method which achieves a better performance. Whereas, the process is still separated into RE and EE. More importantly, the extraction quality is still limited and the annotation cost is not carefully treated.

To address the challenges, we propose an active learning based hybrid neural network for joint IE. The approach integrates the three IE models and shares the same neural network. On the one hand, it considers the relevance of different modules in one framework and avoids the accumulated errors, thus can benefit from the sharing information and overcome their inherent issues. On the other hand, as annotation is hard and costly, our joint model can reduce the cost by completing the three sub-tasks in one round. To further promote the extraction quality, we take all the word and sentence level information into account, and adopt dependency trees with graph attention networks (GATs) to capture wider contexts. Different from existing works, we train our joint model by maximum likelihood estimate (MLE) and fine-tune it by minimum risk training (MRT) [24]. This is because we want to optimize the model in sentence level and avoid the mismatch between the goal in training and testing.

To improve model quality as well as reduce costs up to par, we introduce effective active learning algorithm which strategically chooses the most valuable samples to get better performance with fewer annotations. To save annotation costs, we choose an improved pool-based sampling strategy to generate the most beneficial questions to publish. To reduce latency, we design an iterative batch mode selection of questions to avoid the procrastination from the annotators. We prove the batch mode sentence selection is NP-hard, and devise an efficient algorithms which can exploit the pre-defined joint learning process. To summarize, we make the following contributions.

- We propose an unified human-machine framework on information extraction with high quality and low cost (Sect. 3).
- We present a joint IE method based on a hybrid neural network which takes all kinds of discrete features from the input sentences into consideration (Sect. 4).

- We introduce an effective active learning algorithm to improve model quality, which can select the most beneficial samples for annotating in batch mode (Sect. 5).
- We have conducted extensive experiments to evaluate our methods. Experimental results show the advantages of our method (Sect. 6).

2 Preliminaries

2.1 Problem Formulation

Let $\mathcal{S} = \{w_1, w_2, ...w_n\}$ be a sentence where n is the sentence length and w_i is the $i-th$ token. Let $\mathcal{E} = \{e_1, e_2, e_3, ...\}$ be the set of entity mentions with entity types in the sentence, and $\mathcal{R} = \{r_{11}, r_{12}, ..., r_{ij}, ...\}$ be the relation set between entity e_i and e_j. Similarly, let $\mathcal{G} = \{g_1, g_2, g_3, ...\}$ be the set of triggers with specified event sub-types, and $\mathcal{A} = \{a_{11}, a_{12}, ..., a_{ij}, ...\}$ be the argument set between trigger g_i and its argument e_j. Here, argument e_j can be an entity mention, temporal expression or value with a specific role(type) which can also be deemed as an entity in our model. Let $\mathcal{T}_e = \{te_1, te_2, ...\}$ be the entity type set, $\mathcal{T}_r = \{tr_1, tr_2, ...\}$ be the relation type set, $\mathcal{T}_{tg} = \{tg_1, tg_2, ...\}$ be the trigger subtype set, and $\mathcal{T}_a = \{ta_1, ta_2, ...\}$ be the argument role set. We also consider entity and trigger extractions as sequence labelling tasks and apply the BIO annotation schema to assign a label to each token w_i [13,25].

Given an input $\mathcal{S}$, the NER model predicts the type $\hat{te}_i$ of word w_i by learning from the true type te_i, and output an entity set $\hat{\mathcal{E}} = \{(w_i, \hat{te}_i)|w_i \in \mathcal{S}, \hat{te}_i \in \mathcal{T}_e\}$. Given a set of detected entities $\hat{\mathcal{E}}$ by NER model, the RE model is to predict a relation type $\hat{tr}_i$ for each entity pair, and output a relation set $\hat{\mathcal{R}} = \{(e_i, e_j, \hat{tr}_{ij})|e_i, e_j \in \hat{\mathcal{E}}, i \neq j, \hat{tr}_{ij} \in \mathcal{T}_r\}$. In the EE model, it needs to predict the event subtype $\hat{tg}_i$ for w_i. When a trigger is detected, it then needs to predict the roles $\hat{ta}_{ij}$ that each argument a_{ij} plays in the event. Because the trigger detection can be deemed as an NER process, and the argument classification can be deemed as a RE process, the outputs $\hat{\mathcal{V}}$ can be revised as $\hat{\mathcal{G}} = \{(w_i, \hat{tg}_i)|w_i \in \mathcal{S}, \hat{tg}_i \in \mathcal{T}_g\}$ and $\hat{\mathcal{A}} = \{(g_i, e_j, \hat{ta}_{ij})|g_i \in \hat{\mathcal{G}}, e_j \in \hat{\mathcal{E}}, \hat{ta}_{ij} \in \mathcal{T}_a\}$. By this formulation, the processes can be shared between NER/RE and EE.

Let $\mathcal{L} = \{(x, y)\}_L$ be the dataset of labeled sentences, and $\mathcal{U} = \{(x, y)\}_U$ denotes the unlabeled dataset. Then, $x = \mathcal{S}, y = (\mathcal{E}, \mathcal{R}, \mathcal{V})$, and the output is $\hat{y} = (\hat{\mathcal{E}}, \hat{\mathcal{R}}, \hat{\mathcal{V}})$. Formally, we define the jointly active information extraction problem as follows:

Definition 1 *(Joint Information Extraction) Given a sentence $\mathcal{S} \in \mathcal{U}$, the joint information extraction problem is to find the entity set $\hat{\mathcal{E}}$, relation set $\hat{\mathcal{R}}$, and event set $\hat{\mathcal{V}}$ from $\mathcal{S}$ with the maximal joint probability based on a hybrid neural network.*

Definition 2 *(Jointly Active Information Extraction) Given an unlabeled dataset $\mathcal{U}$, it iteratively selects a set of candidates for labeling in order to maximize the performance of the joint information extraction until the budget is exhausted or the specified quality is reached.*

2.2   Related Work

IE Model. Early research on information extraction has primarily focused on pipelined architectures, where it first identify named entities, then combine the identified entities to classify the relationships. They may suffer from error propagation, information redundancy and inefficiency. Recent work try to develop joint IE models by parameter sharing and tagging scheme. Parameter sharing is a basic strategy and usually implement by neural networks. Miwa and Bansal [16] use sentence level RNN in NER, and BiLSTM to classify relationships according to the shortest path of dependency tree. Katiyar and Cardie [7] build a RE model on RNN using the attention mechanism. Li and Ji [10] develop a joint decoding algorithm based on beam search. MRT investigates minimum risk training to improve the performance. Although the sub-tasks in joint model can interact with each other by parameter sharing, they are also executed in turn. To tackle this problem, Zhang et al. [30] study a globally normalized joint model, and turn the sequence labeling and classification tasks into a sequence labeling tasks. While it can not recognize the overlapping relationships.

In event extraction, Nguyen and Grishman [18] study domain adaptation and event detection via CNNs, and they utilize RNNs and introduce the memory matrix to perform joint EE in [17]. Chen et al. [4] apply dynamic multi-pooling CNNs for EE in a pipelined framework. However, none of these work utilizes parameter sharing of NER, RE, and EE to perform joint IE. Besides, Yang and Li [27] propose a data-augment method applied in EE to tackle the problem of insufficient training data. Although extra synthesis data can improve the performance, the topic is beyond the scope of this article. Based on parameter sharing, our model perform three types of tasks simultaneously, taking into full account the available information of a sentence and the mutual influence among sub-tasks, which can improve efficiency and save costs.

Active Learning. The active learning approaches proposed in recent years are mainly query acquisition/pool-based methods, that is, selecting the sample data with the most valuable information by designing query strategies (sampling rules), including (1) selecting samples with the most uncertainty [2,23,28], (2) selecting an optimal subset based on diversity [22], and their combinations [3]. Unlike the query acquisition method "select" sample, the query synthesizing method "generate" sample. It directly generates samples for model training by using a generation model, such as generation adversarial network (GAN) [15]. These methods are also combined with the uncertainty sampling principle, and usually used in the computer vision field. Because most of the active learning approaches used in IE are still based on query-acquisition, this is the main baseline methods we want to compare. Different from the above approaches, we select samples with larger effect on the model quality when it is mis-predicted. The experiment results show the advantages of our proposed method.

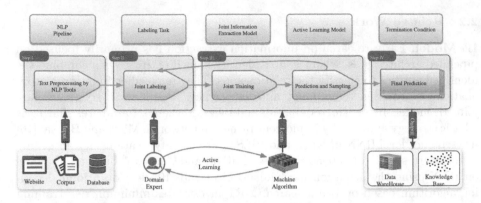

Fig. 1. Overview of the framework.

3   The Framework

We present the framework in Fig. 1. Given an IE task on a specified domain, it takes nature sentences to be annotated from different sources as input, and generates a set of structured entities, relations, and events as output. The steps will be discussed below.

Step I: NLP Pipline. We deploy Stanford CoreNLP to preprocess the data, including tokenizing, sentence splitting, pos-tagging and dependency parsing tree generating. We adopt the features from word and sentence levels which make a trade-off between the quality and the usefulness of the features. These features can be used as the clues for annotation by domain experts, and more importantly, can be the component as the embedding vectors. The dependency parsing tree will drain the dependency syntactic information of a sentence with fewer hops which can capture long-range dependencies.

Step II: Joint Labeling. We deploy an annotation system to label the entity, relation, and event simultaneously. The joint labeling is divided into two periods. The initial labeling will be performed by the domain experts which assure the high quality. The iteration labeling will be performed by domain staffs/students for cost reason. The qualification test and majority voting will be adopted to tolerant human errors in practice.

Step III: Joint Active IE. The joint active IE starts with a joint neural network trained by the initial labeling to select the most informative samples from the unannotated dataset, and the model is continuously fine-tuned by newly annotated samples. We explore a pool-based sampling method by computing entropy and diversity, and devise a batch selection algorithm to select batch samples simultaneously which are most uncertain and dissimilar with each other.

Step IV: Final Prediction. Step II and III repeat until stopping criterion is reached. Usually, the process will stop when the model reaches a satisfactory performance or the budget is exhausted. The performance is hard to verify, and only an approximation guarantee can be given [3]. In this work, we focus on the constraint of budget. After iterative interaction between machine and human, the final prediction will be generated.

4   Joint IE Model

The joint model shares the same embedding layer (Sect. 4.1), bidirectional LSTM encoding layer (Sect. 4.2), GAT layer (Sect. 4.3), and is trained by respective sub-models in joint extraction layer (Sect. 4.4) as shown in Fig. 2.

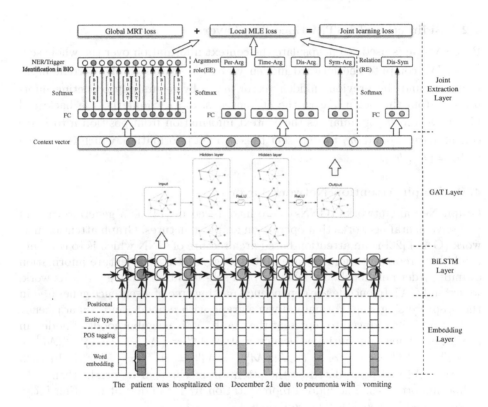

Fig. 2. IE framework.

4.1   Word Embedding Layer

After preprocessing step, we take embedding representations of token w_i into consideration, and transform them to a real-valued vector x_i by concatenating followings: (1) The word embedding vector: We concatenate the word embedding of w_i by looking up a pre-trained embedding matrix Glove [21] and the character embedding which is obtained by sum-up 1-, 2-, and 3-gram character representations of w_i [9]. (2) The POS-tagging embedding: We employ the

Stanford CoreNLP tool to generated POS-tagging labels and transform them to real-valued vectors by looking up the randomly initialized POS-tagging label embedding table. (3) The entity type embedding: We employ the BIO annotation schema [18] to generate type labels of w_i and transform them to real-valued vectors by looking up the randomly initialized type embedding table using the heads of the entity mentions [17]. (4) The positional embedding: We encode the relative distance between w_i as a real-valued vector by looking up the randomly initialized position embedding table [19]. Then, $\mathcal{S}$ will be transformed into a sequence of real-valued vectors $\{x_i\}$, which will be fed into later modules to learn more effective representations.

4.2 Bidirectional LSTM Encoding Layer

BiLSTM [14] is used to encapsulate the context information over the whole sentence. We compute the forward hidden vector $\overrightarrow{h}_i$ based on the current input vector x_i and the previous hidden vector $\overrightarrow{h}_{i-1}$ by recursively collecting information from the beginning to the current position i. Similarly, the backward hidden vector $\overleftarrow{h}_i$ summarizes the context information from position n to i. We concatenate $\overrightarrow{h}_i$ and $\overleftarrow{h}_i$ to represent the encoded information, which is denoted as $\mathbf{h}_i = [\overrightarrow{h}_i, \overleftarrow{h}_i]$.

4.3 Graph Attention Network Layer

Graph Neural Networks (GNNs) were introduced in [14] as a generalization of recursive neural networks that operate on graph structures. Graph attention network (GAT) [26] is an attention-based architecture of GNN which is to compute the hidden representations of each node by aggregating the feature information of multi-order neighborhood, followed by a self-attention strategy. In this work, we extend GAT to encode the dependency parsing tree of a sentence generated in the preprocessing step. Formally, given a graph with n nodes, an $n \times n$ adjacency matrix $\mathcal{M}$ is introduced to represent the graph. $\mathcal{N}_i$ is neighborhoods of node i in the graph. Then, the initial adjacency matrix $\mathcal{M}^0 \doteq \{\mathcal{M}_{ij}^0 \in \{0,1\}^{n \times n} | \mathcal{M}_{ii}^0 = 1;\ \mathcal{M}_{ij}^0 = \mathcal{M}_{ji}^0 = 1\ \ for\ j \in \mathcal{N}_i;\ \mathcal{M}_{ij}^0 = \mathcal{M}_{ji}^0 = 0\ \ for\ j \notin \mathcal{N}_i\}$. In each iteration, $\mathcal{M}$ will be updated by the attention mechanism to weight their individual importances. The hidden representation $\mathbf{h}_i^k$ for node i at iteration k can be computed by the following equations:

$$\mathbf{h}_i^k = f(\sum_{j=1}^{n} \mathcal{M}_{ij} \mathcal{W}_h^k \mathbf{h}_j^{k-1} + \mathbf{b}_h^k) \tag{1}$$

where $f(\cdot)$ is an activation function (e.g., $RELU$), and $\mathcal{W}_h$ and $\mathbf{b}_h$ are the weight matrix and bias vector respectively. $\mathcal{M}_{ij}$ can be calculated by the attention mechanism as the following equation:

$$\mathcal{M}_{ij} = softmax_{j \in \mathcal{N}_i}(\delta(\mathbf{a}^T \cdot [\mathbf{h}_i || \mathbf{h}_j]))$$
$$= \frac{exp(\delta(\mathbf{a}^T \cdot [\mathbf{h}_i || \mathbf{h}_j])))}{\sum_{l \in \mathcal{N}_i} exp(\delta(\mathbf{a}^T \cdot [\mathbf{h}_i || \mathbf{h}_l])))} \tag{2}$$

In our settings, we use a single-layer feedforward neural network to express the attention mechanism parametrized by a attention vector "**a**". $\parallel$ means "concatenate", $(\cdot)^T$ represents transposition, and δ denotes the activation function (e.g., *LeakyReLU*). Multi-head attention is also employed in our model to stabilize the learning process. By stacking the GAT layer, we can not only adaptively accumulate the context information for each token in the sentence, but also capture the shortcut dependencies generated by the dependency parsing tree, which will improve our IE performance.

4.4   Joint Extraction Layer

For NER, each word in the sentence will be assigned an entity type label following the BIO tagging scheme. When detecting the entity type of word w_i, the input h_i obtained from GAT layer will be fed into an FC layer, and the final softmax layer computes normalized probabilities over the possible entity types in $\mathcal{T}_e$:

$$y_{\hat{te}_i} = P_E(\hat{te}_i | \mathcal{S}; \Theta_E) = softmax(\mathcal{W}_E \mathbf{h_i} + \mathbf{b}_E) \tag{3}$$

where $y_{\hat{te}_i}$ is the final output of the $i - th$ entity type label, Θ_E is the parameter set consisting parameters involved from different layers, and $\mathcal{W}_E$ and $\mathbf{b}_E$ are the weight matrix and bias vector for entity detection respectively. Based on these entity tag sequence, we can obtain the entity set $\mathcal{E}$ of the sentence S.

For RE, when detected a set of entities $\mathcal{E}$, we take all possible entity pairs (e_i, e_j) (filtered by relation definition, except tag "O") into consideration to compute the relation set $\hat{\mathcal{R}}$. We aggregate the hidden vectors of token subsequences belong to one entity and feed the concatenation of (e_i, e_j) into a FC layer to predict the relationship as:

$$y_{\hat{tr}_{ij}} = P_R(\hat{tr}_{ij} | \mathcal{S}; \Theta_R) = softmax(\mathcal{W}_R[\mathbf{e_i} || \mathbf{e_j}] + \mathbf{b_R}) \tag{4}$$

where $y_{\hat{tr}_{ij}}$ is the final output for the relation type of (e_i, e_j).

For EE, the processing step of trigger detection is similar to NER and argument role prediction as RE, while executes in another independent sub-process. For convenience of exhibition, we integrate NER and trigger detection in the same part in Fig. 2. The trigger subtype tg_i for w_i also follows the BIO tagging scheme and can be predicted by the following equation:

$$y_{\hat{tg}_i} = P_V^g(\hat{tg}_i | \mathcal{S}; \Theta_V^g) = softmax(\mathcal{W}_V^g \mathbf{h_i} + \mathbf{b_{Vg}}) \tag{5}$$

In the argument role prediction step, we firstly check the trigger subtype tg_i for w_i in the trigger detection sub-process. When detecting a trigger candidate, we will iteratively predict the argument role ta_{ij} for each entity mentions e_j with respect to g_i. When $tg_i = $ "O", we will skip this stage immediately. We aggregate the hidden vectors of token subsequences belong to one entity(trigger) and feed the concatenation of (g_i, e_j) into a FC layer to predict the argument role as:

$$y_{\hat{ta}_{ij}} = P_V^a(\hat{ta}_{ij} | \mathcal{S}; \Theta_V^a) = softmax(\mathcal{W}_V^a[\mathbf{g_i} || \mathbf{e_j}] + \mathbf{b_V^a}) \tag{6}$$

We first train the network by MLE to initialize the model parameters, which minimizes a joint negative log-likelihood function $\mathcal{F}$ as following:

$$
\begin{aligned}
\mathcal{F} = & - logP(\mathcal{E}, \mathcal{R}, \mathcal{V} | \mathcal{S}; \Theta) \\
= & - \sum_i logP_E(te_i | \mathcal{S}; \Theta_E) - \sum_{\substack{e_i, e_j \in \mathcal{E}, i \neq j \\ te_i, te_j \neq \text{``}O\text{''}}} logP_R(tr_{ij} | \mathcal{S}, e_i, e_j; \Theta_R) \\
& - \sum_i logP_V(tg_i | \mathcal{S}; \Theta_G) - \sum_{g_i \in \mathcal{G}, tg_i \neq \text{``}O\text{''}} \sum_{e_j \in \mathcal{E}} P_V^a(ta_{ij} | \mathcal{S}, g_i, e_j; \Theta_A).
\end{aligned}
\tag{7}
$$

where Θ with different subscript are parameter sets of different sub-model. Given the size of the training dataset N, the objective of MLE is to select the optimal parameters $\hat{\Theta}_{\text{MLE}}$ by minimizing function $\mathcal{F}$ on training samples:

$$
\hat{\Theta}_{\text{MLE}} = argmin_\Theta \left\{ \sum_{n=1}^N \mathcal{F}^{(n)} \right\}.
\tag{8}
$$

After that, we train the network by MRT to fine-tune the parameters. MRT uses the loss function $\triangle(\hat{y}, y)$ to describe the divergence between the model prediction $\hat{y}$ and the standard answer y, and try to find a set of parameters to minimize the expected value of the loss on the training set. We define the joint probability of our joint model as:

$$
P(\hat{\mathbf{y}} | \mathcal{S}; \Theta) = P(\hat{\mathcal{E}} | \mathcal{S}; \Theta_E) P(\hat{\mathcal{R}} | \mathcal{S}, \hat{\mathcal{E}}; \Theta_R) P(\hat{\mathcal{V}} | \mathcal{S}, \hat{\mathcal{E}}; \Theta_V).
\tag{9}
$$

Denote $Y(\mathcal{S})$ as the set of all possible outputs of the input sentence $\mathcal{S}$. Similarly, the objective of MRT is to select the optimal model parameters $\hat{\Theta}_{\text{MRT}}$ by minimizing the following expected loss on training samples,

$$
\hat{\Theta}_{\text{MRT}} = argmin_\Theta \left\{ \sum_{k=1}^N \sum_{\hat{\mathbf{y}} \in Y(\mathcal{S})} P(\hat{\mathbf{y}} | \mathcal{S}; \Theta) \triangle(\hat{y}, y) \right\}.
\tag{10}
$$

In fact, MRT is a generalization of MLE with defining the loss function of MRT by 0/1 function. The MRT used in our model has the following advantages: Firstly, MRT does not make any assumptions about the model architecture, which can be applied in any joint learning model like ours flexibly. Secondly, Any evaluation metric can be used as the loss function in MRT which can make the trained model as close to the user's demands as possible, and can alleviate the inconsistent between training and testing stages. Thirdly, MLE only focuses on the correctness of local annotation which will inevitably lose global information, while MRT can easily provide a trade-off among the joint model by designing a global measurement. More importantly, MRT will help our active learning process which will be discussed in the next section.

Enumerating all possible outputs of the input sentence is intractable. Following [25], we select a subset of samples $S(\mathcal{S}) \subset Y(\mathcal{S})$ from the probability distribution of different annotation schemes. The loss function can be approximated as:

$$l_{\text{MRT}}(\Theta) = \sum_{k=1}^{N} \sum_{\hat{\mathbf{y}} \in S(\mathcal{S})} \frac{P(\hat{\mathbf{y}}|\mathcal{S}; \Theta)^{\epsilon}}{\sum_{\mathbf{y}^* \in \mathbf{S}(\mathcal{S})} P(\mathbf{y}^*|\mathcal{S}; \Theta)^{\epsilon}} \triangle(\hat{y}, y). \qquad (11)$$

The hyper-parameter ϵ controls the smoothness of the function [1]. We utilize the F1 score of all the annotations of a sentence to compute $\triangle(\hat{y}, y)$.

5 Active Learning Model

In this section, we first introduce the pool-based active learning, and then discuss how to enhance the pool-based methods. Lastly, we show the overall algorithm.

5.1 Pool-Based Active Learning

Pool-based active learning selects the most valuable samples by designing sampling rules. Our pool-based strategy is a multi-round method. We denote $\mathcal{L}_t$ and $\mathcal{U}_t$ as the training set and the unlabeled set at round t. Let θ^t be the classifier trained on $\mathcal{L}_t$. There exists a small set of labeled data $\mathcal{L}_0$ and a large pool of unlabeled instances $\mathcal{U}_0$. Given budget $\mathcal{B}$, the objective is to select a batch in T training rounds of $\mathcal{Q}$ sentences at a time(i.e. $\mathcal{B} = T \times \mathcal{Q}$) from $\mathcal{U}_t$ in such a way that the classifier w^{t+1} trained on $\mathcal{L}_t \bigcup \mathcal{Q}$ has the maximum performance gain under limited budget $\mathcal{C}$. The newly labeled candidates $\mathcal{Q}$ will be incorporated into $\mathcal{L}_t$ in a continuously fine-tuning way. In our settings, we explore the uncertainty and diversity sampling strategy.

Uncertainty. Uncertainty is defined as the measurement of the informativeness of a sample. Higher uncertainty means more unconfident of the predicted label sequence assigned by the model which we prefer. In our model, the revised version of the output probability of our model $\frac{P(\hat{\mathbf{y}}|\mathcal{S}; \Theta)^{\epsilon}}{\sum_{\mathbf{y}^* \in \mathbf{S}(\mathcal{S})} P(\mathbf{y}^*|\mathcal{S}; \Theta)^{\epsilon}}$ can be used to compute uncertainty, which is denoted as $P'(\hat{\mathbf{y}}|\mathcal{S}; \Theta)$. Then, the uncertainty of $\mathcal{S}_i$ can be formulated by:

$$\mathcal{C}_i = -\frac{1}{|Y|} \max_{\hat{\mathbf{y}}_i \in S(\mathcal{S}_i)} \log P'(\hat{\mathbf{y}}_i|\mathcal{S}_i; \Theta) \qquad (12)$$

Where $|Y|$ is the length of the labeling sequence, and $\mathcal{C}_i \in R^{|\mathcal{U}_t| \times 1}$.

Diversity. Diversity $\mathcal{D}$ is a measure of redundancy (i.e. how similar) between two unlabeled samples. Higher diversity means lower redundancy. In our model, we use text distance between sentences as the measure. We exploit Smooth Inverse Frequency (SIF) as the text distance function. The diversity between two sentences $\mathcal{S}_i$ and $\mathcal{S}_j$ with their embeddings X_i and X_j can be formulated by: $\mathcal{D}_{ij} = Sif(X_i, X_j)$ where $\mathcal{D}_{ij} \in R^{|\mathcal{U}_t| \times |\mathcal{U}_t|}$. By definition, all the entries in $\mathcal{C}_i$ and $\mathcal{D}_{ij}$ are non-negative. Further, $\mathcal{D}_{ii} = 0, \forall i$. We combine $\mathcal{C}_i$ and $\mathcal{D}_{ij}$ into a single uncertainty-diversity matrix $\mathcal{H}$:

$$\mathcal{H}_{ij} = \begin{cases} \mathcal{D}_{ij}, & if \ i \neq j \\ \lambda \mathcal{C}_i, & if \ i = j \end{cases} \tag{13}$$

λ is a trade-off parameter. Then, sentences selection with given batch size $\mathcal{Q}$ can be expressed by the following integer quadratic programming (IQP) problem:

$$\max_k k^T \mathcal{H} k \quad s.t. \ k_i \in \{0,1\}, \forall i \ and \ \sum_{i=1}^{|\mathcal{U}_t|} k_i = \mathcal{Q} \tag{14}$$

Here $k \in \{0,1\}^{|\mathcal{U}_t| \times 1}$ denotes whether the unlabeled sentence will be included in the selected batch ($k_i = 1$) or not ($k_i = 0$).

IQP problem is NP-hard, and we can use the iterative truncated-power algorithm proposed by Yuan and Zhang [29] to solve it. In each round, the sentences in $\mathcal{U}_t$ will be fed into our joint model to obtain $\{\mathcal{C}_i\}$, and every sentence pair in $\mathcal{U}_t$ will be selection to compute $\{\mathcal{D}_{ij}\}$. They compose the matrix $\mathcal{H}$ which can be input into the iterative truncated-power algorithm to get $\mathcal{Q}$ sentences used to train our model in the next round.

5.2   Minimal Expected Prediction Loss

The uncertainty-diversity strategy is a partial solution for sentence selection for only similarity and uncertainty are considered. Moreover, the above algorithm is not efficient (the time complexity is $\mathcal{O}(n^2)$) for large-scale datasets. To overcome these shortages, we design a new criterion which considers not only the uncertainty and similarity, but also the expected prediction loss of our joint model. The intuition is that, the most uncertain sentence is not necessarily the most beneficial one for the model quality. We are prefer the sentence which have larger effect on the model quality when it is mispredicted. We use the expectation of model loss $l_{\hat{\mathbf{y}}_i}(\Theta)$ under different $\hat{\mathbf{y}}_i \in S(\mathcal{S}_i)$ to substitute for uncertainty $\mathcal{C}_i$ as:

$$\mathcal{C}_i' = -\frac{1}{|Y|} \sum_{\hat{\mathbf{y}}_i \in S(\mathcal{S}_i)} \log P'(\hat{\mathbf{y}}_i | \mathcal{S}_i; \Theta) l_{\hat{\mathbf{y}}_i}(\Theta). \tag{15}$$

$l_{\hat{\mathbf{y}}_i}(\Theta)$ can be obtained by incrementally retrain the model with cross validation by classifying $\hat{\mathbf{y}}_i$ of $\mathcal{S}_i$ into different $S(\mathcal{S}_i)$. Then, $\mathcal{H}$ can be revised into $\mathcal{H}'$:

$$\mathcal{H}_{ij}' = \begin{cases} \mathcal{D}_{ij}, & if \ i \neq j \\ \lambda \mathcal{C}_i', & if \ i = j. \end{cases} \tag{16}$$

The quadratic computation complexity is inevitable in solving the above IQP problem. To improve efficiency, we should limit the scale of matrix $\mathcal{H}'$. The common practice is to randomly select a subset of samples from $\mathcal{U}_t$ for each active learning stage [28]. But the effect of different sample size is indeterminate, especially when the budget is very limited. As a substitute for random selection, we exploit least confidence sampling to choose sentences. In our strategy, we sort sentences in descending order according to the uncertainty $\mathcal{C}_i$ of each $\mathcal{S}_i$, and choose $k \times \mathcal{Q}$(usually $2 \leq k \leq 9$ according to $\mathcal{Q}$) sentences to compute $\mathcal{H}'$ in each round. By this way, the sampling quality is guaranteed by the least confidence strategy without extra computational overhead.

Based on the above adjustment, the pool-based active learning process will be more comprehensive and efficient to improve the model quality with given budget. The overall algorithm is shown in Algorithm 1. The computation complexity is $min(t|\mathcal{U}_t|, tk^2Q^2)$, and usually $kQ \ll |\mathcal{U}_t|$. Therefore, our active learning algorithm can be seen as linear to the number of sentences in the unlabeled dataset.

Algorithm 1: ACTIVE LEARNING ALGORITHM.

Input:
  $Model_0$: joint IE model learned from initial training data;
  $\mathcal{U}_0$: initial unlabeled dataset;
  $\mathcal{L}_0$: initial training dataset;
  $\mathcal{Q}$: budget in each round;
  $\mathcal{B}$: total budget;
Output:
  $\mathcal{L}_t$: labeled dataset at round t;
  $Model_t$: fine-tuned model at round t

1 **begin**
2 | **for** *(t = 0 to $\lfloor \mathcal{B}/\mathcal{Q} \rfloor$)* **do**
3 | | **foreach** *sentence $\mathcal{S}_i \in \mathcal{U}_t$* **do**
4 | | | compute $P'(\hat{\mathbf{y}}_\mathbf{i}|\mathcal{S}_i; \Theta)$ by $Model_t$;
5 | | Sort $\mathcal{S}_i$ in descended order wrt. $\mathcal{C}_i$ by Eq. 12;
6 | | Select top $k \times \mathcal{Q}$ sentences, yielding subset $\mathcal{J}$; **foreach** $\mathcal{S}_i \in \mathcal{J}_t$ **do**
7 | | | compute $\mathcal{C}'_i$ by Eq. 15;
8 | | | **foreach** $\mathcal{S}_j \in \mathcal{J}_t$ **do**
9 | | | | compute $\mathcal{D}_{ij}$; compute $\mathcal{H}'_{ij}$ by Eq. 16;
10 | | solve IQP by Eq. 14, yielding $\mathcal{Q}$ unlabeled samples;
11 | | label these sentence, yielding Δ;
12 | | fine-tuning model with Δ, yielding $Model_t$;
13 | | $\mathcal{L}_t = \mathcal{L}_t \cup \Delta$; $\mathcal{U}_t = \mathcal{U}_t \setminus \Delta$; $t = t + 1$;
14 | return $\mathcal{L}_t$, $Model_t$;

---

6 Experiments

6.1 Experiment Setup

Datasets. We evaluate our proposed framework on the ACE 2005 corpus. For the purpose of comparison, we use the same data split as the previous work [4,18]. This data split includes 40 newswire articles (672 sentences) for the test set, 30 other documents (836 sentences) for the development set and 529 remaining documents (14,849 sentences) for the training set. Also, we follow the criteria of the previous work [4] to judge the correctness of the predicted mentions.

Training Details. We obtain initial word embeddings for our IE model using Glove [21]. We use 50 dimensions for pos-tagging embedding, positional embedding and entity type embedding. The dimension of word embedding is 300 and the number of hidden units in LSTM is 220. We apply a three-layer GAT model, and perform mini-batch training with batch size set to 12.

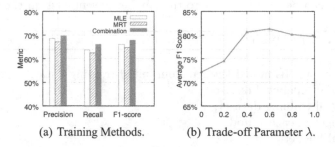

(a) Training Methods.   (b) Trade-off Parameter λ.

Fig. 3. Tuning experiment parameters

6.2   Evaluation on Joint Extraction Models

Competitors. We compare the performance without active learning in this section. For NER and EE joint extraction, we use M&B (2016) [16] and K&C (2017) [7] which are training systems without joint decoding, and L&J (2014) [10] which is a joint decoding algorithm. For event extraction, we use Cross-Ev [11] which uses document level information to improve the performance, DMCNN [4] which uses dynamic multi-pooling to keep multiple information, and JRNN [17] which manually designed features to jointly extraction. ANIE (ATT) is short for our model with self-attention, and ANIE (GAT) is our model with GAT layer instead. We try to add CRF layer [6] but it doesn't improve much. We evaluate the performance by average precision, recall, and F1 score.

Table 1. Resuls of entity and relation extraction

Model	NER			RE		
	P (%)	R (%)	F1 (%)	P (%)	R (%)	F1 (%)
L&J (2014)	85.2	76.9	80.8	**65.4**	39.8	49.5
M&B (2016)	82.9	83.9	83.4	57.2	54.0	55.6
K&C (2017)	84.0	81.3	82.6	55.5	51.8	53.6
MRT (2018)	83.9	83.2	83.6	64.9	55.1	**59.6**
ANIE (ATT)	**85.6**	86.5	86.0	57.6	55.9	56.7
ANIE (GAT)	85.4	**87.6**	**86.5**	59.3	**56.5**	57.9

Evaluating Training Methods. We compare the performance of MLE, MRT and their combination. As shown in Fig. 3(a), training with MRT has the worst performance because the sampling method makes the training result unstable. The combination performs the best because it can leverage the advantages of the two training methods. In our framework we first pre-train the model with MLE, then optimize the local loss and the global loss with MRT. By integrating the two losses, all three extraction models are unaware of the loss from the other side and have a tighter connection among each other. We consider sentence level F1-score, and it characterizes the overall performance of the outputs and make the training objective be consistent with the evaluation metric in testing. In the following experiments, we use the combination as default setting.

Table 2. Resuls of event extraction

Model	Trig-identification			Trig-classification			Arg-identification			Arg-classification		
	P (%)	R (%)	F1 (%)	P (%)	R (%)	F1 (%)	P (%)	R (%)	F1 (%)	P (%)	R (%)	F1 (%)
CROSS-EV	–	–	–	68.7	68.9	68.8	46.5	45.1	45.8	41.2	40.0	40.6
DMCNN	**80.4**	67.7	73.5	**75.6**	63.6	69.1	**62.9**	47.1	53.8	56.8	42.7	48.8
JRNN	68.5	**75.7**	71.9	66.0	**73.0**	69.3	56.1	**58.3**	57.2	49.5	**51.5**	50.5
ANIE (ATT)	75.7	73.6	74.6	71.5	69.5	70.5	59.3	55.8	57.5	55.8	48.2	51.7
ANIE (GAT)	77.6	73.9	**75.7**	72.1	70.7	**71.4**	60.2	56.5	**58.3**	**56.9**	50.3	**53.4**

Comparition with State-of-the-Arts. Table 1 shows the overall performance of NER and RE comparing to the baselines with golden-standard. There is a significant gain with NER, which is 2.9% higher over the best baseline models. The improvement comes from the parameter-sharing framework and sentence-level F1-score optimization by MRT. ANIE with GAT is slightly better than the version of self-attention because the latter can be seen as the single-hope GAT and the former has stronger representation capability and adaptability. Table 2 shows the overall performance of EE comparing to the above baseline methods. We can see that our framework achieves the best F1 scores for both trigger and argument extraction subtasks. The performance reported in argument identification and classification experiments is different with the original paper because we use the entities extracted by our NER model instead of golden-standard. The results demonstrate the benefit of our model with considering all kinds of discrete features from the input sentences and training with the combination of MLE and MRT.

6.3 Evaluation on Active Learning Algorithm

Competitors. For active learning algorithm, we choose Random which samples randomly from the unlabeled data, MNLP [23] which is an uncertainty-based active learning method, and BatchRand [3] which is state-of-the-art method for batch mode active learning with uncertainty-diversity as baselines. ANIE is our active learning algorithm with minimal expected prediction loss. The models are initialized with 10% of training set, and we run 5 trials for each algorithm and record the average results.

Evaluating Trade-Off Between Entropy and Diversity. We check the performance of our active learning algorithm by varying trade-off parameter λ. We choose 30% testing data to fine-tune the IE model, and the result is shown in Fig. 3(b). In this settings, the worst case is $\lambda = 0$ which means only considering diversity. When we only consider entropy (i.e. $\lambda = 1$), the performance is better than $\lambda = 0$. The best case is when we combine entropy and diversity ($\lambda \approx 60\%$). This demonstrates that entropy has greater influence on the model quality than diversity, and the combination will have better performance than considering either side alone. This is because entropy tends to find uncertain samples, but ignores redundancy between unlabeled sentences. We adopt $\lambda = 60\%$ as default.

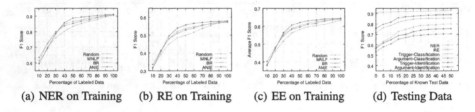

(a) NER on Training   (b) RE on Training   (c) EE on Training   (d) Testing Data

Fig. 4. Comparison of active learning methods

Comparition with State-of-the-Arts. Depicted in Fig. 4(a), 4(b), 4(c), the figures present the similar pattern. Each algorithm performs better than Random baseline. By the time 10% of sentences are labeled, MNLP performs the best because uncertainty-based strategy selects samples with more informativeness and updates model quickly, and ANIE is close to Random because the effect of incrementally retraining with cross validation is not obvious with little training data. ANIE begins to surpass other baselines when 20% ~30% of sentences are labeled, and reaches approximate top performance with 50% labeled data. This is because the proposed algorithm can select samples more effectively. The performance of BR and MNLP is similar while BR performs slightly better than MNLP with the increment of labeled data. This is consistent with the conclusion in Fig. 3(b). Figure 4(d) shows the performance of all sub-tasks when our model is used in testing data. It reaches approximate top performance with 25%–30% testing data which means we can save more than 70% annotation cost.

7   Conclusion

In this paper, we propose a jointly active information extraction framework based on a hybrid neural network. The information extraction model performs NER,

RE, and EE in a unified neural network architecture and takes all kinds of discrete features from the input sentences into consideration. The proposed model is trained by MLE and fine-tuned by MRT which provides stable and consistent training results. We formulate the active information extraction problem as selecting sentences under given budget to maximize the tradeoff between entropy and diversity. We propose a revised pool-based sampling algorithm to address the problem without extra computational overheads. The experiment results on both benchmark dataset show that our framework achieves high quality and can outperform state-of-the-art approaches.

Acknowledgement. This work was supported by Key Projects of Military Logistics Research (BHJ14L010), and medical AI research and development project of PLAGH (2019MBD-046).

References

1. Ayana, S.S., Zhao, Y., Liu, Z., Sun, M.: Neural headline generation with minimum risk training. CoRR, abs/1604.01904 (2016)
2. Beluch, W.H., Köhler, J.M.: The power of ensembles for active learning in image classification. In: CVPR 2018, Salt Lake City, UT, USA, 18–22 June 2018, pp. 9368–9377 (2018)
3. Chakraborty, S., Ye, J.: Active batch selection via convex relaxations with guaranteed solution bounds. IEEE Trans. Pattern Anal. Mach. Intell. **37**(10), 1945–1958 (2015)
4. Chen, Y., Xu, L., Liu, K., Zhao, J.: Event extraction via dynamic multi-pooling convolutional neural networks. In: ACL 2015, 26–31 July 2015, Beijing, China, pp. 167–176 (2015)
5. Ding, X., Liao, K.: ELG: an event logic graph. CoRR, abs/1907.08015 (2019)
6. Greenberg, N., Bansal, T., Verga, P., McCallum, A.: Marginal likelihood training of BILSTM-CRF for biomedical named entity recognition from disjoint label sets. In: EMNLP, Brussels, Belgium, 31 October–4 November 2018, pp. 2824–2829 (2018)
7. Katiyar, A., Cardie, C.: Going out on a limb: joint extraction of entity mentions and relations without dependency trees. In: ACL, Vancouver, Canada, pp. 917–928 (2017)
8. Li, F., Zhang, M., Fu, G., Ji, D.: A neural joint model for entity and relation extraction from biomedical text. BMC Bioinform. **18**(1), 198:1–198:11 (2017)
9. Li, J., et al.: WCP-RNN: a novel RNN-based approach for bio-NER in Chinese EMRs. J. Supercomput. **76**(3), 1450–1467 (2020)
10. Li, Q., Ji, H.: Incremental joint extraction of entity mentions and relations. In: ACL 2014, vol. 1, pp. 402–412, Baltimore, MD, USA, 22–27 June 2014 (2014)
11. Liao, S., Grishman, R.: Using document level cross-event inference to improve event extraction. In: Hajic, J., Carberry, S., Clark, S. (eds.) ACL 2010, pp. 789–797 (2010)
12. Lin, Y., Liu, Z., Sun, M., Liu, Y., Zhu, X.: Learning entity and relation embeddings for knowledge graph completion. In: AAAI, Austin, Texas, USA, 25–30 January 2015, pp. 2181–2187 (2015)
13. Liu, X., Huang, H.: Jointly multiple events extraction via attention-based graph information aggregation. In: EMNLP, Brussels, Belgium, 31 October 2018, pp. 1247–1256 (2018)

14. Marcheggiani, D., Titov, I.: Encoding sentences with graph convolutional networks for semantic role labeling. In: EMNLP, Copenhagen, Denmark, 9 September (2017)
15. Mayer, C., Timofte, R.: Adversarial sampling for active learning. In: WACV 2020, Snowmass Village, CO, USA, 1–5 March 2020, pp. 3060–3068 (2020)
16. Miwa, M., Bansal, M.: End-to-end relation extraction using LSTMS on sequences and tree structures. In: ACL, 7–12 August 2016, Berlin, Germany (2016)
17. Nguyen, T.H., Cho, K., Grishman, R.: Joint event extraction via recurrent neural networks. In: NAACL, USA, 12–17 June 2016, pp. 300–309 (2016)
18. Nguyen, T.H., Grishman, R.: Event detection and domain adaptation with convolutional neural networks. In: ACL, 26–31 July 2015, Beijing, China, pp. 365–371 (2015)
19. Nguyen, T.H., Grishman, R.: Graph convolutional networks with argument-aware pooling for event detection. In: AAAI, New Orleans, Louisiana, USA, pp. 5900–5907 (2018)
20. Ohno-Machado, L.: Natural language processing: algorithms and tools to extract computable information from EHRs and from the biomedical literature. JAMIA **20**(5), 805 (2013)
21. Pennington, J., Socher, R., Manning, C.D.: Glove: global vectors for word representation. In: EMNLP, 25–29 October 2014, Doha, Qatar, pp. 1532–1543 (2014)
22. Sener, O., Savarese, S.: Active learning for convolutional neural networks: a core-set approach. In: ICLR 2018, Vancouver, BC, Canada, 30 April–3 May 2018 (2018)
23. Shen, Y., Yun, H., Lipton, Z.C., Kronrod, Y., Anandkumar, A.: Deep active learning for named entity recognition. In: ICLR 2018, Vancouver, BC, Canada, 30April–3 May 2018 (2018)
24. Smith, D.A., Eisner, J.: Minimum risk annealing for training log-linear models. In: ACL 2006, Sydney, Australia, 17–21 July 2006 (2006)
25. Sun, C., Wu, Y., Lee, K., Wu, K.: Extracting entities and relations with joint minimum risk training. In: EMNLP 2018, Brussels, Belgium, 31 October 2018, pp. 2256–2265 (2018)
26. Velickovic, P., Cucurull, G., Casanova, A., Romero, A., Liò, P., Bengio, Y.: Graph attention networks. In: ICLR, Vancouver, BC, Canada, 30 April–3 May 2018 (2018)
27. Yang, S., Li, D.: Exploring pre-trained language models for event extraction and generation. In: ACL, Florence, Italy, 28 July 2019, pp. 5284–5294 (2019)
28. Yoo, D., Kweon, I.S.: Learning loss for active learning. In: CVPR 2019, Long Beach, CA, USA, 16–20 June 2019, pp. 93–102. IEEE (2019)
29. Yuan, X., Zhang, T.: Truncated power method for sparse eigenvalue problems. J. Mach. Learn. Res. **14**(1), 899–925 (2013)
30. Zhang, M., Zhang, Y., Fu, G.: End-to-end neural relation extraction with global optimization. In: EMNLP, Copenhagen, Denmark, 9–11 September 2017, pp. 1730–1740 (2017)
31. Zheng, S., Hao, Y., Lu, D., Bao, H., Xu, J., Xu, B.: Joint entity and relation extraction based on a hybrid neural network. Neurocomputing **257**, 59–66 (2017)

ABLA: An Algorithm for Repairing Structure-Based Locators Through Attribute Annotations

Iñigo Aldalur$^{(\boxtimes)}$, Felix Larrinaga$^{(\boxtimes)}$, and Alain Perez$^{(\boxtimes)}$

Mondragon University, Arrasate-Mondragon, Spain
{ialdalur,flarrinaga,aperez}@mondragon.edu

Abstract. The growth of the web has been unstoppable in the last decade, which leads to an increasing demand for extracting information from it. Apart from the need to extract information, this growth also has brought the necessity to adapt web pages to user requirements, create annotations or test web applications. Due to the evolution of web pages, the complexity of the implementation of these techniques has increased. Being able to test, annotate, adapt and extract information from web pages correctly and efficiently has become a primary task. In order to perform all these tasks, it is mandatory to have the best mechanisms to effectively and unequivocally locate the desired elements throughout the web page life cycle, especially when a web page evolves. Different mechanisms are used to find web nodes. These mechanisms, called locators, are prone to fail over time owing to changes on websites. Many authors improve life expectancy of locators developing algorithms that use different types of locators. Some others have created algorithms that regenerate locators by saving extra information from the previous structure of the website. These algorithms extend the useful life of locators but their computational and storage cost is much higher. To avoid these problems, we have designed an algorithm that employs an attribute system embedded in the HTML code. The algorithm is able to regenerate the locators based on these attributes every time a single change takes place in a given element attribute. The evaluation of the proposal shows a much lower computational cost than in previous works.

Keywords: Locators · Annotation · Mining · Testing · Web attributes

1 Introduction

Since the creation of the web 30 years ago, one of the objectives has been the extraction of relevant information from this universal source for different purposes. To automatically extract this information locators are used. "A Web locator is a mechanism for uniquely identifying an element on the Web Content i.e. in the Document Object Model (DOM)" [1]. Locators have evolved over time from being fragile and fail to minimal changes of the website to tolerate changes.

© Springer Nature Switzerland AG 2020
Z. Huang et al. (Eds.): WISE 2020, LNCS 12343, pp. 101–113, 2020.
https://doi.org/10.1007/978-3-030-62008-0_7

However, these changes continue to cause locators to stop working, leading to high maintenance costs in terms of both time and money.

Different types of locators have been developed over time. The first ones [22] were based on coordinates and were prone to change at the slightest change. For this reason they have became obsolete. The second generation is based on the structure of the web page (the DOM). These locators are more robust than the previous ones but as soon as there are changes in the structure, they tend to stop working. [22] classifies structure-based locators into XPath locator and attribute locators. Finally, a third generation of image-based locators has been developed (visual locators). These locators use image recognition techniques in order to find a certain element on a web page [22]. Moreover, [1] adds content-based locators to the list. Content-based locators work with the text composition of a website. These locators extract text content even when the desired text is in the middle of a paragraph thanks to its ability to select data within text node content. It is mainly used for web annotation and web harvesting, which is the process of automatically collecting information from the Web [10].

Locators are used in different scenarios: web annotation (establishes a relation between two resources [11]); web augmentation (is the alteration of the original user interface, generally by using scripts running at the client side [12]); web automation (is an activity that can be performed without human intervention [18]); data extraction (retrieves information stored on the Web [10]); testing (are processes designed to verify that a computer code works correctly [17]). For each of these cases you can use different types of locators. Despite the fact that coordinate based locators have almost ceased to be used, some studies still use them for testing and automation [4]. Coordinate-based locators are appropriate when the locations of the tested or automated website nodes are stable. In other words, when nodes are kept in the same position. It is important to highlight that coordinate locators can only be used to locate leaf nodes. Visual locators are appropriate when the structure of the website changes frequently and, as a consequence, structure based locators are less suitable. Additionally, if nodes are relocated on the website, coordinate based locators are useless and visual locators are able to find the element wherever the node is. Hence, visual locators are also suitable for testing and automation [27]. Content based locators are appropriate when the desired web information is in the text [23]. Other types of locators are unable to extract information from a text but structure based locators can be used when the data extract or annotation is the whole node. Visual locators are not suitable because content is not the same all the time. Structure based locator are appropriate for all scenarios [7].

Not only are structure based locators the most extended because of their usefulness in all scenarios, but also they are the most robust type of locator. [21] compares structure based locators to visual locators and concludes that structure based locators are more robust. Furthermore, they compare attribute structure based locators with XPath concluding that the ID attribute is the most robust of all, whereas XPath is the least robust. Taking into account previous works done to compare different types of locators, [1,21] summarize locator robustness classification as follows: structure-ID, structure-attributes, structure-XPath, visual, content, coordinate.

The remainder of the paper is organized as follows. Section 2 analyses the problem we want to solve, its causes and consequences. Section 3 discusses related work in order to give the reader an idea of what has been done in this area. Section 4 describes our approach to solve or mitigate the previously explained problem. Section 5 presents the validation of our approach and its results. Finally, Sect. 6 shows features we would like to enhance about our solution and it concludes the paper.

2   Problem Analysis

The Problem

Different approaches successfully use locators in their daily routines. Nonetheless, after websites upgrades, previously selected web nodes will no longer be found by standard locators. No matter how robust the locators are, they will eventually fail because all websites are updated over time [21].

Causes

One of the main causes is locator fragility. There are several works that have calculated the fragility of the locators [4,9,22]. This fragility is due to the fact that web pages are frequently updated because they are not able to meet users' needs [20]. [16] studied some reasons why locators stop working. The conclusions are that updates on the web may be due to changes in the structure (the DOM), appearance (CSS styles) or content. For example, changes in structure affect locators based on structure, those based on coordinates if the change implies a change of location and those based on content if the change affects the order of the text. The visual locators are those that better support the changes in the structure since they are not affected by these changes. When changes are implemented in styles, the most affected locators are visual and coordinate-based. Conversely, those based on structure and content are not impacted by this type of change since content and structure remain exactly the same as before. Finally, if what changes in a web page is the content, the most affected will be the content-based locators.

Consequences

When a locator breaks, it has to be fixed. Additionally, when there are changes in a website, the more locators there are the more locators will stop working. Therefore, fixing them is time-consuming, which results in high economic cost [14]. That cost rises if such changes are frequent. It is important to emphasize that many of these locator updates are not done automatically but manually, leading to a higher cost [6]. If a locator causes an application to stop working, users could perceive the application as low quality and therefore refuse to continue using it. Therefore, fixing locators and keeping them active is of vital importance. Our goal is to reduce the number of times a developer has to fix failed locators, thus reducing the economic cost and increasing user loyalty.

3   Related Work

First locator implementations focus mainly on generating robust algorithms. Robustness is defined as the ability of a computer system to cope with errors during execution [1]. A good example of development of robust web locators can be found in Robula+ [22]. Robula+ is an algorithm for the automatic creation of structure locators, more precisely XPath expressions. This algorithm prioritizes attributes by recognizing their ability to create robust XPath locators (ID and class attributes). Another similar example is [24] that shows that they can create a robust extractor for product information extraction by combining supervised classification with unsupervised methods. [3] claims that generating locators manually is demanding due to the dynamic nature of the DOM since it involves selecting multiple DOM elements. That is why they have created an automated technique for synthesizing DOM element locators using examples provided interactively by the developer. In spite of the fact that these approaches improve the robustness of previous researches, they end up failing.

In order to extend the useful life of locators, other authors propose the use of multiple locators. [2] uses Floper, a tool developed by their research group that generates XPath expressions that can be adapted manually to obtain a set of alternative queries. In essence, this means that the applications have more than a unique XPath expression to find the desired node. If a locator fails, another one will be utilized to retrieve the same result. [5] presents a web test generation algorithm that selects the most promising alternatives creating more than one locator for each element. [28] uses numerous parallel threads in order to evaluate different XPaths. Nevertheless, even if all these approaches extend life expectancy of locators, they wind up failing.

In recent times, research has changed the approach by trying to regenerate a locator when it failed. There are numerous studies which use XPath expressions to locate UI controls. [25] identifies failed tests and no longer valid XPath expressions. Then, it uses invalid XPath expressions and compares the two DOMs corresponding to a view in the new version and on the original page. Finally, it uses the DOM difference to repair the XPath expressions. In a similar manner, [9] developed a method of automatic wrapper reparation. It is based on comparing stored original structural DOM information for the Web wrapper with the new version and searching for similarities. [26] carries out a visual technique for web testing algorithms, including visual-based test repair for migrating DOM-based tests to visual tests. [15] repairs an invalid locator by using the attributes from the old version of the node. All these articles, in a nutshell, compare previous DOM with the current DOM to regenerate the locator. Nonetheless, it is not the unique technique because other authors use previous node attributes to recreate the new locator. For example, [1] regenerates XPath expressions when they are invalid owing to the fact that they save all structure data of the node and all its ancestors. If there is no single element obtained, then it tries all combinations of ancestor's structure information of the old element gradually. If the created new XPath expression does not locate a unique element using its own attribute combinations, the attribute combinations of the ancestor node of this element are utilized. This process is repeated until a single node is detected. It is considered a failure if no unique node is located when all the ancestor structure data is used. [8] creates a locator based on thirteen attributes and when the locator fails, it uses previously stored node's attributes in order to create a new locator that is able to find the new node. [19] affirms that some "methods tried to repair broken locators by using structural clues, but these approaches usually cannot handle radical changes to page layouts". They determine which node properties (attributes, texts, images, and positions) are trustworthy after comparing old and the new version to create a new locator. Finally, some other works have a manual component when they repair the locator. SITAR [13] is a system to interactively repair test scripts. This reparation is completed by mapping the new version of the interface and creating manually the new version. Unfortunately, all these algorithms are not always able to regenerate the locator despite the fact their performance prolongs locator life. In the next section we present an approach that prolongs locator efficacy in all the above scenarios considering that locators are run frequently.

4   Repairing Structure Attribute Based Locators

Despite all the efforts that have been made to develop the most robust locators, with the passage of time locators end up failing. As we have seen in Sect. 3, certain authors use several of these algorithms in their applications and thus extend locators' useful life. Most recent works try to regenerate broken locators by comparing the old and the new DOM, by using the old attributes. In our research group we have developed numerous web applications for different projects which we test every time we make a change. The problem arises when certain changes made some of the locators used in the test stop working. Then, we had to dedicate time to locate them and generate the test again causing a damage in time and money. Therefore, based on the work of restoring locators, we have developed an algorithm that supported by specific attributes in the DOM, it is able to regenerate the locators based on structure effectively, Annotation-Based Locator Algorithm (ABLA).

To achieve our goal we have defined four different attributes that are used to tag nodes. When the node id, class or another attribute is changed, these especial attributes are used to tag nodes maintaining the previous values:

- **prevId**: used when the id value is changed.
- **prevClass**: used when the class value is modified.
- **preva**: this is used when the value is changed maintaining the attribute name. The attribute name is concatenated to this tag.

– **prevv**: the value name is concatenated to this tag. This is used when the attribute name is changed maintaining the value.

The ABLA algorithm capable of testing a website and restoring locators in case they are broken is shown in Algorithm 1 in which the pseudo code of the algorithm is outlined.

Algorithm 1 performs according to the following steps of the procedure:

Algorithm 1. ABLA algorithm

1: **for** locatorList **do**
2:     attribute = getLocatorAttribute(locatorList, i, 0)
3:     value = getLocatorValue(locatorList, i, 1)
4:     node = executeLocator(attribute, value)
5:     **if** node != **null then**
6:         executeTest(node)
7:     **else**
8:         newNode = findNewNode(attribute, value)
9:         **if** newNode != **null then**
10:             newNodeAttr = getNewNodeAttr(newNode)
11:             newNodeValue = getNewNodeValue(newNode)
12:             restoreLocator(locatorList, attribute, newNodeValue, i)
13:             executeTest(newNode)

– Line 1: We have a list initialized with all nodes we want to find in the website.
– Lines 2 and 3: All attribute types and their values are read from the list one by one. There are three different types of attributes: id, class and the rest.
– Line 4: Each type of attribute has a different mechanism to find the node and this is the reason why we classify them in three categories. "executeLocator" function executes this three different locators in order to find the correct node.
– Line 6: If a node is found, this means that the locator is still correct and the test is executed normally.
– Line 8: Otherwise, "findNewNode" function tries to find the desired node. This function will be explained in Algorithm 2. If the function does not find any node, the test fails and it must be restored manually.
– Line 10 and 11: If a node is found, the new attribute and value are extracted. If the attribute is an id or a class, these attributes will be maintained and what will change is their value. If it is another type of attribute, the change could have occurred in the name of the attribute or its value.
– Line 12: the new attribute and the new values are stored in the list replacing the old values and attributes.
– Line 13: the test is executed.

In line 8 of the Algorithm 1 the function "findNewNode" is executed. It needs the attribute type and its value to find the desired node. The "findNewNode" function is explained in pseudo code in Algorithm 2. Algorithm 2 performs the following step by step procedure:

- Line 1 and 2: the function needs two parameters, the attribute type and its value to carry out its action properly..
- Line 4 and 5: If the attribute is id, the function searches the attribute "prevId" and the value the previous id had. If the "findNode" function does not find any node with this attribute and value, the function will return null. On the contrary, if the "findNode" function finds a node, this element will be returned.
- Line 6 and 7: If the attribute is class, the function looks for the attribute "prevClass" and the value of the previous class. In the same manner, the "findNode" function will return null or it will return the found element.
- Line 9: If the attribute is different from id or class, the function looks for the attribute "preva" concatenating the previous attribute name and the value of the previous attribute. This is because we firstly check if the value has been updated and the attribute is maintained as in the preceding if sentences.
- Line 10 and 11: If the previous step does not find a single node, the "findNode" function checks if the attribute name has been upgraded instead of the value.
- Line 12: the function returns the updated node or null in case no element has been found with the previous attributes and values.

It is also very important to emphasize that the solution of class attributes and the rest of attributes must be unique. If the function provides more than one node it means that it is no longer unique and null will be returned. Since id attributes are unique, they do not need this check.

Algorithm 2. findNewNode function

1: param1: attribute
2: param2: value
3: **if** attribute == "id" **then**
4:     node = findNode("prevId", value)
5: **else if** attribute == "class" **then**
6:     node = findNode("prevClass", value)
7: **else**
8:     node = findNode("preva"+attribute, value)
9:     **if** node == **null then**
10:         node = findNode(attribute, "prevv"+value)
11: **return** node

It is very important to highlight that this method not only serves for small changes that occur continuously in the development of web applications. This method enables, based on annotations, to run test, perform augmentation or data extraction or data mining from web pages that have been completely changed. This can be carried out if we include the old attributes in the new version. The new nodes that have the same representation or aim in the old version website must include previous id, class and other attribute values.

With this work we intend to propose labels which can be created automatically with an editor each time the developer changes any of the attributes. If the creation of these tags in the code is automatically generated, the cost will be practically null for the developer. These labels will be maintained for as long as the developer changes the value of the same attribute when the value of the proposed labels will be updated. This method diminishes in its totality the need to store the information. Unlike [25] that keep the previous version of the DOM or [1] that store all the attributes of the ancestors. In addition, the cost of computation is much lower. This aspect is demonstrated in the following section.

5   Validation

We have defined the following research questions to evaluate our approach:

RQ1 (Efficacy): How effective is ABLA at restoring broken locators?

RQ2 (Efficiency): What is the run-time of ABLA and what is the difference between ABLA and other approaches?

5.1   Approach and Results

The aim of the RQ1 is to evaluate the efficacy of ABLA in different websites. 5 different websites have been utilized to evaluate the algorithm. The criteria to select these websites has been their number of nodes, the number of attributes and their structure complexity which is illustrated in Table 1. The complexity of the HTML structure defines the number of sub-levels existing in the HTML or how far elements are settled down in the structure. We have simulated the behaviour of a web developer. We have identified the nodes that have changed their attribute values and we have added previous attribute values to proposed attributes. WayBackMachine[1] has been applied to obtain old version of the evaluated web pages. The versions of the first day of each month in 2019 has been used for comparison. In this evaluation hidden nodes and removed nodes have not been taken into consideration.

Table 1. A comparison between websites used to test ABLA

Website	#Elements per page	#Attributes per element	Type of structure
IMDB	Many	Many	Moderate
Wikipedia	Moderate	Moderate	Simple
Apple	Moderate	Many	Moderate
Firefox	Few	Few	Simple
BBC	Many	Moderate	Complex

---

[1] https://archive.org/web/.

Table 2 presents the results of the evaluation. The first column shows the websites tested. The next three illustrate the broken IDs, classes and other attributes (first value in each cell) and the number of restored locators by ABLA (second value in the cell). The last column shows the percentage of restored locator in each website. This column shows how ABLA is able to restore always the broken locator and find the desired node even if the website has been updated.

Table 2. ABLA repair results

Website	Broken/Restored IDs	Broken/Restored classes	Broken/Restored attributes	%Restored locators
IMDB	12/12	28/28	267/267	100%
Wikipedia	5/5	3/3	115/115	100%
Apple	0/0	145/145	235/235	100%
Firefox	0/0	6/6	22/22	100%
BBC	1/1	95/95	86/86	100%

Table 3. Regenerate locator algorithms execution order

Algorithm	Exec. order	Success	Test dependency
ABLA	$O(n)$	100%	n: nodes
[25]	$O(m^n)$	87.7%	m: test, n: node children
Weighted Tree Matching [9]	$O(m^n)$	90%	m: test, n: node children and siblings
VISTA [26]	$O(m^n)$	81%	m: test, n: images
WATERFALL [15]	$O(m^n)$	89.3%	m: test, n: node attributes
Regenerative locator [1]	$O(m^n)$	73.3%	m: test, n: node attributes
Genetic Algorithm [8]	$O(m^n)$	87%	m: test, n: node attributes
COLOR [19]	$O(m^n)$	77–93%	m: test, n: nodes or attributes
SITAR [13]	$O(m^n)$	41–89%	m: test, n: nodes

With regard to RQ2, the goal is compare its runtime with other algorithms that restore broken locators too. Besides, we want to evaluate the efficiency of our approach. Figure 1 compares the execution time of ABLA with other algorithms that regenerate broken locators. In particular the same 32 nodes from the same web page where extracted and the time needed to restore each locator was calculated. Answering the research question 2, ABLA has been evaluated individually and it needs 32ms to extract 32 nodes. Furthermore, Regenerative locator [1], VISTA [26], WATERFALL [15], COLOR [19] and SITAR [13] have been evaluated obtaining similar results as those presented in their corresponding researches. Table 3 illustrates the execution order of each algorithm. Their execution order is elevated because all previous algorithms are based on the number of test and the number of nodes, attributes or node siblings and what is more, complex node selection to extract optimal options. All this makes that their execution time is higher comparing with ABLA (see Fig. 1). In case of Regenerative locator [1], a simpler algorithm is used and its execution time is similar to ABLA even if it needs around 100ms to restore each locator. Table 3 also shows the performance of each algorithm. Regenerative locator presents the lowest time cost. In general, these algorithms are excellent restoring their locators with a higher success than 80% but ABLA is superior.

5.2 Limitations

All previous work mentioned in the related work obtain excellent results but all of them have limitations because when certain conditions are met, they all end up failing. In the case of ABLA, if the developer has carried out two updates between two executions on the same node in the same attribute, the algorithm will not find the web node. Given that attribute-based locators are the most robust since they are more resilient against web upgrades, it is very unlikely that they will be updated more than once in a short period of time. To justify this statement, note that in [22] they concluded that in 6 years, less than 2% of the ID-attributes and 18% of the other attributes had failed.

Another limitation is that this regeneration algorithm is intended to be applied on attributes that are unique. Therefore, it cannot be applied to all nodes of a web page since not all nodes have unique attributes.

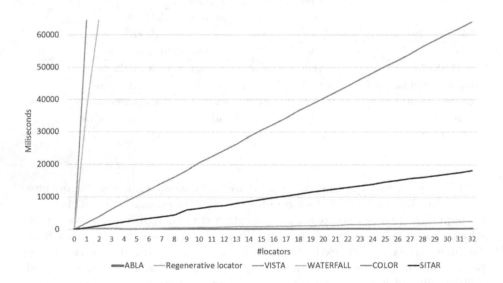

Fig. 1. Comparison between algorithms that regenerate locators

6 Future Work and Conclusions

Locators are sensitive to changes that occur on websites. All developed methods have tried to generate locators as robust as possible and although they are able to find the desired node for a longer time, they tend to fail in the long term when many changes in a page take place. For this reason, recent work tries to regenerate the locators that have stopped working by saving information related to the old code of the web page. All these algorithms are much more robust but they are less efficient because they require a much higher computational cost. Moreover, they need additional memory space to store information related to old versions of the website. The method presented in this article allows to save that extra storage since the data is inserted using special attributes in the HTML code. In addition to this, the algorithm is more efficient since it is not necessary to test a large number of different options until a locator that works is found. If the main locator fails, ABLA only requires to apply another execution, thus considerably reducing regeneration time. Finally, the efficacy of the algorithm reaches 100% if it is executed every time there is a change in the HTML code.

In our future work we plan to investigate and develop an algorithm that is able to regenerate XPaths by using these special tags. XPaths are applicable to all the nodes of a web page to find a specific node. As a consequence, we would avoid one of the limitations of this work since the ALBA algorithm only applies to nodes with unique attributes. In addition to this, our goal is to develop simple mechanisms for detecting updates on third-party websites and be able to keep a 100% effectiveness.

Acknowledgments. This work was carried out by the Software and Systems Engineering research group of Mondragon Unibertsitatea (IT1326-19), supported by the Department of Education, Universities and Research of the Basque Government.

References

1. Aldalur, I., Díaz, O.: Addressing web locator fragility: a case for browser extensions. In: Proceedings of the ACM SIGCHI Symposium on Engineering Interactive Computing Systems, EICS 2017, Lisbon, Portugal, 26–29 June 2017, pp. 45–50 (2017)
2. Almendros-Jiménez, J.M., Luna Tedesqui, A., Moreno, G.: Annotating "fuzzy chance degrees" when debugging XPath queries. In: Rojas, I., Joya, G., Cabestany, J. (eds.) IWANN 2013. LNCS, vol. 7903, pp. 300–311. Springer, Heidelberg (2013). https://doi.org/10.1007/978-3-642-38682-4_33
3. Bajaj, K., Pattabiraman, K., Mesbah, A.: Synthesizing web element locators (T). In: 30th IEEE/ACM International Conference on Automated Software Engineering, ASE 2015, Lincoln, NE, USA, 9–13 November 2015, pp. 331–341 (2015)
4. Bartoli, A., Medvet, E., Mauri, M.: Recording and replaying navigations on AJAX web sites. In: Brambilla, M., Tokuda, T., Tolksdorf, R. (eds.) ICWE 2012. LNCS, vol. 7387, pp. 370–377. Springer, Heidelberg (2012). https://doi.org/10.1007/978-3-642-31753-8_30

5. Biagiola, M., Stocco, A., Ricca, F., Tonella, P.: Diversity-based web test generation. In: 27th ACM Joint European Software Engineering Conference and Symposium on the Foundations of Software Engineering ESEC/FSE, Tallinn, Estonia, 26–30 August 2019, pp. 231–242 (2019)
6. Bures, M., Filipsky, M.: Smartdriver: extension of selenium webdriver to create more efficient automated tests. In: 6th International Conference on IT Convergence and Security, ICITCS 2016, Prague, Czech Republic, 26 September 2016, pp. 1–4 (2016)
7. Chang, C.-H., Lin, Y.-L., Lin, K.-C., Kayed, M.: Page-level wrapper verification for unsupervised web data extraction. In: Lin, X., Manolopoulos, Y., Srivastava, D., Huang, G. (eds.) WISE 2013. LNCS, vol. 8180, pp. 454–467. Springer, Heidelberg (2013). https://doi.org/10.1007/978-3-642-41230-1_38
8. Eladawy, H.M., Mohamed, A.E., Salem, S.A.: A new algorithm for repairing web-locators using optimization techniques. In: 13th International Conference on Computer Engineering and Systems (ICCES), pp. 327–331, December 2018
9. Ferrara, E., Baumgartner, R.: Intelligent self-repairable web wrappers. In: AI*IA 2011: Artificial Intelligence Around Man and Beyond - XIIth International Conference of the Italian Association for Artificial Intelligence, Palermo, Italy, 15–17 September 2011, pp. 274–285 (2011)
10. Ferrara, E., Meo, P.D., Fiumara, G., Baumgartner, R.: Web data extraction, applications and techniques: a survey. Knowl.-Based Syst. **70**, 301–323 (2014)
11. Fiorelli, M., Pazienza, M.T., Stellato, A.: A flexible approach to semantic annotation systems for web content. Int. Syst. Account. Financ. Manag. **22**(1), 65–79 (2015)
12. Firmenich, D., Firmenich, S., Rivero, J.M., Antonelli, L., Rossi, G.: CrowdMock: an approach for defining and evolving web augmentation requirements. Requirements Eng. **23**(1), 33–61 (2018). https://doi.org/10.1007/s00766-016-0257-3
13. Gao, Z., Chen, Z., Zou, Y., Memon, A.M.: SITAR: GUI test script repair. IEEE Trans. Softw. Eng. **42**(2), 170–186 (2016)
14. Guo, J.: Reducing human effort in web data extraction. Ph.D. thesis, University of Oxford, UK (2017)
15. Hammoudi, M., Rothermel, G., Stocco, A.: WATERFALL: an incremental approach for repairing record-replay tests of web applications. In: Proceedings of the 24th ACM SIGSOFT International Symposium on Foundations of Software Engineering, FSE 2016, Seattle, WA, USA, 13–18 November 2016, pp. 751–762 (2016)
16. Hammoudi, M., Rothermel, G., Tonella, P.: Why do record/replay tests of web applications break? In: IEEE International Conference on Software Testing, Verification and Validation, ICST 2016, Chicago, USA, 11–15 April 2016, pp. 180–190 (2016)
17. Herbold, S., Bünting, U., Grabowski, J., Waack, S.: Deployable capture/replay supported by internal messages. Adv. Comput. **85**, 327–367 (2012)
18. Huizinga, D., Kolawa, A.: Automated Defect Prevention: Best Practices in Software Management. Wiley, Hoboken (2007)
19. Kirinuki, H., Tanno, H., Natsukawa, K.: COLOR: correct locator recommender for broken test scripts using various clues in web application. In: 26th IEEE International Conference on Software Analysis, Evolution and Reengineering, SANER 2019, Hangzhou, China, 24–27 February 2019, pp. 310–320 (2019)
20. Lee, T.Y., Bederson, B.B.: Give the people what they want: studying end-user needs for enhancing the web. PeerJ Comput. Sci. **2**, e91 (2016)

21. Leotta, M., Clerissi, D., Ricca, F., Tonella, P.: Visual vs. DOM-based web locators: an empirical study. In: Casteleyn, S., Rossi, G., Winckler, M. (eds.) ICWE 2014. LNCS, vol. 8541, pp. 322–340. Springer, Cham (2014). https://doi.org/10.1007/978-3-319-08245-5_19

22. Leotta, M., Stocco, A., Ricca, F., Tonella, P.: ROBULA+: an algorithm for generating robust Xpath locators for web testing. J. Softw. Evol. Process **28**(3), 177–204 (2016)

23. Lin, A.Y., Ford, J., Adar, E., Hecht, B.J.: VizByWiki: mining data visualizations from the web to enrich news articles. In: Proceedings of the 2018 World Wide Web Conference on World Wide Web, WWW 2018, Lyon, France, 23–27 April 2018, pp. 873–882 (2018)

24. Potvin, B., Villemaire, R.: Robust web data extraction based on unsupervised visual validation. In: Nguyen, N.T., Gaol, F.L., Hong, T.-P., Trawiński, B. (eds.) ACIIDS 2019. LNCS (LNAI), vol. 11431, pp. 77–89. Springer, Cham (2019). https://doi.org/10.1007/978-3-030-14799-0_7

25. Song, F., Xu, Z., Xu, F.: An Xpath-based approach to reusing test scripts for android applications. In: 14th Web Information Systems and Applications Conference, WISA 2017, Liuzhou, China, 11–12 November 2017, pp. 143–148 (2017)

26. Stocco, A., Yandrapally, R., Mesbah, A.: Visual web test repair. In: Proceedings of the 2018 ACM Joint Meeting on European Software Engineering Conference and Symposium on the Foundations of Software Engineering, ESEC/SIGSOFT FSE 2018, Lake Buena Vista, FL, USA, 04–09 November 2018, pp. 503–514 (2018)

27. Yeh, T., Chang, T., Miller, R.C.: Sikuli: using GUI screenshots for search and automation. In: Proceedings of the 22nd Annual ACM Symposium on User Interface Software and Technology, Victoria, BC, Canada, 4–7 October 2009, pp. 183–192 (2009)

28. Zhang, Y., Pan, Y., Chiu, K.: A parallel Xpath engine based on concurrent NFA execution. In: 16th IEEE International Conference on Parallel and Distributed Systems, ICPADS 2010, Shanghai, China, 8–10 December 2010, pp. 314–321 (2010)

Text Mining

Automatic Action Extraction for Short Text Conversation Using Unsupervised Learning

Senthil Ganesan Yuvaraj[1(✉)], Shayan Zamanirad[1], Boualem Benatallah[1], and Carlos Rodriguez[2]

[1] University of New South Wales, Sydney, Australia
{senthily,shayanz,boualem}@cse.unsw.edu.au
[2] Universidad Católica Nuestra Señora de la Asunción, Asunción, Paraguay
carlos.rodriguez@uc.edu.py

Abstract. Collaboration tools are important for workplace communication. The amount of conversation data produced in workplaces are increasing rapidly, placing a burden on workers. There is a necessity to analyze large amounts of data automatically to extract actionable information. Multiple studies were conducted on action extraction to identify actions such as promises and requests. Most of these studies used supervised learning methods. The key problem discussed in this paper are (i) the automatic extraction of action types from short text conversations in collaboration tools such as Twitter and Slack, and (ii) leveraging large amounts of data using unsupervised learning. Data labelling is an important issue when dealing with large datasets for training and extending the corresponding algorithms across different actions and domains. In this paper, we propose an unsupervised learning approach using a combination of relation extraction techniques and word embedding to leverage large amounts of data. The *action2vec* model is created to identify specific actions from short text of conversation data. We have evaluated our unsupervised method against supervised learning and the results are comparable. The action type extractor is integrated with Slack to provide assistance for action type extraction. Thus, the contributions of this paper include an unsupervised learning method to utilize large amounts of data, an automatic extraction of action types from short text and the integration of our approach with state of the art collaboration tools.

Keywords: Action type extraction · Text embedding · Action recognition · Text mining · Short text conversation

1 Introduction

Collaboration tools (e.g., Slack[1]) have become a primary communication means in the workplace and information workers generate large volumes of conversation data in their day-to-day tasks. Email, blogs, wikis and Twitter have

---

[1] https://slack.com/.

© Springer Nature Switzerland AG 2020
Z. Huang et al. (Eds.): WISE 2020, LNCS 12343, pp. 117–128, 2020.
https://doi.org/10.1007/978-3-030-62008-0_8

become increasingly available and accepted in workplace communications [23]. More specifically, email and instant messaging (IM) are considered key tools in this context. Email is typically used on a daily basis across organizations and it is an important tool for non face-to-face communication [8]. Instant messaging (e.g., IBM Lotus Instant Messaging) are multi-tasking tools that allow people to communicate and, at the same time, engage in other activities [16]. It is more immediate than email and communication happens in (near) real time [16].

Despite the advantages provided by these tools, users are typically overwhelmed with the amount of data produced by such communications resulting in a problem known as information overload [4,24]. Workplace productivity can be affected if such information is not properly organized in order to effectively manage ongoing tasks [24]. A key requirement to address this issue is that of automatically identifying actions from email conversations and short texts [10]. Here, an *action* is something that is done by an actor (e.g., a person) or it can be a request, making a commitment to the user interest in natural language conversations [7,15,18]. An *action type* represents a particular action in a process. For example, the e-commerce process includes action types such as *"Deliver, Cancel, Return, Payment and Purchase"*. For instance, *"I will contact customer services if it doesn't arrive on time"* is an example of a particular action type: *"Deliver"*. The identification of such action types are important for the categorization and handling of conversations.

Conversations typically have many actions, which can be identified using manual annotation [11]. This approach can lead to unidentified actions that may result in delays and other issues that affect the organization's activities. The automatic extraction of actions from emails and instant messages can help workers increase their productivity. In this context, multiple studies were conducted in the past to identify actions from emails and collaboration tools [4,7,14,22]. These studies typically use supervised methods for action extraction, which requires labelled data. Such requirement results in important limitations when these supervised approaches need to be extended to various actions. Data labelling is one of the main concerns for action extraction using these approaches and it is important to use large unlabelled data to improve the results [6,12,27]. In this paper, we propose an unsupervised approach and focus on the extraction of *action types* such as *"Deliver, Cancel, Return, Payment and Purchase"* from short text conversation, then we showcase the integration of our approach into a widely used collaboration tool (Slack).

The key contribution of this paper consists in an approach that allows for (i) analyzing conversation data from collaboration tools involving short text conversations, (ii) training large volumes of data without labelling, which can help overcome the limitations of supervised learning approaches, and (iii) seamlessly and effectively adapting to scenarios and domains with multiple actions.

2   Related Work

The problem of action identification from text has been studied extensively in previous research. For example, emails are classified based on the intention of the email sender using nouns *(e.g., information, meeting, task)* and verbs *(e.g., propose, commit, deliver)* called email speech act [7], which identifies the intention in emails such as *"request, propose, amend, commit or deliver"*. Previous research highlighted that users feel overloaded when tracking large number of threads and length of the intervals between messages in email threads [4,24], which involves manual identification of action hidden in the emails. Another study focused on finding the strength of request and commitment in workplace emails using exploratory annotation tasks [15]. The result encourages automatic detection of request and commitment in emails. The classification- or prioritization-based methods are used to group email based on actions [13,21,25]. This method works at the level of the entire email including all statements from the email and do not focus on individual statements [13]. eAssistant [18] learns action verbs and related set of features to identify actionable items such as requests and promises. These studies are focused on identifying generic classes such as request and commitment actions in emails. In our work, instead, we focus on *action type* extraction to identify specific actions such as *"Deliver, Cancel, Return, Payment and Purchase"* from short text conversation data. These short texts have less than 200 characters (i.e., data is sparse) and as a result information extraction becomes harder [9].

The Semanta system [20] was implemented for identifying actions based on consent summarization of emails. It extracts types of actions (e.g., file request, task assignment) from emails. A rule-based method [21] is used to classify actions into categories such as *"Request, Suggest, Assign and Deliver"*. These methods need human intervention to classify emails and involve manual annotation and user reviews [20,21]. In contrast to the text typically found in e-mails, instant messaging texts are fairly short and include idioms and abbreviation [1], making the task of automatic extraction of action types more challenging.

Finally, the work presented in [26] proposes an approach to the problem of event type recognition, which leverages word embedding techniques [17] to find events by averaging the vectors of words in n-grams. Word embedding techniques can help group semantically similar words to find relevant actions. In this paper, we propose an unsupervised method leveraging word embedding for short text conversation.

3   Action Type Framework

The action type extractor framework is illustrated in the Fig. 1. The initial layer has *relation extractor* to extract relation tuples *(subject, relation, object)* from seed data. The open information extraction techniques [19] are used for this task. The relations from these tuples are used to initially identify the action types such as *"Deliver, Cancel, Return, Payment and Purchase"*. The extracted relation includes relevant actions, other actions or no actions. In *Seeds selection*, the relations that are related to action types such as *"Deliver, Cancel, Return, Payment and Purchase"* are selected. If the selected seed relations are used directly for action extraction then it will be able to filter only limited actions

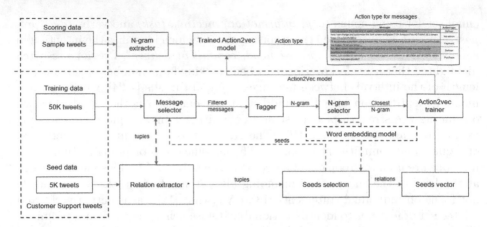

Fig. 1. Action type extractor framework

(those matching the seed relations). We convert the seed relation words to its vector form using word embeddings [17] in order to capture the words semantically similar to each other. This can help us capture actions semantically related to the selected seed relations (i.e., action types such as *"Deliver, Cancel, Return, Payment and Purchase"*). The *message selector* filters the messages that match the seeds for each action type. This is, it picks the relevant message from large datasets to train the action2vec model. The *tagger* extracts nouns, verbs and adjectives from the message, which are used to formulate bi-gram, tri-grams, etc. The n-grams with the closest distance w.r.t. the seeds vector are obtained by the *N-gram selector*. The *Action2vec trainer* improves the vector and produces a final trained *action2vec model* used to find the closest action types for messages.

4   Action Type Extraction

The customer support tweets [3] are considered as data source for training our action2vec model. The action type extractor is used to extract actions from tweets. The nouns, verbs and adjectives are extracted from each tweet to identify actions. Verbs such as *"Deliver, Cancel, etc."* and nouns such as *"Product, Time, etc."* and adjectives such as *"Many, Next, etc."* are extracted to identify actions. In our approach, we extract n-grams and do not restrict ourselves to bi-grams or tri-grams only. This will help us improve the action2vec model to include all keywords from tweets so that we can train our model without such restrictions.

The action type extractor leverages the vector space model (VSM) [17] to extract vectors for each word (nouns, verbs and adjectives) found in a tweet and calculate a *tweet vector* by averaging the vectors of individual words contained in the tweet. The similarity ratio between the *tweet vector* and action2vec model of all action types are then calculated. The action type which has highest similarity ratio with the *tweet vector* and above a given threshold is considered as action type for a given tweet.

4.1   Seed Generation

The seed generation is an initial step to build n-grams for action types. The tuples *(subject, relation and object)* are extracted using Stanford OpenIE [2]. For example, the tweet "I've purchased a gift card for a second time and the first arrived within minutes" has tuples (first, arrive within, minute). The relation "arrive within" related to action type *"Deliver"* is considered as a seed. The *"Deliver"* action type has relations such as *"arrive within", "deliver product",* etc. after excluding stop words. The seeds are extracted from tweet samples for each action type. The word embedding model [17] is used to calculate vectors for seeds in order to create our so-called seeds vectors.

The n-grams vector is created by calculating the average of vectors for each word in the n-grams [26]. For example, *vector("arrive")* and *vector("within")* are extracted for the seed "arrive within". Thus, the individual vectors are averaged to create a vector for the seed *"arrive within"*. Similarly, the average of all seeds of a particular action type *("arrive within", "deliver product", etc.)* is used as seeds vector. Equation 1 and 2 show the calculation of seed and seeds vector.

$$\overrightarrow{seed} = \frac{\sum_{i=1}^{m} \overrightarrow{word_i}}{m} \tag{1}$$

$$\overrightarrow{seeds} = \frac{\sum_{j=1}^{n} \overrightarrow{seed_j}}{n} \tag{2}$$

Here, $\overrightarrow{seed}$ is a vector of a seed by averaging all words for each relation. $\overrightarrow{seeds}$ is a vector of seeds by averaging all seeds for each action type. m is number of words in the relation. n is number of seeds in the action type.

The seeds are created based on specific keywords from a small sample. We need to train seeds vector with larger training sample to find all relevant tweets that belong to the action type. We create action2vec model using a larger sample based on seeds of each action type.

4.2   Action2Vec Model

A large set of tweets are considered to create our action2vec model. The relation phrase are extracted for each tweet. The tweets with relation phrases matching with our seeds are selected for training the action2vec model. The nouns, verbs and adjectives of the selected tweets are extracted for training. The vector is extracted for each word of n-grams *(nouns, verbs and adjectives)* and averaged to create a message vector. This message vector is compared with the seeds vector to find the similarity ratio. The cosine similarity is calculated based on

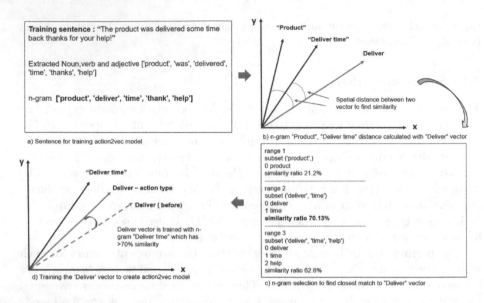

Fig. 2. Training action2vec model

the orientation of two vector (x,y). Equation 3 shows the dot matrix of two vectors (x,y).

$$\cos \theta = \frac{\overrightarrow{x} * \overrightarrow{y}}{|| \overrightarrow{x} |||| \overrightarrow{y} ||} \qquad (3)$$

The threshold 70% is considered (based on experiments) to find the relevant action type for the tweet. If similarity ratio exceeds 70% then the message vector and seeds vector are averaged to train action2vec model. This process is repeated for all tweets in the training data. Figure 2 shows training of action2vec model from training data.

During training, words in tweets representing nouns, verbs and adjectives do not have equal importance. This means that some of the words increase the similarity while others decrease it with the seeds vector. To overcome this issue, we have come up with combinations of up to 5 words for n-grams *(noun, verb and adjective)* to find similarity with the seeds vector. The vector from each combination are compared with the seeds vector and then the highest similarity vector is selected to train action2vec model. This method filters out irrelevant combinations for training the action2vec model and helps reduce noise in the model. Figure 2c) shows the selection of n-grams based on similarity with seeds vector.

5   Dataset and Evaluation

In order to evaluate the feasibility of our approach, we used tweets data to train our action2vec model and compared it with seeds vectors. The large collection of customer support Twitter data [3] was used for the analysis. There are approximately 3 million tweets from multiple brands such as Apple, Amazon, Uber and British Airways. It contains inbound tweets from customers and replies from customer support representatives. The customer support Twitter data has short text conversations and different types of action. The major e-commerce brand Amazon is selected because it has multiple action types such as *"Deliver, Cancel, Return, Payment and Purchase"* for e-commerce processes. A total of 128K Amazon inbound tweets are considered for our analysis.

5.1   Seed Data

The sample data for seeds are extracted from Amazon tweets data. A random sample of 5K tweets (out of 128K tweets) are extracted to find seeds for action types. The Stanford OpenIE [2] is used to find the relations out of these 5K tweets. The extracted relation phrase from tuples are used to identify the relevant seeds for each action type. We focused on action types such as *"Deliver, Cancel, Return, Payment and Purchase"*, which are available in the majority of tweets. For example, the tweet *"I don't want to return all 3 items though, I just want to send one back and have it replaced with the correct one, will that be ok?"* has the tuples *(I, send back, one)*. The seed *"send back"* is relevant to *"Return"* action so it is included in seeds for *"Return"*. The relation that are not relevant to any action types are ignored. For example, *"I tried talking to your chatbot. It wasn't very helpful."* has the tuples (I, talk to, your chatbot) and not related to any of action types. For instance, a total of 57 seeds out of 5K tweets was extracted for the *"Deliver"* action type. Table 1 shows the number of seeds and examples for each action type.

Table 1. Seeds data

Action type	No. of seeds	Example
Deliver	57	Arrive within, deliver thing to
Cancel	21	Cancel order via, ask refund for
Return	16	Send back, replacement
Payment	18	Manage payment through, take money from
Purchase	32	Order package for, place order on

5.2   Training Data

A large set of tweets (50K out of 128K tweets) is used for training action2vec model. The relation phrases are extracted from each tweet in the training data. The relation phrases that match with seeds are filtered for training specific action types. For example, the tweet *"It should arrive in the next couple of days. Still not same day."* has the tuple (it, should arrive in, couple of day), where the relation "should arrive in" matches with seeds in the *"Deliver"* action type. This tweet is considered for training the action2vec model for the *"Deliver"* action type. Using this approach, for instance, a total 1,863 tweets were filtered out of 50K tweets for the *"Deliver"* action type.

Armed with these training data (tweets), we train our action2vec model for each action type. Figure 2 shows examples of the training tweets for action2vec model. The message vector from tweets that have highest similarity with seeds vector are used to train the action2vec model. The number of n-grams selected to train action2vec model are *"Deliver"* (753), *"Cancel"* (148), *"Return"* (167), *"Payment"* (236) and *"Purchase"* (268).

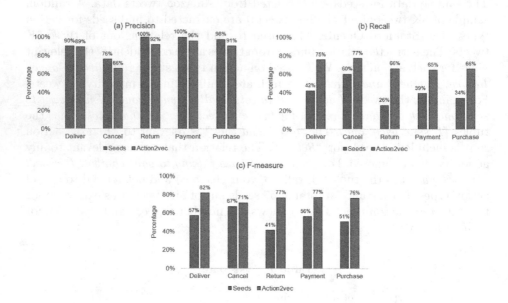

Fig. 3. Seeds vector and action2vec model comparison

5.3   Experiment and Results

There are 5 action types considered for evaluation. We prepared an evaluation dataset with tweets for action types *"Deliver"* (215), *"Cancel"* (217), *"Return"* (211), *"Payment"* (204) and *"Purchase"* (225). Each category contains 50% of tweets related to particular action type and the remaining 50% either have other action types or no actions whatsoever.

The evaluation is done for seeds vector and action2vec model separately. We do this to find out the benefits of using our unsupervised training method. We feed the tweets from our evaluation dataset to both seeds vector and action2vec model to get the respective scores. The threshold is set to 60% based on our experiments as the similarity ratio threshold for each action type. We use precision, recall and F-measure metrics to compare the results of the seeds vector and action2vec model.

The seeds vector achieved better precision than action2vec model for all action types as shown in Fig. 3. For example, for the action type *"Purchase"* seeds vector has 98% precision comparing to 91% precision of action2vec model. The recall is very low for seeds vector comparing to action2vec model. This shows action2vec model is able to identify most of the actions compared to seeds vector. The *"Cancel"* action type has achieved the highest recall (60%) for seeds vector among all action types. The action2vec model has more than 70% in recall for *"Deliver"* and *"Cancel"* and more than 60% for other action types. The action2vec model has higher F-measure for *"Deliver"* (82%), *"Cancel"* (71%), *"Return"* (77%), *"Payment"* (77%) and *"Purchase"* (76%). The results show that a large set of training data has improved action2vec model and it is more effective in finding all action types. Figure 3 shows all the metrics used for comparison of seeds vector and action2vec model results.

We demonstrated how we trained and evaluated our unsupervised method for 5 action types. Our approach can be easily extended to different actions. The relevant seeds from sample data need to be identified initially for each action. Then our unsupervised training method can create the action2vec model for action types by training with large datasets. This unsupervised method can be adapted to different action types and domains with less effort as compared to other approaches that heavily depend on data labelling.

a) Action2vec

Actions	Precision	Recall	F-measure
Deliver	89%	75%	82%
Cancel	66%	77%	71%
Return	92%	66%	77%
Payment	96%	65%	77%
Purchase	91%	66%	76%

b) SVM

Actions	Precision	Recall	F-measure
Deliver	84%	75%	79%
Cancel	65%	60%	63%
Return	95%	75%	84%
Payment	93%	63%	75%
Purchase	86%	69%	77%

Fig. 4. Comparison of action2vec and SVM results

6   Comparison with Machine Learning Method

The unsupervised method action2vec has been evaluated further by comparing with support vector machine (SVM). SVM is a supervised machine learning method that can be used to classify natural language text. We used a SVM classifier from study [5] to find actions from emails.

Given its supervised nature, this classifier requires labelled data. We have prepared labelled tweets for action types such as *"Deliver"* (312), *"Cancel"* (306), *"Return"* (318), *"Payment"* (317) and *"Purchase"* (307) to train the SVM model.

The labelled data is balanced to have unbiased training data. The n-grams from nouns, verbs and adjectives of each tweet sentences were extracted. Stop words were removed and verbs were converted to present tense. The vector is extracted for each word of the n-grams and averaged to calculate the message vector. The message vector has labelled action types that are used to train SVM machine learning model. The SVM model is used to score the evaluation data. The threshold 60% is considered for SVM score to find action types based on experiment to have better precision and recall.

The results show that action2vec model has better F-measure for *"Deliver"* (82%), *"Cancel"* (71%) and *"Payment"* (77%) comparing to the SVM model. This confirms that unsupervised method action2vec has similar results comparing to supervised machine learning approaches. Figure 4 outlines the precision, recall and F-measure numbers for both action2vec and SVM.

7   Integration with Collaboration Tool

The action type extraction from short messages can be helpful in the workplace environment. In this context, workers are typically engaged with multiple, concurrent tasks from different projects and workflows. The automatic extraction of actions from conversations can help workers as an alert/notification mechanism that contribute to avoid unnecessary delays in the corresponding processes and increasing their productivity.

We have integrated our action type extractor in a state of the art collaboration tool: Slack. Slack is a team messaging application widely used nowadays in the workplace to facilitate inter- and intra-communication in organizations. Our action type extractor was implemented as a Slack channel that extracts actions from text. Here, if a user enters a text in the conversation channel it will be compared with our action2vec model for computing a similarity ratio. The action type extractor then selects an action type based on similarity ratio and provides a relevant action type for the given text. For example, if a user asks *"why does amazon take my money but then cancel my order?"*, the action type extractor identifies the action type *"Cancel"*.

8   Conclusion and Future Work

We have explored the action type extraction problem on short text conversation data from collaboration tools. The unsupervised method introduced in this paper for action type extraction helps to train models with large amounts of data. It avoids data labelling, which is one of the major issues found in existing approaches for training large text data. We have evaluated the trained action2vec model with an initial seeds vector. The large set of training data is helpful to improve the model. Our action2vec model is compared with an SVM-based machine learning approach. The results show that action2vec model is comparable to the supervised counterpart. Our unsupervised action type extraction method is adaptable and can be extended to other action types and domains.

To showcase our proposed solution, we integrated the action type extractor with Slack to provide assistance to workers in extracting action types. In future work, we will collect user feedbacks and apply deep learning algorithms to update the model in near real time.

References

1. Almeida, T.A., Silva, T.P., Santos, I., Hidalgo, J.M.G.: Text normalization and semantic indexing to enhance instant messaging and SMS spam filtering. Knowl.-Based Syst. **108**, 25–32 (2016)
2. Angeli, G., Premkumar, M.J.J., Manning, C.D.: Leveraging linguistic structure for open domain information extraction. In: Proceedings of the 53rd Annual Meeting of the Association for Computational Linguistics and the 7th International Joint Conference on Natural Language Processing (Volume 1: Long Papers), pp. 344–354 (2015)
3. Axelbrooke, S.: Customer support on Twitter (2017). https://www.kaggle.com/thoughtvector/customer-support-on-twitter
4. Bellotti, V., Ducheneaut, N., Howard, M., Smith, I.: Taking email to task: the design and evaluation of a task management centered email tool. In: Proceedings of the SIGCHI Conference on Human Factors in Computing Systems, pp. 345–352. ACM (2003)
5. Bennett, P.N., Carbonell, J.: Detecting action-items in e-mail. In: Proceedings of the 28th Annual International ACM SIGIR Conference on Research and Development in Information Retrieval, pp. 585–586 (2005)
6. Blum, A., Mitchell, T.: Combining labeled and unlabeled data with co-training. In: Proceedings of the Eleventh Annual Conference on Computational Learning Theory, pp. 92–100 (1998)
7. Cohen, W.W., Carvalho, V.R., Mitchell, T.M.: Learning to classify email into "speech acts". In: Proceedings of the 2004 Conference on Empirical Methods in Natural Language Processing (2004)
8. Ducheneaut, N., Bellotti, V.: E-mail as habitat: an exploration of embedded personal information management. Interactions **8**(5), 30–38 (2001)
9. Faguo, Z., Fan, Z., Bingru, Y., Xingang, Y.: Research on short text classification algorithm based on statistics and rules. In: 2010 Third International Symposium on Electronic Commerce and Security, pp. 3–7. IEEE (2010)
10. Kateb, F., Kalita, J.: Classifying short text in social media: Twitter as case study. Int. J. Comput. Appl. **111**(9), 1–12 (2015)
11. Khoussainov, R., Kushmerick, N.: Email task management: an iterative relational learning approach. In: Proceedings of the Conference on Email and Anti-Spam (2005)
12. Kiritchenko, S., Matwin, S.: Email classification with co-training. In: Proceedings of the 2011 Conference of the Center for Advanced Studies on Collaborative Research, pp. 301–312. IBM Corp. (2011)
13. Kushmerick, N., Lau, T., Dredze, M., Khoussainov, R.: Activity-centric email: a machine learning approach. In: Proceedings of the National Conference on Artificial Intelligence, vol. 21, p. 1634. AAAI Press, Menlo Park, MIT Press, Cambridge, London (2006)

14. Lampert, A., Dale, R., Paris, C.: Detecting emails containing requests for action. In: Human Language Technologies: The 2010 Annual Conference of the North American Chapter of the Association for Computational Linguistics, pp. 984–992. Association for Computational Linguistics (2010)

15. Lampert, A., Paris, C., Dale, R., et al.: Can requests-for-action and commitments-to-act be reliably identified in email messages. In: Proceedings of the 12th Australasian Document Computing Symposium, pp. 48–55. Citeseer (2007)

16. Maina, T.M.: Instant messaging an effective way of communication in workplace. arXiv preprint arXiv:1310.8489 (2013)

17. Mikolov, T., Sutskever, I., Chen, K., Corrado, G.S., Dean, J.: Distributed representations of words and phrases and their compositionality. In: Advances in Neural Information Processing Systems, pp. 3111–3119 (2013)

18. Nezhad, H.R.M., Gunaratna, K., Cappi, J.: eAssistant: cognitive assistance for identification and auto-triage of actionable conversations. In: Proceedings of the 26th International Conference on World Wide Web Companion, WWW 2017 Companion, International World Wide Web Conferences Steering Committee, Republic and Canton of Geneva, Switzerland, pp. 89–98 (2017). https://doi.org/10.1145/3041021.3054147

19. Niklaus, C., Cetto, M., Freitas, A., Handschuh, S.: A survey on open information extraction. arXiv preprint arXiv:1806.05599 (2018)

20. Scerri, S., Davis, B., Handschuh, S., Hauswirth, M.: Semanta – semantic email made easy. In: Aroyo, L., et al. (eds.) ESWC 2009. LNCS, vol. 5554, pp. 36–50. Springer, Heidelberg (2009). https://doi.org/10.1007/978-3-642-02121-3_7

21. Scerri, S., Gossen, G., Davis, B., Handschuh, S.: Classifying action items for semantic email. In: LREC (2010)

22. Shim, E., Singla, V., Krishnaswamy, V.: Application frameworks and methods for assisting tracking of actionable items. In: Proceedings of the 6th International Conference on Ubiquitous Information Management and Communication, p. 5. ACM (2012)

23. Turner, T., Qvarfordt, P., Biehl, J.T., Golovchinsky, G., Back, M.: Exploring the workplace communication ecology. In: Proceedings of the SIGCHI Conference on Human Factors in Computing Systems, pp. 841–850 (2010)

24. Whittaker, S., Sidner, C.: Email overload: exploring personal information management of email. In: Proceedings of the SIGCHI Conference on Human Factors in Computing Systems, pp. 276–283 (1996)

25. Yang, Y., Yoo, S., Lin, F., Moon, I.C.: Personalized email prioritization based on content and social network analysis. IEEE Intell. Syst. **4**, 12–18 (2010)

26. Zamanirad, S., Benatallah, B., Barukh, M.C., Rodriguez, C., Nouri, R.: Dynamic event type recognition and tagging for data-driven insights in law-enforcement. Computing **102**(7), 1627–1651 (2020). https://doi.org/10.1007/s00607-020-00791-z

27. Zelikovitz, S., Hirsh, H.: Improving short text classification using unlabeled background knowledge to assess document similarity. In: Proceedings of the Seventeenth International Conference on Machine Learning, vol. 2000, pp. 1183–1190 (2000)

Topic Analysis by Exploring Headline Information

Rong Yan[1,2]([✉]) and Guanglai Gao[1,2]

[1] College of Computer Science, Inner Mongolia University,
Hohhot, People's Republic of China
{csyanr,csggl}@imu.edu.cn
[2] Inner Mongolia Key Laboratory of Mongolian Information Processing Technology,
Hohhot, People's Republic of China

Abstract. As for the topic representation in standard topic models, the words that appear in a document are considered with the same weight under the assumption of 'bag of words'. The word-topic assignment will lean to the high-frequency words and ignore the influence of the low-frequency words. As a result, it will ultimately impact on the performance of topic representation. Generally, the statistical information obtained from the whole document collection can be used to improve this situation. In addition, headlines of some kind of documents, such as news articles, usually summarize the important elements in the document, and the words in headlines are more appropriate to represent the topics. However, few previous studies consider the headline rich information, which is significant for topic modeling. In this paper, we propose a new headline-based topic model in order to accomplish a well-formed topic description. Experimental results on three widely used datasets show that the proposed headline-based modeling scheme achieves lower perplexity.

Keywords: Headline information · Topic analysis · Latent Dirichlet Allocation

1 Introduction

Probabilistic topic modeling family has a class of statistical techniques that bring out the latent representation in the collection, like LDA (Latent Dirichlet Allocation) [1,3,5–10,13], which is able to find new data correlations in a low-dimensional latent topic space. The quality of topic model affects the performance of several content analysis tasks in NLP (Natural Language Processing) [4,8,11,12,14,15,18], such as opinion recognition, text summarization, word sense disambiguation, text classification and information retrieval. It assumes that there are K latent variables or latent topics in a collection of documents that are used as mixtures to generate or model each document, where each latent variable is characterized by a distribution over words. The modeling manner

© Springer Nature Switzerland AG 2020
Z. Huang et al. (Eds.): WISE 2020, LNCS 12343, pp. 129–142, 2020.
https://doi.org/10.1007/978-3-030-62008-0_9

makes it easy to assign a probability to a previously unseen document. Generally, the possibility of a word assigned to each topic is effected by two probability distributions. One is the word-topic probability distribution, which is used to represent topic at the collection level; another is the topic-document probability distribution, which is used to represent document at the document level. Thus, latent topic acts as a middle chain to connect the word and the document. Based on the research of the evolution process of topic modeling, it is obvious to find that suitable descriptions for these two probability distributions will be the key factor directly affecting the performance of topic analysis. Obviously, the word in a document is usually closely associated with its frequency. Especially in LDA, the frequency is described as a multinomial distribution with *Dirichlet* prior, and it is estimated using variational Bayes inference [3] or Gibbs sampling [1,7]. Generally, the word frequency in a document collection always follows Zipf's law, but not for an individual document. LDA models the collection by employing the multinomial distribution with *Dirichlet* prior. However, the *Dirichlet* multinomial setting can't capture Zipf's law of the word distribution [10], nor to provide well-formed topic descriptions [17]. Actually, the truth is that the few high-frequency words outweigh the majority of low-frequency words. As a result, it will let the word-topic assignment lean to the high-frequency words that are more popular or general, and make them be a much stronger topic-specific indicator. In this way, the quality of topic description can't be guaranteed because of no constraints on the inference of word-topic distribution in the common topic modeling algorithms [17].

The remainder of this paper is organized as follows: Sect. 2 gives the motivation of this paper. Section 3 describes the proposed headline-based topic model in detail. Section 4 discusses the experiments and result analysis. Finally, Sect. 5 provides a short conclusion and possible future work.

2   Motivation

Generally, removing the word with high frequency is the primary tactics to avoiding the influence of the word frequency before topic modeling [3,7,15]. Meanwhile, some low-frequency words that are supposed to have few impacts on the topic, are also removed directly. Both of them are regarded as stop-words in many applications of NLP research. We think this manner is too coarse for the quality of topic description.

In fact, it ignores two facts. One is that some low-frequency terms indeed have certain practical significance, which is possibly the research focus or innovation, especially in the fields of Science, technical literature and Medical research study. Generally, the term with bigger probability will have more opportunity to be the topic description term. However, the truth is that the term with low probability will not appear in word-topic distribution. Consider the term 'AIDS', appear 77 times in text contents in XINHUA[1] four years news collection. We model texts in XINHUA using LDA with 60 topics, and the probability of 'AIDS' in

---

[1] http://research.nii.ac.jp/ntcir/index-en.html.

a specific topic is 0.001435402536324239 with ranked 126, whose probability is too low to become the part of description for the topic.

In common topic modeling, different document is forced to represent by the same specific topics so that the same term would be shared to appear in different topics. Thus, another fact is that some high-frequency terms indeed can't act as a topic-specific indicator role, and are too general to many topics in the collection, which will restrict the difference between each topic pair. Actually, this kind of term always helps to give background to represent a specific topic. Just removing these terms will have little impact to the description of the topic. However, there are still some high-frequency terms, which just appear in some finite topics, have strong semantic representation for the topic. Thus, it will lead to the semantic loss for the topic after removing them directly. In a word, it is not a good option to removing all high-frequency and low-frequency terms directly before modeling.

Of course, some other works support not removing any terms before modeling. Xia et al. [16] tried to find the topic-specific indicator features of the word in advance before document topic modeling. Wilson et al. [15] proposed a different term weighting scheme for LDA, which is more concerned about the influence of the high-frequency words in text modeling than the low-frequency words. Conversely, some works [9,15] assign words with different weights without removing any word during the process of document modeling. However, all these works ignore the frequency influence between words in topic modeling. In this paper, we focus on improving the model performance in the generation of topics description during the iterative process of topic modeling by making use of the rich headline information in the collection.

In topic modeling, such as LDA, due to the 'bag-of-words' assumption, the order of words that appear in a document is not important, because the document is modeled by maximizing the likelihood. Then, the word-topic assignment will tend to the general or high-frequency terms, and ignore the low-frequency terms. In reality, some low-frequency terms are the strong topic-specific indicators, whose role in the topics should be enhanced, especially in some specific domains.

It is worth noting that some certain documents have the specific headline information, such as news article, which is an important component for this kind of document because it is strong relevant to the content of the document. But in fact, the headline information has few attentions in topic modeling. Just for this reason, we do some statistical analysis for the actual news collection. Figure 1 exhibits the analysis on XINHUA news collection, and it shows the frequency comparison of the same word in the headline and in the text content respectively.

From Fig. 1, we notice that for part of the words, the word with higher frequency in the headline (Fig. 1(a)) also has more opportunity to appear in the text contents (Fig. 1(b)), which means that the frequency of words in headlines are highly consistent with that in text contents. At the same time, we are glad to find the truth is that some words in text contents with low frequency but

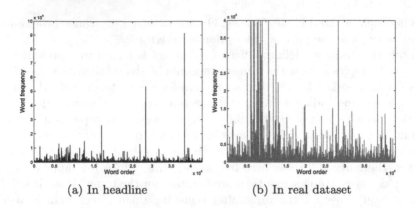

(a) In headline   (b) In real dataset

Fig. 1. Comparison of the same word frequency in headline and dataset.

have high frequency in headlines. Enlightened by this observation, we propose an approach which makes it possible to use the headline information to re-weighting the word in the topic, in order to improve and remedy the influence of word frequency for topic modeling.

In this work, we focus on using headline information for topic model that named HTM (Headline-based Topic Model) on three real datasets in order to overcome the dilemma of word frequency as well as achieve a well-formed topic description. To the best of our knowledge, this is the first work to improve the description of topic modeling using headline information.

3   Our Proposed Model

Fig. 1 sends us a good message that the headline information in news collection would have the positive effect for alleviating the word frequency for topic modeling. We depict our proposed HTM model in Fig. 2 using a graphical model in plate notation.

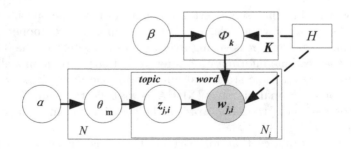

Fig. 2. Our proposed model HTM in plate notation.

In the graphical model shown in Fig. 2, N denotes the number of documents in a collection with V unique words. K denotes the number of latent variables.

N_i denotes the number of word in the document i. $w_{j,i}$ denotes the jth word in the document i. $z_{j,i}$ denotes the topic of the word $w_{j,i}$. α is the T-dimensional parameter vector of the *Dirichlet* distribution over θ_j ($j=1,\cdots,K$), and β is the parameter of the *Dirichlet* distribution over ϕ_k ($k=1,\cdots,V$). $\theta_{z=j}^{(d)}$ represents the topic distribution of the document d. $\phi_w^{(z=j)}$ specifies the probability of generating the word w given the topic j. H denotes the whole headline in collection.

There are many text feature weighting methods, such as Information Gain (IG), Odds ratio, χ^2 Statistics (CHI) and document frequency, which are different from the weight calculation algorithm. In this work, we focus on dealing with unlabeled data, so we choose the well-known unsupervised method TF-IDF (term frequency-inverse document frequency).

In topic modeling, the probability distribution of a word implies its importance to the topic. Unfortunately, some important topical words will not be considered as a strong topic-specific indicator due to their low frequency. According to our experimental results in Fig. 1, here, we employ a simple rule to increase and remedy the probability distribution of the low-frequency words, so as to obtain a well-formed topic description. The technique assumes that if a word w in the document d has a high $weight(w,H)$ value in H, then it will have more opportunity to appear in topic representation and more suitable for describing the topic. Thus, the $weight(w, H)$ of word w is computed by Eq. (1)

$$weight(w_{i,j}, H) = TF(w_{i,j}) \times IDF(w_{i,j}, H) = \frac{N(w_{i,j})}{|d_j'|} \times log\frac{V_H+1}{df_H(w_{i,j})} \quad (1)$$

where $TF(w_{i,j})$ is the frequency of the ith word $w_{i,j}$ in the headline j, $IDF(w_{i,j}, H)$ denotes the word $w_{i,j}$ inverse headline frequency in H. d_j' is the jth headline in H, $|d_j'|$ is the number of word tokens in d_j'. $N(w_{i,j})$ is the count of $w_{i,j}$ appearing in d_i'. V_H is the number of the headline in H, $df_H(w_{i,j})$ is the headline frequency of word $w_{i,j}$.

In this work, we adopt collapsed Gibbs sampling [7,9] for approximation to compute the posterior probability of the hidden variables. In Gibbs sampling, multivariate probability distribution problem is tackled with a sequential process, and posterior probability of the new topic sample $z_{i,j}$ can be estimated by $\phi_w^{(z=j)}$ and $\theta_{z=j}^{(d)}$.

Actually, the posterior probability of the new topic sample $z_{i,j}$ can be estimated by Eq.(2).

$$P(z_{j,i} = k|w_{j,i}, Z^{-j,i}, W^{-j,i}) \propto P(w_{i,j}|z_{i,j} = k, z_{-i,j}, w_{-i,j})P(z_{i,j} = k|z_{-i,j}) \quad (2)$$

In the right side of Eq. (2), the first item likelihood result can be calculated by Eq. (3). Then, we use Bayes rule to calculate Eq. (3), and the prior result in the right side of Eq. (2) can be calculated by Eq. (5).

$$P(w_{i,j}|z_{i,j} = k, z_{-i,j}, w_{-i,j}) = \int P(w_{i,j}|z_{i,j} = k, \phi_k)P(\phi_k|z_{-i,j}, w_{-i,j})d\phi_k \quad (3)$$

$$P(w_{i,j}|z_{i,j} = k, z_{-i,j}, w_{-i,j}) = \int P(w_{i,j}|z_{i,j} = k, \phi_k)P(w_{-i,j}|\phi_k, z_{-i,j})P(\phi_k)d\phi_k \quad (4)$$

$$P(z_{i,j} = k | z_{-i,j}) = \int P(z_{i,j} = k | \theta_i) P(\theta_i | z_{-i,j}) d\theta_i \qquad (5)$$

Here, Eq. (4) is used to realize the estimation of $\phi_w^{(z=j)}$, and Eq. (5) is used to realize the estimation of $\theta_{z=j}^{(d)}$. Equation (6) give the specific implementation of these two distributions.

$$\phi_w^{(z=j)} = \frac{N_{k,w}^{-j,i} + \beta}{N_{k,\cdot}^{-j,i} + V\beta}, \quad \theta_{z=j}^{(d)} = \frac{N_{j,k}^{-j,i} + \alpha}{N_{j,\cdot}^{-j,i} + K\alpha} \qquad (6)$$

where $N_{j,k}^{-j,i}$ denotes the number of times that the topic k appears in the document j except of the current instance, $N_{k,w}^{-j,i}$ denotes the number of times that the word w appears in the topic k except of the current instance.

In particular, Gibbs sampling is a strictly sequential process, but the sampling of $z_{i,j}$ can obtain independently given by $\theta_{z=j}^{(d)}$ or $\phi_w^{(z=j)}$. Thus, in our sampling procedure, we can carry out another naive scheme to estimate each word in V: if the word w that obeys the above rule often appears in H, it will add the value of $(weight(w, H)+1)$ to $N_{j,k}^{-j,i}$ and $N_{k,w}^{-j,i}$ instead of plusing 1 directly. For doing this, the probability distribution of some low-frequency words that appeared with high frequency in H would be increased. In the end, they would be promoted the probability in the topic, as well as having more opportunity to describe the topic.

4   Experiment and Analysis

4.1   DataSets

- **NTCIR 8:** It is a widely used news dataset for information retrieval research. It is a simplified Chinese dataset and contains several different topics with 308,845 documents for four years (2002–2005) XINHUA news. We use the '*TEXT*' field for topic modeling, and the '*HEADLINE*' field as the headline information. We first do some pre-processing for this dataset, including automatic word segmentation and standard stop-word removal.
- **the Associated Press (AP):** It is also a widely used news dataset[2] for information retrieval research. It contains 242,918 documents for three years (1988–1990) the Associated Press news. We use the '*TEXT*' field for topic modeling, and the '*HEAD*' field as the headline information.
- **Reuters-10:** Reuters-21578 is a widely used dataset[3] for document classification research. It contains 21,578 documents in 135 categories. But this dataset is very imbalanced and the variation of category size is quite large. In the experiment, we only used the 10 largest categories (**Reuters-10**) with a total of 7,285 documents, which categories are including acq, coffee, crude, earn, gold, interest, money-fx, ship, sugar and trade. We use the '*BODY*' field for topic modeling. Because the '*TITLE*' is full of rich description information for each document, we adopt this field as the headline information.

---

[2] http://trec.nist.gov/.
[3] http://kdd.ics.uci.edu/database/reuters21578/reuters21578.html.

For both English datasets, we remove a standard English stoplist with 418 stop words and all terms are processed by Porter's Stemmer[4].

In addition, we find that not all the headline information in XINHUA and AP datasets are useful, approximately 99.8 percent is related to the contents, which we consider useful, and the rest is non useful, which is only related to the editing of the article. Typically, the latter cases a single or few words, such as '*rewrite*' and '*please attention*' in XINHUA, '*EDITORS*' and '*04-06-88 02:24*' in AP. Therefore, we remove this kind of information that has little representative meaning during pre-processing.

4.2   Comparison Metric

To verify the generalization performance of our proposed approach, we use *perplexity* [2] value as a comparison criterion for topic analysis evaluation, which is generally used to describe the ability of generalization performance of the model. In general, a lower *perplexity* represents a higher performance. Formally, given a test set R_{test} with M' documents, the *perplexity* is defined as Eq. (7):

$$Perplexity(R_{test}) = \exp\left\{ -\frac{\sum_{m=1}^{M'} log P(r_m)}{\sum_{m=1}^{M'} N_m} \right\} \qquad (7)$$

where N_m is the word number in document r_m. $P(r_m)$ is the probability of document r_m.

Experiment Setting. We apply directly standard LDA to obtain the topic description results from each dataset. In the beginning, we initialize the parameters as follows: $\alpha = 50/K$, $\beta = 0.01$, the number of iterations adopt a fixed value is 100. We use JGibbLDA[5] toolkit to get *perplexity* value while the topic number K varies from 10 as beginning. We use fivefold cross-validation in the experiments.

4.3   Results and Analysis

We test the experiments on three datasets respectively. Figures 3, 4 and 5 exhibit the comparisons of the *perplexity* value on four facets with different topic numbers on XINHUA (Fig. 3(a)–3(d)), AP (Fig. 4(a)–4(c)) and Reuters-10 (Fig. 5(a)–5(j)) respectively ('WithLow' denotes the model with low-frequency words; 'WithoutLow' denotes the model without low-frequency words; 'WithH_WithLow' denotes the model use H and with low-frequency words; 'WithH_NoLow' denotes the model use H and without low-frequency words).

From Fig. 3, 4 and 5, we observe that our proposed model always produces lower *perplexity* value in all datasets. That means that our scheme produced

---

[4] https://tartarus.org/martin/PorterStemmer/.
[5] http://sourceforge.net/projects/jgibblda/.

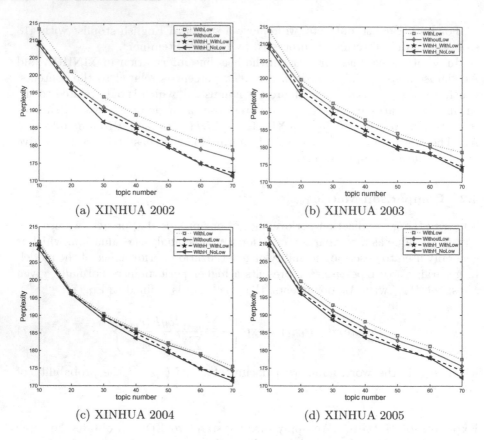

(a) XINHUA 2002

(b) XINHUA 2003

(c) XINHUA 2004

(d) XINHUA 2005

Fig. 3. Perplexity on four years XINHUA dataset.

much better generalization performance. At the same time, this work gives empirical evidence that it can get lower *perplexity* value, even though the low-frequency words are not removed, i.e., by making use of the headline information within the standard LDA setting. Furthermore, we can conclude that the ability of model generalization will become stronger when the headline information is adopted even if removing the low-frequency words in advance. Especially for Reuters-10 dataset, the generalization performance is better than another two datasets because the '*TITLE*' field information is more concise and appropriate for the description of the document.

In this work, our goal is to obtain a well-formed topic description and better topic representation for a given document collection. Table 1 and Table 2 respectively display some examples of the top-20 word-topic assignment in XINHUA 2002, AP 1988 and 'coffee' category in Reuters-10 with $K = 50$, where '*No_headline*' denotes the results that we do not adjust by using H. In this experiment, we do not remove the low-frequency words. The word-topic assignments have been manually classified in different categories in advance.

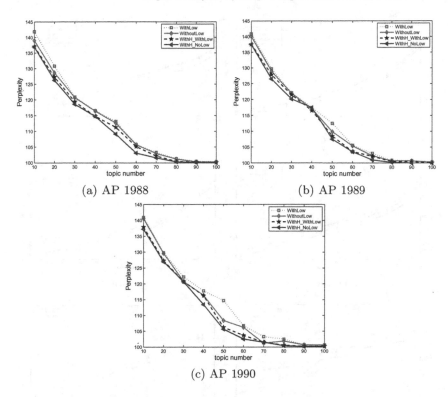

Fig. 4. Perplexity on three years AP dataset.

As shown in Table 1 and Table 2, we can observe that our proposed model can obtain well topic description. The top-ranked terms in our method can better describe the specific content of the topic, and the location of the terms with well speciality for describing the intrinsic semantic of the topic promotes to some extent. Furthermore, we notice that some terms which appear both in **Our method** and **No_headline** with different rank order, and some terms firstly appear in topic description in **Our method** but have stronger peculiarity, such as '*square meter*' and '*ecology*' in Table 1, '*tobacoo*' and '*sharp*' in Table 2. In third column of Table 2, the terms '*Contras*' and '*Contra*', both appeared in the '*Military*' category, have the same meaning, but only appeared once in **Our method** that make the topic description much more sense. Especially, the term '*American*' firstly appears in '*Military*' category, which can help us to better understand the topic. We can conclude that not only our proposed model can accomplish the stronger ability for the description of the topic, but also enhance the difference between the topics.

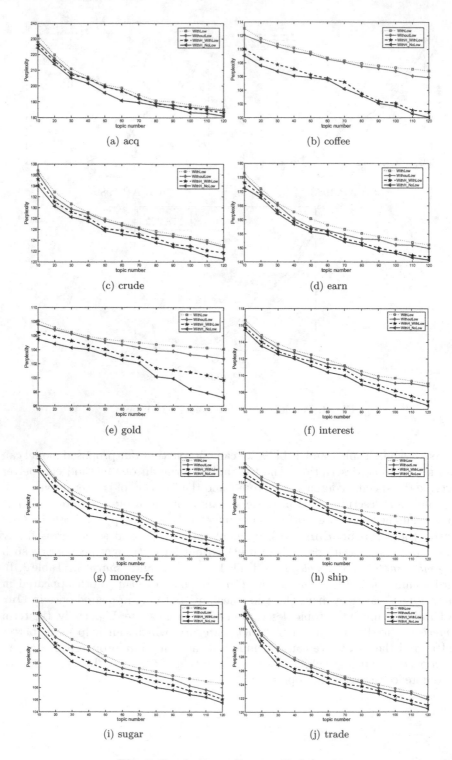

(a) acq

(b) coffee

(c) crude

(d) earn

(e) gold

(f) interest

(g) money-fx

(h) ship

(i) sugar

(j) trade

Fig. 5. Perplexity on Reuters-10 dataset.

Table 1. Example results of LDA in XINHUA 2002 with K=50

XINHUA 2002			
City construction		Politics	
No_headline	Our method	No_headline	Our method
city	environment	management	management
environment	area	department	society
construction	engineering	regulations	regulations
engineering	ecology	institution	investment
project	pollution	personnel	implementation
facility	protection	investment	criteria
have	square meter	revolution	institution
completed	a unit of area	Beijing	specification
foundation	ton	manner	requirement
quality	water	method	method
pollution	most	society	medical
emphasis	project	enterprise	conform
geology	synthesis	implementation	carry out
protection	foundation	formulate	legal
environment protection	buildup	legal	supervision
major	governance	requirement	government
region	reach	business	this year
plan	improvement	foreign-funded	personnel
launch	hectare	today	direct
traffic	residence	further	mechanism

Finally, we do some further statistical analysis for our re-weighting scheme results. In XINHUA, about 29.2% words, with low occurrence frequency in the collection but have relatively high weights in the headlines, firstly appear in the top 20 word-topic assignments, while about 19.7% in AP and 9.6% in Reuters-10. Meanwhile, about 18.9% words in XINHUA, whose probability value in the word-topic assignment, have been gained different degree of ascension, while about 20.0% in AP and 23.0% in Reuters-10. Obviously, our method will give the low-frequency words much more chances to appear or promote the probability in word-topic assignment to represent the topics. The results demonstrate the effectiveness of our re-weighting scheme in improving the topic representation.

Table 2. Example results of LDA in AP 1988 and Reuters-10 (coffee) with $K = 50$

AP 1988				Reuters-10 (coffee)	
Medical		Military		Finance	
No_headline	Our method	No_headline	Our method	No_headline	Our method
bill	medical	aid	aid	debt	price
AIDS	heart	military	troops	foreign	current
drug	health	troops	military	prices	debt
people	treatment	Contra	rebels	external	billion
medical	disease	rebels	Contra	Gaviria	Gaviria
legislation	blood	government	forces	told	sharp
veto	patients	Contras	congress	markets	markets
Dr	tobacco	Nicaraguan	border	total	total
death	care	Honduras	United	revenues	coffee
disease	AIDS	congress	Nicaragua	state	collapse
blood	hospital	United	package	rescheduled	bank
federal	cancer	Central	violence	said	state
health	director	forces	administration	affect	loss
patients	researchers	arrested	government	major	later
testing	smoking	Sandinista	American	Colombia	country
drugs	doctors	Honduran	President	price	journalists
women	cocaine	border	plan	immediate	mean
care	Dr	package	Sandinista	position	decline
doctors	association	authorities	Honduran	situation	immediate
child	institute	violence	Wright	finance	hope

5   Conclusion and Future Work

In order to make the topic modeling results better reflect the essence of the collection and reduce the influence of high-frequency words, we propose a headline-based topic model. In this paper, we focus on exploring a new topic analysis thread to characterize and reveal the profound semantic and organization of the topic for the text that are behind in statistical results, as well as provide portray abundant internal relationship within the topics and between the topics. Experimental results show the effectiveness of our model that can better capture the important words in each topic by using the headline rich information.

Our work will have positive effect to improve the quality of topic content analysis. Based on these, we plan to deploy a new document representation space by using a set of words and the co-relationships in them that are abstracted from the topics, so as to refine the semantic representation for short document expansion.

Acknowledgment. This research is jointly supported by the Natural Science Foundation of China (Grant No. 61866029, 61763034) and Natural Science Foundation of Inner Mongolia Autonomous Region (Grant No. 2018MS06025).

References

1. Blei, D.M.: Probabilistic topic models. Commun. ACM **55**(4), 77–84 (2012)
2. Blei, D.M., Lafferty, J.D.: Correlated topic models. In: Proceedings of the 18th International Conference on Neural Information Processing Systems (NIPS 2005), pp. 147–154. MIT Press, MA (2005)
3. Blei, D.M., Ng, A.Y., Jordan, M.I.: Latent Dirichlet allocation. J. Mach. Learn. Res. **3**, 993–1022 (2003)
4. Chen, X.Y., Xia, Y.Q., Jin, P., Carroll, J.: Dataless text classification with descriptive LDA. In: Proceedings of the 29th AAAI Conference on Artificial Intelligence (AAAI 2015), pp. 2224–2231. AAAI (2015)
5. Gao, Y., Xu, Y., Li, Y.F.: Discovering evolutionary theme patterns from text: an exploration of temporal text mining. In: Proceeding of International Conference on Web Information Systems Engineering (WISE 2014), pp. 186–201. Springer International Publishing, Cham, October 2014
6. Gao, Y., Xu, Y., Li, Y., Liu, B.: A two-stage approach for generating topic models. In: Pei, J., Tseng, V.S., Cao, L., Motoda, H., Xu, G. (eds.) PAKDD 2013. LNCS (LNAI), vol. 7819, pp. 221–232. Springer, Heidelberg (2013). https://doi.org/10.1007/978-3-642-37456-2_19
7. Griffiths, T.L., Steyvers, M.: Finding scientific topics. Proc. Natl. Acad. Sci. **101**(suppl 1), 5228–5235 (2004)
8. Mei, Q., Zhai, C.X.: Topical pattern based document modelling and relevance ranking. In: Proceedings of the 11th ACM SIGKDD International Conference on Knowledge Discovery in Data Mining (KDD 2005), pp. 198–207. ACM, New York, August 2005
9. Petterson, J., Smola, A., Caetano, T., Buntine, W., Narayanamurthy, S.: Word features for latent Dirichlet allocation. In: Proceedings of the 23rd International Conference on Neural Information Processing Systems (NIPS 2010), vol. 2, pp. 1921–1929. Curran Associates Inc., New York, December 2010
10. Sato, I., Nakagawa, H.: Topic models with power-law using pitman-Yor process. In: Proceedings of the 16th ACM SIGKDD International Conference on Knowledge Discovery and Data Mining (KDD 2010), pp. 673–682. ACM, New York, July 2010
11. Tang, G., Xia, Y., Sun, J., Zhang, M., Zheng, T.F.: Statistical word sense aware topic models. Soft. Comput. **19**(1), 13–27 (2014). https://doi.org/10.1007/s00500-014-1372-z
12. Trabelsi, A., Zaïane, O.R.: A joint topic viewpoint model for contention analysis. In: Natural Language Processing and Information Systems, pp. 114–125. Springer International Publishing, Cham (2014). https://doi.org/10.1007/978-3-319-07983-7_16
13. Wang, C., Blei, D.M.: Collaborative topic modeling for recommending scientific articles. In: Proceedings of the 17th ACM SIGKDD International Conference on Knowledge Discovery and Data Mining (KDD 2011), pp. 448–456. ACM, New York, August 2011

14. Wei, X., Croft, W.B.: LDA-based document models for ad-hoc retrieval. In: Proceedings of the 29th Annual International ACM SIGIR Conference on Research and Development in Information Retrieval (SIGIR 2006), pp. 178–185. ACM, New York, August 2006
15. Wilson, A.T., Chew, P.A.: Term weighting schemes for latent Dirichlet allocation. In: Proceedings of the 2010 Annual Conference of the North American Chapter of the Association for Computational Linguistics (HLT 2010), pp. 465–473. Association for Computational Linguistics, Stroudsburg, June 2010
16. Xia, Y.Q., Tang, N., Hussain, A., Cambria, E.: Discriminative Bi-term topic model for headline-based social news clustering. In: Proceedings of the 28th Florida Ariticial Intelligence Research Society Conference, pp. 311–316. AAAI, April 2015
17. Zeng, J.P., Duan, J.J., Cao, W.J., Wu, C.R.: Topics modeling based on selective Zipf distribution. Expert Syst. Appl. **49**(7), 6541–6546 (2012)
18. Zhai, C.X., Velivelli, A., Yu, B.: A cross-collection mixture model for comparative text mining. In: Proceedings of the 10th ACM SIGKDD International Conference on Knowledge Discovery and Data Mining (KDD 2004), pp. 743–748. ACM, New York, August 2004

A Text Mining Approach to Extract and Rank Innovation Insights from Research Projects

Francesca Maridina Malloci[2(✉)] [iD], Laura Portell Penadés[1] [iD],
Ludovico Boratto[1] [iD], and Gianni Fenu[2] [iD]

[1] Data Science and Big Data Analytics Unit, EURECAT - Centre Tecnògic
de Catalunya, Carrer de Bilbao 72, 08005 Barcelona, Spain
`laura.portell@eurecat.org, ludovico.boratto@acm.org`
[2] Department of Mathematics and Computer Science, University of Cagliari,
Via Ospedale 72, 09124 Cagliari, Italy
`{francescam.malloci,fenu}@unica.it`

Abstract. Open innovation is a new paradigm embraced by companies to introduce transformations. It assumes that firms can and should use external and internal ideas to innovate. Recently, commercial and research projects have undergone an exponential growth, leading the open challenge of identifying possible insights on interesting aspects to work on. The existing literature has focused on the identification of goals, topics, and keywords in a single piece of text. However, insights do not have a clear structure and cannot be validated by comparing them with a straightforward ground truth, thus making their identification particularly challenging. Besides the extraction of insights from previously existing initiatives, the issue of how to present them to a company in a ranking also emerges. To overcome these two issues, we present an approach that extracts insights from a large number of projects belonging to distinct domains, by analyzing their abstract. Then, our method is able to rank these results, to support project preparation, by presenting first the most relevant and timely/recent insights. Our evaluation on real data coming from all the Horizon 2020 European projects, shows the effectiveness of our approach in a concrete case study.

Keywords: Information extraction · Ranking · Text mining

1 Introduction

Open innovation is a strategy that combines internal knowledge and skills with those of other external entities, for sharing, integrating, and acquiring ideas in collaboration with other interested parties (consumers, users, employees, other

---

All authors equally contributed to this research.

© Springer Nature Switzerland AG 2020
Z. Huang et al. (Eds.): WISE 2020, LNCS 12343, pp. 143–154, 2020.
https://doi.org/10.1007/978-3-030-62008-0_10

companies, technology or research centers, universities, etc.) [5,15,21]. The key role of this process is represented by the consultants, who act as facilitators and accelerators for the whole process. However, consultants spend a lot of time reading similar projects to shape their idea until the product is launched.

A key role in these activities is represented by the manual extraction of *insights*, i.e., aspects that haven been identified in previous projects has open issues, challenges, and interesting aspects to work on. Relying on previous initiatives is a central aspect to provide foundations that an aspect is worth working on. Indeed, these insights have the dual role of providing (i) consultants with ideas and (ii) references of existing projects in the same area, to highlight and contextualize the work. Given the incredible amount of both private and public projects being created and approved on a daily basis, a manual detection of insights is clearly neither efficient nor effective.

Related Work. Text mining is the task of automatically extracting patterns from text [2]. Information extraction techniques have been used for analyzing company data, to identify goals, topics, and keywords. Alabdulkareem et al. [1] identified the preferences and goals through natural language statements. Each statement is first matched through regular expressions to distinguish between the preference component and the goal component. The former is mapped to a preferential strength measure, while the latter is used to identify the relevant goal, through a statistical semantic similarity. Related work on topic and keyword extraction was also conducted. Aras et al. [3] developed a text-mining tool that extracts a list of keywords that are relevant and reveal the invention in the patent text. They adopted a noun-phrase extraction and analysis and phrase weighting based on features such as length, position, term frequency–inverse document frequency (TF-IDF). Larraaga et al. [12] designed a Computer Supported Learning System able to extract didactic resources from electronic documents using ontologies and NLP techniques as part-of-speech. Kathait et al. [11] suggested an unsupervised methodology that uses the noun words and phrases, their occurrence and co-occurrences to extract key-phrases from text articles.

Open Issues. While we acknowledge the work done to extract information from a single piece of text, getting insights in a given area/topic of interest is a fundamentally different and more challenging problem. The first difference is at conceptual level. Indeed, an insight is a much more complex concept than a goal or a keyword. It can be a challenge (e.g., for the autonomous vehicle industry, it could be to react to unexpected signals, or to increment sensing capabilities), or an open issue (e.g., car recycling, or automate car disassembly process). Basically, it should give consultants an idea of what are the relevant aspects to work on, but it is definitely more challenging to recognize w.r.t. a goal, a keyword, or a named entity. The other main difference is on the algorithmic approach needed to extract insights, which drastically departs from those existing in the literature. Goal recognition, as defined in [1], assumed the existence of a goal model that describes a domain of intention for a particular stakeholder, such as a goal decomposition model for achieving the goal "schedule meeting". This technique implied the identification of a preference component and the goal component inside a text. Similarly, Aras et al. [3] designed a model to extract specific keywords for a goal, or rather, to all those that are relevant and reveal the invention in the patent text. Hence, the existing models are *domain dependent*. In our work, the issue is to develop a non-circumscribed model for an objective queried by the user, thus being able to recognize insights in various domains through the analysis of abstracts written in formal language. The formal register also presents challenges, as by definition it is aseptic and leaves no room for ambiguity or interpretation. It has variable structures dictated by sectoral terminology that can assume a scientific, legal, administrative, artistic formalism.

The second open issue we face is that, given the multitude of insights coming from previous projects, a consultant should be provided with clear information about what are the most relevant insights to work on. Hence, we need to generate a *ranking* of the extracted insights, which should give consultants the opportunity to consider first the most relevant aspects that were extracted.

Our Contribution. In this paper, we propose an insight extraction approach that mainly identifies challenges and open issues from a project's abstract. To make our problem useful in a real-world scenario, we embed it in a Web platform (not public at the time of the publishing of this paper). The tool receives as input a query (e.g., "autonomous and industrial robotics") and extracts insights from the retrieved abstracts. Given the extracted insights, we also propose a ranking algorithm that presents them in an ordered way to consultants, based on the relevance for the query and the timeliness of the extracted insights (i.e., how recent are the projects that mentioned those insights). As proof of concept, we validate our algorithms in a concrete use case, considering European projects written in English and provided in the CORDIS database (the Community Research and Development Information Service) as a training for our approach.

Roadmap. The rest of the paper is structured as follows. Section 2 introduces our insight extraction approach, and in Sect. 3 we present our ranking algorithm. The details of our evaluation framework are given in Sect. 4. Finally, we conclude the paper in Sect. 5, with some final remarks and future developments.

2   Insight Extraction

Given a piece of text (specifically, a project's abstract), our approach extracts insights from it by following three main steps:

1. **Text pre-processing.** We pre-process the abstract text, to remove parts that are not linked to insights (e.g., numbers and links).
2. **Candidate insight extraction.** We extract candidate insights from the text, with two subtasks based on (i) word frequency and (ii) syntax analysis.
3. **Sentence embedding extraction.** Each candidate insight is mapped into a sentence embedding, for their aggregation in the subsequent ranking task.

These steps are now presented in detail.

2.1 Text Pre-Processing

Given as input a project's abstract, we removed punctuation, white spaces, numbers, round brackets, and link removal, by using regex expressions.

2.2 Candidate Insight Extraction

As mentioned in the Introduction, insights are challenging aspects and open issues that a company has to face when starting a new project. The size of the key phrase that forms an insight depends upon its intended application [20]. Based on domain knowledge provided by tens of consultants we interviewed prior this work, we identified an insight as a *noun phrase or key phrase, consisting of two to four words*. Clearly, this is a parameter, that could be easily adapted in case of different needs and domains.

It should be clear that not all of the key phrases that present these characteristics are insights. In this section, we present our approach to extract *candidate insights*, i.e., key phrases that are likely to be insights. A candidate insight has the role of shaping the definition of the actual insights, in case the same or very similar concepts appear several times. This goal of merging candidate insights into actual insights will be done by our ranking algorithm.

In the literature, several key-phrase extraction techniques were presented, which can be summarized into five categories [13]: Rule Based Linguistic approaches [9], Statistical approaches [11], Machine Learning approaches [7,8,10], Graph-based approaches [6], and Hybrid approaches [22]. We employed rule-based approaches, where a key phrase is extracted based on word frequency and syntax analysis. The motivation behind the choice of these techniques is that they allowed us to identify more insights and minimize the noise that can be present due to the adoption of an unsupervised method. Our approaches to extract candidate insights are now presented in detail.

Insight Extraction Based on Word Frequency. The intuition behind our use of word-frequency technique to extract candidate insights is that very relevant aspects are usually underlined several times in a piece of text. Hence, simple world frequency can help us identify what might be particularly relevant in a given project. Our word frequency based approach to candidate-insight extraction employs the *Rapid Automatic Keyword Extraction* (RAKE) algorithm [17]. A scored-weight is calculated for every candidate phrase using a score-weight matrix. The score-weight matrix is obtained by calculating:

- word frequency: the number of times the word appears in the document;
- word degree: the sum of the number of co-occurrences the word has with any other content word in the text;
- ratio of degree to frequency: score of candidate framework.

Keywords are considered candidate phrases if their score falls within a specific range, which is necessary to reduce the presence of noise. We assumed that an insight is an item s that achieved a score in a range from β and γ i.e., $\beta \leq s \leq \gamma$. The choice of a range was done to remove too high/low scores. High scores capture information that goes beyond the insight to be identified, thus including noise (e.g., an insight, plus words used to introduce it). Similarly, keywords with a low score are usually composed of a single term that is not explanatory enough to express a challenge or an open issue (hence, not detecting an insight).

Insight Extraction Based on Syntax Analysis. Our second approach considers candidate insights as noun phrases in the text. Our intuition is that, besides frequences, project writing follows specific patterns to remark relevant aspects. With this second approach, we want to identify these patterns.

Noun phrases are flat phrases that have a noun followed or anticipated by a word describing the noun, e.g., "cost-effective car". In order to implement an approach to identify noun phrases, also known in the literature as *noun phrases chunking*, we used the model implemented in the spaCy library[1], which allowed us to extract part-of-speech tags. From them, candidate insights are extracted as named entities, by defining a *chunk grammar*, which is a set of rules that indicates how sentences should be chunked. After considering several combinations, we outline five rules that are typically used inside a text to express insights (for each rule, we also present an example of it):

1. NN-VBG (e.g., Car recycling);
2. NN-RB (e.g., Car disassembly);
3. NN-VBG-(NN—NNS) (e.g., Vehicle sensing capabilities);
4. JJ-JJ-(NNS—NN) (e.g., Reliable robust solutions);
5. JJ-NN-(NNS—NN) (e.g., Flexible motion planning).

We use Rule 1 to explain how rules should be read. A noun chunk should be formed whenever the chunker finds a noun (NN) followed by a verb/gerund/present participle taking (VBG). In the other rules, adverbs (RB), adjectives (JJ) and plural nouns (NNS) can be part of our candidate insights.

2.3 Sentence Embedding Extraction

Sentence embeddings are phrases from a vocabulary that are mapped to vectors of real numbers. They are capable of capturing the context of a word in a document, semantic and syntactic similarity, and relation with other words. Theoretically, the space with many dimensions per word is mapped to a continuous vector space with a lower dimension. In recent years, this method was exploited in the literature for several NLP tasks in information retrieval. Given that we extract candidate insights with two approaches that lead to sentences of different length, in our study we extracted sentence embeddings, to allow the subsequent ranking activity to filter the insights extracted in the previous task.

---

[1] https://spacy.io/.

At the state of the art, many algorithms were developed to this objective, such as Word2vec, GloVe, ELMo and BERT [4]. A modified BERT (Bidirectional Encoder Representations from Transformers) algorithm called Sentence-BERT (SBERT) [16] was adopted in our approach, due to its ability to build a context-dependent, and therefore instance-specific, embedding.

3   Ranking

In the previous section, we proposed an approach to extract candidate insights from the text. Each insight is represented as a sentence embedding. In this section, we show how to group together the candidate insights and present the detected insights in a ranking.

We assume as input a query q, containing the topic of interest for which we want to detect the insights. A set of n abstracts A is returned and, for each abstract $a \in A$, we run the candidate-insight extraction approaches, who return a set S_a of embeddings.

Our approach to group and rank these insights follows two main steps:

1. **Clustering.** This steps merges together highly similar candidate insights.
2. **Ranking.** We assign a score to each cluster based on the cohesiveness of the insights it contains and how recent are the abstracts that contain them.

3.1   Clustering

Clustering was employed to classify similar candidate insights. Given that the insights S_a associated to an abstract are obtained with two different techniques and that insights can be repeated several times (either inside an abstract or in different abstracts), in this step we cluster together the candidate insights $S = \bigcup S_a, \forall a \in A$, with the goal of providing consultants with relevant insights.

Specifically, we employ the Universal Sentence Encoder (USE) to distinguish those groups of insights that are very similar to the user's query q and those that are not, but which are equally related to the sector in question. This technique due to its ability to discover semantic similarity from formal texts, such the ones we consider in this work (abstracts from publicly founded projects). Similarity was determined by comparing sentence embedding vectors and taking the pairs that achieved a score greater than or equal to θ. Cosine similarity was calculated to determinate the distance between the embeddings. The output is a set of clusters of insights, denoted as C.

3.2   Ranking

We developed and introduced a ranking system to present the most relevant insights, based on (i) the relevance for the query q, and (ii) the timeliness of the insight (i.e., how recently it was proposed).

Let $c \in C$ be a cluster of insights generated from different abstracts. We measure the timeliness of a cluster c with a score α_c, attributed according to the closing date of the project. Considering the application scenario of our platform (Horizon 2020 projects), the covered period is from 2014 to 2020. Each year y, was assigned a weight β_y, in ascending order (where $1 \leq \beta_y \leq 6$). For each cluster, we count how many insights are associate to each year, considering the year in which the project will be/was closed. Then, the score α_a is the β_y value of the year to which most insights belong, to give an idea of how timely they are (i.e., how recently most of the insights were proposed).

The total score associated to the cluster is computed as follows:

$$rscore_c = \alpha_a + \sum_{i=1}^{|c|} \frac{score_i}{|c|}$$

where $score_i$ represents the score obtained by the USE model, by comparing word vectors during the step described in Sect. 3.1, and $|c|$ is the cluster cardinality expressed as number of insights contained cluster. Insights with an high $rscore$ are suggested as the first results. In case of parity, the cluster with the highest cardinality $|c|$ is suggested first, because it certainly contains more relevant insights, as it is present in several abstracts.

4 Experimental Framework

As we mentioned, the presented algorithms are part of a Web platform that is about to go into production for a consulting company. This means that, at this stage, no live user evaluation could be performed. Before delivering the platform, we conducted a case study to validate our algorithms. In this section, we present the experimental environment and the case study performed to evaluate the proposed approach.

4.1 Experimental Environment

The experimental framework exploits the Python scikit-learn 0.19.1 library, executed on a computer equipped with a 2,3 GHz Intel Core i5 processor and 8 GB of RAM. The candidate-insight extraction algorithms were implemented with the support of the spaCY[2] open-source library for advanced NLP in Python. The dataset were stored into Elastich Search[3], a full-text, distributed NoSQL database, which was also used to retrieve our abstracts given the query q.

Regarding the parameter setting, given a query, we considered the first $n = 50$ returned abstracts. When extracting the candidate insights based on word frequency, we employed a range where β was 9 and γ was 15. The similarity threshold θ, for cluster formation, is set to 0.7.

---

[2] https://spacy.io/.
[3] https://www.elastic.co/.

4.2 Dataset

Training Dataset. We trained our algorithms based on a CORDIS dataset that contains projects and related organizations funded by the European Union under the Horizon 2020 program. The dataset was retrieved by the CORDIS website itself[4]. It is composed of 25,363 instances (projects). Each project is represented by the following information: Record Control Number (RCN), project ID (grant agreement number), project acronym, project status, funding programme, topic, project title, project start date, project end date, project objective, project total cost, EC max contribution (commitment), call ID, funding scheme, coordinator, coordinator country, participants, participant countries. In our case study, we considered the *project objective*, i.e., the abstract that summarizes the project.

Test Dataset. From the full dataset we removed the project of a company (a private organization heavily involved in Horizon 2020 initiatives), in a leave-one-out experimental setting. The test set is composed of 231 projects won by the company, characterized by the same attributes of the rest of the datasets.

4.3 Case-Study Description

A preliminary case study was conducted to evaluate the unsupervised approach we proposed. The case study was based on the query *autonomous and industrial robotics*, which is a topic covered both in the training and test datasets. To generate a ground truth that serves as validation to our approach, three independent consultants manually extracted the insights from the abstract of the projects in the test dataset and ranked them. The company associated to the test dataset, who executed these project, revised the ranking, as a validation of what are the relevant topics in this area they work on. In a final stage, we joined the independent consultants and those of the company, to reach a final consensus (i.e., a final list of ranked insights).

To present the results we obtained during this case-study, we considered *qualitative results* (i.e., examples of candidate insights and of similarities returned by our approach) and *quantitative results*, to measure ranking quality by comparing it against the ground truth, with standard metrics.

4.4 Quantitative Evaluation Metrics

Evaluation metrics for ranking systems can be categorized into four classes [19]: *predictive accuracy metrics, classification accuracy metrics, rank accuracy metrics* and *non-accuracy metrics*. Given the nature of our work, which extracts insights and ranks them, we evaluated the proposed approach in terms of *classification accuracy metrics* and *rank accuracy metrics*.

Classification accuracy metrics measure the amount of correct and incorrect classifications as relevant or irrelevant items that are ranked and are actually useful for the user [19]. We measured the following metrics:

---

[4] https://data.europa.eu/euodp/en/data/dataset/cordisref-data.

- *Precision* is defined as $TP/(TP + FP)$;
- *Recall* is defined as $TP/(TP + FN)$;
- F_2 - *measure* is defined as: $5 \cdot ((precision \cdot recall)/(4 \cdot precision + recall))$. The choice to use $F2$, rather than $F1$, was made because in our context it is more important to classify correctly as many positive insights as possible, rather than maximizing the number of correctly classified insights;
- *Fallout* is computed as $FP/(FP + TN)$;
- *MissRate* is defined as $FN/(TP + FN)$.

Ranking accuracy metrics. In our study, we aimed to measure the correlation ranking between the list recommended by the algorithm, described in Sect. 3, against the ranked ground truth, built by considering the insights extracted from the test dataset. To this end, we computed two standard metrics, *Kendall's* τ and *Spearman's* ρ.

Kendall's τ is a non-parametric measure of the degree of correlation. It measures the strength of the relationship between two ordinal level variables. Let $C = c_1 \ldots c_m$ be a set of clusters containing insights. Let π and σ denote two distinct orderings of C, and $T(\pi, \sigma)$ the minimum number of adjacent transpositions needed to bring π to σ. Kendall's τ is defined as:

$$\tau = 1 - \frac{2T(\pi, \sigma)}{N(N - 1)/2}$$

where N is the number of objects (i.e., items) being ranked.

Spearman's ρ is a non-parametric version of the Pearson correlation coefficient. It measures the strength of a monotonic relationship between paired data. The association between two variables is expressed in a single value between -1 and +1, which is calculated as follows (d_i is the difference in paired rank):

$$\rho = 1 - \frac{6 \sum d_i^2}{N(N^2 - 1)}$$

A value of +1 indicates a perfect positive correlation, −1 denotes a perfect negative correlation, and 0 represents a no correlation between ranks.

4.5   Experimental Results

This section presents the results obtained in our evaluation.

Candidate Insights' Examples and Similarity Assessment. Figure 1 presents a heatmap with the similarity for a subset of candidate insights extracted from the abstracts related to our query. The first focus is on the effectiveness of our candidate-insight extraction algorithm, which is able to highlight concrete and relevant aspects to work on in the autonomous and industrial robotics area. The heatmap also highlights that our sentence embeddings are a very effective means to detect the similarity between candidate insights. Indeed, when computing the cosine similarity between the vectors, we can see that we can

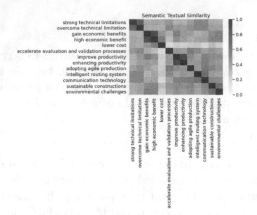

Fig. 1. Heatmap providing an example of extracted insights and their similarity.

Table 1. Classification accuracy metrics.

Precision	Recall	F_2 - measure	Fallout	MissRate
0.90	0.86	0.86	0.4	0.13

detect insights at higher levels (e.g., by putting together "sustainable construc-tions" and "environmental challenges"). At the same time, unrelated topics are very dissimilar (e.g., "accelerate evaluation and validation process" and "lower cost"). This confirms that both our candidate insight extraction process and similarity comparison lead to clusters of similar candidate insights, thus leading us to an effective insight presentation to the user.

Classification Accuracy. Table 1 summarizes the results obtained by com-puting different accuracy metrics. The precision is 0.90, which means that we are effectively extracting insights that correspond to the end-user's need. The recall value is slightly lower 0.86, but still remains extremely high, considering that the maximum is 1 and that this result can be achieved in combination with a very high precision. Indeed, the 0.86 obtained in terms of F_2 - measure confirmed the robustness of our estimator, which does not miss a significant number of instances. However, the two measures, precision and recall, are typ-ically inversely related. Because of this mutual dependence, it makes sense to evaluate the ranking algorithm in conjunction with the other two metrics we introduced, i.e., fallout and miss rate. The obtained fallout value of 0.4 outlines that the probability of an irrelevant insight to be provided is low. Similarly, the miss rate which is the probability that a relevant insights is not provided is 0.13.

Rank Accuracy Metrics. Kendall's τ and Spearman's ρ metrics were com-puted by comparing the list generated by the algorithm against the ranked hand-built list considering the insights manually extracted from the test dataset (our ground truth). The τ coefficient comparing the ranked lists is 0.83 and the ρ coefficient is 0.97. The results achieved are very close and thus invariably lead to the same inferences. Between the rankings there is a strong relationship, which means that the algorithm is able to suggest in the list a set of challenges that is very close to the user and, at the same time, it is also very close to the related and relevant insights considering the sector in question.

5   Conclusions and Future Work

In this paper, we proposed a novel approach to automatically extract and rank insights from a wide range of abstract projects belonging to distinct domains. By the term insight, we define the challenges and open issues that a company faces when starting a new project. Its identification by an automatic tool allows simplifying the consultancy process for shaping new project ideas. The approach was developed by combining different NLP techniques and ranking approaches connecting the relevance and timeliness of the topic. Given a topic as input, the method extracts a list of insights related to the user's topic of interest. The preliminary case-study conducted shows the effectiveness of our method in terms of rankings. The approach can suggest the more relevant challenges based on the user's query and the challenges faced by the user's company. As future work, we will extend our approach to introduce a personalization perspective, allowing us to present a different ranking to each company, based on their preferred topics and expertise [14,18]. This will be done through a mix of content-based approaches, content information mining, and collaborative filtering, to identify possibly relevant insights based on peer companies' expertise.

References

1. Alabdulkareem, F., Cercone, N., Liaskos, S.: Goal and preference identification through natural language. In: 23rd IEEE International Requirements Engineering Conference, RE, pp. 56–65. IEEE Computer Society (2015)
2. Allahyari, M., et al.: A brief survey of text mining: Classification, clustering and extraction techniques (2017). CoRR abs/1707.02919
3. Aras, H., Hackl-Sommer, R., Schwantner, M., Sofean, M.: Applications and challenges of text mining with patents. In: Proceedings of the First International Workshop on Patent Mining and Its Applications (IPaMin 2014). CEUR Workshop Proceedings, vol. 1292. CEUR-WS.org (2014)
4. Bavier, A., Peterson, L., Mosberger, D.: Bert: A scheduler for best effort and realtime tasks. Technical Report (1999)
5. Bogers, M., Chesbrough, H., Moedas, C.: Open innovation: research, practices, and policies. Calif. Manag. Rev. **60**(2), 5–16 (2018)
6. Boudin, F.: Unsupervised keyphrase extraction with multipartite graphs (2018). arXiv preprint arXiv:1803.08721
7. Dessì, D., Fenu, G., Marras, M., Reforgiato Recupero, D.: COCO: semantic-enriched collection of online courses at scale with experimental use cases. In: Rocha, Á., Adeli, H., Reis, L.P., Costanzo, S. (eds.) WorldCIST'18 2018. AISC, vol. 746, pp. 1386–1396. Springer, Cham (2018). https://doi.org/10.1007/978-3-319-77712-2_133

8. Dessì, D., Reforgiato Recupero, D., Fenu, G., Consoli, S.: A recommender system of medical reports leveraging cognitive computing and frame semantics. In: Tsihrintzis, G.A., Sotiropoulos, D.N., Jain, L.C. (eds.) Machine Learning Paradigms. ISRL, vol. 149, pp. 7–30. Springer, Cham (2019). https://doi.org/10.1007/978-3-319-94030-4_2

9. Gorinski, P.J., et al.: Named entity recognition for electronic health records: a comparison of rule-based and machine learning approaches (2019). arXiv preprint arXiv:1903.03985

10. Hasan, H.M., Sanyal, F., Chaki, D.: A novel approach to extract important keywords from documents applying latent semantic analysis. In: 2018 10th International Conference on Knowledge and Smart Technology (KST), pp. 117–122. IEEE (2018)

11. Kathait, S.S., Tiwari, S., Varshney, A., Sharma, A.: Unsupervised key-phrase extraction using noun phrases. Int. J. Comput. Appl. **162**, 1–5 (2017)

12. Larrañaga, M., Elorriaga, J.A., Arruarte, A.: A heuristic NLP based approach for getting didactic resources from electronic documents. In: Dillenbourg, P., Specht, M. (eds.) EC-TEL 2008. LNCS, vol. 5192, pp. 197–202. Springer, Heidelberg (2008). https://doi.org/10.1007/978-3-540-87605-2_22

13. Loukam, M., Hammouche, D., Mezzoudj, F., Belkredim, F.Z.: Keyphrase extraction from modern standard Arabic texts based on association rules. In: Smaïli, K. (ed.) ICALP 2019. CCIS, vol. 1108, pp. 209–220. Springer, Cham (2019). https://doi.org/10.1007/978-3-030-32959-4_15

14. Ramos, G., Boratto, L.: Reputation (in)dependence in ranking systems: demographics influence over output disparities. In: Proceedings of the 43rd International ACM SIGIR Conference on Research and Development in Information Retrieval, SIGIR '20, pp. 2061–2064. Association for Computing Machinery, New York (2020). https://doi.org/10.1145/3397271.3401278

15. Rauter, R., Globocnik, D., Perl-Vorbach, E., Baumgartner, R.J.: Open innovation and its effects on economic and sustainability innovation performance. J. Innov. Knowl. **4**(4), 226–233 (2019)

16. Reimers, N., Gurevych, I.: Sentence-bert: sentence embeddings using siamese bert-networks (2019). arXiv preprint arXiv:1908.10084

17. Rose, S., Dave, E., Nick, C., Wendy, C.: Automatic keyword extraction from individual documents. Text Min. Appl. Theory **1**, 1–20 (2010)

18. Saúde, J., Ramos, G., Caleiro, C., Kar, S.: Reputation-based ranking systems and their resistance to bribery. In: 2017 IEEE International Conference on Data Mining, ICDM 2017, pp. 1063–1068. IEEE Computer Society (2017)

19. Schröder, G., Thiele, M., Lehner, W.: Setting goals and choosing metrics for recommender system evaluations. In: UCERSTI2 workshop at the 5th ACM Conference on Recommender Systems, vol. 23, p. 53 (2011)

20. Sifatullah, S., Sharan, A.: Keyword and keyphrase extraction techniques: a literature review. Int. J. Comput. Appl. **109**(2), 18–23 (2015)

21. West, J., Bogers, M.: Open innovation: current status and research opportunities. Innovation **19**(1), 43–50 (2017)

22. Wu, J., Choudhury, S.R., Chiatti, A., Liang, C., Giles, C.L.: Hesdk: a hybrid approach to extracting scientific domain knowledge entities. In: 2017 ACM/IEEE Joint Conference on Digital Libraries (JCDL), pp. 1–4 (2017)

Automatic Complex Word Identification Using Implicit Feedback from User Copy Operations

Ilan Kirsh[✉][iD]

The Academic College of Tel Aviv-Yaffo, Tel Aviv, Israel
kirsh@mta.ac.il

Abstract. Complex Word Identification (CWI) is one of the key components of lexical text simplification. This paper proposes a new approach to CWI on websites, based on tracking what web users copy to their clipboards. Users may copy to the clipboard words that they are not familiar with or that make the text difficult to understand, in order to search for more information on the internet. Accordingly, this study examines the hypothesis that word copying on a website is an indicator of word complexity. Copied words on a sample website are compared to uncopied words using three simple word complexity indicators: number of syllables, number of characters, and general word frequency. The results show that copied words are more likely to be evaluated as complex than uncopied words and words that are copied more frequently are more likely to be evaluated as complex than words that are copied less frequently, by all three indicators. Consequently, word copying on a website can be considered a novel CWI indicator. Unlike traditional CWI indicators, which are based on static word features, this new indicator provides a different approach by targeting complex words based on dynamic user behavior. Therefore, simplifying these complex words might be particularly helpful to the readers. Further work should evaluate using this word copying indicator in complete CWI and text simplification implementations.

Keywords: Complex Word Identification (CWI) · Lexical text simplification · Clipboard copy and paste · Web usage mining · Web pages

1 Introduction

Shardlow defined text simplification as "the process of modifying natural language to reduce its complexity and improve both readability and understandability" [17]. Similarly to text translation and text summarization, text simplification can also benefit from automation. But translation and summarization could be more tolerant of errors than simplification, at least in some applications. Errors in text simplification could lead to output text that is more complex than the input, making the "simplification" result unusable [17].

© Springer Nature Switzerland AG 2020
Z. Huang et al. (Eds.): WISE 2020, LNCS 12343, pp. 155–166, 2020.
https://doi.org/10.1007/978-3-030-62008-0_11

Automatic text simplification is a challenging task. Many studies focus on the more modest goal of lexical text simplification, which involves replacing individual complex words with simpler words with similar meanings, without changing sentence structures and grammar [6,14,19,20]. Lexical text simplification can be performed in stages, where the first stage is Complex Word Identification (CWI) [17]. Any word that reduces the readability or understandability of the text may be considered complex. Word complexity is audience dependent. For example, a word can be simple for native speakers of a language and complex for non-native speakers. Words that are identified as complex (for the prospective audience) are candidates for substitution with simpler words in the next stages of the text simplification process.

Simple features of words can be used as indicators of word complexity. Three of the most commonly used word complexity indicators are:

1. **Syllable Count** - Complex words tend to be longer and have more syllables. Readability tests such as the Gunning fog index [5] and the SMOG grade [12] classify words with three syllables or more as complex.
2. **Character Count** - An alternative readability test, the Coleman–Liau index [2], uses the number of characters in a word as a complexity measure. On average, complex words tend to have more characters.
3. **Frequency** - Less frequent words are usually less familiar and therefore more complex than more frequent words [11].

These three indicators have been found to be strongly correlated with word complexity [16]. Other indicators, sense count and synonym count, which may indicate potential word ambiguity and therefore complexity, have been found to have weaker correlations with word complexity [16].

Many CWI implementations use a combination of indicators, including the indicators described above, as features in machine learning models. Various machine learning methods, including SVM classifiers, Random Forests, Neural Networks, and Bayesian Ridge classifiers, have been examined in the SemEval 2016 task 11 [15] and the CWI 2018 shared task [21]. Taking into account the context in which words appear in the text can improve the results [4].

Eye gaze tracking is commonly used in research on reading behaviors. It may be used in the context of CWI to identify complex words, because encountering complex words may be reflected in the user eye gaze, for example, as extended reading time [1]. Identifying words that are complex for real users, using eye tracking methods, could be more reliable than using static word complexity indicators. However, collecting eye tracking data requires special equipment and user collaboration, so the scope of this approach is limited, and it is usually impractical to collect eye gaze data from ordinary users on public websites.

This study proposes a new approach to automatic identification of complex words on web pages by tracking web users' copy operations. Users copy strings of various types to the clipboard [8], including words and phrases to look up elsewhere [10], key sentences for citations and text summaries [9], and programming code fragments [10]. This paper shows that copying words to the clipboard is a word complexity indicator. Consequently, tracking word copying on websites should be considered a new technique in CWI and text simplification. Similarly to eye gaze tracking, it has the benefit of tracking real users and finding their real needs, but it is not constrained by the limitations that make eye tracking impractical on most websites.

2 Implementation

The architecture of the CWI implementation that was developed and explored as part of this study is shown in Fig. 1.

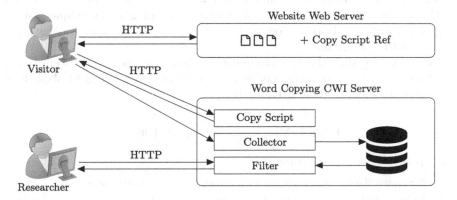

Fig. 1. Architecture of the word copying CWI implementation

The pages of the website are modified to include a reference to a *Copy Script*. When these pages are loaded into a visitor's browser, the browser follows the reference and loads the Copy Script. The Copy Script includes JavaScript code that tracks copy events and reports them to the *Collector* component in the server, which stores the data (following anonymization) in a dedicated database.

The *Filter* component is responsible for distinguishing copied words from all the other copied strings, and its output is the copied words along with their corresponding copying frequencies. Only copy operations of single text words are counted, copy operations of multiple words are filtered out, as well as copy operations of text strings contained in HTML PRE elements, which are often code fragments. A valid word is defined in this context (of CWI for texts in English), as a string consisting of lower case letters, except for the first character that can also be an upper case letter. All copied strings are converted to lower case for case-insensitive counting.

This basic Filter implementation is not perfect. For example, it rejects some possible types of complex words (e.g. abbreviations), as well as complex phrases made up of several words. It is also adjusted for the specific website being tested. Nevertheless, this basic implementation is satisfactory for the purposes of this study. Further work is required in order to develop more advanced and general filtering methods that can be used on other types of websites.

This CWI implementation can be used as a standalone system, where the most frequently copied words (above a given threshold) are identified as complex, or in combination with other word complexity indicators, providing an additional source of information and indication of word complexity in a multi-indicator CWI system.

3   Experiments and Results

The CWI implementation was run on 231 documentation pages of the ObjectDB website (www.objectdb.com). Usage data were collected for a period of several months, ending in March 2020. 654,399 page views of 241,644 unique visitors (estimated) and 53,131 copy operations were recorded and used as the input dataset for the experiment.

Table 1 shows that the most frequently copied strings on this website are code fragments. Programmers may copy code fragments to their clipboards in order to paste them in their IDEs [8,10].

Table 1. Most frequently copied strings

#	Text (long strings are truncated)	Count
1	@SequenceGenerator(name = "seq", initialValue = 1, allocationSi...	1,143
2	@GeneratedValue(strategy = GenerationType.AUTO)	1,069
3	@GeneratedValue(strategy = GenerationType.SEQUENCE, gene...	686
4	@GeneratedValue	464
5	@Embeddable	454
6	@IdClass(ProjectId.class)	391
7	ParameterExpression<Integer> p = cb.parameter(Integer.class);	364
8	@Transient	358
9	CriteriaBuilder cb = em.getCriteriaBuilder();	350
10	@GeneratedValue(strategy = GenerationType.IDENTITY)	315

Table 2 shows the most frequently copied text strings, obtained by filtering out copy operations of code fragments that are wrapped in HTML PRE tags.

Most of the strings in Table 2 are technical terms, code fragments (embedded in the text), and long sentences, rather than complex words. Users may copy technical terms to the clipboard in order to search for more information about them [8] and complete sentences for citations and text summaries [9]. None of these strings is a valid word according to the definition in Sect. 2, so they are all filtered out by the Filter component.

Table 3 shows the 30 most frequently copied text words, produced by the Filter. For a proficient English speaking user, the words in Table 3 may not seem

Table 2. Most frequently copied text strings

#	Text (long strings are truncated)	Count
1	JPQL	109
2	Composite Primary Key	85
3	EntityManager	67
4	The IDENTITY strategy also generates an automatic value ...	63
5	ObjectDB	58
6	EntityManagerFactory	49
7	persistence.xml	47
8	The sequence strategy consists of two parts - defining a named ...	46
9	Marking a field with the @GeneratedValue annotation specifies ...	39
10	Embedded Primary Key	37

very complex, but they may be complex relative to the website vocabulary level, which consists of mainly simple English words. Very simple and frequent words (e.g. "the", "of", etc.) are not on the list, even though they are very common and appear many times on every page. This is the first indication that users are more likely to copy complex words than simple words. Section 4 presents statistical evidence that supports this hypothesis.

Table 3. Most frequently copied words

#	Word	Count	#	Word	Count	#	Word	Count
1	criteria	36	11	entity	10	21	detached	7
2	transient	24	12	persistable	10	22	allocation	6
3	embeddable	21	13	hollow	9	23	retrieves	6
4	embedded	20	14	explicitly	9	24	query	6
5	persistence	17	15	pessimistic	9	25	identity	6
6	composite	16	16	polymorphic	9	26	equivalent	6
7	redundant	14	17	persistent	8	27	sequence	6
8	retrieved	14	18	dangling	8	28	persist	6
9	explicit	11	19	instantiation	8	29	ascending	6
10	cascaded	11	20	retrieval	7	30	orphan	6

The significant differences in copying frequency (the "Count" columns) in Tables 1, 2, and 3 show that on this website code fragments are copied much more frequently than text words. Due to these differences, complex words are only exposed in Table 3, after filtering out the other elements.

Figures 2 and 3 show some of the resulting complex words in the context of the website text. The words that were copied by users are framed.

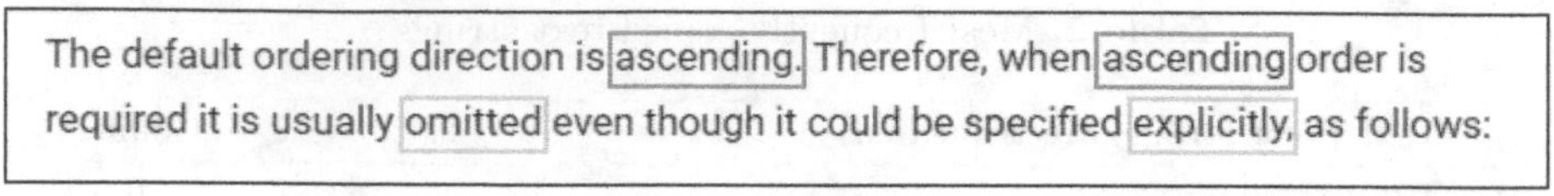

The default ordering direction is ascending. Therefore, when ascending order is required it is usually omitted even though it could be specified explicitly, as follows:

Fig. 2. Copy visualization: "ascending" and other words

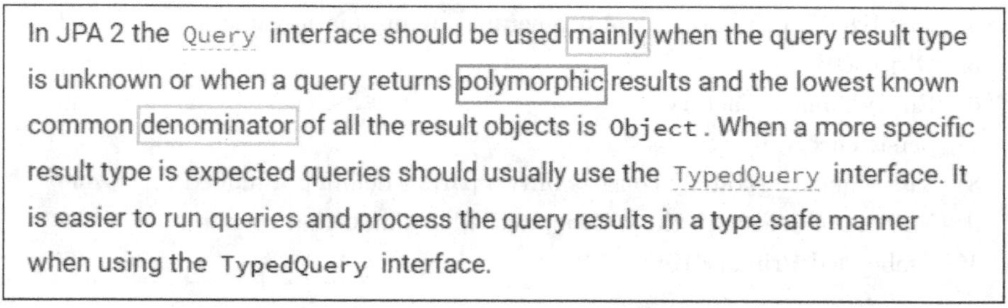

In JPA 2 the `Query` interface should be used mainly when the query result type is unknown or when a query returns polymorphic results and the lowest known common denominator of all the result objects is `Object`. When a more specific result type is expected queries should usually use the `TypedQuery` interface. It is easier to run queries and process the query results in a type safe manner when using the `TypedQuery` interface.

Fig. 3. Copy visualization: "polymorphic" and other words

The word "ascending" (#29 in Table 3) was copied 6 times, and the word "polymorphic" (#16 in Table 3) was copied 9 times. Each of the other framed words (with the yellow border) was copied once in the text shown above (the word "explicitly" was copied 9 times in total on all the tracked web pages).

4 Evaluation

The dataset contains 53,131 copy operations resulting from 654,399 page views, though most of the copy operations are related to code. Only 823 copy operations are accepted by the Filter as related to valid text words (based on the strict definition of valid words in Sect. 2), and these copy operations relate to 326 different words. For evaluation purposes, words not included in the list of the 333,333 most frequent words in the Google's Trillion Word Corpus [13] (e.g. "persistable") were excluded, so the evaluation focused on 316 distinct copied words in 801 copy operations.

These relatively small numbers may not be sufficient for a complete evaluation of the proposed CWI approach as a standalone implementation (e.g. compared to other CWI methods), but as shown in this section, they are sufficient to conclude that words that are copied by users are more likely to be complex words than words that are not copied. In other words, copying words to the clipboard on a website can be considered a CWI indicator.

The null hypothesis is that there is no difference in complexity between copied and uncopied words. To test the null hypothesis we can use the three CWI indicators that have been described in Sect. 1: syllable count, character count, and frequency. In addition to simplicity, using these three indicators rather than testing against human tagging of complex words, which is often used in full evaluation and comparison of CWI implementations (e.g. in [15, 21]), has the advantage of objectivity and avoiding biases. Human complex word tagging was proved to be subjective and inconsistent among different taggers [15,18].

Google's Trillion Word Corpus can be used to estimate word frequency. Given the list of 333,333 most frequently used words in this corpus ordered by decreasing frequency [13], we can define the frequency rank of a word as its position in the list (e.g. #1, the highest, for the word "the", which is the most frequent word in the corpus) and expect words that are less frequent to be generally more complex. Syllables in words have been counted using a Java library[1], which although not 100% accurate, is sufficient for the purpose.

Table 4 shows the values of these three indicators for the most frequently copied words in the dataset (words with at least 8 copy operations). We can reject the null hypothesis by showing that word complexity as evaluated by these indicators is significantly different for copied and uncopied words, i.e. copied words are more likely to be evaluated as complex than uncopied words, with a significant statistical difference.

For the evaluation, the words on the website are divided into two sets: 3,234 uncopied words, which were not copied at all in a single-word copy operation, and 311 copied words, which were copied at least once. Table 5 presents a comparison of these two sets using eight binary criteria for word complexity, based on the

Table 4. Complexity indicators values for the most frequently copied words

#	Word	Copies	Syllables	Characters	Frequency rank
1	criteria	36	3	8	2,468
2	transient	24	2	9	12,548
3	embeddable	21	4	10	89,240
4	embedded	20	3	8	5,356
5	persistence	17	3	11	14,474
6	composite	16	3	9	6,414
7	redundant	14	3	9	12,423
8	retrieved	14	2	9	7,609
9	explicit	11	3	8	6,371
10	cascaded	11	3	8	70,361
11	entity	10	3	6	4,067
12	hollow	9	2	6	9,566
13	oxplicitly	9	4	10	8,551
14	pessimistic	9	4	11	32,253
15	polymorphic	9	4	11	37,407
16	persistent	8	3	10	9,645
17	dangling	8	2	8	25,694
18	instantiation	8	5	13	42,186

---

[1] https://github.com/wfreitag/syllable-counter-java.

three CWI indicators with ranges of values that indicate complexity (relative to the average values for the website's words). Each row in the table shows how many words in these two sets meet each criterion, and so are identified as complex.

As shown in the table, copied words are more likely than uncopied words to be classified as complex by all of the examined complexity criteria. Note that these complexity criteria are based on simple and generic indicators, which cannot determine complexity precisely. For example, the words "Wikipedia" and "electricity" are not necessarily complex, despite the numbers of syllables and characters. In addition, the division into copied and uncopied words is highly dependent on the dataset (uncopied words could become copied words if more copy operations were recorded for more users). Therefore, the exact percentages of words that are classified as complex by these complexity criteria are not expected to be accurate or important on their own. In fact, it might be reasonable to expect the new word copying indicator to be more accurate than the three indicators that are used in this evaluation, as discussed in Sect. 5.

Table 5. Uncopied words vs. copied words

Complexity criterion		Uncopied words	Copied words	p-value
Syllables	≥3	1,378/3,234 (42.6%)	184/311 (59.2%)	0.000000
	≥4	536/3,234 (16.6%)	79/311 (25.4%)	0.000160
Characters	≥8	1,420/3,234 (43.9%)	198/311 (63.7%)	0.000000
	≥9	979/3,234 (30.3%)	140/311 (45.0%)	0.000000
	≥10	623/3,234 (19.3%)	85/311 (27.3%)	0.001046
	≥11	356/3,234 (11.0%)	48/311 (15.4%)	0.024497
Frequency rank	≥5,000	1,369/3,234 (42.3%)	166/311 (53.4%)	0.000198
	≥10,000	782/3,234 (24.2%)	105/311 (33.8%)	0.000347

The results in Table 5 reject the null hypothesis, as they show a significant statistical difference between copied words and uncopied words for each of the eight word complexity criteria. Such differences are not expected under the null hypothesis. For each complexity criterion (a row in the table, representing a 2×2 contingency table) the p-value is calculated using the two-tailed Fisher's exact test.

To examine if the frequency of copying each word matters, Table 6 compares two subsets of the set of copied words: the words that have been copied exactly once and the words that have been copied at least 8 times (shown in Table 4). Table 6 shows that words in the second subset are more likely than words in the first subset to be classified as complex by each of the eight complexity criteria. Due to the sizes of the sets, high statistical significance for these differences (p-value ≤ 0.05) is obtained only for the more inclusive criteria, $Syllables \geq 3$, $Characters \geq 8$, and $FrequencyRank \geq 5,000$.

Table 6. Words copied once vs. words copied at least 8 times

Complexity criterion		Number of times copied		p-value
		Exactly once	At least 8 times	
Syllables	≥3	88/167 (52.7%)	14/18 (77.8%)	0.048170
	≥4	38/167 (22.8%)	6/18 (33.3%)	0.381191
Characters	≥8	93/167 (55.7%)	16/18 (88.9%)	0.005636
	≥9	64/167 (38.3%)	11/18 (61.1%)	0.077752
	≥10	39/167 (23.4%)	7/18 (38.9%)	0.158121
	≥11	26/167 (15.6%)	4/18 (22.2%)	0.500274
Frequency rank	≥5,000	80/167 (47.9%)	16/18 (88.9%)	0.000879
	≥10,000	49/167 (29.3%)	9/18 (50.0%)	0.105954

5 Discussion

Section 4 shows that copied words are more likely to be evaluated as complex than uncopied words and words that are copied more frequently are more likely to be evaluated as complex than words that are copied less frequently, by three different CWI indicators. This has been shown for one website, and further experiments on other websites are required in order to establish these findings.

A reasonable explanation of the connection between complex words and the user behavior of copying words to the clipboard is that users search for definitions and translations of complex words. It is also possible that some users copy familiar but complex words in order to paste them instead of typing them while writing text. It is difficult to think of other convincing reasons as to why users copy to the clipboard regular words, such as "composite", "redundant", "explicit", and "ascending". Copying of very simple words (e.g. "the", "of", etc.) was not observed in the dataset.

Consequently, word copying on a website can be considered a novel CWI indicator. It may be used in a standalone CWI implementation, as described in Sect. 2, where the most frequently copied words are identified as complex, or in combination with other CWI indicators in a multi-indicator CWI implementation. In both cases, the output of the CWI implementation can be used for text simplification (automatic or manual).

Using copy operations for CWI requires large amounts of web usage data, as shown by the demonstration of the filtering process in Sect. 3. The dataset used in this study is based on web usage data collected over several months from a medium traffic website. On low traffic websites, copy operations may be less effective for CWI. On high traffic websites (e.g. Wikipedia) they could be significantly more effective. Collecting data for longer periods may help.

Tracking copy operations is related to *session recording*, which is a common practice in modern web analytics where user activity on websites, including mouse movements and keystrokes, is recorded. It raises interesting questions regarding user privacy and personal data protection, due to the risk of collecting sensitive personal information intentionally or unintentionally [3]. However, session recording does not necessarily require prior user consent under personal data protection regulations, such as GDPR (under certain terms, as discussed by the IT and privacy lawyer Arnoud Engelfriet [3]). Sensitive personal data, which are not required for CWI, should not be collected. If the collected data are completely anonymized, which is a standard practice in web analytics, then they are no longer considered personal data (e.g. according to GDPR). In some sense, counting copy operations is similar to counting page-views, which has always been considered a legitimate web analytics practice.

The evaluation in Sect. 4 shows significant statistical evidence that word copying is a CWI indicator and that this indicator is stronger for words that are copied more frequently. Note that the three indicators that have been used for evaluation in Sect. 4 are not very accurate and cannot be used to evaluate the effectiveness of other CWI indicators. It is reasonable to expect that the word copying indicator (assuming it reflects users' need for assistance with complex words) is more accurate than the three indicators that have been used to test it. However, being well known tested, objective word complexity indicators, they are useful for showing that word copying is also a CWI indicator. Further work is required to assess the effectiveness of the word copying indicator, both as a standalone indicator and in combination with other CWI indicators.

The word copying indicator reflects the collective experience of the website audience by considering copy operations as implicit user "votes". Words identified by these votes as complex might be the biggest barriers to understanding the text and may mostly appear in paragraphs that are read more frequently. Therefore, simplifying these complex words might be particularly helpful to the readers. Accordingly, this new approach might have the potential of being more effective, reliable, and accurate than other CWI indicators, and it might also be more accurate than human tagging of complex words, which is subjective and inconsistent among different experts [15,18]. Further evaluation is required in order to assess this potential.

Other user activities on websites might also indicate word complexity. An interesting hypothesis that has to be tested in this context, is that slower mouse movements near words also indicate word complexity. Some users move the mouse cursor during reading to mark the reading position, so slowing or stopping near words might indicate difficulties in reading or understanding [7].

6   Conclusions and Further Work

This study introduces a new approach to automatic CWI on websites, based on tracking copy operations of users. An experiment on a sample website shows that copied words are more likely to be evaluated as complex than uncopied words and words that are copied more frequently are more likely to be evaluated as complex than words that are copied less frequently, by three different word complexity indicators. Consequently, word copying on a website can be considered a novel CWI indicator, which targets complex words based on real user behavior.

Further work should investigate using this word copying indicator in complete CWI and text simplification implementations, and evaluate the effectiveness of using copy operations in CWI on various types of websites.

References

1. Bingel, J., Barrett, M., Klerke, S.: Predicting misreadings from gaze in children with reading difficulties. In: Proceedings of the Thirteenth Workshop on Innovative Use of NLP for Building Educational Applications, pp. 24–34. Association for Computational Linguistics, New Orleans, June 2018. https://doi.org/10.18653/v1/W18-0503, https://www.aclweb.org/anthology/W18-0503
2. Coleman, M., Liau, T.L.: A computer readability formula designed for machine scoring. J. Appl. Psychol. **60**, 283–284 (1975)
3. Gilliam Haije, E.: Are session recording tools a risk to internet privacy, March 2018. https://mopinion.com/are-session-recording-tools-a-risk-to-internet-privacy/
4. Gooding, S., Kochmar, E.: Complex word identification as a sequence labelling task. In: Proceedings of the 57th Annual Meeting of the Association for Computational Linguistics, pp. 1148–1153. Association for Computational Linguistics, Florence, July 2019. https://doi.org/10.18653/v1/P19-1109, https://www.aclweb.org/anthology/P19-1109
5. Gunning, R.: The Technique of Clear Writing. McGraw-Hill, New York (2019)
6. Horn, C., Manduca, C., Kauchak, D.: Learning a lexical simplifier using Wikipedia. In: Proceedings of the 52nd Annual Meeting of the Association for Computational Linguistics, vol. 2, pp. 458–463. Association for Computational Linguistics, Baltimore, June 2014. https://doi.org/10.3115/v1/P14-2075, https://www.aclweb.org/anthology/P14-2075
7. Kirsh, I.: Using mouse movement heatmaps to visualize user attention to words. In: Proceedings of the 11th Nordic Conference on Human-Computer Interaction (NordiCHI 2020), Tallinn, Estonia, forthcoming. Association for Computing Machinery, New York, October 2020
8. Kirsh, I.: What web users copy to the clipboard on a website: a case study. In: Proceedings of the 16th International Conference on Web Information Systems and Technologies (WEBIST 2020), forthcoming. INSTICC, SciTePress, Setúbal, November 2020
9. Kirsh, I., Joy, M.: An HCI approach to extractive text summarization: selecting key sentences based on user copy operations. In: Proceedings of the 22nd HCI International Conference (HCII 2020), Communications in Computer and Information Science. Springer, Cham, July 2020. https://doi.org/10.1007/978-3-030-60700-5_43
10. Kirsh, I., Joy, M.: Splitting the web analytics atom: from page metrics and KPIs to sub-page metrics and KPIs. In: Proceedings of the 10th International Conference on Web Intelligence, Mining and Semantics (WIMS 2020), Biarritz, France, pp. 33–43. Association for Computing Machinery, New York, June 2020. https://doi.org/10.1145/3405962.3405984
11. Leroy, G., Kauchak, D.: The effect of word familiarity on actual and perceived text difficulty. J. Am. Med. Inform. Assoc. JAMIA **21**, 169–172 (2013). https://doi.org/10.1136/amiajnl-2013-002172
12. McLaughlin, G.H.: SMOG grading: a new readability formula. J. Read. **12**, 639–646 (1969)

13. Norvig, P.: Natural language corpus data. In: Segaran, T., Hammerbacher, J. (eds.) Beautiful Data, pp. 219–242. O'Reilly Media, Inc., USA (2009). https://norvig. com/ngrams/
14. Paetzold, G., Specia, L.: Benchmarking lexical simplification systems. In: Chair, N.C.C., et al. (eds.) Proceedings of the Tenth International Conference on Language Resources and Evaluation (LREC 2016). European Language Resources Association (ELRA), Paris, May 2016
15. Paetzold, G., Specia, L.: SemEval 2016 task 11: complex word identification. In: Proceedings of the 10th International Workshop on Semantic Evaluation (SemEval-2016), pp. 560–569. Association for Computational Linguistics, San Diego, June 2016. https://doi.org/10.18653/v1/S16-1085, https://www.aclweb.org/anthology/S16-1085
16. Shardlow, M.: A comparison of techniques to automatically identify complex words. In: 51st Annual Meeting of the Association for Computational Linguistics Proceedings of the Student Research Workshop, pp. 103–109. Association for Computational Linguistics, Sofia, August 2013. https://www.aclweb.org/anthology/P13-3015
17. Shardlow, M.: A survey of automated text simplification. Int. J. Adv. Comput. Sci. Appl. (IJACSA) Spec. Issue Nat. Lang. Process. 4(1) (2014). https://doi.org/10.14569/SpecialIssue.2014.040109
18. Specia, L., Jauhar, S.K., Mihalcea, R.: Semeval-2012 task 1: English lexical simplification. In: Proceedings of the First Joint Conference on Lexical and Computational Semantics, SemEval 2012, pp. 347–355. Association for Computational Linguistics, USA (2012)
19. Stajner, S., Saggion, H., Ponzetto, S.: Improving lexical coverage of text simplification systems for Spanish. Expert Syst. Appl. **118**, 80–91 (2019). https://doi.org/10.1016/j.eswa.2018.08.034
20. Swain, D., Tambe, M., Ballal, P., Dolase, V., Agrawal, K., Rajmane, Y.: Lexical text simplification using WordNet. In: Singh, M., Gupta, P.K., Tyagi, V., Flusser, J., Ören, T., Kashyap, R. (eds.) ICACDS 2019. CCIS, vol. 1046, pp. 114–122. Springer, Singapore (2019). https://doi.org/10.1007/978-981-13-9942-8_11
21. Yimam, S.M., et al.: A report on the complex word identification shared task 2018. In: Proceedings of the Thirteenth Workshop on Innovative Use of NLP for Building Educational Applications, pp. 66–78. Association for Computational Linguistics, New Orleans, June 2018. https://doi.org/10.18653/v1/W18-0507, https://www.aclweb.org/anthology/W18-0507

A Densely Connected Encoder Stack Approach for Multi-type Legal Machine Reading Comprehension

Peiran Nai[1], Lin Li[1(✉)], and Xiaohui Tao[2]

[1] School of Computer Science and Technology, Wuhan University of Technology,
Wuhan, China
{neng245547874,cathylilin}@whut.edu.cn
[2] School of Sciences, University of Southern Queensland, Toowoomba, Australia
Xiaohui.Tao@usq.edu.au

Abstract. Legal machine reading comprehension (MRC) is becoming increasingly important as the number of legal documents rapidly grows. Currently, the main approach of MRC is the deep neural network based model which learns multi-level semantic information with different granularities layer by layer, and it converts the original data from shallow features into abstract features. Owing to excessive abstract semantic features learned by the model at the top of layers and the large loss of shallow features, the current approach still can be strengthened when applying to the legal field. In order to solve the problem, this paper proposes a Densely Connected Encoder Stack Approach for Multi-type Legal MRC. It can easily get multi-scale semantic features. A novel loss function named multi-type loss is designed to enhance the legal MRC performance. In addition, our approach includes a bidirectional recurrent convolutional layer to learn local features and assist in answering general questions. And several fully connected layers are used to keep position features and make predictions. Both extensive experiments and ablation studies in the biggest Chinese legal dataset demonstrate the effectiveness of our approach. Finally, our approach achieves 0.817 in terms of F1 in CJRC dataset and 83.4 in the SQuAD2.0 dev.

Keywords: Multi-type question answering · Legal machine reading comprehension · Dense encoder stack

1 Introduction

The ability to comprehend text and answer questions is crucial for natural language processing (NLP). Due to the creation of various large-scale datasets [1–4], remarkable advancements have been made in the machine reading comprehension (MRC) tasks. It could be widely applied in many domain related applications. The most-used MRC benchmark datasets are the Stanford SQuAD v1.1 and v2.0 datasets [5,6]. Recently, large pre-trained language models such as BERT

© Springer Nature Switzerland AG 2020
Z. Huang et al. (Eds.): WISE 2020, LNCS 12343, pp. 167–181, 2020.
https://doi.org/10.1007/978-3-030-62008-0_12

and XLNET based on transfer learning make a great progress in MRC [7,8] and speed up its applications in other domains. MRC in legal field attracts the attentions from both of industry and academia since it can assist judges to sum up the evidence from legal documents and make judgments.

In a legal document, a case consists of two parts: fact description and corresponding judgment result. Each document meets the legal norms and includes fact description, court view, parties, judgment result and other information. Therefore, semanteme and structure are clearer than other documents, for example, from Wikipedia. Previous models, such as R-net [9], use abstract feature expressions from top layers to answer questions. Automatic feature learning mechanism of most deep learning model thinks that the deeper layer it reaches, the more abstract knowledge it learns. We observe that as information about the input context of law text passes through layer by layer, the effect of features in lower layers on the result will downgrade. Directly using information learned by the top layer of a deep neural network will not lead to a solid result.

Domain: civil

Context: "After examination and verification, it is found that the Audi car with the license plate of SUAXXX owned by Gong X6 have been covered by motor vehicle loss insurance at PICC for a period from 15:00 on May 12, 2011 to 15:11 on May 12, 2012, and the amount of insurance was 394500 yuan. When Gong X6 drove SUAXXX car along XiKang road from south to north, the well cover managed by TongDa company protrudes from the road surface, resulted in the vehicle accident caused by its collision with the car. ...... On July 18, 2012, Gong X6 issued a transfer of motor vehicle insurance rights and agreed to transfer the corresponding rights and interests claimed by him to the responsible party of PICC "

Question 1: "What is the cause of the accident?"

Answer 1: "The well cover managed by TongDa company protrudes the road surface"

Question 2: "Will the applicant completely transfer the right of recourse to the third party?"

Answer 2: "YES"

Wrong Answer: "Unknown"

Question 3: "What is the sum insured by the person at the plaintiff?"

Answer 3: "394500 yuan"

Fig. 1. Three questions with its answers from the dataset. Relevant keywords in context and question are shown in different colors. (Color figure online)

This paper works on the biggest Chinese legal dataset called CJRC which is collected from first-instance judgment document published by the Supreme People's Court of China[1]. Figure 1 shows an example. As first-instance verdict consists of statements of the parties, cause of action and judgments. Moreover,

---

[1] http://wenshu.court.gov.cn/.

it has a strictly legal structure, and related narrations clearly show the fact. Accordingly, documents of legal cases have multi-scale semantic features. Questions in CJRC are multi-type. Models of legal MRC need to not only extract a sentence, phrase or an entity to answer questions based on the document, but also determine to abstain from answering when the question is not mentioned in the context. Furthermore, as Question 2 in Fig. 1, it should answer "yes" or "no" when it is a general question. Specifically, like SQuAD 2.0 dataset [6], each case in CJRC consists of a judgment context, five questions and corresponding answering results. To deal with multi-scale semantic features and multi-type questions, we design a densely encoder stack approach.

First, in the CAIL2019 competition, most of the contestants use BERT as their baselines [7]. However, directly using BERT does not always work well. According to the problems aforementioned, documents of legal cases have multi-scale semantic features. Both semantic features at the top and features learned by previous layers are needed to answer different style questions. Jawahar et al. show that different BERT encoder stacks capture different hierarchy of linguistic information [10]. On the image processs field, there exists similar observation. For example, Huang et al. [11] proposed a DenseNet which has been used in image segmentation widely with good performacne. Inspired by these researches, we propose a densely connected encoder stack approach. Our approach can learn multi-scale semantic features from legal documents and can be used for multi-type questions in MRC.

Second, in order to solve multi-type questions, a convolutional layer is at the top of our approach to learn local features. Although BERT can deal with the multi-type questions task, legal judgement documents are closely related. Local features are strengthened in our approach, which results in a better understanding on contexts and questions, and thus, better accuracy. Moreover, several fully connected layers are added after densely connected encoder block.

Third, as questions in CJRC are multi-type, if models focus on one of them, the accuracy of the others may decrease. R-net and BERT can deal with multi-type MRC [7,9,14], however, most of the researches just add a threshold to adjust the distribution of the answer in the dataset. Our DeCES designs a multi-type loss function to deal with the yes-or-no and span prediction at the same time. In addition, it can directly be transferred to other languages as verified by experimental results on a benchmark data.

To summarize, our main contributions are listed as follows:

1. A Densely Connected Encoder Stack Approach (DeCES) is proposed, contributing to learn multi-scale semantic features. Different encoders of BERT encode a rich hierarchy of linguistic information.
2. For the sake of multi-type questions, a convolution layer are applied to get local features. Moreover, a new loss function is formulated by combining the logit of different tasks and using subtraction. It can decrease the error rate of different answer styles.

3. Eextensive experiments are conducted in the largest Chinese legal dataset and SQuAD2.0 dataset. Results show that our DeCES significantly improves the performance of both of legal MRC and general MRC compared with popular baselines.

The rest of the paper is organized as follows. Section 2 surveys the related works. Section 3 introduces the detailed descriptions of our approach. The experiment is given in Sect. 4. Section 5 concludes the paper and looks forward to the future work.

2 Related Work

Machine reading comprehension tasks have been widely investigated recently. So far, depending on the answer type, we can divide MRC task into three categories: cloze-style, span prediction and yes-or-no. MRC mainly focus on span prediction task and MRC general questions can be seen as classification task.

Recently big progress on MRC is mainly due to the availability of large-scale datasets [2,5,9], since they make it possible to train large end-to-end neural network models. Hermann et al. have proposed a method for obtaining large quantities of triples through news articles and its summary [1]. Along with the release of cloze-style reading comprehension dataset, they also proposed an attention-based neural network to handle this task. Experimental results show that the proposed neural network is more effective than traditional baselines.

Since SQuAD has been released [5], question style turns to span prediction and free-form text. Wang and Jiang built the match of question-answer with match-LSTM, and predict answer boundaries in the passage with pointer networks [15,16]. The method is being widely used now. Seo et al. introduced bidirectional attention flow networks and Xiong et al. proposed dynamic co-attention networks [17,18]. Wang et al. present R-Net [9]. Google Brain and Carnegie Mellon University combined all these above and proposed QANet [19], while the ingenious model made a great success. However, as mentioned in the introduction, existing textual deep learning models are not compatible with legal field. With the network becoming increasingly deep [20,21], information about the input or gradient passes through many layers, it can vanish or be ignored by the time it reaches the end of the network [22,23]. As structured legal document is easy to learn, the problem above become more serious. Compared with our work, these previous studies pay more attention to learning on deep semantics features and ignored the position features and lower semantics features.

Text classification is a solution for yes-or-no MRC. The category can be semantic (e.g., political and economic) or sentimental (e.g., positive and negative) [15]. Nearly all classical (shallow) classifiers have been used in text classification, such as support vector machine, logistic regression. Recently deep neural network behaves well in text classification task. Legal documents' structures are clearer than other documents (e.g. Wikipedia), and thus, we should pay more attention to local features. We add a bidirectional recurrent convolutional layer to learn local features and assist in answering general questions.

BERT is the most widely used model for MRC dataset, which is designed to pre-train deep bidirectional representations by jointly conditioning on both left and right context in all layers of Transformers [7]. BERT is the first unsupervised, deeply bidirectional system for pre-training NLP. Unsupervised means that BERT was trained using only a plain text corpus, which is important because an enormous amount of plain text data is publicly available on the web in many languages. BERT obtains new state-of-the-art results on eleven natural language processing tasks, which include machine reading comprehension and classification. Many researchers try to improve BERT. Pre-training task like Roberta [24], uses more dataset and more parameter, which makes it extremely expensive. Other researchers add additional components on the top of BERT. Inspired by DenseNet that is mainly applied in computer vision area by using convolution [11], a densely connected encoder stack approach is proposed in legal MRC. Different from DenseNet, our work is answering multi-type questions in MRC domain. It introduces direct connections between any two layers and can scale naturally to many layers.

3   DeCES Framework

3.1   Overview of DeCES

In this section, dense encoder is described firstly since the model our system is built upon it. And several algorithmic improvements on top of it are introduced. Figure 2 shows the overview network structure of our DeCES.

1. Dense encoder block in Fig. 2 learns both deep and superficial semantics which makes considerable reduction on the loss of feature learned in the beginning.
2. A convolution layer and attention mechanism in the bottom of in Fig. 2 aggregate evidence from the whole documents and refine the passage representation, which is then fed into the output layer to answer the general questions whether it is true, false or not mentioned in the documents.
3. A multi-type loss function in the right of in Fig. 2 combines span prediction loss and classification loss to train our model.

3.2   DenseEncoder Stack

BERT Base. This paper enhances BERT to match a legal dataset and add Yes mask, No mask and Unknown mask. Five new learnable parameters are added: a Start vector $S \in \mathbb{R}$, an End vector $E \in \mathbb{R}$, a Yes vector $Y \in \mathbb{R}$, a No vector $N \in \mathbb{R}$ and an Unknown vector $U \in \mathbb{R}$. These five logits can be calculated in Eq. (1):

$$Vector\_log(i) = Vector_i \cdot T_i, \tag{1}$$

where T_i denotes the final hidden vector from BERT for the ith input token, and $Vector$ represents the five vectors aforementioned. During training, the probabilities of word i being the start and end of the answer can be calculated in Eq. (2):

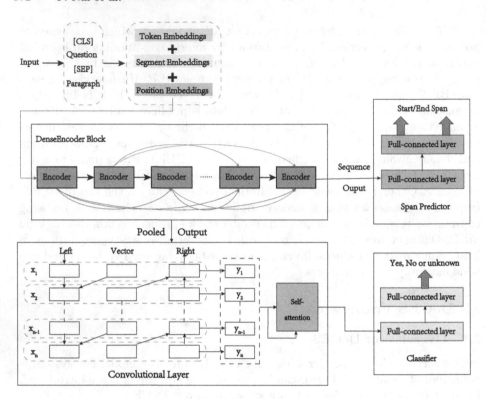

Fig. 2. An overview of our DeCES

$$P_i^{start} = start\_log(i)\Phi[start\_log(i)], \quad P_i^{end} = end\_log(i)\Phi[end\_log(i)] \quad (2)$$

where the probabilities of the answer is "yes", "no" or "unknown" are calculated separately in Eqs. (3) and (4):

$$P_i^{yes} = yes\_log(i)\Phi[yes\_log(i)], \quad P_i^{no} = no\_log(i)\Phi[no\_log(i)] \quad (3)$$

$$P_i^{unknown} = unknown\_log(i)\Phi[unknown\_log(i)], \quad (4)$$

where $\Phi(x)$ is the probability density function of the normal distribution.

DenseEncoder Block. As shown in Fig. 2, the encoder output $[h_{l1}, ..., h_{ln}]$ is used as an input to downstream layers and then design a densely connected encoder stack approach. Following Peters et al. [7], using the last layer is not always the best way. According to DenseNet [11], convolutional networks can be substantially deeper, more accurate and efficient to train if they contain shorter connections between different layers. DenseEncoder block tends to yield consistent improvement in accuracy with a growing number of parameters. Each layer has access to all the preceding features in its block, which means that most of the information learned before would remain. Different decode layers of BERT learn different representations of language [10]. Legal documents are made up by the detail of the case with sound evidence of each event. According to the researches above, features at low levels are still useful for prediction. We argue that linking different layers can help model learn different levels of the legal document, such as the suspect and the reason why the case happens. To further improve the information transmission between different layers, a different connectivity pattern is given. Direct connections from any layer to all subsequent layers are introduced. Consequently, the lth layer receives the feature with a tunable weight of all preceding layers $[h_1, h_2, ..., h_{l-1}]$ as input H:

$$h_l = P_l([h_1, h_2, ..., h_{l-1}]), \tag{5}$$

where $[h_1, h_2, ..., h_{l-1}]$ refers to the concatenation of the feature produced in the layers above. We define P_l as a multi-head self-attention method as:

$$Attention\,(Q, K, V) = softmax\left(\frac{QK^T}{\sqrt{d_k}}\right)V \tag{6}$$

$$head_i = Attention\,(QW_i{}^Q, KW_i{}^K, VW_i{}^V) \tag{7}$$

$$MultiHead = Concat\,(head_i, ..., head_n)\,W^o, \tag{8}$$

where Q, K, and V are linear projections from different encoders, representing attention queries, keys and values [13].

An extra full-connected layer is overlied to strengthen the learning ability of the model, as shown in predictor and classifier modules of Fig. 2. Using an extra full-connected layer is effective and $F1$ result improves further.

3.3   Bidirectional Recurrent Convolution Layer

This paper combines word and its context to present a word's local feature. Our approach lets a bidirectional recurrent convolutional layer to capture the contexts and get more precise semantics than directly using BERT.

$t_l(w_i)$ is defined as the left context of word w_i, which is calculated by Eq. (9). As for $e(w_{i-1})$, it is the output vector of word w_{i-1} from DenseEncoder block. The left-side context for the first word in document uses the same shared parameters $t_l(w_i)$. W^l is a weight matrix that transforms the lth layer into the next layer. W^{sl} is a weight matrix that is used to combine the semantic of the current word with the next word's left context. f is a non-linear activation function. The right-side context $t_r(w_r)$ is calculated in a similar manner, as defined in Eq. (10).

$$t_l(w_i) = f[W^{(l)}t_l(w_{i-1}) + W^{sl}e(w_{i-1})] \tag{9}$$

$$t_r(w_i) = f[W^{(r)}t_r(w_{i+1}) + W^{sr}e(w_{i+1})] \tag{10}$$

As shown in Eqs. (9) and (10), the context vector captures the semantics of all left-side and right-side contexts. Then, the representation x_i of word w_i is defined in Eq. (11).

$$x_i = [t_l(w_i); e(w_i); t_r(w_i)] \tag{11}$$

The recurrent structure can obtain all t_l in a forward scan of the text and t_r in a backward scan of the text. $tanh$ is used to activate the model and apply attention layer.

$$y_i^{(2)} = tanh(w^2 x_i + b^2) \tag{12}$$

$y_i^{(2)}$ is a latent semantic vector, in which local semantic factor will be analyzed to determine the most useful factor.

3.4 Multi-type Loss Function

Although previous machine reading comprehension models are capable of learning answer extraction, they mainly use thresholds to adjust the distribution of whether it can be answered or not. For this adjustment method, there is still room for improvement. In the CJRC dataset, there are not only span answers but also general questions. Like BERT-base approach, the loss function is at high complexity. For example, softmax function is used to deal with the prediction value and then multiply the tagged value. Without reducing the performance of the system, this papers improves a loss function by using subtraction to reduce complexity, and pinpoints the error prediction to make it more reasonable.

This loss is designed to concentrate on both answer extraction and classification. For each question, predictions are made for the possibility of answering "yes", "no" or "unknown" as yes_log, no_log and unk_log. For each word in the document, $start\_log$ and end_log are defined as possible start and end of span prediction. The predictions all above are connected as:

$$S_{start} = [start\_log, yes\_log, no\_log, unk\_log] \tag{13}$$

$$S_{end} = [end\_log, yes\_log, no\_log, unk\_log] \tag{14}$$

Similarly, the correct mask is used in the document to define $Mask_{start}$ and $Mask_{end}$ as:

$$Mask_{start} = [start\_mask, yes\_mask, no\_mask, unk\_mask] \tag{15}$$

$$Mask_{end} = [end\_mask, yes\_mask, no\_mask, unk\_mask] \tag{16}$$

In order to reduce the computational complexity, the logit of span prediction and classification are combined and a negative infinity number is added to the score of the wrong mask. The multi-type loss is defined in Eqs. (17) to (21):

$$Start_{normal} = log \sum_{i=1}^{n} exp(S_{start}) \tag{17}$$

$$Start_{score} = log \sum_{i=1}^{n} exp[S_{start} - n(1 - Mask_{start})], \qquad (18)$$

where we set n to e^{29}.

$$Start_{loss} = -(Start_{score} - Start_{normal})$$
$$= log \sum_{j=1}^{n} exp(S_{start\_wrong}) \qquad (19)$$

Equally, End_{loss} is defined and finally the loss is calculated as:

$$End_{loss} = -(End_{score} - End_{normal})$$
$$= log \sum_{j=1}^{n} exp(S_{end\_wrong}) \qquad (20)$$

$$loss = \frac{Start_{loss} + End_{loss}}{2} \qquad (21)$$

3.5   The Learning Process of Our Approach

Algorithm 1 Training DeCES

Input: contexts: $T = [T_1, T_2, ..., T_n]$
          questions:$Q = [Q_{11}, Q_{12}, Q_{13}, ..., Q_{n3}]$
          true answers:$A = [A_{11}, A_{12}, A_{13}, ..., A_{n3}]$
Output: classification lable:$C' = [C'_{11}, C'_{12}, C'_{13}, ..., C'_{n3}]$
          span start prediction:$S' = [S'_{11}, S'_{12}, S'_{13}, ..., S'_{n3}]$
          span end prediction:$E' = [E'_{11}, E'_{12}, E'_{13}, ..., E'_{n3}]$
1: **if** $A! = 'Yes', 'No'$ or $'Unknown'$ **then**
2:      span start S, span end E, classification label $C = 2$
3: **else**
4:      span start $S = -1$, span end $E = -1$, classification label C
5: **end if**   // Format the true answers
6: **repeat**
7:      **for** i=1 to n **do**
8:          $t_i = f_1(T_i)$ // context representation
9:          **for** j=1 to 3 **do**   // one context has three questions
10:             $q_{ij} = f_2(q_{ij})$ // question representation
11:             $C'_{ij} = f_p(t_i, q_{ij})$ // use pooled out put to do classify
12:             $S'_{ij}, E'_{ij} = f_s(t_i, q_{ij})$ // use sequence out put to do span prediction
13:         **end for**
14:     **end for**
15:     $Loss = \sum_{i=1}^{n} \sum_{j=1}^{3} [e(C_{ij}, C'_{ij}), e(S_{ij}, S'_{ij}), e(E_{ij}, E'_{ij})]$
16: **until** minimize Loss

Algorithm 1 describes the trainning procedure of our approach. After initialization, an iterative process for MRC begins. Firstly, each word of context or query is mapped to a vector space. Secondly, all the encoders are connected to couple multi-scale semantic features. Model makes a preliminary analyze under the guidance of context T and the question Q. Pooled output helps answer general questions, Sequence output helps answer other questions. And then a bidirectional recurrent convolution layer is employed to scan the context and help answer general questions C'_{ij}. Sequence output is directly used to do span prediction S'_{ij}, E'_{ij}. Lastly, both classification results and span predictions are considered to minimize loss.

4   Experiment

4.1   Data Description and Experiment Configuration

Experiments have been conducted with CJRC (China Judicial Reading Comprehension) dataset. The input to the model is a question with a context paragraph, and the output should be the span of text in the paragraph, "yes", "no" or "unknown" based on the question. 25,000 examples are used as the training set, 5,000 examples as the validation set, and 10,000 examples as the test set, which follows the CAIL2019. Table 1 shows the detail of the dataset.

Table 1. Detail of the CJRC dataset

Answer type	Span prediction	Yes	No	Unknown
Train and validation set	25409	2664	1185	742
Test set	8249	901	358	492

TensorFlow implementation of BERT model and pre-trained weights are used here released by Cui et al. [26]. The hidden size is 768 and the number of layers is 12. Our model is fine-tuned with the data described in Sect. 4.1 and search for the optimal combination of hyperparameters. Due to the condition limitation, the batch size is set to 8 and the max seq length to 512. The stride in the sliding window for passages is 128, the max question length to 64, and the max answer length to 55. Our model is trained with an initial learning rate of 2e-5 by training for 5 hours on a TITANX GPU.

4.2   Evaluation Measures

Our evaluation measures is macro-average F1 which is used by the CoQa [25]. Considering that our work based on legal document which are expected to be inferred according to the fact descriptions of cases, ExactMatch(EM) is also considered. Macro-average F1 is the harmonic average of precision and recall defined in Eqs. (22) to (26).

$$Lg = len(gold), Lp = len(pred), Lc = InterSec(gold, pred) \qquad (22)$$

$$precision = \frac{Lc}{Lp}, recall = \frac{Lc}{Lg} \tag{23}$$

$$F1 = \frac{2 \times precision \times recall}{precision + recall} \tag{24}$$

$$Avef1 = \frac{\sum_{i=1}^{Count_{ref}} max(f1(gold_i, pred))}{Count_{ref}} \tag{25}$$

$$F1_{macro} = \frac{\sum_{i=1}^{N} Avef1_i}{N}, \tag{26}$$

where *gold* is the ground truth in the dataset and *pred* is the prediction answer. *InterSec* function calculate the intersection of *gold* and *pred*, while *gold* and *pred* are word level. $Count_{ref}$ is the amount of standard answer. More details can be refered to CoQa [25].

EM measures whether a predicted result matches the correct answer exactly:

$$EM = \begin{cases} 1 & \text{if } pred \equiv gold \\ 0 & otherwise \end{cases} \tag{27}$$

4.3 Baselines

Some state-of-the-art baslines in the machine reading comprehension area are listed below and we adjust these models to fit our dataset.

- BIDAF [17]: a classifier module is added to answer true or false and delete the char embedding.
- BERT [7]: a question is marked with yes, no and unknown before training the model directly.
- BERT+AOA [12]: an attention-over-attention model is applied on BERT to answer question.
- BERT+ Self-attention [13]: a self-attention model after the CLS output is utilized to assist classifying yes, no and unknown.
- Match-LSTM [15]: a classifier module is added to deal with the basic Match-LSTM model.
- R-Net [9]: Microsoft has proposed an MRC model that is composed of gated attention-based recurrent networks and self-matching attention mechanism.
- QANet [19]: Yu et al. have proposed the QA architecture without recurrent networks.

4.4 Overall Results

Our approach is evaluated and the result is displayed on Table 2. Our DeCES is better than seven baselines mentioned in Sect. 4.3. Baseline results are cited from competition leaderboard and Cui et al. [26]. It leads to an improvement of 7.2% over the base BERT model and 11.5% over BIDAF, which demonstrates its effectiveness. We find that adding AoA does not improve performance. This is because Transformer has already used bidirectional self-attention, so no more column attention is needed. Experiments show that the extra self-attention layer is not helpful for our task. In the judgement documents, each case consists of two parts: fact description and corresponding judgment result. Deep neural network will learn the information excessively which is not mentioned in the document and make a wrong decision. Experiment shows DeCES can learn more accurate features than the other models and get a better result.

Table 2. Performance comparison on CJRC datasets (highlighted vaules are the best)

Model	F1 (%)	EM (%)
Match-LSTM [15]	58.3	38.9
BIDAF [17]	60.2	41.3
R-Net [9]	62.8	44.5
QANet [19]	64.6	47.3
BERT-base [7]	74.5	54.7
BERT+AoA [12]	75.6	54.3
BERT+Self-Attention [13]	74.8	54.9
DeCES (ours)	**81.7**	**61.7**

4.5 SQuAD2.0

Moreover, the results on SQuAD2.0, a general-purpose dataset, verify the generality of our approach. SQuAD2.0., i.e., Stanford Question Answering Dataset (SQuAD) is a benchmark dataset in machine reading comprehension, consisting of questions posed by crowd workers on a set of Wikipedia articles [6].

Table 3. Performance comparison on SQuAD2.0 dev (highlighted vaules are the best)

Model	F1 (%)	EM (%)
BIDAF [17]	68.9	65.7
R-Net [9]	69.8	67.1
QANet [19]	70.6	68.2
BERT-base [7]	74.5	70.8
BERT-large [7]	81.8	79.0
DeCES (ours)	**83.4**	**80.4**

Table 3 reports the performance of DeCES and baseline models. Baseline results are cited from SQuAD2.0 leaderboard. Experiments have been conducted as single model without ensemble. As shown in Table 3, DeCES achieves significantly better results than several baseline models. In detail, it improves overall F1 by 8.9%, compared with base BERT model. Experiment shows DeCES also works in general-purpose dataset.

4.6 Ablation Study

Next, an ablation study has been conducted on the CJRC dataset to show the effects of our proposed approach. Table 4 shows the F1 and EM on CJRC dataset by removing each key components of our work separately.

Table 4. Result with and w/o our methods (highlighted vaules are the best)

Model	F1 (%)	EM (%)
W/o Multi-type Loss	80.3	60.2
W/o Bi-RCNN layer	80.1	60.4
W/o Dense module	78.2	57.3
BERT + last 2 encoders	76.2	55
DeCES (ours)	**81.7**	**61.7**

Loss function focusing on independent task will result in a performance drop for the whole work. Our loss balances the both answer boundary and classification while result shows that it performs better than the other. And then we remove our work on CLS and find that adding a convolution layer on the top of the BERT improve the classification result. If BERT is used alone, model will answer "YES" to nearly 1% questions erroneously and the EM decrease obviously. When two encoders are added on BERT, experiment result increases 1.7%. We can observe that the performance drops obviously after removing the connection in DenseEncoder block. The F1 decreases 3.5%. It shows that linking different layers can really help model learn different levels of the legal document. Therefore, ablation study verifies that densely connection module plays the most important role in our approach and other modules are also indispensable parts.

4.7 Case Study

We have several observations in Table 5 through the following case study. Firstly, for such simple questions like question1, DeCES can find the related sentence in the document and give the correct answer. Secondly, we observe that if there are more than one possible answer, DeCES can combine the multi-scale semantic features to understand the relationship between different possibility with question. Therefore, it will give a more appropriate answer at last. Finally, about general questions that are not clearly noted in the paragraph, owing to DeCES have a Bidirectional Recurrent Convolution Layer, the correct answer are more likely to be extracted than others.

Table 5. Case Study. The top several rows show an example of the legal document and questions, while the questions and answers are shown at the following lines.

Context: On April 18th, 2014, the plaintiff parked his Jeep Wrangler Rubicon in the parking lot managed by the defendant. ...... Judging from the video of the teahouse' monitoring probe, the car was stolen at 3 a.m. on April 19th, 2017, and the thief drove the car away from other outlets. And more truths are as followed: 1. The plaintiff Zhang bought the Jeep Wrangler Rubicon 3604CC, a four-door SUV, on January 14th, 2014 with a total price of 461900 yuan (including Value Added Tax 67113.68 yuan)2. To prove that the plaintif zhang did consume and park the involved car in West Garden run by the defendant from April 18th to the early morning of 19th,2017, the plaintiff Zhang submitted a machine-printing invoice issued by the defendant. The defendant had a parking booth under the West Garden, on which he posted a warning: There's no guard after 1 a.m., and if issues happen with cars parking overnight, no responsibility is accepted by us. The parking lots are for consumers of West Garden with no charge. We offer no safekeeping service of cars. Parking time: 9:00-24:00			
Question 1	When did the plaintiff purchase the vehicle?		
Question 2	Who is the parking lot under the West Garden open to?		
Question 3	Do the parking lot offer safekeeping service at night?		
Question 4	What is the invoice amount of the parking lot?		
	Correct answer	DeCES (ours)	BERT-base
Answer 1	January 14, 2014	January 14, 2014	January 14, 2014
Answer 2	consumers of West Garden	consumers of West Garden	the defendant
Answer 3	NO	NO	YES
Answer 4	NULL	NULL	461900 yuan

5   Conclusions and Future Work

This paper proposes a densely connected encoder stack approach for multi-type legal MRC. Both DenseEncoder block and bidirectional recurrent convolution layer are key through ablation studies. Prediction performance is further improved by our novel loss function. Experiments show that our approach is effective not only in legal area but also in common area. In the future, leveraging legal knowledge database to guide the learning process can be further studied.

References

1. Hermann, K.M., Kocisky, T., et al.: Teaching machines to read and comprehend. In: Advances in Neural Information Processing Systems, pp. 1693–1701 (2015)
2. Nguyen, M., et al.: Ms Marco: a human generated machine reading comprehension dataset. arXiv preprint arXiv:1611.09268 (2016)
3. Joshi, M., Choi, E., Weld, D., et al.: TriviaQA: a large scale distantly supervised challenge dataset for reading comprehension. In: The 55th Annual Meeting of the Association for Computational Linguistics, pp. 1601–1611 (2017)
4. Kočiský, T., et al.: The narrativeqa reading comprehension challenge. Trans. Assoc. Comput. Ling. **6**, 317–328 (2018)
5. Rajpurkar, P., et al.: SQuAD: 100,000+ questions for machine comprehension of text. In: EMNLP, pp. 2383–2392 (2016)

6. Rajpurkar, P., et al.: Know what you don't know: unanswerable questions for SQuAD. In: ACL (Volume 2: Short Papers), vol. 1, pp. 784–789 (2018)
7. Devlin, J., et al.: BERT: pre-training of deep bidirectional transformers for language understanding. In: NAACL, pp. 4171–4186 (2019)
8. Yang, Z., ei al.: XLNet: generalized autoregressive pretraining for language understanding. arXiv preprint arXiv:1906.08237 (2019)
9. Wang, W., et al.: Gated self-matching networks for reading comprehension and question answering. In: ACL, pp. 189–198 (2017)
10. Jawahar, Ganesh, et al.: What does BERT learn about the structure of language? In: The 57th Conference of the Association for Computational Linguistics, pp. 3651–3657 (2019)
11. Huang, G., Liu, Z., et al.: Densely connected convolutional networks. In: The IEEE Conference on Computer Vision and Pattern Recognition, pp. 4700–4708 (2017)
12. Cui, Y., et al.: Attention-over-attention neural networks for reading comprehension. In: The 55th Annual Meeting of the Association for Computational Linguistics, pp. 593–602 (2017)
13. Vaswani, A., et al.: Attention is all you need. In: Advances in Neural Information Processing Systems, pp. 6000–6010 (2017)
14. Hu, M., et al.: Read+ verify: Machine reading comprehension with unanswerable questions. In: The AAAI Conference on Artificial Intelligence, pp. 6529–6537 (2019)
15. Wang, S., Jiang, J.: Machine comprehension using match-LSTM and answer pointer. In: International Conference on Learning Representations (2017)
16. Vinyals, O., Fortunato, M., Jaitly, N.: Pointer networks. In: Advances in Neural Information Processing Systems, pp. 2692–2700 (2015)
17. Seo, M., Kembhavi, A., Farhadi, A., Hajishirzi, H.: Bidirectional attention flow for machine comprehension. In: ICLR (2017)
18. Xiong, C., Zhong, V., Socher, R.: Dynamic coattention networks for question answering. In: ICLR (2017)
19. Yu, Adams Wei, et al.: Qanct: combining local convolution with global self-attention for reading comprehension. arXiv preprint arXiv:1804.09541 (2018)
20. Gheisari, M., et al.: NSSSD: a new semantic hierarchical storage for sensor data. In: 2016 IEEE 20th International Conference on Computer Supported Cooperative Work in Design (CSCWD), Nanchang, pp. 174–179 (2016)
21. Peng, M., Xie, Q., Wang, H., Zhang, Y., Tian, G.: Bayesian sparse topical coding. IEEE Trans. Knowl, Data Eng (2018)
22. Peng, M., et al.: Mining event-oriented topics in microblog stream with unsupervised multi-view hierarchical embedding. TKDD **12**(3), 38 (2018)
23. Peng, M., et al.: Neural sparse topical coding. In: Proceedings of the 56th Annual Meeting of the Association for Computational Linguistics, vol. 1, Long Papers, pp. 2332–2340, July 2018
24. Liu, Y., Ott, M., et al.: Roberta: a robustly optimized bert pretraining approach. arXiv preprint arXiv:1907.11692 (2019)
25. Reddy, S., et al.: Coqa: a conversational question answering challenge. Trans. Assoc. Comput. Linguist. **7**, 249–266 (2019)
26. Cui, Y., Che, W., Liu, T., et al.: Pre-training with whole word masking for Chinese BERT. arXiv preprint arXiv:1906.08101 (2019)

Security and Privacy

Privacy-Preserving Data Generation and Sharing Using Identification Sanitizer

Shuo Wang[1,2]([✉]), Lingjuan Lyu[3], Tianle Chen[2], Shangyu Chen[4],
Surya Nepal[1], Carsten Rudolph[2], and Marthie Grobler[1]

[1] CSIRO's Data61, Melbourne, Australia
{shuo.wang,Surya.Nepal,Marthie.Grobler}@csiro.au
[2] Monash University, Melbourne, Australia
{shuo.wang,Carsten.Rudolph}@monash.edu, tche119@student.monash.edu
[3] National University of Singapore, Singapore, Singapore
lyulj@nus.edu.sg
[4] University of Melbourne, Melbourne, Australia
shangyuc@student.unimelb.edu.au

Abstract. In this paper, we propose a practical privacy-preserving generative model for data sanitization and sharing, called Sanitizer-Variational Autoencoder (SVAE). We assume that the data consists of identification-relevant and irrelevant components. A variational autoencoder (VAE) based sanitization model is proposed to strip the identification-relevant features and only retain identification-irrelevant components in a privacy-preserving manner. The sanitization allows for task-relevant discrimination (utility) but minimizes the personal identification information leakage (privacy). We conduct extensive empirical evaluations on the real-world face, biometric signal and speech datasets, and validate the effectiveness of our proposed SVAE, as well as the robustness against the membership inference attack.

Keywords: Generative model · Data sharing · Privacy-preserving · Deep learning · Variational autoencoder

1 Introduction

The major challenges faced by both the data mining and security communities include: (1) Datasets shared to public use often contain sensitive information that can reveal identification, e.g. electroencephalogram or voice signals have been successfully used as a novel biometric technique in security and authentication applications for person identification [2,3,14]; (2) Data availability remains limited in a lot of domains. On the other hand, generative models, such as variational autoencoder (VAE) [13] and Generative Adversarial Nets (GANs) [16,19,20] have brought the practical potential to relieve the data availability limitation. However, it has been demonstrated that releasing only the generative distribution derived from generative models may reveal private information

© Springer Nature Switzerland AG 2020
Z. Huang et al. (Eds.): WISE 2020, LNCS 12343, pp. 185–200, 2020.
https://doi.org/10.1007/978-3-030-62008-0_13

of the original training samples [4]. Thus, it is necessary to embed privacy-preserving mechanisms into the generative models to break privacy barriers that hamper data sharing. Differential privacy [7], has been implemented as a state-of-the-art privacy principle in the training of generative models [5,8,12,18,25,26]. In this line of work, data curator publishes a synthetic sample generation model trained on the original data in a differentially private manner instead of releasing a sanitized version of the original data. Based on the released generative model, it is feasible to produce an unlimited amount of synthetic sample from distributed parties without privacy concerns. However, the utility of existing differentially private GANs is still limited due to the tight privacy-preserving level of the ordinary Gaussian mechanism [26].

In this paper, we make a reasonable assumption that data consists of identification-relevant and irrelevant components. For example, speech recognition, where voice can be used to recognize text as well as explicit speaker identity information. For a specific task (e.g. arrhythmia diagnosis or speech-to-text recognition), it would be ideal if there is a sanitizer which can provide task-relevant information only, while ignoring (generalizing) identification-relevant features. Namely, the sanitizer can maximize task-relevant performance (utility) and at the same time minimize identity-relevant information (privacy). Consequently, in this paper, we propose a privacy-preserving generation scheme, to generate and sanitize data using identity sanitizer, by perturbing the identification-irrelevant latent space only.

Our contributions can be summarized as follows:

- We apply a variational autoencoder (VAE) network that maps an instance into low-dimensional and disentangled latent representation (a.k.a. latent codes) by the encoder and reconstructs the latent codes into a new reconstructed instance by the decoder. It is then enhanced with the generative adversarial network (GAN), aiming to boost the performance of disentanglement (a change in one dimension of the latent vector corresponds to a change in one feature only) and reconstruction (generation).
- We incorporate the VAE with a classifier-based identity sanitizer to minimize the identity-relevant features while only maintaining the identity-irrelevant information, called Sanitizer-Variational Autoencoder (SVAE). Besides, we apply differential privacy to the SVAE by perturbing disentangled latent codes or perturbing the training of SVAE. Additionally, by perturbing latent codes, it is feasible to generate and share unlimited task-relevant but identity-irrelevant synthetic instances for the identity-irrelevant tasks.
- We conduct an extensive empirical evaluation using real-world face image, ECG signals as well as speech datasets, to validate the efficiency and privacy-preserving performance of SVAE. We demonstrate the ability of SVAE to effectively support augmentation tasks for data from distributed parties while providing privacy protection and maintaining a reasonable utility.

2   Sanitizer-Variational Autoencoder Scheme

2.1   Overview of the SVAE

The purpose of this work is to convert and augment sensitive instances into a large volume of sanitized synthetic instances that are safe from privacy concerns. The identification relevant information is stripped at first and then the perturbation is conducted on the latent representation instead of the input space. Therefore, as illustrated in Fig. 1, SVAE consists of: (1) an identification-irrelevant and disentangled latent representation learning model with an identification sanitizer (i.e. inference stage), and (2) differential privacy-based perturbation and sanitized generation (i.e. perturbation and generation stage).

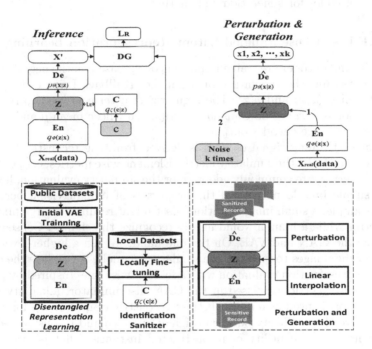

Fig. 1. Frameworks of SVAE.

For the first component, an encoder En conducts feature extraction which transforms the high-dimensional input X into the low-dimensional latent reorientation Z (a.k.a. codes), which can be convolutional neural network (CNN) based or recurrent neural network (RNN) based. The latent codes are then passed to the decoder De to reconstruct the Z back to high-dimensional output X. Further, to improve the reconstruction quality, a generative adversarial network (GAN) is jointly trained with VAE on clean training instances, which provides a discriminator DG to improve the reconstruction error evaluation so that it enhances the quality of reconstructed instances. The initial VAE-based disentangled latent representation learning model can be trained on public-released relevant datasets as the inputs, and then shared to multi-parties to fine-tune using local data. For the identification sanitizer, an identification classifier C is jointly trained (fine-tuning) with the pre-trained encoder En on the local sensitive data to filter identification-relevant features and only retain the identification-irrelevant components of latent vector for synthetic instances reconstruction and generation. Namely, the identification classifier C intercepts the feature vector Z with which it estimates the identity class labels' probabilities $P(I)$, aiming to strip the identification-relevant features. After locally fine-tuning the inference model $\hat{En}$ and $\hat{De}$, we can simply conduct linear interpolation on $\hat{En}$ and feed it to decoder net to generate new data points. To further address the privacy concerns, differential privacy is applied to the locally fine-tuning procedure or the learned latent codes for generation and sharing.

2.2 VAE-based Disentangled Latent Representation Learning

The encoder and decoder are trained simultaneously based on the negative reconstruction error and the regularization term, i.e. Kullback-Leibler (KL) divergence between $q_\phi(z|x)$ and $p(z)$. The regularization term is used to regularize the distribution $q_\phi(z|x)$ to be Gaussian distribution whose mean μ and diagonal covariance $\sum$ are the encoder output.

The major training signal of VAE is derived from the reconstruction error objective evaluated by the similarity metric. Element-wise metrics, e.g., the pixel-wise squared error, are commonly adopted for the similarity evaluation, however, they are simple but do not model the properties of human visual perception [15]. For example, a small image rotation offset might result in a large pixel-wise squared error while a human would hardly recognize the change. Consequently, we combine the GAN and VAE. On the one hand, we use a higher-level representation of the images to measure the similarity, namely, we replace the default pixel-wise reconstruction evaluation with a feature-wise one. Inspired by [10, 15], the feature-wise metric expressed in the GAN discriminator, a.k.a. style error or content error, is used as a more abstract reconstruction error L_R for VAE to better measure the similarity between original instance and reconstructed ones, aiming to improve the utility of reconstructed instance. That is, we can use learned feature representations in the discriminator of a pre-trained GAN as the basis for the VAE reconstruction objective. Formally, let x and $\hat{x}$ be the original image and the reconstructed image and $F^l(x)$ and $F^l(\hat{x})$ be their respective hidden feature representation of the lth layer of the discriminator, the representation loss to reveal reconstruction error is defined in Eqs. 1 and 2. A Gaussian observation model is applied for $F^l(x)$ with mean $F^l(\hat{x})$ and identity covariance $p(F^l(x)|z) = N(F^l(x)|F^l(\hat{x}), I)$.

$$L_{rec}^{F_l} = -E_{q(z|x)}[log \; p(F^l(x)|z))] \tag{1}$$

$$L_{rec}^{content}(x, \hat{x}, l) = \frac{1}{2}\sum_{i,j}(F_{i,j}^l - \hat{F}_{i,j}^l)^2 \tag{2}$$

Simply, a GAN is trained on original clean instances. A single layer l is then chosen from the discriminator network and used to obtain representation similarity according to $F^l(x)$. The representation error is then considered as L_R, leveraging the learned feature representations in the discriminator as the basis for the VAE reconstruction objective, instead of l_2 loss between the original instance and the reconstructed instance.

Note that another variant of GAN, such as info-GAN [6], can be directly incorporated into our framework.

2.3 Identification Sanitizer

We assume that the latent codes of instances consist of an identity-invariant component (e.g. arrhythmia diagnosis or speech-to-text recognition) and an identity-relevant component (essential attributes of identification). Then, it is desired to obtain better disentangled and identification-irrelevant latent codes, by stripping identification information. Consequently, we propose SVAE, derived from VAE. It is incorporated with a classifier on the top of the latent code to achieve disentangle the latent codes z that follow a fixed prior distribution while being unrelated to the label, i.e. to extract class irrelevant z. The classifier and En will be trained in an adversarial manner.

To obtain identification-irrelevance for good disentanglement in z, we apply the irrelevance term L_E, estimated by a classifier parameterized by ς on the latent codes z, derived from the encoder $q_\phi(z|x)$, to classify the label of z. Adversarial learning approach, similar to [19], is used to train the classifier. An additional adversarial classifier is added after z to distinguish its label, while encoder tries to fool it, as the C component demonstrated in Fig. 1 left. The objective of the classifier can be defined as cross-entropy loss:

$$L_C = -E_{q_\phi(z|x)} \sum_c I(c = y) log\ q_\varsigma(c|z) \tag{3}$$

Here $I(c = y)$ is the indicator function, and $q_\varsigma(c|z)$ is softmax probability output of the classifier. Simply, we assume the labels are distributed uniformly across all inputs, i.e. class probabilities $\pi = 1/C$. To peel the effect of labels and to extract class irrelevant z, the encoder $q_\phi(z|x)$ is simultaneously trained to fool the classifier with loss added by a cross entropy loss defined as follows:

$$L_E = E_{q_\phi(z|x)} \sum_c \frac{1}{C} log\ q_\varsigma(c|z) \tag{4}$$

2.4 Perturbation and Generation of SVAE

For each dimension of the latent vector z_i, DP is implemented to find the optimal noise scale n_i to perturb z_i to be $\hat{z}_i = z_i + n_i$. Specifically, there are two types of perturbation we can apply to locally fine-tune $\hat{En}$ and $\hat{De}$: the inference procedure perturbation and generation procedure perturbation. For the inference procedure perturbation, we apply perturbation mechanisms during fine-tuning the $\hat{En}$ and $\hat{De}$ via Stochastic gradient descent (SGD) perturbation. For the generation procedure perturbation, we apply perturbation mechanisms after fine-tuning the $\hat{En}$ and $\hat{De}$. In this case, the inputs that are considered to be the learned latent codes.

SGD-Based Inference Procedure Perturbation. Motivated by deep learning with differential privacy [1], we add random noise to the mini-batch stochastic gradient descent (SGD) [1,24] during training the training process of $\hat{En}$ and $\hat{De}$. At every stage of the SGD, when the gradient $\nabla_\theta L(\theta, x_i)$ is calculated for a random subset of examples, the l_2-norm of gradients are clipped by a threshold T at first to bound the sensitivity by a constant T. We then obtain the average of these gradient values, followed by injecting random noise sampled from a Gaussian distribution that has a variance proportional to the gradient clipping to protect the privacy. Note that a privacy accountant A is adopted in this work, similar to [1], to track the cumulative privacy loss in each step. These procedures are iterated until convergence or exceeding the privacy budget. As the gradient clipping bound T is a hyper-parameter, small T leads to excessive truncation of the gradients while large T leads to overestimating the sensitivity, both causing slow convergence and poor utility. On the other hand, the magnitudes of gradients of the weights and biases vary significantly across different layers. To improve the utility, we assume a small amount of public relevant data Pb is available. During each training iteration, a batch of examples from Pb is randomly sampled to estimate the clipping bound of each parameter using the average gradient norm w.r.t the current batch. Additionally, we fine-tune the classifier used for identification.

Latent Codes-Based Generation Perturbation. In this case, we apply Laplace or Gaussian mechanisms to the output of En, i.e. the latent vector z. The scale of noise can be decided by *sensitivity S/privacy budget ϵ* and $\sqrt{2log(1.25/\sigma)}\frac{S}{\epsilon}$ respectively, where S is the approximate sensitivity of En. It is impossible to figure out the exact sensitivity of the unknown En during the training. As observed from our empirical studies, each factor of latent vector z can be bounded within a value range, in which the reconstructed instances retain the original features. Consequently, the latent vector z can be clipped by the threshold of T. This allows us to assume that the upper bound of En is the norm of T, i.e. the sensitivity $S = T$. Alternatively, the perturbation can be only applied to one or more randomly selected latent codes, instead of applying on all the latent codes, which can further improve the utility.

Synthetic Generation with Anonymity. After training the identification sanitizer, we use the fine-tuned $\hat{En}$ and the fixed $\hat{De}$ as a generic generative model. Based on the locally fine-tuning the inference model $\hat{En}$ and $\hat{De}$, we can simply conduct linear interpolation on $\hat{En}$ and feed it to decoder net to generate new data points. Specifically, as shown in Fig. 1 right, given a sensitive instance x as seed, k times perturbations are iteratively conducted on the latent $\hat{z} \leftarrow En(x)$ (adding k times noise), which will be re-construed back to k synthetic instances via the decoder $\hat{De}$. The norm of perturbation is restricted within a norm γ. This not only contributes to data augmentation but also provide (k, γ)-anonymity for each sensitive instance, defined as follows:

Theorem 1. *For a given sensitive instance x, locally fine-turned $\hat{E}n$ and $\hat{D}e$, and an identification classification mechanism IC, we state that x is releasable with (k, γ)-anonymity, if there exist at least k distinct records $x_1, \cdots, x_k$ from $\hat{D}e(\hat{E}n(x) + noise)$ such that*

$$IC(x_i) \neq IC(x) \text{ and } ||\hat{E}n(x_i) - \hat{E}n(x)||_p \leq \gamma, \ \forall k \qquad (5)$$

Therefore, besides ensuring the differential privacy based sensitivity of the model $\hat{E}n$ with respect to input data records, we provide enough indistinguishable records using the reconstruction ability of SVAE. Note that the γ can be considered as S, the approximate sensitivity of En. k is set at 10 in this work.

3 Experimental Results

In this section, we present empirical evaluations of the proposed SVAE over three benchmark datasets (CelebA, EGG, and Speech). These experiments aim to address two key questions: (1) Does SVAE enable the synthesization of realistic datasets with differential privacy perturbation? (2) Does the generated data retain satisfactory quality and utility for various data analysis tasks?

3.1 Datasets

CelebA. CelebA (cropped version) [17]: consists of 202,599 RGB $64 \times 64 \times 3$ images of various celebrities, including 10,177 unique identities and 40 binary attribute annotations per image.

ECG-ID Database [11]. The database contains 310 ECG recordings, obtained from 90 persons. The records were obtained from volunteers (44 men and 46 women aged from 13 to 75 years who were students, colleagues, and friends of the author). The number of records for each person varies from 2 (collected during one day) to 20 (collected periodically over 6 months).

Speech Database. The TIMIT acoustic-phonetic corpus [9,27] contains abundant recordings of phonetically-balanced read speech. There are 6300 utterances presented with 10 sentences from each of 630 speakers, of which about 70% are male and 30% are female. Each utterance comes with manually time-aligned phonetic and word transcriptions, as well as a 16-bit, 16 kHz speech waveform file. Kaldi's TIMIT recipe is used to split train/dev/test sets and exclude dialect sentences (SA), with 462/50/24 non-overlapping speakers in each set respectively. We select 50% male and female speakers respectively, and randomly split every individual's audio data into training (80%) and evaluation (20%) sets. The data from the remaining speakers are used for the gender testing set.

We split all samples into the training set (80%) and hold out the rest as a test set (20%). The former 85% of the dataset is used as the training data and the remainder 15% is used as the validation data (e.g. as the public-released relevant datasets).

3.2   Metrics

The performance of the proposed SVAE is evaluated in terms of three different aspects: the privacy-based evaluation of perturbation mechanism, the data utility-based evaluation of the synthesis mechanism and the resistibility of membership attacks [23]. The detailed evaluation metrics are given as follows:

Privacy-Related Evaluation. Distance-based metrics are widely used in many data privacy works. Euclidean distance is used to evaluate how similar generated samples from the SVAE with SGD-based and Latent code-based perturbation mechanisms (SP-SVAE and LP-SVAE), the generated samples without perturbation (O-SAVE) with ordinary samples (Original) respectively. By comparing the distances between samples of Original and O-SVAE/LP-SVAE/SP-SVAE, we can investigate the distortion of generated samples from the perturbed/reconstructed generator. Here, we use the L_2 loss as privacy gain, i.e., Euclidean distance between a sample s_p of LP-SVAE/SP-SVAE or s_s of O-SVAE to the corresponding s_o in Original.

We also measure the percentage of re-identified records given a certain attacker model. We assume a practical scenario where the attacker assumed in our case is that the attacker does not have specific targets but does his/her best to re-identify as many records as possible using available background information. Here, we adopt a pre-trained identification classifier on the entire original samples with identification labeled as a strong attack. Thus, the successful re-identification ability is simply calculated as re-identification ratio: $RR = \frac{re-id \ \#}{total \ id \ \#}$.

Data Utility-Related Evaluation. First, statistic distributions distribution of each attribute is used to compare the statistical similarity between the original data and the corresponding attribute in perturbed/sanitized data. In addition, the inception score (IS) [21] is adopted to evaluate the quality of data generated by SVAE.

$$IS = exp(\mathbb{E}_x KL(Pr(y|x)||Pr(y))) \qquad (6)$$

where x is a sample generated by SVAE, $Pr(y|x)$ is the conditional distribution of a pre-trained classifier to predict its label y. A small entropy of $Pr(y|x)$ reveals that x is similar to a real sample. Note that baseline classifiers to predict $Pr(y|x)$ can be trained on an entire training set. The inception score IS can describe both the quality and diversity of the synthetic sample. However, since the inception score only relies on the final probabilities of the classifier, it is sensitive to noise and is not able to detect mode collapse. Thus, a pre-trained Deep4 model [22] is adopted as a replacement for the inception model. Furthermore, Machine learning score similarity is used to evaluate the compatibility of the model. We apply different attribute-wise classifiers to evaluate the machine learning score similarity. After fixing a classification or regression algorithm and its parameter, we train with the original data or the synthesized data generated with and without the perturbation mechanism. We use F-1 values for classification tests and mean relative error (MRE) for regression tests. Such evaluation can reveal the ability of the synthetic sample to be used for real applications.

Membership Attacks Resistibility. The membership attack is practical to common machine learning algorithms [23], aiming to infer whether a target sample is in training samples or not after observing the outputs of a target machine learning algorithm. It is assumed that attackers have black-box access to the target model and know its detailed algorithm design and architecture. Using the black-box access, attackers can mimic training samples for malicious models. To attack the proposed SVAE, we customize the membership attack method presented in [23] that was designed to attack classification models. Attackers are assumed to enable to obtain as many outputs as they want from our SVAE model and know the algorithm and architecture of SVAE. Note that black-box access to the decoder of SVAE is allowed and access to other neural networks of SVAE is blocked because they are not related to releasing synthetic samples after being trained. We use Multilayer Perceptron, DecisionTree, AdaBoost, Random-Forest, and SVM classifiers to build identification classifiers as attack models and their best parameters are recorded. We evaluate the attack performance based on the F-1 score.

Note that the experiments were performed on three classical differential privacy levels, i.e., $(\epsilon, \delta) = \{high : (1, 10^{-5}); medium : (4, 10^{-5}); low : (10, 10^{-5})\}$. Each experiment was tested 10 times and the average result was reported.

3.3   Quantitative Evaluation

In this section, a set of quantitative evaluations are conducted to evaluate the performance of SVAE, to address the second question: *Does the generated data retain satisfactory quality and utility for various data analysis tasks?*

Inception Score Evaluation. Table 1 summarizes the Inception scores of original and synthetic sample (generated by SVAE with and without perturbation mechanism) for the CelebA, ECG and speech datasets.

We found that LP-SVAE enables to synthesize data with Inception scores close to the original data, similar to O-SVAE and better than SP-SVAE. For CelebA, the difference between the original sample and the synthetic sample by O-SVAE/LP-SVAE/SP-SVAE is around 1/1.5/2, respectively. For ECG, the difference between the original sample and the synthetic sample by O-SVAE/LP-SVAE/SP-SVAE is around 1.1/1.5/2.5, respectively. For Speech, the difference between the original sample and the synthetic sample by O-SVAE/LP-SVAE/SP-SVAE is around 1.2/1.3/2.2, respectively. In all cases, synthetic samples from LP-VAE and O-VAE are similar to the original samples. LP-SVAE with the high-privacy setting performs better than SP-SVAE. The latent perturbation strategy shows better synthesis performance than SGD-based perturbation.

Table 1. IS of original and synthetic sample on CelebA, ECG and Speech dataset on different settings of SVAE

Database	Model	n(10^4)	(ϵ, δ)	IS
CelebA	Original	500	-	8.12 ± 0.06
	O-SVAE	400	-	7.12 ± 0.06
	SP-SVAE	320	$(1, 10^{-5})$	6.15 ± 0.06
	LP-SVAE	320	$(1, 10^{-5})$	6.78 ± 0.06
ECG	Original	300	-	6.56 ± 0.04
	O-SVAE	200	-	5.42 ± 0.04
	SP-SVAE	180	$(1, 10^{-5})$	4.11 ± 0.04
	LP-SVAE	180	$(1, 10^{-5})$	5.07 ± 0.04
Speech	Original	300	-	5.27 ± 0.02
	O-SVAE	300	-	4.01 ± 0.04
	SP-SVAE	280	$(1, 10^{-5})$	3.09 ± 0.02
	LP-SVAE	280	$(1, 10^{-5})$	3.89 ± 0.02

Statistics Distribution Evaluation. The statistic distributions of specific sensitive attributes for the original and synthetic sample are demonstrated in Figs. 2 and 3, based on the sample label in the CelebA, ECG, Speech using original samples, samples derived from O-SVAE without perturbation, samples derived from LP-SVAE and samples derived from SP-SVAE under various privacy-preserving level perturbation. We find that the statistics results for each label using SVAE with the low-privacy setting are more realistic than regular SVAE under the same setting. And the performances of LP-SVAE in all cases are better than SP-SVAE, which is similar to O-SVAE.

Machine Learning Score Similarity Classification. The classification score similarity is demonstrated in Fig. 4. Several conventional classification algorithms, e.g., decision tree, random forest, and multilayer perception classifiers, are conducted on the original dataset and synthetic datasets from SVAE models

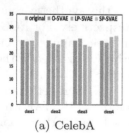
(a) CelebA

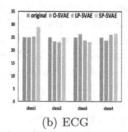
(b) ECG

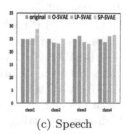
(c) Speech

Fig. 2. Label distribution of the CelebA, ECG and Speech datasets derived from O-SVAE, LP-SVAE and SP-SVAE with low privacy-preserving setting $\epsilon = 10, \delta \leq 10^{-5}$.

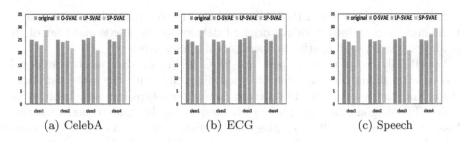

(a) CelebA   (b) ECG   (c) Speech

Fig. 3. Label distribution of the CelebA, ECG and Speech datasets derived from O-SVAE, LP-SVAE and SP-SVAE with high privacy-preserving setting $\epsilon = 1, \delta \leq 10^{-5}$.

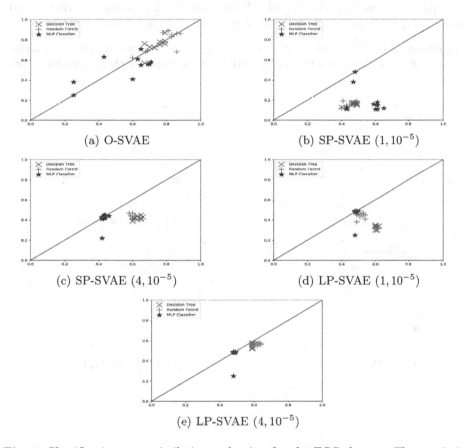

(a) O-SVAE   (b) SP-SVAE $(1, 10^{-5})$

(c) SP-SVAE $(4, 10^{-5})$ (d) LP-SVAE $(1, 10^{-5})$

(e) LP-SVAE $(4, 10^{-5})$

Fig. 4. Classification score similarity evaluation for the ECG dataset. The x-axis is F-1 score of the classifiers trained with the original dataset and y-axis is that of the classifiers trained with synthesized samples.

(O-SVAE, SP-SVAE, and LP-SVAE) respectively. We then plot the F-1 score of the classifiers trained with the original dataset as the x axis and that of the classifiers trained with synthesized samples as the y axis.

Specifically, we plot (x, y) pairs for ECG in Fig. 4, where each pair is calculated after fixing the same classification algorithm with the same parameter setup. Note that we conduct 10 different parameter setups for each classification model. The diagonal line (i.e., x = y) is used as the baseline to denote perfect classification compatibility, i.e., the classification score similarity of the model trained on synthetic samples is the same as that of the model trained on the original samples. Figures 4(a–e) displays the F-1 score similarity in the ECG dataset using O-SVAE, SP-SVAE, and LP-SVAE with low and high privacy-preserving level perturbation respectively. LP-SVAE with the low-privacy setting in (e) shows good F-1 score similarity with very small differences to O-SVAE.

Regression. In Fig. 5, we show the results of the regression score similarity evaluation. The mean relative error (MRE) is adopted as the regression score

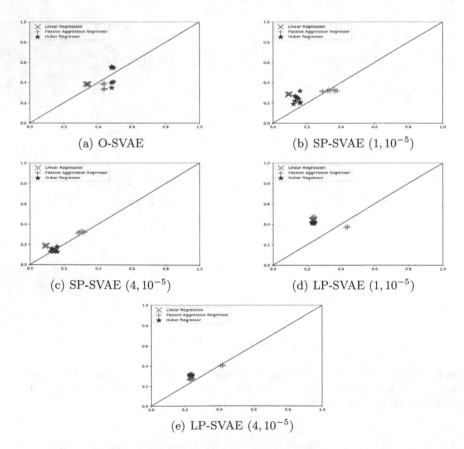

(a) O-SVAE

(b) SP-SVAE $(1, 10^{-5})$

(c) SP-SVAE $(4, 10^{-5})$

(d) LP-SVAE $(1, 10^{-5})$

(e) LP-SVAE $(4, 10^{-5})$

Fig. 5. Regression score similarity evaluation for the ECG dataset. The x axis is F-1 score of the classifiers trained with the original dataset and y axis is that of the classifiers trained with synthesized samples.

similarity metric. We also conduct 10 different parameter setups for linear regression, Lasso regression, passive-aggressive regression, Huber regression models. The diagonal line (i.e., x = y) is used as the baseline to denote perfect regression compatibility, i.e., the regression score similarity of the model trained on synthetic samples is the same as that of the model trained on the original samples. Figures 5(a–e) display the MSE-based regression score similarity in the ECG dataset using O-SVAE, SP-SVAE, and LP-SVAE with low and high privacy-preserving level perturbation respectively. LP-SVAE with the low-privacy setting in (e) shows good MSE-based regression score similarity with very small differences to O-SVAE (a).

Time-Based Evaluation. At last, the efficiency of LP-SVAE is evaluated, compared with SP-SVAE, as shown in Table 2. We first evaluate the latent code perturbation on the number of allowed iterations given the same privacy constraints, compared with SGD-based perturbation during training. We illustrate the maximum number of iterations under the same privacy constraint. The number of allowed iterations is significantly increased by LP-SVAE, resulting in improved utility in the generators.

Table 2. Effect on allowed iterations #.

Itereations	# SP-SVAE	# LP-SVAE
CelebA	100000	180000
ECG	3300	4500
Speech	15000	22000

3.4 Privacy Evaluation

Distance-Based Privacy Gain. We evaluate the proposed and baseline methods for privacy using the re-identification ratio, distance-based privacy-gain and membership attack, as shown in Table 3. The average and standard deviation of distances of (x, x') pairs as well as the re-identification ratio are shown in Tables 3, where x is an original record and x' is the corresponding sample in terms of x via O-SVAE/ LP-SVAE/SP-SVAE. Ideally, the average distance is large, and the standard division is small, i.e. x and x' are very close but the identification is hard to be determined. O-SVAE, LP-SVAE and SP-VAE show very stable average and standard deviation values. In contrast, LP-SVAE with the high-privacy setting shows longer average distance values (i.e. substantially lower probabilities of privacy leakage) than SP-SVAE. As expected, it is almost impossible to re-identity original values from synthetic values.

Table 3. Distance-based privacy gain (Distance) and re-identification ratio (RR)

Distance	O-SVAE	SP (low)	SP (high)	LP (low)	LP (high)
CelebA	1.24 ± 0.31	2.85 ± 0.43	3.56 ± 0.52	2.53 ± 0.42	2.78 ± 0.41
ECG	0.53 ± 0.12	0.88 ± 0.24	0.96 ± 0.32	0.76 ± 0.19	1.93 ± 0.22
Speech	0.96 ± 0.42	1.46 ± 0.32	1.58 ± 0.41	1.22 ± 0.22	1.26 ± 0.28
RR%	O-SVAE	SP (low)	SP (high)	LP (low)	LP (high)
CelebA	15	6	3	5	0
ECG	11	4	2	3	0
Speech	9	2	1	1	0

Table 4. Membership attack resistibility evaluations

F-1	O-SVAE	LP-SVAE (low)	LP-SVAE (high)	SP-SVAE (low)	SP-SVAE (high)
CelebA	0.62	0.45	0.32	0.59	0.43
ECG	0.42	0.32	0.24	0.36	0.34
Speech	0.58	0.41	0.21	0.50	0.42
AUC-ROC	O-SVAE	LP-SVAE (low)	LP-SVAE (high)	SP-SVAE(low)	SP-SVAE (high)
CelebA	0.71	0.58	0.46	0.64	0.62
ECG	0.60	0.5	0.45	0.48	0.51
Speech	0.58	0.5	0.5	0.49	0.52

Membership Attack Resistance. The attack performance, averaged over three classes, is illustrated in Table 4. Overall, the low-privacy setting allows some information leakage to the attacker, resulting in higher the F-1 and AUC-ROC values. As privacy-preserving level increases, the attack performance decreases. In these three datasets, F-1 is dropped after implementing LP-SVAE, which means unsuccessful membership attacks. The attack performance is decreased by 12% in the high-privacy configuration, compared to that in the low-privacy configuration, in all O-SVAE, LP-SVAE and SP-SVAE. For instance, F-1 of Speech is dropped from 0.58 to 0.21, which means unsuccessful membership attacks. In many other cases, the attack performance is decreased by 10% in the high-privacy configuration, compared to that in the low-privacy configuration.

4   Conclusion

In this paper, we proposed SVAE, a generic scheme of publishing to synthesize data in a practical privacy-preserving manner. Instead of releasing sanitized datasets, SVAE establishes differentially private generative models, which can be used by analysts to synthesize an unlimited amount of data for identification-irrelevant analysis tasks. SVAE integrates the VAE framework with identification sanitizer to strip identification-relevant features and applies differential privacy mechanisms to protect the privacy data generation. Additionally, we employ a suite of optimization strategies to address the utility and training stability challenges. To our knowledge, our method is the first attempt to synthesize and sanitized the realistic real-world image and time-series databases in a practical differential privacy-preserving manner, combining the utility and stability optimizations of privacy-preserving generation into a single task. In the future, we plan enhance the generation quality, and extend our method to more complex synthetic sample scenarios.

References

1. Abadi, M., et al.: Deep learning with differential privacy. In: Proceedings of the 2016 ACM SIGSAC Conference on Computer and Communications Security, pp. 308–318. ACM (2016)
2. Abdulkader, S.N., Atia, A., Mostafa, M.S.M.: Brain computer interfacing: applications and challenges. Egypt. Inf. J. 16(2), 213–230 (2015)
3. Alyasseri, Z.A.A., Khader, A.T., Al-Betar, M.A., Papa, J.P., Alomari, O.A.: Eeg feature extraction for person identification using wavelet decomposition and multi-objective flower pollination algorithm. IEEE Access 6, 76007–76024 (2018)
4. Arjovsky, M., Chintala, S., Bottou, L.: Wasserstein gan. arXiv preprint arXiv:1701.07875 (2017)
5. Beaulieu-Jones, B.K., Wu, Z.S., Williams, C., Greene, C.S.: Privacy-preserving generative deep neural networks support clinical data sharing. BioRxiv, p. 159756 (2017)
6. Chen, X., Duan, Y., Houthooft, R., Schulman, J., Sutskever, I., Abbeel, P.: Infogan: interpretable representation learning by information maximizing generative adversarial nets. In: Advances in Neural Information Processing Systems, pp. 2172–2180 (2016)
7. Cynthia, D.: Differential privacy. Automata, languages and programming, pp. 1–12 (2006)
8. Esteban, C., Hyland, S.L., Rätsch, G.: Real-valued (medical) time series generation with recurrent conditional gans. arXiv preprint arXiv:1706.02633 (2017)
9. Garofolo, J.S., Lamel, L.F., Fisher, W.M., Fiscus, J.G., Pallett, D.S.: Darpa timit acoustic-phonetic continous speech corpus cd-rom. nist speech disc 1–1.1. NASA STI/Recon technical report n 93 (1993)
10. Gatys, L.A., Ecker, A.S., Bethge, M.: A neural algorithm of artistic style. arXiv preprint arXiv:1508.06576 (2015)
11. Goldberger, A.L., et al.: Physiobank, physiotoolkit, and physionet: components of a new research resource for complex physiologic signals. Circulation 101(23), e215–e220 (2000)
12. Guibas, J.T., Virdi, T.S., Li, P.S.: Synthetic medical images from dual generative adversarial networks. arXiv preprint arXiv:1709.01872 (2017)
13. Kingma, D.P., Welling, M.: Auto-encoding variational bayes. arXiv preprint arXiv:1312.6114 (2013)
14. Kumari, P., Vaish, A.: Brainwave based authentication system: research issues and challenges. Int. J. Comput. Eng. Appl. 4(1), 2 (2014)
15. Larsen, A.B.L., Sønderby, S.K., Larochelle, H., Winther, O.: Autoencoding beyond pixels using a learned similarity metric. arXiv preprint arXiv:1512.09300 (2015)
16. Li, Y., Swersky, K., Zemel, R.: Generative moment matching networks. In: International Conference on Machine Learning, pp. 1718–1727 (2015)

17. Liu, Z., Luo, P., Wang, X., Tang, X.: Deep learning face attributes in the wild. In: Proceedings of the IEEE International Conference on Computer Vision, pp. 3730–3738 (2015)
18. Lyu, L., et al.: Towards fair and privacy-preserving federated deep models. IEEE Trans. Parallel Distrib. Syst. 31(11), 2524–2541 (2020)
19. Makhzani, A., Shlens, J., Jaitly, N., Goodfellow, I., Frey, B.: Adversarial autoencoders. arXiv preprint arXiv:1511.05644 (2015)
20. Mescheder, L., Nowozin, S., Geiger, A.: Adversarial variational bayes: unifying variational autoencoders and generative adversarial networks. arXiv preprint arXiv:1701.04722 (2017)
21. Salimans, T., Goodfellow, I., Zaremba, W., Cheung, V., Radford, A., Chen, X.: Improved techniques for training gans. In: Advances in Neural Information Processing Systems, pp. 2234–2242 (2016)
22. Schirrmeister, R.T., et al.: Deep learning with convolutional neural networks for EEG decoding and visualization. Hum. Brain Mapp. 38(11), 5391–5420 (2017)
23. Shokri, R., Stronati, M., Song, C., Shmatikov, V.: Membership inference attacks against machine learning models. In: 2017 IEEE Symposium on Security and Privacy (SP), pp. 3–18. IEEE (2017)
24. Song, S., Chaudhuri, K., Sarwate, A.D.: Stochastic gradient descent with differentially private updates. In: 2013 IEEE Global Conference on Signal and Information Processing (GlobalSIP), pp. 245–248. IEEE (2013)
25. Xie, L., Lin, K., Wang, S., Wang, F., Zhou, J.: Differentially private generative adversarial network. arXiv preprint arXiv:1802.06739 (2018)
26. Zhang, X., Ji, S., Wang, T.: Differentially private releasing via deep generative model. arXiv preprint arXiv:1801.01594 (2018)
27. Zue, V., Seneff, S., Glass, J.: Speech database development at mit: Timit and beyond. Speech Commun. 9(4), 351–356 (1990)

Robust Blockchain-Based Cross-Platform Audio Copyright Protection System Using Content-Based Fingerprint

Juan Zhao$^{(\boxtimes)}$, Tianrui Zong$^{(\boxtimes)}$, Yong Xiang , Longxiang Gao ,
and Gleb Beliakov

Deakin University, Burwood, VIC 3125, Australia
{zhaojua,tianrui.zong,yong.xiang,longxiang.gao,
gleb.beliakov}@deakin.edu.au

Abstract. Copyright infringement is a serious problem in the digital media industry. The traditional watermarking methods can neither protect copyright against attacks, especially the de-synchronization attacks, nor achieve cross-platform copyright protection. The existing blockchain-based methods have the potential to protect copyright across multiple online platforms. However, they are vulnerable to attacks. In this paper, we propose a novel Ethereum blockchain-based robust copyright protection system (RobustCPS) for audio signals, which can not only resist attacks but also achieve cross-platform copyright protection. In the proposed RobustCPS, the host audio signal is first segmented into blocks. Then the singular value decomposition (SVD) is performed on each block to generate the content-based fingerprint. After that, a database query is executed to search whether a similar fingerprint has existed in the Ethereum blockchain. If so, the rightsholders will be informed that their copyright might have been violated. Otherwise, the generated fingerprint will be stored in the Ethereum blockchain for future copyright protection actions. Simulation results demonstrate the superiority of the proposed RobustCPS.

Keywords: Audio signal · Fingerprinting system · Singular value decomposition · De-synchronization attacks · Ethereum blockchain

1 Introduction

The rapid development of the Internet and multimedia technology facilitates the easy distribution of multimedia files, which may seriously violate the interests and rights of the rightsholder. According to the IFPI digital music report [1] released by the recording industry association of America (RIAA) in 2015, 20% of worldwide fixed-line users regularly accessed the services offering pirated music

This work was supported in part by the Australian Research Council under grant LP170100458.

and 4 billion songs were downloaded by BitTorrent alone in 2014. Besides, Game of Thrones (Season 7) was pirated more than 1 billion times [2] within 7 weeks since it was released on 16 July 2017. Since copyright violation has caused huge economic loss, there is a strong demand to protect the copyright of multimedia.

Digital watermarking is a popular method to protect the copyright, which can be applied on cloud-based applications [3] and is applicable to image [4–6], video [7–9], and audio [10–12]. It embeds the watermark information (e.g., signature, ID, author information, logo, etc.) into the host signal. When necessary, the watermark can be extracted to protect the copyright. In this paper, we focus on copyright protection for audio files. Effective audio watermarking methods should exhibit traits such as high robustness, high embedding capacity, imperceptibility, and security. So far, many audio watermarking methods have been proposed. Cox *et al.* introduced a spread spectrum (SS) based watermarking method [13], which utilizes the correlation between the watermarked signal and the watermark sequence to extract the embedded watermark. However, the host signal interference (HSI) weakens the robustness of the method. To reduce the HSI, Malvar *et al.* proposed an improved spread spectrum (ISS) [14] based method to improve the robustness of the method in [13] by introducing a controlling factor. Xiang *et al.* [15] improved the embedding capacity of the SS-based methods by constructing a group of pseudonoise (PN) sequences to represent the watermark sequences rather than the watermark bit. Later, Xiang *et al.* [16] further reduced the computational complexity of [15] without compromising the embedding capacity.

The SS-based watermarking methods are robust against the common signal processing attacks, such as Gaussian noise addition, low-pass filtering, amplitude scaling, and compression. However, they are vulnerable to the de-synchronization attacks, such as jittering, pitch invariant time scaling, and time invariant pitch scaling. To tackle the de-synchronization attacks, Wang *et al.* proposed a blind audio watermarking algorithm based on support vector regression (SVR) and quantization index modulation (QIM) in which the extracted template features are adopted to locate the synchronization points [17]. Xiang *et al.* [18] proposed a method, without compromising the embedding capacity, to defend the de-synchronization attacks by embedding the synchronization codes in the logarithmic DCT (LDCT) domain of the watermarked signals. Liu *et al.* [19] presented a patchwork-based audio watermarking algorithm to resist the de-synchronization attacks by embedding the synchronization codes.

Although the aforementioned watermarking methods are designed to resist the de-synchronization attacks, their robustness is limited. Moreover, the existing watermarking methods are designed to protect the copyright within a single platform rather than multiple platforms. Blockchain technology, which emerges in recent years, has the potential to solve cross-platform copyright protection problem. However, the existing blockchain-based systems, such as DotBlockchain [20], Ujo [21], Bittunes [22], and Mycelia [23], cannot resist the common signal process attacks and the de-synchronization attacks.

In this paper, we propose a robust copyright protection system (RobustCPS) based on Ethereum blockchain, which achieves cross-platform copyright protection under the common signal processing attacks and the de-synchronization attacks. In the RobustCPS, the audio signal is firstly segmented into blocks. Then singular value decomposition (SVD) is performed on each block to generate the content-based fingerprint (CF). After that, a database query is executed to determine the result of the request. If there is a similar fingerprint in the Ethereum blockchain, a warning will be sent to the corresponding rightsholder and then the rightsholder can determine the result of the request, which can effectively prevent the copyright violation. Otherwise, the generated fingerprint will be stored in the Ethereum blockchain for future copyright protection actions. Furthermore, different from the traditional watermarking methods, there is no embedding process in the proposed RobustCPS. Therefore, the audio signal will not suffer from perceptual quality degradation.

The reminder of this paper is organized as follows. Section 2 demonstrates the proposed RobustCPS in detail. The robustness of the proposed RobustCPS is illustrated in Sect. 3. Finally, Sect. 4 concludes this paper.

2   The Proposed RobustCPS

The framework of the proposed RobustCPS is illustrated in Fig. 1. Firstly, the user sends an upload request r and the audio signal s to the server. Secondly, the server transmits s to the smart contract deployed on the Ethereum blockchain. Thirdly, the smart contract will be executed to determine the result of the request r. The result is determined by the output of the CF extraction and the similarity detection steps. In the CF extraction step, a fingerprint is extracted from s, which can uniquely prove the ownership of s. In the similarity detection step, a database query is executed to search whether there is a similar fingerprint existed in the Ethereum blockchain and then determine the result of r. Finally, the result will be returned to the server and the user.

2.1   CF Extraction

Assume the length of the audio signal s is L_s, then s can be expressed as

$$s = \{s(i)|1 \leq i \leq L_s\}, \tag{1}$$

where $s(i)$ represents the i-th sample of s. Split s into T blocks with equal length, and then the original audio signal s can be expressed as

$$s = [s_1, s_2, \cdots, s_T]. \tag{2}$$

Note the value of T controls the length of the CF. The i-th block of s can be denoted as

$$s_i = [s_i(1), s_i(2), \cdots, s_i(L)], \tag{3}$$

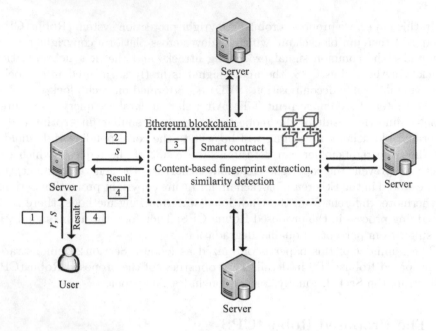

Fig. 1. The framework of the RobustCPS.

where $s_i(j)$ refers to the j-th sample in the i-th block, $1 \le i \le T$, and L is the length of each block, which can be calculated by

$$L = L_s/T. \tag{4}$$

Apply discrete cosine transform (DCT) [24] on s_i to obtain the corresponding DCT coefficients by

$$S_i(j) = l(j) \sum_{n=1}^{L} s_i(n)\cos\frac{\pi(2n-1)(j-1)}{2L}, \tag{5}$$

where $S_i(j)$ is the j-th DCT coefficient of s_i, $j = 1, 2, \ldots, L$, and

$$l(j) = \begin{cases} \frac{1}{\sqrt{L}}, & \text{if } j = 1; \\ \frac{2}{\sqrt{L}}, & \text{otherwise.} \end{cases} \tag{6}$$

Obviously, the DCT coefficients of s can be expressed as

$$S = [S_1, S_2, \cdots, S_T]. \tag{7}$$

Since the main energy of a signal concentrates on the low-frequency part, the low-frequency components of each block are chosen to perform the SVD. Set the boundary of the selected frequency coefficient as f. Then the selected low-frequency band $\bar{S}_i$ for the i-th block can be denoted as

$$\bar{S}_i = [S_i(1), S_i(2), \cdots, S_i(f)]. \tag{8}$$

Decompose $\bar{S}_i$ via SVD, then $\bar{S}_i$ can be expressed as

$$\bar{S}_i = U_i \Sigma_i V_i^T, \tag{9}$$

where U_i and V_i are unitary matrices and Σ_i is the diagonal matrix with singular values. We adopt Σ_i to extract the CF as the singular values in Σ_i are stable. Denote the maximum value in Σ_i as c_i. Then build C to accommodate these maximum values as

$$C = [c_1, c_2, \cdots, c_T]. \tag{10}$$

Calculate the average value of the elements in C by

$$\bar{c} = \text{avg}(C), \tag{11}$$

where $\text{avg}(\cdot)$ is the average operator and $\bar{c}$ is the average value of C. Then generate the content-based fingerprint H for s based on the following principle

$$H(i) = \begin{cases} 1, & \text{if } c_i \geq \bar{c}; \\ 0, & \text{otherwise}, \end{cases} \tag{12}$$

where $H(i)$ is the i-th element of the CF.

2.2 Similarity Detection

In the similarity detection step, H will be compared with the existing fingerprints stored in the Ethereum blockchain to determine the result of r. Assume at this moment, there are M fingerprints stored in the Ethereum blockchain. Denote Q with size $T \times M$ as the set of these existing fingerprints. The similarity $P(k)$ between H and the k-th existing fingerprint Q_k can be calculated by

$$P(k) = \frac{1}{T} \sum_{i=1}^{T} Q_k(i) \oplus H(i), \tag{13}$$

where $\oplus$ is the XOR operator, and $k = 1, 2, ..., M$. If $\exists\, j \in [1, 2, ..., M]$ satisfies

$$P(j) \leq \alpha, \tag{14}$$

where α is a predefined threshold, the copyright is considered as violated. In this case, a warning will be sent to the rightsholder of the j-th existing fingerprint and the result of r will be determined by the rightsholder of the j-th fingerprint. Otherwise, H will be stored in the Ethereum blockchain and r will be approved. A smaller value of α implies the determination is more strict and a larger value of α indicates the determination is looser. In this paper, we set $\alpha = 0.1$. Figure 2 shows the flowchart of CF extraction and similarity detection processes.

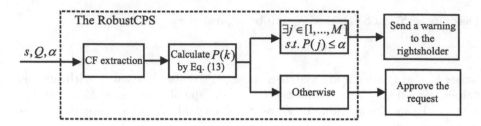

Fig. 2. The flowchart of CF extraction and similarity detection.

3   The Robustness of the CF

In the real-world application, the audio signals are easy to suffer from attacks. Denote the CF of the host signal after attacks as H'. Similar to Eq. (13), the similarity P' between H and H' can be formulated as

$$P' = \frac{1}{T} \sum_{i=1}^{T} H(i) \oplus H'(i). \tag{15}$$

Since the value of P' determines the result of request r, in this paper, the robustness the RobustCPS is evaluated by P'. A larger value of P' indicates less robustness against attacks while a smaller value of P' implies stronger robustness against attacks. All the simulations are conducted on Matlab R2018b via a computer featured a 2.11 GHz Intel Core i7 processor with 16 GB RAM. The test audio signals include 5 genres of music. Each genre contains 20 audio clips. The duration, sample rate, and resolution of the audio clips are 60 s, 44.1 kHz, and 16 bits, respectively. Table 1 shows the types of the test audio signals used in the simulation.

Table 1. The test audio signals used in the simulation.

Type	Genres	Host signals	Sample rate	Bits per sample	Duration
Type I	Claasic music	$s_{001} \sim s_{020}$	44.1 kHz	16 bits	60 s
Type II	Folk music	$s_{021} \sim s_{040}$			
Type III	Hip hop	$s_{041} \sim s_{060}$			
Type IV	Pop music	$s_{061} \sim s_{080}$			
Type V	Rock music	$s_{081} \sim s_{100}$			

3.1   The Selection of L

As mentioned in Sect. 2.1, the segment length L and the boundary of the selected DCT coefficients f are two parameters used to extract the CF of an audio signal. To guarantee the generated CF has resilience against the low-pass filtering attack, we set $f = L/3$. Figure 3 shows the relationship between L and P'. The values of P' in Fig. 3 is calculated by averaging the similarities between the CF of the host signal and the CFs of the host signals after the common signal processing attacks and the de-synchronization attacks. From Fig. 3, it is easy to observe that the robustness of the CF increases with the increase of L. One the other hand, a larger value of L influences the detection accuracy in the similarity detection step as a larger value of L implies a smaller T. To balance the robustness and the detection accuracy, in this paper we set $L = 6615$ in this paper.

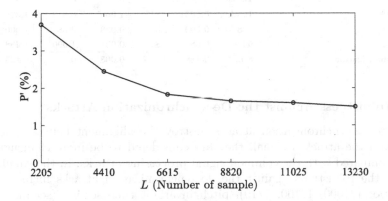

Fig. 3. The relationship between the L and the P'.

3.2   Robustness Against the Common Signal Processing Attacks

The common signal processing attacks include Gaussian noise addition, low passing filtering, amplitude scaling, ACC compression, MP3 compression, and requantization. Table 2 shows the performance of the RobustCPS under different common signal processing attacks. According to Table 2, the RobustCPS has great resilience against the Gaussian noise addition (signal-to-noise ratio (SNR): 0 dB, 10 dB, 20 dB) attacks. Even when SNR $= 0$ dB, the P' values are less than 3%. When the SNR is larger than 10 dB, the P' values are less than 1%. For the amplitude scaling attacks (scaling factor: 80%, 90%, 110%, 120%), they have no impact on the CF as the DCT transform is linear.

Since the RobustCPS extracts the CF by using the low-frequency DCT coefficients, the RobustCPS is robust against the low-pass filtering attack. Similarly, the RobustCPS is also resistance to the ACC and MP3 compression attacks, because both these two attacks keep the low-frequency part of the signal. Besides, the RobustCPS also has great resilience against the 8 bits requantization attack as P' is less than 0.2% for all types of audio signal.

Table 2. Performance ($P'(\%)$) of the RobustCPS under the common signal processing attacks.

Attacks		P' (%)				
		Type I	Type II	Type III	Type IV	Type V
Gaussian noise addition (SNR)	0 dB	1.500	1.563	2.625	1.813	1.438
	10 dB	0.063	0.438	0.625	0.750	0.563
	20 dB	0.063	0.125	0.063	0	0.125
Low-pass filtering	8 kHz	0	0	0	0	0
Amplitiude scaling	80%	0	0	0	0	0
	90%	0	0	0	0	0
	110%	0	0	0	0	0
	120%	0	0	0	0	0
AAC	128 kps	0.563	0.813	1.063	0.813	1.375
MP3	128 kps	0.500	1.063	0.938	0.688	1.000
	64 kps	0.688	1.813	0.938	1.000	1.438
Requantization	8 bits	0.188	0	0.063	0	0.125

3.3 Robustness Against the De-synchronization Attacks

Since the de-synchronization attacks destroy the alignment between the host signal and the attacked signal, they are considered to be more challenging to tackle compared with the common signal processing attacks. In the simulation, we test the robustness against the de-synchronization attacks include jittering (ratio: 1/1000, 1/100, 1/10), pitch invariant time scaling (scaling factor: 80%, 90%, 110%, 120%), and time invariant pitch scaling (scaling factor: 80%, 90%, 110%, 120%). Table 3 shows the performance of the RobustCPS under the de-synchronization attacks. Since the extraction of CF is based on the statistical attribute of the audio signal, the RobustCPS is robust against the de-synchronization attacks. According to Table 3, all of the P' values are less than 9% even though severe de-synchronization attacks are performed on the audio signal.

3.4 RobustCPS vs Robust Traditional Watermarking Method

Since the proposed RobustCPS focuses more on the robustness against attacks, we compare our system with a patchwork-based watermarking method in [19], which is one of the latest robust watermarking methods. In this comparison, we set the embedding capacity as 10 bps. The perceptual quality of the method in [19] is evaluated using the objective difference grade (ODG) [25], which ranges in the interval of $[-4, 0]$. When the ODG value is larger than -1, the distortion is imperceptible. Since our proposed RobustCPS does not include an embedding procedure, which means the perfect perceptual quality can be achieved, for a fair comparison, the performance of the method in [19] is evaluated under the ODG value of -0.5. Table 4 shows the performance of the proposed RobustCPS and the method in [19] under the common signal processing attacks and the de-synchronization attacks. According to Table 4, our proposed RobustCPS is more

Table 3. Performance $(P'(\%))$ of the RobustCPS under the de-synchronization attacks.

Attacks		$P'(\%)$				
		Type I	Type II	Type III	Type IV	Type V
Jittering	1/1000	0.313	0.563	1.000	0.438	0.438
	1/100	0.438	1.063	1.250	0.563	0.500
	1/10	0.563	1.375	1.625	0.938	0.750
Pitch invariant time scaling	80%	1.125	4.125	2.875	1.313	2.313
	90%	0.938	3.000	2.500	1.250	1.938
	110%	0.750	2.688	1.813	1.313	1.813
	120%	0.813	3.875	2.750	1.563	1.938
Time invariant pitch scaling	80%	4.375	8.563	8.688	4.750	7.500
	90%	1.625	7.688	7.875	4.625	6.250
	110%	1.813	3.250	2.250	4.375	2.313
	120%	4.818	8.063	7.688	2.000	6.313

Table 4. Performance comparison $(P'(\%))$ of the RobustCPS with the method in [19] under common signal processing attacks and the de-synchronization attacks.

Attacks		$P'(\%)$	
		The method in [19]	The proposed RobustCPS
Gaussian noise addition (SNR)	0 dB	39.667	1.788
	10 dB	31.000	0.488
	20 dB	22.500	0.075
Low-pass filtering	8 kHz	40.667	0
Amplitiude scaling	80%	0	0
	90%	0	0
	110%	0	0
	120%	0	0
AAC	128 kps	0	0.925
MP3	128 kps	4.636	0.838
	64 kps	8.167	1.175
Requantization	8 bits	9.833	0.075
Jittering	1/1000	5.967	0.550
	1/100	6.167	0.763
	1/10	7.695	1.050
Pitch invariant time scaling	80%	10.898	2.350
	90%	8.950	1.925
	110%	7.959	1.675
	120%	11.564	2.188
Time invariant pitch scaling	80%	12.858	6.775
	90%	8.565	2.613
	110%	8.959	2.800
	120%	12.564	5.776

robust under the common signal processing attacks and the de-synchronization attacks compared with the method in [19].

3.5   RobustCPS vs Blockchain-Based Copyrigtht Protection Systems

In this section, we compare our proposed RobustCPS with two existing blockchain-based systems, which are the DotBlockchain system in [20] and the Ujo system in [21], from three aspects, including the applicability to cross-platform, the robustness against attacks, and no unique file format. The comparison results are shown in Table 5. The DotBlockchain, Ujo as well as our proposed RobustCPS are applicable to multiple platforms as all of them utilize the blockchain technology. However, the DotBlockchain and the Ujo become ineffective under attacks. As for the proposed RobustCPS, it is resistant to attacks, which is mentioned in Sect. 3.2 and Sect. 3.3. Furthermore, the DotBlockchain requires a unique file format, which further restricts its usage.

Table 5. The effectiveness of the RobustCPS compared with the DotBlockchain system in [20] and the Ujo system in [21].

	Applicability to cross-platform	Robust against the attacks	No unique file format requirement
DotBlockchain in [20]	√	×	×
Ujo in [21]	√	×	√
The proposed RobustCPS	√	√	√

4   Conclusions

In this paper, we present an Ethereum blockchain-based copyright protection system called RobustCPS, which achieves multiple platforms copyright protection under the common signal processing attacks and the de-synchronization attacks. Compared with the traditional watermarking methods designed for a single platform copyright protection, our proposed RobustCPS is more robust and achieves no perceptual quality degradation as there is no embedding process. Compared with the existing blockchain-based systems, the proposed RobustCPS is also superior as none of the existing blockchain-based systems are effective under attacks.

References

1. IFPI Digital music report (2015). https://www.riaa.com/wp-content/uploads/2015/09/Digital-Music-Report-2015.pdf. Accessed 11 May 2020
2. 'Game of Thrones' season 7 was pirated more than 1 billion times. https://www.businessinsider.com.au/game-of-thrones-season-7-pirated-1-billion-times-torrents-streaming-muso-2017-9. Accessed 11 May 2020

3. Kabir, E., Mahmood, A., Wang, H., Mustafa, A.: Microaggregation sorting framework for k-anonymity statistical disclosure control in cloud computing. IEEE Trans. Cloud Comput. **8**(2), 408–417 (2015)
4. Zong, T., Xiang, Y., Guo, S., Rong, Y.: Rank-based image watermarking method with high embedding capacity and robustness. IEEE Access **4**, 1689–1699 (2016)
5. Zong, T., Xiang, Y., Elbadry, S., Nahavandi, S.: Modified moment-based image watermarking method robust to cropping attack. Int. J. Autom. Comput. **13**(3), 259–267 (2016). https://doi.org/10.1007/s11633-015-0926-6
6. Zong, T., Xiang, Y., Natgunanathan, I., Guo, S., Zhou, W., Beliakov, G.: Robust histogram shape-based method for image watermarking. IEEE Trans. Circuits Syst. Video Technol. **25**(5), 717–729 (2014)
7. Asikuzzaman, M., Alam, M.-J., Lambert, A.-J., Pickering, M.-R.: Robust DT CWT-based DIBR 3D video watermarking using chrominance embedding. IEEE Trans. Multimedia **18**(9), 1733–1748 (2016)
8. Joshi, A.M., Gupta, S., Girdhar, M., Agarwal, P., Sarker, R.: Combined DWT–DCT-based video watermarking algorithm using arnold transform technique. In: Satapathy, S.C., Bhateja, V., Joshi, A. (eds.) Proceedings of the International Conference on Data Engineering and Communication Technology. AISC, vol. 468, pp. 455–463. Springer, Singapore (2017). https://doi.org/10.1007/978-981-10-1675-2_45
9. Mareen, H., Praeter, J.-D., Wallendael, G.-V., Lambert, P.: A novel video watermarking approach based on implicit distortions. IEEE Trans. Consum. Electron. **64**(3), 250–258 (2018)
10. Xiang, Y., Peng, D., Natgunanathan, I., Zhou, W.: Effective pseudonoise sequence and decoding function for imperceptibility and robustness enhancement in time-spread echo-based audio watermarking. IEEE Trans. Multimedia **13**(1), 2–13 (2010)
11. Xiang, Y., Natgunanathan, I., Peng, D., Zhou, W., Yu, S.: A dual-channel time-spread echo method for audio watermarking. IEEE Trans. Inf. Forensics Secur. **7**(2), 383–392 (2011)
12. Natgunanathan, I., Xiang, Y., Hua, G., Beliakov, G., Yearwood, J.: Patchwork-based multilayer audio watermarking. IEEE/ACM Trans. Audio Speech Lang. Process. **25**(11), 2176–2187 (2017)
13. Cox, I.J., Kilian, J., Leighton, F.T., Shamoon, T.: Secure spread spectrum watermarking for multimedia. IEEE Trans. Image Process. **6**(12), 1673–1687 (1997)
14. Malvar, H.S., Florêncio, D.A.: Improved spread spectrum: a new modulation technique for robust watermarking. IEEE Trans. Signal Process. **51**(4), 898–905 (2003)
15. Xiang, Y., Natgunanathan, I., Rong, Y., Guo, S.: Spread spectrum-based high embedding capacity watermarking method for audio signals. IEEE/ACM Trans. Audio Speech Lang. Process. **23**(12), 2228–2237 (2015)
16. Xiang, Y., Natgunanathan, I., Peng, D., Hua, G., Liu, B.: Spread spectrum audio watermarking using multiple orthogonal PN sequences and variable embedding strengths and polarities. IEEE/ACM Trans. Audio Speech Lang. Process. **26**(3), 529–539 (2017)
17. Wang, X., Qi, W., Niu, P.: A new adaptive digital audio watermarking based on support vector regression. IEEE Trans. Audio Speech Lang. Process. **15**(8), 2270–2277 (2007)
18. Xiang, Y., Natgunanathan, I., Guo, S., Zhou, W., Nahavandi, S.: Patchwork-based audio watermarking method robust to de-synchronization attacks. IEEE/ACM Trans. Audio Speech Lang. Process. **22**(9), 1413–1423 (2014)

19. Liu, Z., Huang, Y., Huang, J.: Patchwork-based audio watermarking robust against de-synchronization and recapturing attacks. IEEE Trans. Inf. Forensics Secur. **14**(5), 1171–1180 (2018)
20. Ambili, K.N., Sindhu, M., Sethumadhavan, M.: On federated and proof of validation based consensus algorithms in blockchain. In: IOP Conference Series: Materials Science and Engineering, vol. 225, p. 012198. IOP Publishing, India (2017). https://doi.org/10.1088/1757-899x/225/1/012198
21. The blockchain phenomenon-the disruptive potential of distributed consensus architectures. https://www.etla.fi/wp-content/uploads/ETLA-Working-Papers-38.pdf. Accessed 11 May 2020
22. Bittunes homepage. http://bittunes.co.uk/. Accessed 11 May 2020
23. Mycelia for music. http://myceliaformusic.org/. Accessed 11 May 2020
24. Wu, J.L., Shin, J.: Discrete cosine transform in error control coding. IEEE Trans. Commun. **43**(5), 1857–1861 (1995)
25. Thiede, T., et al.: PEAQ-the ITU standard for objective measurement of perceived audio quality. J. Audio Eng. Soc. **48**(1), 3–29 (2000)

Distributed Differential Evolution for Anonymity-Driven Vertical Fragmentation in Outsourced Data Storage

Yong-Feng Ge[1], Jinli Cao[1], Hua Wang[2], Yanchun Zhang[2],
and Zhenxiang Chen[3]([✉])

[1] Department of Computer Science and Information Technology,
La Trobe University, Melbourne 3083, Australia
[2] Institute for Sustainable Industries and Liveable Cities, University of Victoria,
Melbourne 3011, Australia
[3] School of Information Science and Engineering, University of Jinan,
Jinan 250022, China
czx@ujn.edu.cn

Abstract. Vertical fragmentation is a promising technique for outsourced data storage. It can protect data privacy while conserving original data without any transformation. Previous vertical fragmentation approaches need to predefine sensitive associations in data as the optimization objective, therefore unavailable for the data lacking related prior knowledge. Inspired by the anonymity measurement in anonymity approaches such as k-anonymity, an anonymity-driven vertical fragmentation problem is defined in this paper. To tackle this problem, a set-based distributed differential evolution (S-DDE) algorithm is proposed. An island model containing four sub-populations is adopted to improve population diversity and search efficiency. Two set-based update operators, i.e., set-based mutation operator and set-based crossover operator, are designed to transfer the calculation of discrete values to corresponding sets in vertical fragmentation. Extensive experiments are carried out, and the performance of S-DDE on anonymity-driven vertical fragmentation is verified. The computation efficiency of S-DDE is investigated, and the effectiveness of the generated vertical fragmentation solution by S-DDE is confirmed.

Keywords: Data security and privacy · Vertical fragmentation · Distributed differential evolution

1 Introduction

Due to the advantages of high scalability and low cost, outsourced data storage is widely adopted by web-based applications [7,8,22,23]. Meanwhile, when adopting outsourced data storage, data security and privacy are big concerns

© Springer Nature Switzerland AG 2020
Z. Huang et al. (Eds.): WISE 2020, LNCS 12343, pp. 213–226, 2020.
https://doi.org/10.1007/978-3-030-62008-0_15

[13, 17, 18, 20]. Encryption techniques [6] can protect the sensitive information of web-based applications from releasing. However, encryption of data reduces data query efficiency and distorts the patterns in original data [10, 15, 21]. Separate data models based on fragmentation can split the data of sensitive associations into multiple fragments and conserve the original data patterns.

In the previous literature, various approaches have been proposed for vertical fragmentation. To break sensitive association between data attributes, encryption-based vertical fragmentation [2] was proposed. In [3], the optimization of vertical fragmentation was identified as NP-hard. Two greedy heuristic approaches were proposed to optimize the number of fragments. Subsequently, a graph search approach [19] was designed to satisfy confidentiality constraints. Although these approaches are effective in vertical fragmentation with predefined constraints, they cannot be applied to the data that lacks prior knowledge about the sensitive association between attributes. With the same objective of privacy protection, the performance of anonymity approaches [9, 14] can be measured by the value of anonymity degree. Accordingly, the performance of vertical fragmentation can also be evaluated by anonymity degree. Unlike sensitive associations and constraints in vertical fragmentation, anonymity degree can be directly calculated and do not need prior knowledge or association mining. In this way, if the anonymity degree is set as the optimization objective, vertical fragmentation can be automatically optimized. Based on the above consideration, a problem of anonymity-driven vertical fragmentation is defined.

Besides the predefined optimization objective, previous vertical fragmentation approaches have also shown a disadvantage in search efficiency. Heuristic greedy strategy and graph search algorithm are very likely to get trapped by local optimum in problems of complex search space. In the last decades, various differential evolution (DE) algorithms were proposed [5, 11, 12] and shown efficiency and robustness in complex optimization problems such as protein structure prediction [25], microwave circuit design [24], and seismic inversion [4]. Unlike the heuristic greedy approach and graph search algorithm, population-based DE is more likely to achieve the trade-off between explorative search and exploitative search. Considering the effectiveness of DE in these applications, it is worthy of applying DE in vertical fragmentation and investigating its performance.

To tackle the anonymity-driven vertical fragmentation problem, a set-based distributed differential evolution algorithm (S-DDE) is proposed. In S-DDE, each vertical fragmentation is indicated by an individual, and the fitness value is set as the anonymity degree of each vertical fragmentation. The update of individuals and increase of fitness value reflects the enhancement of anonymity degree in vertical fragmentation. In S-DDE, an island model containing four islands is utilized to improve the population diversity as well as enhance the search efficiency of DE. Two set-based operators, i.e., set-based mutation and set-based crossover, are proposed to transfer the calculation of discrete values to the corresponding sets in vertical fragmentation. According to the experimental results, the optimization performance of S-DDE is verified, and the vertical fragmentation result of S-DDE is effective in actual public datasets.

The rest of this paper is organized as follows. In Sect. 2, related work of vertical fragmentation for privacy protection is reviewed. Afterward, a brief description of operators and related strategies in DE are outlined. Subsequently, the proposed S-DDE, including an island mode, set-based mutation, and set-based crossover, are illustrated in Sect. 4. Extensive experiments and the corresponding discussion are given in Sect. 5. Finally, Sect. 6 concludes.

2   Related Work

In [1], vertical fragmentation for data privacy was firstly introduced, in which attributes are divided into two groups. After that, encryption embedded database fragmentation was utilized in [2], which is effective in breaking the sensitive association between attributes. Authors in [3] proved that the vertical fragmentation problem is NP-hard. Moreover, two heuristic strategies were designed to optimize the number of fragments. A graph search approach was designed in [19] to achieve optimal fragmentation according to the given confidentiality constraints. In [16], loose association between fragments in vertical fragmentation was investigated. In these approaches, vertical fragmentation is designed according to the predefined sensitive association between attributes and additional constraints. When facing a data fragmentation problem that lacks prior knowledge, these approaches cannot be directly applied.

To protect data privacy from being released, various anonymity approaches including k-anonymity [14], l-diversity [9] have been proposed. In these approaches, the performance of privacy protection can be measured by anonymity degree. This kind of measure can also be applied in the fragmentation problem and indicate the performance of vertical fragmentation in privacy protection.

3   Differential Evolution (DE)

To update individuals, DE contains three operators, i.e., mutation, crossover, and selection. An elementary description of these three operators is given as follows.

Mutation: In mutation, evolution information of different individuals is exchanged and construct mutation individuals. At generation g, the ith mutant individual $\mathbf{v}_i^g$ is generated. Frequent mutation strategies are outlined as follows:

DE/rand/1

$$\mathbf{v}_i^g = \mathbf{x}_{r1}^g + F \cdot (\mathbf{x}_{r2}^g - \mathbf{x}_{r3}^g) \tag{1}$$

DE/current-to-best/1

$$\mathbf{v}_i^g = \mathbf{x}_i^g + F \cdot (\mathbf{x}_{best}^g - \mathbf{x}_i^g) + F \cdot (\mathbf{x}_{r1}^g - \mathbf{x}_{r2}^g) \tag{2}$$

DE/best/1

$$\mathbf{v}_i^g = \mathbf{x}_{best}^g + F \cdot (\mathbf{x}_{r1}^g - \mathbf{x}_{r2}^g) \tag{3}$$

DE/best/2

$$\mathbf{v}_i^g = \mathbf{x}_{best}^g + F \cdot (\mathbf{x}_{r1}^g - \mathbf{x}_{r2}^g) + F \cdot (\mathbf{x}_{r3}^g - \mathbf{x}_{r4}^g) \tag{4}$$

DE/rand/2

$$\mathbf{v}_i^g = \mathbf{x}_{r1}^g + F \cdot (\mathbf{x}_{r2}^g - \mathbf{x}_{r3}^g) + F \cdot (\mathbf{x}_{r4}^g - \mathbf{x}_{r5}^g) \tag{5}$$

where $\mathbf{x}_{best}^g$ represents the best individual in current population; $r1$, $r2$, $r3$, $r4$, and $r5$ are indexes of five randomly chosen individuals; F indicates the differential factor.

Crossover: Crossover operator is helpful in maintaining the population diversity. The mutant individual $\mathbf{v}_i^g$ is combined with the current individual $\mathbf{x}_i^g$ and generate a trial individual $\mathbf{u}_i^g$. This process is formulated as:

$$\mathbf{u}_{i,j}^g = \begin{cases} \mathbf{v}_{i,j}^g, & if\ rand(0,1) \leq Cr\ or\ j = j_{rand} \\ \mathbf{x}_{i,j}^g, & otherwise \end{cases} \tag{6}$$

where $rand(0,1)$ is a random float number between 0 and 1; j_{rand} indicates a random integer to guarantee at least one bit of $\mathbf{u}_i^g$ is different from $\mathbf{x}_i^g$; Cr represents the crossover rate, which determines the ratio of components from mutant individual $\mathbf{v}_i^g$.

Selection: Selection is carried out compare the fitness of target individual and the generated trail individual by mutation and crossover operators. For a minimization problem $f(x)$, the selection process is expressed as follows:

$$\mathbf{x}_i^{g+1} = \begin{cases} \mathbf{u}_i^g, & if\ f(\mathbf{u}_i^g) \leq f(\mathbf{x}_i^g) \\ \mathbf{x}_i^g, & otherwise \end{cases} \tag{7}$$

where $f(x)$ is the objective function of optimization problem and $\mathbf{x}_i^{g+1}$ represents the target individual in the next generation.

In DE, each solution for vertical fragmentation can be represented by an individual. The update of individuals in DE is based on the fitness of individuals. In this problem, the fitness of an individual is defined as the anonymity degree of vertical fragmentation. In this way, the update of individuals in DE can help obtain better solutions in anonymity-driven vertical fragmentation.

4 Set-Based Distributed Differential Evolution (S-DDE)

In this section, the island model of distributed differential evolution (DDE) is firstly described in detail. Then, two set-based DE operators, i.e., set-based mutation and set-based crossover, are introduced. Finally, the overall process of proposed S-DDE is illustrated.

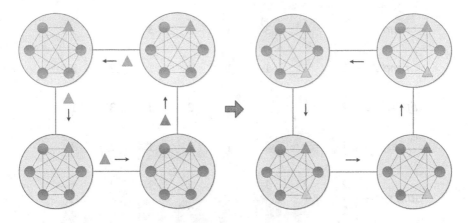

Fig. 1. Illustration of island model.

4.1 Island Model

Direct distributed implementation is one advantage of DE. At the beginning of DDE, the entire population is uniformly divided into multiple sub-populations, which are also referred as islands. Afterward, each island evolves independently. With a given interval, elite individuals of sub-populations are migrated according to the topology of islands. Each sub-population sends one or more elite individuals to its neighborhood on the topology and receives migrated individuals from the neighborhood. One or more individuals in the sub-population will be then replaced by the migrated individuals.

As shown in Fig. 1, an island model containing four sub-populations is given. Each big circle represents a sub-population. Small triangles and circles in the sub-population indicate the best individuals and other individuals, respectively. In this example, a unidirectional ring topology is plotted. With a given frequency, migration between islands is carried out. The best individual in each sub-population is sent to the neighbor sub-population. When receiving the best individual from the neighbor island, an individual in the current sub-population will be chosen by random and replaced.

The island model can enhance the performance of DE from two angles. On the one hand, the population diversity of DE is maintained in the separate sub-populations. Migration between sub-populations can help exchange evolutionary information in different sub-populations. In this way, a trade-off between exploration and exploitation is achieved by the island model. On the other hand, since each island evolves independently and only individuals are exchanged between sub-populations. DE based on the island model can be easily implemented on a distributed computation platform and obtain ideal computation efficiency.

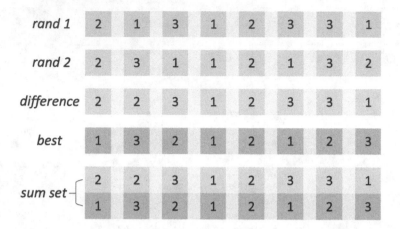

Fig. 2. An example of set-based mutation operator. (Color figure online)

4.2 Set-Based Mutation

In the original mutation of DE, the calculation of values on each bit can be transferred to the corresponding vector. For the values of discrete values, this kind of transfer cannot work. Based on this consideration, a set-based mutation is proposed for vertical fragmentation, which can help transfer the calculation of discrete values to the corresponding set variation. The process of set-based mutation is listed as follows. Firstly, the best individual and two random individuals are chosen. Then, the difference between two random individuals is calculated. The rules for calculating the difference between two individuals are defined as follows. For each bit, if the values of two individuals are the same, the same value is kept. If the values of two individuals are different, a new value is generated by random. Subsequently, the sum of the best individual and the difference is calculated. On each bit, the value of the best individual and the difference are combined together as a sum set, which means the values in these two individuals are both alternatives for the subsequent crossover.

An example of the set-based mutation is given in Fig. 2. The blue bar and the purple bar indicate the first random individual and the second random individual, respectively. Then, the difference between two individuals is calculated. On the first bit, since the value of two random individuals is 2, this value is directly transferred to difference. On the second bit, since the values of two random individuals are different, a new value is generated by random in difference individual. With the same rules, values of difference on all the other bits are generated. Then, the difference individual and the best individual construct a sum set. On each bit, the set contains the values from two individuals.

With the help of set-based mutation, the calculation of discrete values in the individuals is transferred to set variation. For the difference calculation, if two individuals are of the same value, this value will be kept. On the contrary, if two individuals are of different values, a new value is generated by random, which can help improve the diversity of solutions. For the sum calculation, the sum set can keep the information from two individuals and make it an alternative for crossover operator.

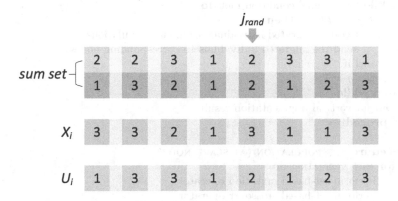

Fig. 3. An example of set-based crossover operator.

4.3 Set-Based Crossover

In the proposed set-based crossover, the crossover of mutant individual and current individual is based on the generated value set. For each bit, one value from the sum set is firstly selected by random. Then, this value is combined with the corresponding value in the current individual x_i and form a new set. With the predefined crossover rate, the final value in u_i utilizes the value from the sum set. Otherwise, the value in the current individual x_i is directly copied to u_i. Moreover, to guarantee at least one bit comes from the sum set, a bit "j_{rand}" is chosen, and the value of sum set on this bit is directly copied to u_i.

As shown in Fig. 3, the sum set and the current individual x_i are given. Then, for each bit, one value from the sum set is randomly chosen and combined with x_i. On the first bit, with the crossover rate, value 1 in the sum set is chosen and copied to u_i. In the second bit, value 3 is copied from the current individual. To be noted, the 5th bit is chosen as the j_{rand} bit, and the value in u_i is copied from the sum set. In the same manner, values of u_i on other bits are extracted from the sum set and x_i.

In the set-based crossover, information from the mutant set and current individual is combined together. Unlike original crossover, a sum set instead of a single mutant individual is provided to set-based crossover. In this way, information from the sum set is sufficiently considered and inputted to the final individual. In a random manner, in the set-based crossover, values in the sum set are firstly selected by random and combined with the current individual. In this way, the weights of the sum set and current the current individuals are distinguished, and the information from two sources is considered.

Algorithm 1. Pseudo-code of S-DDE

1: **procedure** GLOBAL CONTROLLER (AT MASTER NODE)
2:     $g = 0$
3:     Spawn N sub-populations on N computation nodes
4:     **while** no terminal condition met **do**
5:         **if** $g \% MI = 0$ **then**
6:             Receive migrated individuals from sub-populations
7:             Send the migrated individuals to corresponding nodes
8:         **end if**
9:         $g = g + 1$
10:     **end while**
11:     Output vertical fragmentation result
12: **end procedure**
13:
14: **procedure** SUB-POPULATION (AT SLAVE NODE)
15:     **for** each generation **do**
16:         Perform set-based mutation operator
17:         Perform set-based crossover operator
18:         Perform selection operator
19:         **if** a migrated individual is received **then**
20:             Replace a random individual
21:         **end if**
22:     **end for**
23: **end procedure**

4.4 Overall Process

As shown in Algorithm 1, pseudo-code of proposed S-DDE is given. The island model of DDE is implemented by a master-slave manner. The master node takes charge of global controlling and communication between sub-populations. At the beginning of evolution, the entire population is spawned to N sub-populations. While the terminal condition is not met, each sub-population evolves independently in each slave node. Individuals in each node are updated according to the order of set-based mutation operator, set-based crossover operator, and selection operator. With a predefined migration interval MI, migration is executed. Once the global controller receives the migrated individuals, it sends the migrated individuals to the corresponding sub-populations according to the unidirectional ring topology.

5 Experimental Result

5.1 Experimental Setup

To evaluate the performance of proposed S-DDE, 16 test cases of various properties are utilized. These 16 test cases are generated based on the public dataset

Table 1. Property of 16 test cases

Test cases	NR	NSR	NA	NS
T_1	100000	1000	15	3
T_2	100000	2000	15	3
T_3	100000	3000	15	3
T_4	100000	4000	15	3
T_5	200000	1000	20	4
T_6	200000	2000	20	4
T_7	200000	3000	20	4
T_8	200000	4000	20	4
T_9	300000	1000	25	5
T_{10}	300000	2000	25	5
T_{11}	300000	3000	25	5
T_{12}	300000	4000	25	5
T_{13}	400000	1000	30	6
T_{14}	400000	2000	30	6
T_{15}	400000	3000	30	6
T_{16}	400000	4000	30	6

provided by New York State Department of Health[1]. Properties of 16 test cases including number of records NR, number of sample records NSR, number of attributes NA, and number of sites NS are outlined in Table 1.

The parameters of S-DDE are set as follows. The entire population size NP is set as 40, and the number of sub-populations is set as 4; migration interval MI is set as 10; crossover rate Cr is set as 0.5. For S-DDE and all the compared approaches, the maximum number of fitness evaluations is set as $NS \times 10^3$.

The island model of S-DDE is implemented based on the MPI framework. Each island is assigned to a single computation core and evolves independently. The proposed S-DDE and all the compared approaches are performed on a computer cluster using Intel 8-core i7 CPU.

5.2 Comparisons with Competitive Approaches

To verify the competitiveness of proposed S-DDE on vertical fragmentation optimization, two competitive approaches, i.e., heuristic algorithm [3] and differential evolution [12] are utilized in the comparison. These two competitive approaches are listed as follows:

1. HA [3]: This is a state-of-the-art heuristic algorithm for vertical fragmentation.

---

[1] https://health.data.ny.gov/Health/Hospital-Inpatient-Discharges-SPARCS-De-Identified/82xm-y6g8.

Table 2. Comparisons with competitive approaches

Approaches	HA		DE		S-DDE	
	Avg	Std	Avg	Std	Avg	Std
T_1	1.25E+00	1.68E-01	2.02E+00	3.01E-01	**2.38E+00**†	9.26E-02
T_2	1.46E+00	2.83E-01	2.72E+00	3.78E-01	**3.14E+00**†	9.89E-02
T_3	1.67E+00	3.30E-01	3.08E+00	4.71E-01	**3.87E+00**†	1.75E-01
T_4	1.89E+00	4.75E-01	3.67E+00	5.52E-01	**4.46E+00**†	2.70E-01
T_5	1.31E+00	2.32E-01	1.88E+00	1.87E-01	**2.13E+00**†	2.60E-01
T_6	1.49E+00	2.74E-01	2.31E+00	3.07E-01	**2.59E+00**†	4.32E-01
T_7	1.52E+00	3.43E-01	2.71E+00	5.21E-01	**3.09E+00**†	4.42E-01
T_8	1.55E+00	3.81E-01	2.96E+00	4.94E-01	**3.46E+00**†	4.84E-01
T_9	1.41E+00	3.22E-01	1.60E+00	1.34E-01	**2.08E+00**†	3.12E-01
T_{10}	1.55E+00	3.48E-01	2.02E+00	3.35E-01	**2.62E+00**†	4.50E-01
T_{11}	1.85E+00	3.98E-01	2.26E+00	2.80E-01	**3.05E+00**†	4.79E-01
T_{12}	1.96E+00	6.20E-01	2.64E+00	4.62E-01	**3.59E+00**†	5.45E-01
T_{13}	1.09E+00	6.37E-02	1.19E+00	2.97E-02	**1.37E+00**†	1.28E-01
T_{14}	1.14E+00	8.38E-02	1.32E+00	6.93E-02	**1.68E+00**†	2.33E-01
T_{15}	1.23E+00	1.28E-01	1.40E+00	1.37E-01	**1.82E+00**†	3.43E-01
T_{16}	1.34E+00	1.59E-01	1.47E+00	1.41E-01	**1.96E+00**†	3.56E-01

2. DE [12]: This approach acts as the baseline of DE for vertical fragmentation. The difference between DE and S-DDE indicates the effectiveness of the proposed operators.

Furthermore, to guarantee the objectiveness of the comparison, S-DDE and all the compared approaches are performed in 25 independent runs.

As shown in Table 2, the average and standard deviation value of each approach on each test case are calculated and outlined, where the best results are labeled in boldface. Overall, S-DDE can outperform the other approaches on all the test cases. Compared with HA, S-DDE can outperform since it can achieve a comprehensive search in the solution space. Different from DE and S-DDE, HA can only search in a narrow direction, which is led by the predefined heuristic strategy. For test cases of complex solution space, the search based on predefined heuristic strategy is more likely to get trapped by local optima and cannot obtain ideal optimization results. Compared with DE, the effectiveness of the island model and proposed set-based operators is verified. With the help of the island model, population diversity is enhanced. Moreover, the design of set-based operators can transfer the calculation of continuous values to discrete set variation, which makes it efficient for the discrete vertical fragmentation.

To investigate the performance of S-DDE in a statistical sense, Wilcoxon rank-sum at a significance level 0.05 is employed. In Table 2, symbol † indicates the marked best results can achieve a significant difference from the other compared results. In total, proposed S-DDE is significantly better than the other competitive algorithms on all the 16 test cases.

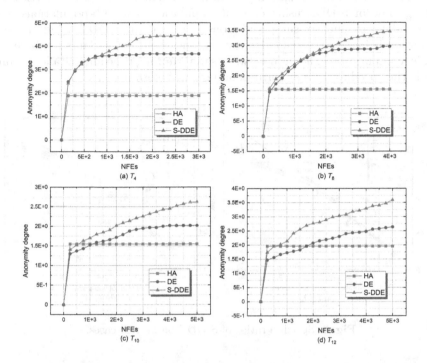

Fig. 4. Convergence curves of competitive approaches on four typical test cases.

The convergence curves of these three algorithms on four typical test cases are plotted in Fig. 4. The convergence curves of HA, DE, and S-DDE are indicated by the blue line, red line, and green line, respectively. The value of each point on the line is obtained by calculating the average value in 25 independent runs. At the beginning of optimization, all these three approaches can achieve a quick convergence. After then, HA quickly stagnates since it lacks high diversity to avoid local optima. Compared with DE, S-DDE exhibits much stronger global search ability, which is obtained from the island model and set-based operators. Overall, in all these four typical test cases, S-DDE can achieve the highest solution accuracy and convergence speed.

5.3   Speedup Ratio

Speedup ratio is an important indicator of distributed algorithms, which can directly reflect the distributed computation efficiency. A distributed algorithm of a high speedup ratio is of higher scalability and better application value. In proposed S-DDE, since each sub-population is assigned to a single computation node, the parallel granularity of S-DDE is directly decided by the number of sub-populations.

To obtain the value of speedup ratio in different parallel granularyties, the running time of S-DDE using one sub-population, two sub-populations, four sub-populations, and six sub-populations is measured. S-DDE using one sub-population is based on serial computation, and the other variants are based on parallel computation. The ratio between serial running time and parallel running time is the speedup ratio at corresponding parallel granularity.

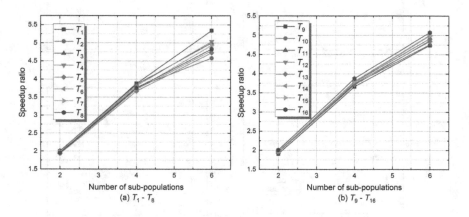

Fig. 5. Speedup ratios of S-DDE on all test cases.

The speedup ratios obtained by S-DDE on 16 test cases are plotted in Fig. 5. With the increase of parallel granularity, speedup ratios also increase. In different test cases, the speedup ratio curves also vary. Speedup ratio is affected by the relationship between computation time and communication time. In general, a test case of higher complexity needs more computation time. Thus, a test case of high complexity can achieve a higher speedup ratio. Also, the balance between different computation nodes is also important. The Unbalance of computation tasks in each computation node can also affect the value of the speedup ratio.

6   Conclusion

In this paper, an anonymity-driven vertical fragmentation problem has been defined, which is valuable for outsourced data storage. To tackle this problem, a set-based distributed differential evolution algorithm is proposed. An island model is utilized to improve the search efficiency as well as maintain the population diversity. Two set-based operators, i.e., set-based mutation operator and set-based crossover operator, are proposed to transfer discrete value calculation to problem-dependent set variation. Experimental results have shown the advantage of the proposed algorithm in optimization performance.

References

1. Aggarwal, G., et al.: Two can keep a secret: a distributed architecture for secure database services. In: CIDR 2005 (2005)
2. Ciriani, V., De Capitani di Vimercati, S., Foresti, S., Jajodia, S., Paraboschi, S., Samarati, P.: Fragmentation and encryption to enforce privacy in data storage. In: Biskup, J., López, J. (eds.) ESORICS 2007. LNCS, vol. 4734, pp. 171–186. Springer, Heidelberg (2007). https://doi.org/10.1007/978-3-540-74835-9_12
3. Ciriani, V., Vimercati, S.D.C.D., Foresti, S., Jajodia, S., Paraboschi, S., Samarati, P.: Combining fragmentation and encryption to protect privacy in data storage. ACM Trans. Inf. Syst. Secur. **13**(3), 22 (2010)
4. Gao, Z., Pan, Z., Zuo, C., Gao, J., Xu, Z.: An optimized deep network representation of multimutation differential evolution and its application in seismic inversion. IEEE Trans. Geosci. Remote Sens. **57**(7), 4720–4734 (2019). https://doi.org/10.1109/tgrs.2019.2892567
5. Ge, Y.F., et al.: Distributed differential evolution based on adaptive mergence and split for large-scale optimization. IEEE Trans. Cybern. **48**(7), 2166–2180 (2018). https://doi.org/10.1109/tcyb.2017.2728725
6. Köhler, J., Jünemann, K., Hartenstein, H.: Confidential database-as-a-service approaches: taxonomy and survey. J. Cloud Comput. **4**(1), 1 (2015)
7. Li, J., Yao, W., Zhang, Y., Qian, H., Han, J.: Flexible and fine-grained attribute-based data storage in cloud computing. IEEE Trans. Serv. Comput. **10**(5), 785–796 (2017). https://doi.org/10.1109/tsc.2016.2520932
8. Li, M., Sun, X., Wang, H., Zhang, Y., Zhang, J.: Privacy-aware access control with trust management in web service. World Wide Web **14**(4), 407–430 (2011). https://doi.org/10.1007/s11280-011-0114-8
9. Machanavajjhala, A., Gehrke, J., Kifer, D., Venkitasubramaniam, M.: L-diversity: privacy beyond k-anonymity. In: 22nd International Conference on Data Engineering. IEEE (2006). https://doi.org/10.1109/icde.2006.1
10. Peng, M., Zeng, G., Sun, Z., Huang, J., Wang, H., Tian, G.: Personalized app recommendation based on app permissions. World Wide Web **21**(1), 89–104 (2017). https://doi.org/10.1007/s11280-017-0456-y
11. Price, K., Storn, R.M., Lampinen, J.A.: Differential evolution: a practical approach to global optimization. Springer Science & Business Media, Heidelberg (2006)
12. Price, K.V.: Differential evolution. In: Handbook of Optimization, pp. 187–214. Springer, Heidelberg (2013)
13. Rani, K., Sagar, R.K.: Enhanced data storage security in cloud environment using encryption, compression and splitting technique. In: 2017 2nd International Conference on Telecommunication and Networks (TEL-NET). IEEE (2017). https://doi.org/10.1109/tel-net.2017.8343557
14. SWEENEY, L.: K-anonymity: a model for protecting privacy. Int. J. Uncertainty, Fuzziness Knowl. Based Syst. **10**(05), 557–570 (2002). https://doi.org/10.1142/s0218488502001648
15. UbaidurRahman, N.H., Balamurugan, C., Mariappan, R.: A novel dna computing based encryption and decryption algorithm. Procedia Comput. Sci. **46**, 463–475 (2015)
16. De Capitani di Vimercati, S., Foresti, S., Jajodia, S., Livraga, G., Paraboschi, S., Samarati, P.: Loose associations to increase utility in data publishing. J. Comput. Secur. **23**(1), 59–88 (2015)

17. Wang, H., Wang, Y., Taleb, T., Jiang, X.: Special issue on security and privacy in network computing. World Wide Web **23**(2), 951–957 (2020). https://doi.org/10.1007/s11280-019-00704-x
18. Wang, H., Zhang, Z., Taleb, T.: Special issue on security and privacy of IoT. World Wide Web **21**(1), 1–6 (2018)
19. Xu, X., Xiong, L., Liu, J.: Database fragmentation with confidentiality constraints: a graph search approach. In: 2015 ACM Conference on Data and Application Security and Privacy, pp. 263–270 (2015)
20. Yu, Y., Au, M.H., Ateniese, G., Huang, X., Susilo, W., Dai, Y., Min, G.: Identity-based remote data integrity checking with perfect data privacy preserving for cloud storage. IEEE Trans. Inf. Forensics Secur. **12**(4), 767–778 (2017)
21. Zhang, F., Wang, Y., Liu, S., Wang, H.: Decision-based evasion attacks on tree ensemble classifiers. World Wide Web (2020). https://doi.org/10.1007/s11280-020-00813-y
22. Zhang, J., Tao, X., Wang, H.: Outlier detection from large distributed databases. World Wide Web **17**(4), 539–568 (2014)
23. Zhang, Y., Chen, X., Li, J., Wong, D.S., Li, H., You, I.: Ensuring attribute privacy protection and fast decryption for outsourced data security in mobile cloud computing. Inf. Sci. **379**, 42–61 (2017). https://doi.org/10.1016/j.ins.2016.04.015
24. Zheng, L.M., Zhang, S.X., Zheng, S.Y., Pan, Y.M.: Differential evolution algorithm with two-step subpopulation strategy and its application in microwave circuit designs. IEEE Trans. Industr. Inf. **12**(3), 911–923 (2016). https://doi.org/10.1109/tii.2016.2535347
25. Zhou, X.G., Peng, C.X., Liu, J., Zhang, Y., Zhang, G.J.: Underestimation-assisted global-local cooperative differential evolution and the application to protein structure prediction. IEEE Trans. Evol. Comput., 1 (2019). https://doi.org/10.1109/tevc.2019.2938531

A Graph Data Privacy-Preserving Method Based on Generative Adversarial Networks

Aiping Li$^{(\boxtimes)}$, Junbin Fang, Qianye Jiang, Bin Zhou, and Yan Jia

School of Computer Science, National University of Defense Technology, Changsha, China
liaiping@nudt.edu.cn

Abstract. We proposed a graph anonymization method which is based on a feature learning model of Generative Adversarial Network (GAN). We used the differential privacy to ensure the privacy and take both anonymity and utility into consideration. The method consists of the following two parts: Firstly, we designed a graph feature learning method based on GAN. The method used the bias random walk strategy to sample the node sequence from graph data, and trained the GAN model. After training, the GAN generated a set of simulation sequences that are highly like the real sampled sequence. Secondly, we proposed an anonymous graph construction method based on the simulation node sequence. We calculated the number of edges in the node sequences and constructed a probability adjacency matrix. The differential privacy noise is added to get the anonymous probability adjacency matrix. Then we extract the edges from the anonymous matrix and then constructed the anonymous graph. We evaluate our methodology, showing that the model had good feature learning ability through embedding visualization and link prediction experiments, compared with other anonymous graphs. Through experiments such as metric evaluation, community detection, and de-anonymization attack, we proved that the anonymous method we proposed is better than the current mainstream anonymous method.

Keywords: Graph data · Generative Adversarial Networks · Differential privacy · Privacy preservation

1 Introduction

With the rapid development of research and application of graph data, private information such as individual existence and associations in graph data also faces huge risks of leakage and abuse [1]. When we publish or share data with others, we need to anonymize the data by anonymization methods. But the anonymity is usually obtained by giving up data utility.

In the scenario of data sharing and publishing, in order to ensure that the value of the published information is retained as much as possible, and the data accepter cannot obtain the specific individual information on the datasets, the datasets are usually anonymized by anonymization technology, and then the anonymized data is published to public. The common privacy preservation algorithms include naive ID removal, k-anonymity [2], l-diversity [3], t-closeness [4], ε-differential privacy [5], etc., which achieve data

© Springer Nature Switzerland AG 2020
Z. Huang et al. (Eds.): WISE 2020, LNCS 12343, pp. 227–239, 2020.
https://doi.org/10.1007/978-3-030-62008-0_16

anonymization mainly by hiding node attributes and the network structure features, adding edges to perturb the data.

In order to verify the effect of the anonymization algorithm and to assess the risk of reidentification of personal privacy information in the published dataset, many researchers have proposed various of de-anonymization methods. Some seed-based deanonymization algorithms apply the reidentification process by using a set of known node pairs between anonymized graph and auxiliary graph, and it can accurately infer other node pairs by using the seed node pairs. But, unfortunately, the algorithms would be serious misleading by the wrong seeds and achieve a bad reidentification result. The seedless methods could reidentify the node pairs without the pre-knowledge of seeds, and using the structure information to define the similarity of node pairs. But the method required a well definition of node feature and an accurate similarity. It often has a poor performance when the noise of the fake edges increased.

This paper proposes a graph data privacy-preserving method using Generative Adversarial Network, named GDPGAN, to achieve excellent anonymity and utility balance on graph data publishing. we designed a graph feature learning method based on GAN. The method used the bias random walk strategy to sample the node sequence from graph data, and trained the GAN model. After training, the GAN generated a set of simulation sequences that are highly like the real sampled sequence. Then proposes an anonymous graph construction method based on the simulation node sequence. We calculated the number of edges in the node sequences and constructed a probability adjacency matrix. The differential privacy noise is added to get the anonymous probability adjacency matrix. Then we extract the edges from the anonymous matrix and then constructed the anonymous graph. Our main contributions are as follow.

- Using the GAN to perform feature learning on the graph data. In the zero-sum game between the generator and discriminator, the model can learn the statistic distribution of the input graph and generate nearly true features.
- The method constructs anonymous graph using previously learned features and differential privacy noise. So that, anonymous graphs can guarantee individual privacy while accomplishing data mining tasks with minimal loss.
- We define an evaluation module for the privacy and utility of anonymous data. The experiments show that, the evaluation module has strong practicability, accuracy and operability.

The rest is organized as follows: in Sect. 2, we summarize the related works. In Sect. 3, a graph feature learning method based on GAN is proposed. Then our anonymous graph construction method is shown in section IV. In Sect. 5 experimental results are presented. In Sect. 6, we conclude our work.

2 Related Work

Nowadays, privacy preservation technology for graph data sharing has been extensively studied. Liu et al. designed an access system, named LinkMirage, that controls the privacy of social networks [6]. LinkMirage clusters social networks to super nodes and then anonymize the links between clusters. Applications can only access anonymous data by the interface of LinkMirage's for various data mining tasks. Bhagat et al. used an interaction-oriented anonymization algorithm to cluster nodes with similar attributes or labels [7]. Yi et al. proposed a k-anonymity model for privacy preserving data publication, which called (p, a)-sensitive k-anonymity and (p^{+}, a)-sensitive k-anonymity model, can reduce the privacy breach [8]. Cluster-based anonymization methods can effectively protect the structural information, but the utility loss is large.

Rossi et al. proposed a new anonymization framework for time-varying multi-level graphs, by perturbing the structure of time-varying graphs [9]. The degree sequence of each node is compared with at least other k-1 nodes and cannot be distinguished. Because over-generalized data may render data of little value or useless, Min et al. proposed a generalization boundary techniques to maximize data usability while, minimizing disclosure of privacy [10, 11]. By solving the equivalence graph process, a set of k-anonymous methods to ensure high availability of data are proposed. This approach provides better usability while providing the same privacy guarantees.

Sala et al. obtains a dK-2 sequence by counting the number of edges between different degrees [12]. Gao et al. propose an anonymity method of generating differential privacy graph based on the dk-1, dk-2 and dk-3 sequences [13]. They uses low-dimensional features to retain more features and high-dimensional features are more sensitive to noise. Two different graph generation methods CAT and LTH are designed to generate graphs and reduce data loss under differential privacy. Because the relationships of graph data encoded are often sensitive, people begin to find effective ways to release representative graphs which nevertheless protect the privacy of the data subjects. Jorgensen et al. proposed an approach to release such graphs under the strong guarantee of differential privacy [14]. They introduce a new model and show how to augment them with meaningful privacy.

3 Graph Feature Learning Method

In this section, the learning model of graph data feature is built by using the Generated Adversarial Network. The GAN was proposed by Goodfellow et al. [15]. Once it was proposed, it attracted extensive attention and research. GAN has a strong ability of feature learning. We use GAN to learn the potential feature distribution of the input real sampling random walk sequence, and use the trained generator to generate a highly simulated random walk sequence. The simulation sequence can further use word2vec to generate node embedding representation, and compare with the original embedding representation to verify whether it learns the potential relationship distribution in the graph. Graph representation is aimed at mapping nodes into a low-dimensional vector, which can quantitatively to measure the feature of nodes. At present, the research on graph representation is widely studied, including Deepwalk [16], GraphGAN [17], etc. In this section, we utilize the biased second-order random walk strategy proposed by node2vec [18] to getting length T random walks. The GAN model consists of two parts: the generator and the discriminator.

3.1 Generative Model

The design goal of the generator is to generate a simulation node sequence that is close to the real random walk sequence, so that the discriminator cannot distinguish the true and the false of the input sequence correctly. Therefore, the generator uses the LSTM (Long Short-Term Memory) [19] as the basic unit, and generates a node as the output according to the parameter state of the generator at each time. Then generates a node sequence of length t as the output simulation node sequence after t time. The construction of generator and the generation process of simulation node sequence are shown in Fig. 1.

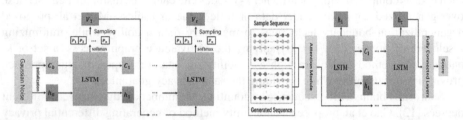

Fig. 1. The design of generator   **Fig. 2.** The design of discriminator

In this model, the nodes in the input and output sequences are encoded in *one-hot* format, that is, for the graph data containing n nodes, each node is represented by a vector of length n, only one element in the vector is 1, and the rest n-1 elements are 0. In the initial stage of the model, Gaussian noise (that is, noise satisfying normal distribution) is used to initialize the state parameters C_0 and h_0 of the LSTM unit randomly, and the original state information of the model is obtained. At the time of t_1, the input data is the original state parameter. According to the parameter settings and various gate structures of the current LSTM model, an output vector h_1 and a state vector C_1 are obtained. For h_1, the *Sotfmax* is used to obtain the probability distribution vector of length n, and each element in the vector represents the occurrence probability of the corresponding numbered node. Then, a node number is selected from the probability distribution as the output node of LSTM at that time, which is represented by *one-hot* form.

At the time of t_2, the *one-hot* vector of the output node v_1 of t_1 and the state parameters C_1 and h_1 are used as input. As the operation process of t_1, after passing through the LSTM and *Sotfmax* layers, the probability distribution vector of t_2 is obtained, and the output node v_2 of t_2 is selected according to the probability. In this way, through the t-round iteration, the node sequence $<v_1, v_2, ..., v_t>$ is obtained. Each node is represented in the form of *one-hot*, that is, the simulation random walk sequence generated by the generator.

3.2 Discriminative Model

The design goal of the discriminator is to correctly distinguish the feature difference between the real random walk sequence and the simulated random walk sequence generated by the generator. The discriminator also uses LSTM as the basic unit, combined with the attention layer to improve the operation effect of the discriminator. For the input real or simulation node sequence, each time takes one of the nodes, gets the output value of the last node in the corresponding sequence after t times, and transforms it into a value.

For the output values corresponding to different input sequences, different objective functions are used to optimize the parameters of the generator and the discriminator respectively, to improve the simulation ability of the generator and the judgment ability of the discriminator. In the training process of game confrontation, the discriminator and generator promote each other, and achieve their own design goals. The construction and operation flow of the discriminator is shown in Fig. 2.

3.3 Training and Verification

Based on the offset walk sampling strategy of the above node sequence, the implementation method of LSTM basic unit and attention mechanism, we build a graph data feature learning model based on the generated countermeasure network. The structure of the model is shown in Fig. 3. In model training, the original graph uses node sampling strategy to obtain a set of random walk sequences. At the same time, the generator also generates a series of simulation nodes according to the initial parameter settings. Two sets of sequences are used as the input of the discriminator. The loss function and the optimizer of the generator and the discriminator are defined respectively by the discriminator, and their model parameters are adjusted continuously in the iterative training. Finally, the generator can generate a set of simulation sequences with high similarity to the sampling sequence as the feature representation of the original graph. The process of training and verification of feature learning model is as follows:

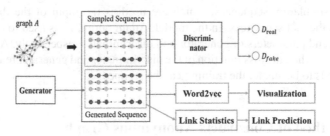

Fig. 3. Graph data feature learning model based on GAN

1) Build training data. Before the model training, the offset random walk strategy is used to generate a set of real random walk sequences as the feature of the original map. This group of node sequence needs to have enough number to ensure the feature distribution covering the whole original map data. In each round of training, a certain number of sequences are extracted from the sequences as a batch of training data.

2) Build feature learning model. Using the above-mentioned design method of generator and discriminator, the feature learning model of the GAN is built. The simulation sequence generated by the generator and the real sampling node sequence are used as the input of the discriminator to start the training process of the model.

3) The loss function of generator and discriminator is defined. Assuming that the size of each batch of data is m, for the generator, to make the simulation sequence cheat the discriminator as much as possible, even if the output D fake of the discriminator is as large as possible. Therefore, the loss function of the generator is defined as:

$$G_{loss} = -\frac{\sum_m D_{fake}}{m} \tag{1}$$

For the discriminator, its training goal is to make it correctly identify whether the input data is from the real sequence or the simulation sequence. Therefore, for the input of the real node walk sequence, the output d real value of the discriminator should be as large as possible, while for the input of the simulation sequence, the output D fake of the discriminator should be as small as possible. According to the optimization method for GAN model training in Wasserstein [20], the loss function of discriminator is defined as:

$$D_{loss} = \frac{\sum_m D_{fake}}{m} - \frac{\sum_m D_{real}}{m} \tag{2}$$

Then the Lipschitz-Penalty term is applied to limit the gradient of the discriminator to solve the problems of gradient explosion and convergence. Finally, L_2 regular terms are applied to the generator and the discriminator to obtain the final loss function.

4) Training and parameter optimization In each round of training, the generator also generates m simulation sequences, which are used as the input of the discriminator together with the real sequences. In the model, Adam optimizer is used to optimize the loss function and parameters of generator and discriminator respectively. After enough training rounds, when the loss function of the discriminator and generator reaches a good effect and tends to be stable, the training can be stopped.

4   The Method of Constructs Anonymous Graph

Over the training steps, generator can generate many near real random walks and we can construct the anonymous graph data onto it. The process of anonymous graph construction needs several steps.

Step 1. By using the near real random walks, the number of the edges of all nodes is counted and the probability distribution of all edges is calculated.
Step 2. Adding Laplacian noise that satisfies the differential privacy requirement in the probability distribution of the edges obtained in the previous step. Then we can obtain an anonymous probability distribution of edges.
Step 3. According to number of edges E specified, we can sample the most top-E possible edges of the probability distribution, and construct an anonymous graph.

4.1 Anonymous Graph Building Model

In order to construct anonymous graphs, simulation sequences need to be transformed into computable feature distributions. First, all the links in the simulation sequence are counted. Assuming that there are n nodes in the graph, an $n \times n$ count adjacency matrix can be constructed.

The value of x_{ij} in the matrix represents the number of times of the link between nodes i and j in the simulation sequence. The larger the value is, the more likely the edge will exist in the original graph according to the feature state of model learning. Therefore, if the probability of link between each node i and any node j in the graph is recorded as e_{ij}, then all elements of the row in which node i is in the count adjacency matrix e_{ij} are normalized according to the following:

$$e_{ij} = \frac{x_{ij}}{\sum_{j=1}^{n} x_{ij}} \tag{3}$$

The probability matrix is obtained by calculating the probability distribution of node i in each row. We add differential privacy noise to the probability distribution matrix to ensure the anonymity of the constructed graph. Differential privacy anonymity technology has become one of the most widely used anonymity technologies because of its powerful privacy protection ability. The implementation of differential privacy is based on probability, so it has certain randomness. At the same time, because the added noise is controlled, the degree of privacy protection can be calculated quantitatively.

For the probability distribution matrix, each row represents the probability distribution of the edge between node i and any other node. The larger the probability value of an element, the greater the probability that there is a link relationship between corresponding nodes, that is, the association characteristics of graph data are reflected in these probability distributions. When anonymous operation is performed on the probability distribution, in the probability distribution of each line, random noise value satisfying Laplace noise is added to each probability component. According to the probability range of each line, the added noise sensitivity and privacy budget parameters are determined, and finally an anonymous probability distribution matrix satisfying differential privacy is obtained. Controlled noise makes the probability of edge occurrence contain more randomness, thus ensuring the effect of privacy protection.

After obtaining the anonymity probability distribution matrix satisfying the differential privacy, the probability distribution in the matrix includes the feature distribution of the anonymity graph. In order to construct an anonymous graph, we need to extract a certain number of elements from the anonymous probability distribution matrix as the edge of the graph. The construction strategy of anonymous graph is as shown in Algorithm 1. First, we need to ensure that each row of node i has a connected node j, to ensure that there will be no isolated points in the anonymous graph, and then extract a specified number of edges according to the set requirements to build the anonymous graph.

Algorithm 1. The algorithm of anonymous graph building

   Input:

      Anonymous probability distribution matrix M

   Output:

      The anonymous graph data G_a

1: **for** each row in M **do**

2:     Randomly select an edge and add to G_a ;

3:     Ensure that there are no isolated points in the graph

4: **end for**

5: **while** the count of edges of G_a <= Threshold **do**

6:     Calculate the remaining required edge m;

7: **if** (m >0) **then**

8:     Take one edge from M according prob and add to G_a;

9: **end if**

10: **end while**

11: **return** G_a

The data availability of the constructed anonymous graph mainly depends on the feature learning ability of the generated adversary network. The simulated node sequence generated by the generator from the learned feature distribution comprehensively includes the statistical characteristics and potential node correlation characteristics of the original graph. There are two main sources of anonymity of anonymous graphs: one is that the generation process of simulation sequence is random, which can make the edge constructed different from that in the original graph while retaining the statistical characteristics of graph data; the other is to add differential privacy mechanism to the edge probability distribution, which theoretically guarantees that the constructed anonymous graphs can provide effective privacy protection ability.

4.2 Evaluation Module

The privacy and utility evaluation module is used to measure the privacy and data utility of anonymous graph. Meanwhile, the module can be used to evaluate the performance of the generated graph after each training iteration.

Algorithm 2. The ranking strategy of evaluation module
___
Input:

The results of metrics for original and anonymous graphs

Output:

Performance score RS

1: **for** each evaluation metric result **do**

2:    $rank = 0$;

3:    **for** each m in M **do**

4:      $m = |m - m'|$; // m' is the corresponding utility metric in original graph.

5:      $m = sorted(m)$; //rank(m) is the ranking of m in all m values

6:    **end for**

7:    **for** each n in N **do**

8:      $n = sorted(n)$; //rank(n) is the ranking of m in all n values

9:      $rank = rank + n$;

10:   **end for**

11:   $RS = rank \cdot |M + N|^{-1}$

12: **end for**

13: **return** RS
___

Evaluation module considers 10 utility metrics and 2privacy metrics. The utility metrics includes max degree (the maximum degree of nodes in the graph, abbreviated as md), wedge count (the number of nodes which connect with two nodes, abbreviated as wc), claw count (the number of nodes which connect with three nodes, abbreviated as clc), triangle count (the number of three-nodes connected with each other, abbreviated as tc), square count (the number of quads composed by four nodes, abbreviated as sc), $gini$ (the Gini coefficient of the graph), relative edge distribution entropy(Compute the relative edge distribution entropy of the input graph, abbreviated as re), associativity (Calculate the degree of classification of the graph, abbreviated as as), Clustering coefficient (the aggregation coefficient, abbreviated as cc), cpl (the characteristic path length). The privacy metric includes dd(The difference between degree of the same nodes) and odd (The difference between degree of two-hop node). The strategy of Relative Score (RS) calculation is formally described in Algorithm 2.

5   Experiments

In this section, we verify the performance of GPAGAN anonymization method on a real world dataset and compare the performance with 2 other typical anonymization methods.

This experiment uses the Cora-ML, which is a citation dataset containing 2,810 nodes and 7,981 edges. We applied GPAGAN, K-anonymous, and Differential Privacy to the Cora-ML dataset and get corresponding anonymous graphs. Finally, the evaluation module was used to verify the performance of each method.

All experiments were carried out on a Server with Intel Corei7 CPU@3.2 Hz and 120 GB RAM. The operating system is Ubuntu 18.04. The GPAGAN algorithm was implemented in Python 3.6, the and K-anonymous was implemented in Java8. Our experiments are conducted on both real and synthetic datasets.

We use the same k-anonymous and differential privacy algorithm [1], and generate a set of k-anonymous graphs and a set of differential privacy anonymity graphs according to different parameters (the sensitivity uses the default value) as the comparison method. In the process of task evaluation, the community partition method proposed in [21] and the anonymous attack method proposed in [22] are used to verify the original graph and each anonymous graph respectively.

This experiment shows two problems by measuring the performance of different anonymous graphs. One is that the current mainstream anonymous methods cannot take into account both anonymity and availability. The other is that the GAN anonymity method can achieve the same degree of anonymity as the K anonymity and differential anonymity in the high anonymity state with the same parameters. The other is that the GAN anonymity method can achieve the same degree of anonymity as the k-anonymous and differential anonymity in the high availability state with the same parameters Anonymity and differential privacy achieve the same degree of availability, that is, they can guarantee high anonymity and high availability at the same time.

Used the best anonymous graph with 115% edges generated by GPAGAN to compare with anonymous graphs generated by k-anonymous and differential privacy method respectively. The parameter ε of differential privacy is set to 0.05, 0.1, 0.2, 0.5, and 1. The parameter k of k-anonymous is set to 1, 5, 10, 20 and 50. The larger the k value is, the higher the anonymity is, but the lower the data availability is. In the differential privacy anonymity algorithm, the privacy parameter $\varepsilon = 5, 10, 50, 100$. The smaller the ε value is, the higher the anonymity is. The original graph, Gan anonymity graph, k-anonymous graph and differential privacy anonymity graph under different parameters are calculated respectively. Among them, GAN represents GAN anonymous graph, $k5$ represents k-anonymous graph with parameter $k = 5$, DP5 represents differential privacy anonymous graph with parameter $\varepsilon = 5$, and so on. The visual comparison of indicators in the table is shown in Fig. 4. In Fig. 4, the visual comparison results of indicators are more obvious. In indicator measurement, the larger the value is, the higher the anonymity is, and all indicators of Gan anonymity graph are kept at a higher value. From the calculation results of anonymity synthetical score, the lower the score value is, so the anonymity of Gan anonymity method is as strong as that of k 50 and DP5 with strong anonymity parameters.

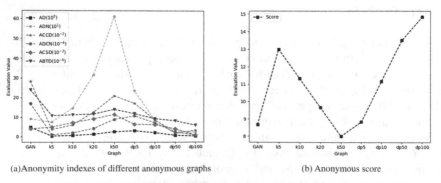

(a)Anonymity indexes of different anonymous graphs   (b) Anonymous score

Fig. 4. Comparison of anonymity indexes and anonymity ranking

Figure 5 and Fig. 6 show the comparison of availability indicators. The abscissa is different graphs, and the first scale is the original graph. Ordinates are measures. Among the anonymous graphs, the closer they are to the original graph, the higher their availability. The anonymity graph of GAN is very close to the original graph in all indexes and has strong data availability. This result shows that the existing mainstream anonymity methods can achieve better data availability by setting loose privacy parameters (for example, K value in K anonymity is set to 5 or lower, ε in differential privacy is set to 50 or higher). And GAN anonymity algorithm can guarantee the same degree or even better data availability as the mainstream algorithm of these high availability state parameters.

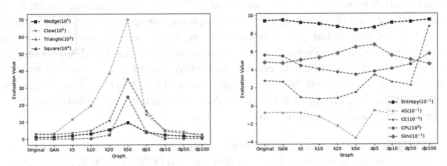

Fig. 5. Comparison of availability indexes of different anonymous graph

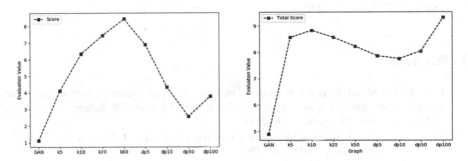

Fig. 6. Score of availability   **Fig. 7.** Synthetical score and ranking

In the above evaluation results of anonymity and availability for different anonymity graphs, we can see that the anonymity graph of GAN has the same degree of availability as the mainstream anonymity algorithm under the high availability parameter, and also has the same degree of anonymity as the anonymity algorithm under the high anonymity parameter, that is, both anonymity and availability can be achieved in the anonymity graph of GAN. While the current mainstream anonymous methods need to give up the other side when they get the effect of one side.

Finally, the synthetical score of all usability and anonymity evaluation indicators is calculated, and the score and ranking of indicators are shown in Fig. 7. The lower the score, the higher the ranking, the better the synthetical performance. According to the calculation results of the evaluation indexes, the performance of the anonymity and availability indexes of GAN method is the best, and its score is about 4.89. After that, the scores of DP10, DP5 and K50 are obviously behind. This result further proves that compared with the current mainstream anonymous graph, GAN anonymous graph has a strong advantage in both anonymity and usability.

6 Conclusions

In this paper, we propose a GPAGAN method to perform anonymization operations on the graph data which can achieve good performance both in privacy and utility. The use of GAN represents the most advanced method of graph learning and makes it possible to fully learn the characteristics of the graph without specifying specific features. The addition of differential privacy noise in the graph generation process ensures the privacy performance of anonymous graph. Experiments show that the GPAGAN method has a significant improvement in privacy and utility retention compared to existing privacy preservation methods. In future research, we will focus on optimizing of training efficiency, expanding the applicable types of data, and verifying the performance of the data onto actual machine learning tasks.

Acknowledgements. The work described in this paper is partially supported by the National Key Research and Development Program of China (No. 2017YFB0802204, 2016QY03D0603, 2016QY03D0601, 2017YFB0803301, 2019QY1406), the Key R&D Program of Guangdong Province (No. 2019B010136003), and the National Natural Science Foundation of China (No. 61732004, 61732022, 61672020).

References

1. Ji, S., Li, W., Mittal, P., et al.: Secgraph: a uniform and open-source evaluation system for graph data anonymization and de-anonymization. In: 24th USENIX Security Symposium USENIX Security 15), pp. 303–318 (2015)
2. Sweeney, L.: k-anonymous: a model for protecting privacy. Int. J. Uncertainty, Fuzziness Knowl. Based Syst. **10**(05), 557–570 (2002)
3. Sun, X., Wang, H., Li, J., Zhang, Y.: Satisfying privacy requirements before data anonymization. Comput. J. **55**(4), 422–437 (2012)
4. Li, N., Li, T., Venkatasubramanian, S.: T-closeness: privacy beyond k-anonymous and l-diversity. In: 2007 IEEE 23rd International Conference on Data Engineering, pp. 106–115. IEEE (2007)
5. Dwork, C.: Differential privacy. In: Encyclopedia of Cryptography and Security, pp. 338–340 (2011)
6. Liu, C., Mittal, P.: LinkMirage: enabling privacy-preserving analytics on social relationships. In: NDSS 2016 (2016)
7. Bhagat, S., Cormode, G., Krishnamurthy, B., et al.: Class-based graph anonymization for social network data. Proc. VLDB Endowment **2**(1), 766–777 (2009)
8. Yi, X., Zhang, Y.: Privacy-preserving distributed association rule mining via semi-trusted mixer. Data Knowl. Eng. **63**(2), 550–567 (2007)
9. Rossi, L., Musolesi, M., Torsello, A.: On the k-anonymization of time-varying and multi-layer social graphs. In: Ninth International AAAI Conference on Web and Social Media (2015)

10. Li, M., Sun, X., Wang, H., et al.: Privacy-aware access control with trust management in web service. World Wide Web-internet Web Inf. Syst. **14**(4), 407–430 (2011)
11. Wang, H., Zhang, Z., Taleb, T.: Editorial: special issue on security and privacy of IoT. In: World Wide Web-Internet & Web Information Systems, pp. 1–6 (2017)
12. Sala, A., Zhao, X., Wilson, C., et al.: Sharing graphs using differentially private graph models. In: Proceedings of the 2011 ACM SIGCOMM Conference on Internet Measurement Conference, pp. 81–98. ACM (2011)
13. Gao, T., Li, F.: Sharing social networks using a novel differentially private graph model. In: 2019 16th IEEE Annual Consumer Communications & Networking Conference (CCNC), pp. 1–4. IEEE (2019)
14. Jorgensen, Z., Yu, T., Cormode, G.: Publishing attributed social graphs with formal privacy guarantees. In: The 2016 International Conference. ACM (2016)
15. Goodfellow, I., Pouget-Abadie, J., Mirza, M., et al.: Generative adversarial nets. In: Advances in Neural Information Processing Systems, pp. 2672–2680 (2014)
16. Kipf, T.N., Welling, M.: Semi-supervised classification with graph convolutional networks. arXiv preprint arXiv:1609.02907 (2016)
17. Wang, H., Wang, J., Wang, J., et al.: Graphgan: graph representation learning with generative adversarial nets. In: Thirty-Second AAAI Conference on Artificial Intelligence (2018)
18. Grover, A., Leskovec, J.: node2vec: scalable feature learning for networks. In: Proceedings of the 22nd ACM SIGKDD International Conference on Knowledge Discovery and Data Mining, pp. 855–864. ACM (2016)
19. Sundermeyer, M., Schlüter, R., Ney, H.: LSTM neural networks for language modeling. In: Interspeech, pp. 601–608 (2012)
20. Arjovsky, M., Chintala, S., Bottou, L.: Wasserstein generative adversarial networks. In: International Conference on Machine Learning, pp. 214–223 (2017)
21. Blondel, V.D., Guillaume, J.-L., Lambiotte, R., et al.: Fast unfolding of communities in large networks. J. Stat. Mech. Theory Exp. **2008**(10), P10008 (2008)
22. Ji, S., Gu, Q., Weng, H., et al.: De-Health: all your online health information are belong to us. arXiv preprint arXiv:1902.00717 (2019)

Channel Correlation Based Robust Audio Watermarking Mechanism for Stereo Signals

Tianrui Zong[1]($\boxtimes$)(iD), Yong Xiang[1](iD), Iynkaran Natgunanathan[1](iD),
Longxiang Gao[1](iD), and Wanlei Zhou[2](iD)

[1] School of Information Technology, Deakin University, Burwood, VIC 3125,
Australia
{tianrui.zong,yong.xiang,iynkaran.natgunananthan,
longxiang.gao}@deakin.edu.au
[2] School of Software, University of Technology Sydney, Ultimo, NSW 2007, Australia
wanlei.zhou@uts.edu.au

Abstract. The music industry, especially the Internet-based platforms, is confronting the challenge of protecting the copyright. The traditional watermarking algorithms have limited robustness against the desynchronization attacks, such as time-invariant-pitch-scaling (TIPS) and pitch-invariant-time-scaling (PITS). Furthermore, only a very small part of the existing methods is designed for stereo signals, which is one of the most widely used formats in web-based music platforms. In this paper, we propose a robust watermarking mechanism for stereo signals for web-based applications. We first transform both of the left channel signal and the right channel signal into the frequency domain. Then the watermark bits are embedded by adjusting the Pearson correlation coefficient (PCC) between the frequency coefficients of the left channel and the right channel in the selected frequency band. Since TIPS and PITS can be approximated as stretching the signal in the frequency domain, the proposed channel correlation based watermarking mechanism is resilient to these attacks. Besides, the proposed method is also robust to other common attacks. Theoretical analysis and simulation results show the validity of the proposed method.

Keywords: Audio watermarking · Stereo signal · Correlation coefficient · Desynchronization attacks · Online copyright protection

1 Introduction

Thanks to the rapid development of the World Wide Web technology, nowadays the multimedia files are produced, stored, traded, and transmitted mostly

This work was supported in part by the Australian Research Council under grant LP170100458.

Z. Huang et al. (Eds.): WISE 2020, LNCS 12343, pp. 240–251, 2020.
https://doi.org/10.1007/978-3-030-62008-0_17

in digital formats. However, the multimedia files in digital formats can be easily manipulated, duplicated, and distributed via the Internet, which leads to severe piracy issues. For instance, the U.S. economy loses $12.5 billion every year because of music theft [1] and $71 billion annually for film and TV industry as a result of global digital piracy [2]. Therefore, a great concern has been raised for copyright protection for web-based applications. Digital watermarking is a promising technology to tackle copyright infringement. In digital watermarking, the copyright information, such as signatures, logos, and IDs, is embedded directly into the multimedia signals. When the violation of copyright happens, the watermark can be extracted for copyright protection. Digital watermarking can be applied to cloud-based applications [3,4], and is applicable to audio [5–7], image [8–10], and video [11–13]. In this paper, we narrow our focus on audio signals.

In the past, a lot of effort has been made to cope with copyright infringement of audio signals for online-based applications. Cox et al. proposed a spread spectrum (SS) based watermarking method in [14] where the watermark signal and the host audio signal are orthogonal. Xiang et al. [15] further increased the embedding capacity of the SS based watermarking strategy by utilizing the orthogonal watermark sequences. In [16], the authors proposed an echo hiding based watermarking method which has better perceptual quality than the SS based methods. Later on Xiang et al. [17] improved the method in [16] by splitting the host signal into two channels, which reduced the host signal interference and consequently enhanced the robustness. Although the SS and echo hiding based methods show excellent robustness against common signal processing attacks such as noise addition, compression, and filtering, they are vulnerable to desynchronization attacks, which include jittering, time-invariant-pitch-scaling (TIPS), and pitch-invariant-time-scaling (PITS).

Desynchronization attacks are challenging attacks for audio watermarking, as they destroy the synchronization between the embedder and the receiver. Nowadays, desynchronization attacks can be easily implemented using free audio processing software. In order to tackle desynchronization attacks, Feng et al. proposed a quantization index modulation (QIM) based watermarking method in [18] which modified the mean value of the host signal. In [19], the authors utilized the log coordinate mapping feature to cope with PITS. Liu et al. [20] tackled TIPS and PITS by constructing the frequency domain coefficients logarithmic mean feature.

Despite the above-mentioned methods can resist desynchronization attacks to some extent, their robustness is still limited. Furthermore, most of the research outcomes focus on mono audio signals. However, the majority of the audio signals used by web-based applications are stereo signals. Since the left channel and the right channel of a stereo signal is positive highly correlated, in this paper we propose a channel correlation based audio watermarking algorithm for stereo audio signals for online-based applications. We first segment both channels of the stereo signal, and transform them into the discrete cosine transform (DCT) domain. Then within the selected frequency range, we modify the

Pearson correlation coefficient (PCC) between the left channel signal and the right channel signal. As the PCC between both channels is robust to common attacks, including the challenging desynchronization attacks, the proposed method shows excellent resistance against attacks. The contribution of this paper can be summarised as follows:

- Our proposed method designs a robust watermarking algorithm for the widely used stereo audio signals, while most of the existing methods focus on the mono audio signals.
- Our proposed method is resistant to desynchronization attacks, while the exisiting methods are not effective against desynchronization attacks.

The remainder of this paper is organized as follows. Section 2 introduces the proposed method. The robustness of the proposed method and the comparison with the traditional watermarking schemes are demonstrated in Sect. 3. Finally, Sect. 4 concludes the paper.

2   Proposed Method

Instead of vintage records, cassettes and CDs, it is more and more popular to purchase the digital files of songs from online platforms. After the online order is processed and the payment is confirmed, the details of the transaction could be embedded into the host audio signal by our proposed watermarking method prior to the download procedure. As a result, each sold digital copy will contain a unique watermark sequence, which later can be used for copyright protection. The flowchart of the proposed method is shown in Fig. 1.

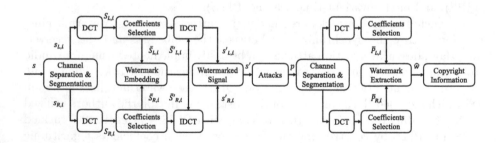

Fig. 1. The flowchart of the proposed method.

2.1   PCC

PCC is an indicator of the linear correlation between two vectors. Denote the host stereo signal as s with size $L \times 2$, the left channel signal as s_L, and the right channel signal as s_R. Then S could be expressed as

$$s = [s_L, \ s_R].\tag{1}$$

The PCC between s_L and s_R can be calculated as

$$\rho = \frac{1}{L} \sum_{k=1}^{L} \frac{(s_L(k) - E(s_L))(s_R(k) - E(s_R))}{\sigma_L \sigma_R}, \tag{2}$$

where ρ is the PCC between s_L and s_R, $E(.)$ is the mean value operator, σ_L is the standard deviation of s_L, and σ_R is the standard deviation of s_R. According to [19], the TIPS attack and the PITS attack can be formulated as

$$\begin{cases} P_L(k) = S'_L(\beta k); \\ P_R(k) = S'_R(\beta k), \end{cases} \tag{3}$$

where $P_L(k)$ is the kth DCT coefficient in the left channel of the received signal, $P_R(k)$ is the kth DCT coefficient in the right channel of the received signal, $S'_L(k)$ is the kth DCT coefficient in the left channel of the watermarked signal, $S'_R(k)$ is the kth DCT coefficient in the right channel of the watermarked signal, and β is the frequency scaling factor introduced by TIPS or PITS. Similar to (2), the PCC $\hat{\rho}$ between $P_L(k)$ and $P_R(k)$ can be expressed as

$$\hat{\rho} = \frac{1}{L'} \sum_{k=1}^{L'} \frac{(P_L(k) - E(P_L))(P_R(k) - E(P_R))}{\hat{\sigma}_L \hat{\sigma}_R}, \tag{4}$$

where L' is the length of the received signal, $\hat{\sigma}_L$ is the standard deviation of P_L, and $\hat{\sigma}_R$ is the standard deviation of P_R. By substituting (2) and (3) into (4), we can have

$$\hat{\rho} = \frac{1}{L'} \sum_{k=1}^{L'} \frac{(S'_L(\beta k) - E(P_L))(S'_R(\beta k) - E(P_R))}{\hat{\sigma}_L \hat{\sigma}_R}$$

$$= \frac{1}{L} \sum_{k=1}^{L} \frac{(S'_L(k) - E(S'_L))(S'_R(k) - E(S'_R))}{\sigma'_L \sigma'_R}$$

$$= \rho', \tag{5}$$

where ρ' is the PCC between S'_L and S'_R, σ'_L is the standard deviation of S'_L, and σ'_R is the standard deviation of S'_R. Therefore, PCC is robust to TIPS and PITS.

2.2   Segmentation

To utilize PCC in watermark embedding, firstly divide s_L into N segments with length M, where $M = L/N$. The ith segment of s_L can be formulated as

$$s_{L,i} = [s_L((i-1) \cdot M + 1), \ s_L((i-1) \cdot M + 2), \dots, s_L(i \cdot M)], \tag{6}$$

where $s_{L,i}$ is the ith segment of s_L and $i \in [1, N]$. Secondly, transform $s_{L,i}$ into DCT domain using [21] as

$$S_{L,i}(m) = l(m) \sum_{n=1}^{M} s_{L,i}(n) \cdot \cos\frac{\pi(2n-1)(m-1)}{2M}, \tag{7}$$

where $S_{L,i}(m)$ is the mth DCT coefficient of $s_{L,i}$, $m = 1, 2, \ldots, M$, and

$$l(m) = \begin{cases} \frac{1}{\sqrt{M}}, & \text{if } m = 1; \\ \frac{2}{\sqrt{M}}, & \text{otherwise.} \end{cases} \tag{8}$$

Since most of the audible components of an audio signal are in the low and mid frequency [19], in order to tackle filtering and compression operations, we choose the first K DCT coefficients for watermark embedding. Denote $\bar{S}_{L,i}$ as the selected DCT coefficients in the ith segment. Then $\bar{S}_{L,i}$ can be expressed as

$$\bar{S}_{L,i} = [S_{L,i}(1), \ S_{L,i}(2), \ldots, S_{L,i}(K)]. \tag{9}$$

Thirdly, similar to the left channel, by following (6)–(9), we can also obtain $s_{R,i}$, $S_{R,i}$, and $\bar{S}_{R,i}$ for the right channel.

2.3   Watermark Embedding

In our proposed method, the watermark bits are embedded by adjusting the PCC between $\bar{S}_{L,i}$ and $\bar{S}_{R,i}$. Similar to (2), the PCC for the ith segment ρ_i can be expressed as

$$\rho_i = \frac{1}{K} \sum_{k=1}^{K} \frac{(\bar{S}_{L,i}(k) - E(\bar{S}_{L,i}))(\bar{S}_{R,i}(k) - E(\bar{S}_{R,i}))}{\sigma_{L,i}\sigma_{R,i}}, \tag{10}$$

where $\sigma_{L,i}$ and $\sigma_{R,i}$ are the standard deviations of $\bar{S}_{L,i}$ and $\bar{S}_{R,i}$, respectively. Denote the watermark sequence as w and the ith watermark bit as $w(i)$, where $w(i) \in \{0, 1\}$. $w(i)$ will be embedded using the following procedures:

$w(i)$ is "0" A watermark bit "0" is embedded by modifying the PCC to satisfy

$$\rho_i' \geq T_0, \tag{11}$$

where ρ_i' is the modified PCC for the ith segment and T_0 is a predefined threshold. If $\rho_i \geq T_0$, no modifications are needed. Otherwise, introduce a mean value sequence D_i as

$$D_i(k) = \frac{\bar{S}_{L,i}(k) + \bar{S}_{R,i}(k)}{2}, \tag{12}$$

where $k = 1, 2, \ldots, K$. Then modify $\bar{S}_{L,i}$ and $\bar{S}_{R,i}$ as follows

$$\begin{cases} \bar{S}_{L,i}' = \bar{S}_{L,i} + \alpha_0 \cdot (D_i - \bar{S}_{L,i}); \\ \bar{S}_{R,i}' = \bar{S}_{R,i} + \alpha_0 \cdot (D_i - \bar{S}_{R,i}), \end{cases} \tag{13}$$

where $\bar{S}'_{L,i}$ and $\bar{S}'_{R,i}$ are the watermarked counterparts of $\bar{S}_{L,i}$ and $\bar{S}_{R,i}$, respectively, and α_0 is the parameter which controls the strength of the watermark. Substitute (13) into (10), ρ'_i can be calculated as

$$\rho'_i = \frac{1}{K} \sum_{k=1}^{K} \frac{(\bar{S}'_{L,i}(k) - E(\bar{S}'_{L,i}))(\bar{S}'_{R,i}(k) - E(\bar{S}'_{R,i}))}{\sigma'_{L,i}\sigma'_{R,i}}, \tag{14}$$

where $\sigma'_{L,i}$ and $\sigma'_{R,i}$ are the standard deviations of $\bar{S}'_{L,i}$ and $\bar{S}'_{R,i}$, respectively.

From (14) we can see that a larger α_0 will result in a higher ρ'_i. When $\alpha_0 = 1$, from (13) it is obvious that $\bar{S}'_{L,i} = \bar{S}'_{R,i}$, which leads to $\rho'_i = 1$. However, a larger α_0 also cause more perceptual quality degradation. Since the PCC may vary for different segments, in order to preserve the perceptual quality of the watermarked signal, α_0 is selected using an adaptive manner. Algorithm 1 shows the procedures of embedding a bit "0" for the ith segment, where α_{int0} is the initial value for α_0 and $\Delta\alpha$ is the step size.

Algorithm 1. The embedding algorithm for bit "0" for the ith segment.

Input: $\bar{S}_{L,i}$, $\bar{S}_{R,i}$, T_0, α_{int0}, $\Delta\alpha$
Output: $\bar{S}'_{L,i}$, $\bar{S}'_{R,i}$
1: Calculate D_i using Equation (12).
2: Set $\alpha_0 = \alpha_{int0}$.
3: Set $\bar{S}'_{L,i} = \bar{S}_{L,i}$, $\bar{S}'_{R,i} = \bar{S}_{R,i}$.
4: Calculate ρ'_i using Equation (14).
5: **while** $\rho'_i < T_0$ **do**
6:    Update $\alpha_0 = \alpha_0 + \Delta\alpha$.
7:    Calculate $\bar{S}'_{L,i}$ and $\bar{S}'_{R,i}$ using Equation (13).
8:    Calculate ρ'_i using Equation (14).
9: **end while**

$w(i)$ **is "1"** To embed a bit "1", we will reduce the PCC so that

$$\rho'_i \leq T_1, \tag{15}$$

where T_1 is a predefined threshold. To achieve this, we generate two additive white Gaussian noise (AWGN) sequences $n_{L,i}$ and $n_{R,i}$ with length K. The PCC between $n_{L,i}$ and $n_{R,i}$ are close to 0. Then modify $\bar{S}_{L,i}$ and $\bar{S}_{R,i}$ as follows

$$\begin{cases} \bar{S}'_{L,i} = \bar{S}_{L,i} + \alpha_1 \cdot n_{L,i}; \\ \bar{S}'_{R,i} = \bar{S}_{R,i} + \alpha_1 \cdot n_{R,i}, \end{cases} \tag{16}$$

where α_1 is the parameter which controls the strength of the watermark. A larger α_1 will result in a lower ρ'_i. When α_1 is large enough to make $\bar{S}'_{L,i} \approx n_{L,i}$ and $\bar{S}'_{R,i} \approx n_{R,i}$, we will have $\rho'_i \approx 0$. Similar to the case when $w(i) = 0$, in order to preserve the perceptual quality, α_1 is also selected by an adaptive manner.

Algorithm 2. The embedding algorithm for bit "1" for the ith segment.

Input: $\bar{S}_{L,i}$, $\bar{S}_{R,i}$, T_1, α_{int1}, $\Delta\alpha$, $n_{L,i}$, $n_{R,i}$
Output: $\bar{S}'_{L,i}$, $\bar{S}'_{R,i}$
1: Set $\alpha_1 = \alpha_{int1}$.
2: Set $\bar{S}'_{L,i} = \bar{S}_{L,i}$, $\bar{S}'_{R,i} = \bar{S}_{R,i}$.
3: Calculate ρ'_i using Equation (14).
4: **while** $\rho'_i > T_1$ **do**
5:    Update $\alpha_1 = \alpha_1 + \Delta\alpha$.
6:    Calculate $\bar{S}'_{L,i}$ and $\bar{S}'_{R,i}$ using Equation (16).
7:    Calculate ρ'_i using Equation (14).
8: **end while**

Algorithm 2 demonstrates the steps to embed a watermark bit "1", where α_{int1} is the initial value for α_1.

Once $w(i)$ is embedded and $\bar{S}'_{L,i}$ and $\bar{S}'_{R,i}$ are obtained, we can have

$$\begin{cases} S'_{L,i} &= [\bar{S}'_{L,i},\ S_{L,i}(K+1),\ S_{L,i}(K+2),\dots,\ S_{L,i}(M)]; \\ S'_{R,i} &= [\bar{S}'_{R,i},\ S_{R,i}(K+1),\ S_{R,i}(K+2),\dots,\ S_{R,i}(M)], \end{cases} \quad (17)$$

where $S'_{L,i}$ and $S'_{R,i}$ are the watermarked DCT coefficients for the ith segment for the left channel and the right channel, respectively. Then by applying the inverse DCT (IDCT) to $S'_{L,i}$ and $S'_{R,i}$, the ith segment of the watermarked signal can be calculated as

$$\begin{cases} s'_{L,i} &= \text{IDCT}(S'_{L,i}); \\ s'_{R,i} &= \text{IDCT}(S'_{R,i}), \end{cases} \quad (18)$$

where $s'_{L,i}$ and $s'_{R,i}$ are the watermarked left channel signal and right channel signal of the ith segment, respectively, and IDCT(.) is the IDCT operator.

Finally, by combining the watermarked signals of all N segments, we can have the watermarked left channel signal as s'_L and right channel signal as s'_R. Then the watermarked stereo signal s' can be obtained as

$$s' = [s'_L,\ s'_R]. \quad (19)$$

2.4   Watermark Extraction

The watermarked stereo signal may encounter intentional or unintentional attacks. Denote the received stereo signal as p with size $L' \times 2$. Similar to (6)–(9), the left channel and the right channel of p will be segmented into N blocks and transformed into DCT domain. Denote $\bar{P}_{L,i}$ and $\bar{P}_{R,i}$ as the selected DCT coefficients of the left channel and the right channel with length K', respectively. According to (10), we can calculate the PCC $\hat{\rho}_i$ between $\bar{P}_{L,i}$ and $\bar{P}_{R,i}$ as

$$\hat{\rho}_i = \frac{1}{K'} \sum_{k=1}^{K'} \frac{(\bar{P}_{L,i}(k) - E(\bar{P}_{L,i}))(\bar{P}_{R,i}(k) - E(\bar{P}_{R,i}))}{\hat{\sigma}_{L,i}\hat{\sigma}_{R,i}}, \quad (20)$$

where $\hat{\sigma}_{L,i}$ and $\hat{\sigma}_{R,i}$ are the standard deviations of $\bar{P}_{L,i}$ and $\bar{P}_{R,i}$, respectively. Then the ith extracted watermark bit $\hat{w}(i)$ can be obtained by

$$\hat{w}(i) = \begin{cases} 0, & \text{if } \hat{\rho}_i > \frac{T_0+T_1}{2}; \\ 1, & \text{otherwise.} \end{cases} \tag{21}$$

2.5 Synchronization

In order to tackle attacks which can modify the duration of the watermarked stereo signal, such as jittering and PITS, similar to [20], we divide w into three groups, which are the 1st synchronization codes x_1, the 2nd synchronization codes x_2, and the information bits y, where $x_1 = x_2$. Therefore, the watermark sequence w can be rewritten as

$$w = [x1, \ x2, \ y]. \tag{22}$$

During the watermark extraction step, firstly the 1st synchronization codes $\hat{x}_1$ and the 2nd synchronization codes $\hat{x}_2$ will be extracted and compared. The comparison process can be expressed as

$$Q = \sum \hat{x}_1 \oplus \hat{x}_2, \tag{23}$$

where $\oplus$ is the XOR operator. If $Q = 0$, the current frame window is considered as synchronized, and the information bits $\hat{y}$ will be extracted. Otherwise, we will slightly move the frame window and extract $\hat{x}_1$ and $\hat{x}_2$ again until the frame window is synchronized.

3 Experimental Results

In this section, we will compare the performance of the proposed method with [18,20,22] under common signal processing attacks such as closed-loop, noise addition attack, LP filtering, MP3 compression, AAC compression, and amplitude scaling, and common desynchronization attacks including jittering, TIPS, and PITS. [18] and [20] are two robust watermarking methods for mono signals. [22] is a singular value decomposition (SVD) based watermarking scheme for stereo signals. We used 100 audio clips for the performance comparison. The testing audio clips include pop, folk, jazz, funk, and rock with a duration of 15 seconds each. The perceptual quality of the watermarked signals is measured by the objective difference grade (ODG) [23], which ranges from -4 to 0. When the ODG value is greater than -1, the distortion caused by the watermark embedding procedure can be considered as imperceptible. In order to maintain the perceptual quality of the watermarked signals, the performance is evaluated under ODG ≈ -0.7 for all of the methods. The embedding rate is set to 10 bps for all of the methods. The performance is evaluated by bit-error-rate (BER), which can be formulated as

$$\text{BER} = \frac{\text{Number of error bits}}{\text{Number of watermark bits}} \times 100\%. \tag{24}$$

A lower BER indicates a better robustness. The parameters used in the proposed method are as follows: $\alpha_{int0} = 0.5$, $\alpha_{int1} = 0.1$, $\Delta\alpha = 0.01$, $T_0 = 0.99$, and $T_1 = 0.88$.

3.1 Common Signal Processing Attacks

Table 1 shows the performance of the method in [18], the method in [20], the method in [22], and the proposed method under common signal processing attacks. It can be observed that the proposed method achieves excellent robustness against all kinds of signal processing attacks. The closed-loop means no attacks present. Therefore, all 4 methods can detect the watermark without any errors. The LP filtering removes the frequency components higher than the cut-off frequency of the audio signal. The MP3 compression and the AAC compression reduce the size of the audio file by compressing the high frequency components while maintaining the low frequency components of the audio signal. Since the proposed method embeds the watermark bits into the low frequency components of the host audio signal, it is resilient against the LP filtering, the MP3 compression, and the AAC compression. On the other hand, the method in [22] cannot resist LP filtering and compression attacks. The noise addition attack introduces AWGN to the audio signal, while the quantization attack reduce the resolution of the audio signal. Since the PCC is a statistical property of the audio signal, it is robust to minor modification caused by the noise addition attack and the quantization attack. Quite the constrary, the method in [20] and the method in [22] suffer from the noise addition attack. The amplitude scaling attack modifies the amplitude of the audio signal by a scaling factor. Since the PCC will not be affected by amplitude scaling, the proposed method is invariant to the amplitude scaling attack. However, the method in [18] is based on QIM, which is vulnerable to the amplitude scaling attack.

Table 1. BER (%) for the method in [18], the method in [20], the method in [22], and the proposed method under common signal processing attacks.

Attacks	Method in [18]	Method in [20]	Method in [22]	Proposed method
Closed-loop	0	0	0	0
LP filtering 4 kHz	3	51	15	0
MP3 128 kbps	0	0	25	0
MP3 64 kbps	0	0	50	0
AAC 128 kbps	0	0	13	0
Noise addition SNR 20 dB	0	23	14	0
Noise addition SNR 10 dB	0	35	59	0
Quantization 12 bits	0	0	0	0
Quantization 8 bits	0	0	0	0
Amplitude scaling 90%	9	0	0	0
Amplitude scaling 80%	77	0	0	0

3.2   Common Desynchronization Attacks

Table 2 demonstrates the robustness of all 4 methods against common desynchronization attacks. We can see that the proposed method outperforms the other 3 methods under common desynchronization attacks. The jittering attack takes out a number of samples from random positions of the audio signal. As the PCC is a statistical property of the audio signal, which will not be significantly modified by removing a small portion of samples, the proposed method is robust against the jittering attack. The TIPS modifies the pitch of the audio signal without changing the duration, while the PITS modifies the duration of the signal without shifting the pitch. As analysed in Sect. 2.1, the proposed method is resilient against TIPS and PITS. However, the method in [22] shows little resistance against TIPS and PITS.

Table 2. BER (%) for the method in [18], the method in [20], the method in [22], and the proposed method under common desynchronization attacks.

Attacks	Method in [18]	Method in [20]	Method in [22]	Proposed method
Jittering 1%	0	0	1	0
Jittering 5%	0	0	1	0
Jittering 10%	0	1	1	0
TIPS 80%	8	2	45	0
TIPS 90%	4	1	40	0
TIPS 110%	1	0	41	0
TIPS 120%	1	1	42	0
PITS 80%	5	3	44	0
PITS 90%	3	1	42	0
PITS 110%	2	1	39	0
PITS 120%	6	2	42	0

4   Conclusion

In this paper, we introduced a channel correlation based audio watermarking algorithm for stereo signals for web-based applications. The watermarking embedding is achieved by modifying the PCC between the left channel signal and the right channel signal in the frequency domain. In order to preserve the perceptual quality, the parameters which control the watermark strength is selected using an adaptive manner for each segment. The effectiveness of PCC against common desynchronization attacks, especially TIPS and PITS, has been theoretically analysed. The simulation results show that the proposed method outperforms the existing audio watermarking methods under common signal processing attacks and common desynchronization attacks.

References

1. The True Cost of Sound Recording Piracy to the U.S. Economy. https://www.riaa.com/reports/the-true-cost-of-sound-recording-piracy-to-the-u-s-economy, Accessed 18 May 2020
2. Digital video piracy costs movie and TV industry at least $29 billion a year. https://www.cnet.com/news/digital-video-piracy-costs-the-movie-and-tv-industry-at-least-29-billion-study-says, Accessed 18 May 2020
3. Wang, H., Yi, X., Bertino, E., Sun, L.: Protecting outsourced data in cloud computing through access management. Concurrency Comput. Pract. Exp. **28**(3), 600–615 (2014)
4. Cheng, K., et al.: Secure k-NN query on encrypted cloud data with multiple keys. IEEE Trans. Big Data **14**(8), 1–14 (2017)
5. Xiang, Y., Peng, D., Natgunanathan, I., Zhou, W.: Effective pseudonoise sequence and decoding function for imperceptibility and robustness enhancement in time-spread echo-based audio watermarking. IEEE Trans. Multimedia **13**(1), 2–13 (2011)
6. Xiang, Y., Natgunanathan, I., Guo, S., Zhou, W., Nahavandi, S.: Patchwork-based audio watermarking method robust to de-synchronization attacks. IEEE/ACM Trans. Audio Speech Lang. Process. **22**(9), 1413–1423 (2014)
7. Xiang, Y., Natgunanathan, I., Peng, D., Hua, G., Liu, B.: Spread spectrum audio watermarking using multiple orthogonal PN sequences and variable embedding strengths and polarities. IEEE/ACM Trans. Audio Speech Lang. Process. **26**(3), 529–539 (2018)
8. Su, P., Chang, Y., Wu, C.: Geometrically resilient digital image watermarking by using interest point extraction and extended pilot signals. IEEE Trans. Inf. Forensics Secur. **8**(12), 1897–1908 (2013)
9. Zong, T., Xiang, Y., Natgunanathan, I., Guo, S., Zhou, W., Beliakov, G.: Robust histogram shape-based method for image watermarking. IEEE Trans. Circuits Syst. Video Technol. **25**(5), 717–729 (2015)
10. Zong, T., Xiang, Y., Guo, S., Rong, Y.: Rank-based image watermarking method with high embedding capacity and robustness. IEEE Access **4**(1), 1689–1699 (2016)
11. Lee, M., Kim, K., Lee, H.: Digital cinema watermarking for estimating the position of the pirate. IEEE Trans. Multimedia **12**(7), 605–621 (2010)
12. Stütz, T., Autrusseau, F., Uhl, A.: Non-blind structure-preserving substitution watermarking of H.264/CAVLC inter-frames. IEEE Trans. Multimedia **16**(5), 1337–1349 (2014)
13. Asikuzzaman, M., Alam, M.J., Lambert, A.J., Pickering, M.R.: Imperceptible and robust blind video watermarking using chrominance embedding: a set of approaches in the DT CWT domain. IEEE Trans. Inf. Forensics Secur. **9**(9), 1502–1517 (2014)
14. Cox, I.J., Kilian, J., Leighton, F.T., Shamoon, T.: Secure spread spectrum watermarking for multimedia. IEEE Trans. Image Process. **6**(12), 1673–1687 (1997)
15. Xiang, Y., Natgunanathan, I., Rong, Y., Guo, S.: Spread spectrum based high embedding capacity watermarking method for audio signals. IEEE/ACM Trans. Audio Speech Lang. Process. **23**(12), 2228–2237 (2015)
16. Ko, B., Nishimura, R., Suzuki, Y.: Time-spread echo method for digital audio watermarking. IEEE Trans. Multimedia **7**(2), 212–221 (2005)
17. Xiang, Y., Natgunanathan, I., Peng, D., Zhou, W., Shui, Y.: A dual-channel time-spread echo method for audio watermarking. IEEE Trans. Inf. Forensics Secur. **7**(2), 383–392 (2012)

18. Feng, H., Jiang, W., Huang, X., Jing, X.: Robust audio watermarking algorithm against TSM. In: International Conference Intelligence Computer and Integrated System, pp. 76–78. IEEE, Guilin (2010)
19. Kang, X., Yang, R., Huang, J.: Geometric invariant audio watermarking based on an LCM feature. IEEE Trans. Multimedia **13**(2), 181–190 (2011)
20. Liu, Z., Huang, Y., Huang, J.: Patchwork-based audio watermarking robust against de-synchronization and recapturing attacks. IEEE Trans. Inf. Forensics Secur. **14**(5), 1171–1180 (2019)
21. Wu, J., Shin, J.: Discrete cosine transform in error control coding. IEEE Trans. Commun. **43**(5), 1857–1861 (1995)
22. Hwang, M., Lee, J., Lee, M., Kang, H.: SVD-based adaptive QIM watermarking on stereo audio signals. IEEE Trans. Multimedia **20**(1), 45–54 (2018)
23. Thiede, T., et al.: PEAQ-the ITU standard for objective measurement of perceived audio quality. J. Audio Eng. Soc. **48**(1), 3–29 (2000)

Adaptive Online Learning
for Vulnerability Exploitation
Time Prediction

Jiao Yin[1,2], MingJian Tang[3], Jinli Cao[1(✉)], Hua Wang[4], Mingshan You[5],
and Yongzheng Lin[6]

[1] Department of Computer Science and Information Technology,
La Trobe University, Melbourne, VIC 3083, Australia
j.cao@latrobe.edu.au
[2] School of Artificial Intelligence, Chongqing University of Arts and Sciences,
Chongqing 402160, China
[3] Huawei Technologies Co. Ltd, Shenzhen 518129, China
[4] Institute for Sustainable Industries and Liveable Cities, Victoria University,
Melbourne, VIC 3083, Australia
[5] School of Computer Science and Electrical Engineering, Hunan University,
Changsha 410082, China
[6] School of Information Science and Engineering, University of Jinan, Jinan
250022, China

Abstract. Exploitation analysis is vital in evaluating the severity of
software vulnerabilities and thus prioritizing the order of patching.
Although a few methods have been proposed to predict the exploitability
of vulnerabilities, most of them treat this problem as an offline binary
classification problem. To suit the real-world data stream applications
and provide more fine-grained results for vulnerability evaluation, we
believe that it is better to treat the exploitation time prediction problem
as a multiclass online learning problem. In this paper, we propose an
adaptive online learning framework for exploitation time prediction to
tackle the combined challenges posed by online learning, multiclass learning and dynamic class imbalance. Within this framework, we design a
Sliding Window Imbalance Factor Technique (SWIFT) to capture the
real-time imbalanced statuses and thus to handle the dynamic imbalanced problem. Experimental results on real-world data demonstrate
that the proposed framework can improve the predictive performance
for both the minority class and the majority class.

Keywords: Exploitation time prediction · Online learning · Multiclass
imbalance

1 Introduction

The world is increasingly linked with internet-connected devices [10,20]. At the
same time, software vulnerabilities and risks, as well as cyber threats and attacks

© Springer Nature Switzerland AG 2020
Z. Huang et al. (Eds.): WISE 2020, LNCS 12343, pp. 252–266, 2020.
https://doi.org/10.1007/978-3-030-62008-0_18

are soaring, with a much faster speed than can be patched in time [14, 29]. Therefore, vulnerability evaluation and risk assessment become the most cost-effective way and strategic choices for security vendors to make more flexible decisions [1, 12, 18, 28].

In industry, the de facto standard—Common Vulnerability Scoring System (CVSS)—provides a score between 0 to 10 as the overall vulnerability assessment from multiple perspectives, such as attack complexity, required privileges, user interaction and confidentiality [16]. However, CVSS has been criticized as being inconsistent with the real-world situation in terms of exploitability [4, 7]. Therefore, vulnerability exploitability evaluation and prediction has long been considered a research hotspot in academia.

Most researchers have focused on the binary-class exploitability prediction problem, that is, to predict whether a vulnerability will be exploited or not [2, 4]. Obviously, multiclass exploitation time prediction makes much more sense to decision-makers, because it can predict when a vulnerability will be exploited for consideration. However, to the best of our knowledge, no previous research has specifically investigated the exploitation time prediction problem, except a few papers on Zero-Day Exploit detection [3], which is also a binary classification problem. Furthermore, previous studies all belong to offline learning strategies [19, 21]. The main problem of offline learning for data steam applications is that it involves using knowledge learnt from later-appeared samples to predict earlier-appeared samples and thus causing an inflated performance.

To suit the real-world situation and provide more fine-grained results for vulnerability evaluation, we treat the exploitation time prediction problem as a multiclass imbalanced online learning problem. In this paper, there are threefold challenges posed by multiclass learning, imbalanced learning and online learning to deal with.

Firstly, multiclass classification tasks have been proved to be more challenging than binary classification in offline learning [24], because of the increased data complexity and the imbalanced one-vs-all data distribution for each class. Predictably, multiclass learning could be even more tricky when learning online because of dynamic data distributions [5, 9].

The second challenging issue is the dynamic class imbalance. When the data distributions across known classes are biased or skewed over time, it is an imbalanced online learning problem [17]. Because of lacking prior knowledge of the whole data, some specific techniques should be employed to detect and track the class imbalanced pattern over time [13]. Corresponding training strategies are also required to adapt the online learner over time and thus mitigate the negative impact caused by imbalanced data. Otherwise, models may achieve poor predictive performance, especially for the minority class [11].

Thirdly, compared with offline learning, online learning may involve new concepts when new samples arrive [27]. The classification model should be deliberately updated over time to adapt new samples and improve the final performance.

To dress the above-mentioned pains, we propose an adaptive online learning framework for exploitation time prediction. Specifically, we design a Sliding Window Imbalance Factor Technique (SWIFT) within the framework to tackle the dynamic imbalanced learning problem.

The rest of this paper is organized as follows. Section 2 firstly formulates the multiclass imbalanced online learning problem and then introduces some related works. Section 3 presents the proposed adaptive learning framework and how SWIFT algorithm works. Section 4 gives the exploitation time prediction results on real-word data from 1988 to 2020. Section 5 concludes the paper and points out our future work.

2   Problem Formulation and Related Work

In this paper, we focus on the task of multiclass imbalanced online learning problem. Let $x^{(t)} \in \mathbb{R}^n$ is a vector in n dimensional feature space $\mathcal{X}$ observed at time step t. $y^{(t)}$ is the corresponding label and $y^{(t)} \in \{c_{[1]}, c_{[2]}, \cdots, c_{[N]}\} (N \geqslant 3)$, where $c_{[1]}, c_{[2]}, \cdots, c_{[N]}$ are the N class labels that have appeared so far. We call $x^{(t)}$ an instance of a data stream at time step t and a pair $(x^{(t)}, y^{(t)})$ a labelled instance.

To keep track of the real time imbalanced statuses of all classes, paper [22] defined an imbalance factor $w^{(t)} = \{w_{[1]}^{(t)}, w_{[2]}^{(t)}, \cdots, w_{[N]}^{(t)}\}$ to indicate the overall class proportions in time step t. Specifically, the data proportion of class $c_{[k]}$, denoted as $w_{[k]}^{(t)}$, is updated by (1).

$$w_{[k]}^{(t)} = \frac{(t-1) * w_{[k]}^{(t-1)} + [(x^{(t)}, c_{[k]})]}{t}, (k = 1, 2, \cdots, N \text{ and } t \geq 1) \qquad (1)$$

where $w_{[k]}^{(0)} = 0$, $[(x^{(t)}, c_{[k]})] = 1$ if the true label of $x^{(t)}$ equals to $c_{[k]}$, otherwise 0. When the class distributions keep stable as time goes by, $w_{[k]}^{(t)}$ will gradually get closer to the real imbalanced status over time. However, the imbalanced status in exploitation time prediction scenario, as well as many other data streams, often changes dynamically and irregularly. Equation (1) is insensitive to reflect the new imbalanced status, due to the great influence of old data. To weaken the effect of old data, a time decay factor θ is introduced to update the imbalance factor $w_{[k]}^{(t)}$, as shown in (2) [23].

$$w_{[k]}^{(t)} = \theta w_{[k]}^{(t-1)} + (1 - \theta)[(x^{(t)}, c_{[k]})], (k = 1, 2, \cdots, N \text{ and } t \geq 1) \qquad (2)$$

After obtaining $w_{[k]}^{(t)}$, a Multiclass Oversampling (MO) method and a Multiclass Undersampling (MU) method were used to handle the multiclass imbalanced online learning problem in [23]. Both MO and MU resample data by training the current labelled instance $(x^{(t)}, y^{(t)})$ K times when updating classifier from $f^{(t)}$ to $f^{(t+1)}$, where K obeys a Poisson Distribution, namely,

$K \sim Poisson(\lambda)$ and λ is defined by (3).

$$\lambda = \begin{cases} w_{\max}^{(t)}/w_{[k]}^{(t)}, (k = 1, 2, \cdots, N \text{ and } t \geq 1), \text{ for MO method} \\ w_{\min}^{(t)}/w_{[k]}^{(t)}, (k = 1, 2, \cdots, N \text{ and } t \geq 1), \text{ for MU method} \end{cases} \tag{3}$$

where $w_{\min}^{(t)} = \min_{k=1}^{N} w_{[k]}^{(t)}$ is the minimum class proportion at time step t and $w_{\max}^{(t)} = \max_{k=1}^{N} w_{[k]}^{(t)}$ is the maximum class proportion. Obviously, $\lambda \geq 1$ for MO method. Therefore, according to the properties of Poisson Distribution, MO algorithm can increase the training epochs for the minority samples, equivalent to oversample instances from the minority class. Similarly, $\lambda \leq 1$ for MU method and the chance of learning majority classes will be reduced.

Time decay Imbalance Factor calculated by (2) assumes that all classes decay in the same rate θ and finally the first item in Eq. (2) will gradually converge to 0 over time. However, this assumption can hardly hold true in real-world applications, especially for data steam situations like the exploitation time prediction problem.

3   Adaptive Online Learning Framework

In this section, we first set forth the proposed adaptive online learning framework for multiclass imbalanced online learning problems. It is in particular designed for exploitation time prediction, but can also be used for similar data stream applications. Then, we go into detail about the principle and implementation of the SWIFT algorithm.

3.1   Work Flow of the Proposed Framework

The work flow of the proposed adaptive online learning framework is shown in Fig. 1. This is an algorithm-agnostic framework, which means any feature extraction algorithms and classifiers can be used to fill in this framework. In Fig. 1, $D = \{d^{(1)}, d^{(2)}, \cdots, d^{(t)}, \cdots\}$ is the raw input data stream and $Y = \{y^{(1)}, y^{(2)}, \cdots, y^{(t)}, \cdots\}$ is the corresponding labels; $F = \{f^{(1)}(\cdot), f^{(2)}(\cdot), \cdots, f^{(t)}(\cdot), \cdots\}$ is a status sequence of the same classifier with different parameters, in other words, $f^{(t)}(\cdot)$ is the sequential status updated from $f^{(t-1)}(\cdot)$. When there is no prior knowledge or data available, $f^{(1)}(\cdot)$ is initialized with random parameters. However, when some pre-obtained data is available, $f^{(1)}(\cdot)$ could be initialized with a pre-trained model.

There are two steps to be done at each time step t ($t \geq 1$) in this adaptive framework, namely, prediction and classifier updating.

Prediction. When raw datum $d^{(t)}$ arrives, the first thing to do is transforming $d^{(t)}$ into numerical features $x^{(t)}$ by some feature extraction and selection algorithms. Then, using the current classifier $f^{(t)}(\cdot)$, we can predict the output related to the current sample for downstream applications by calculating

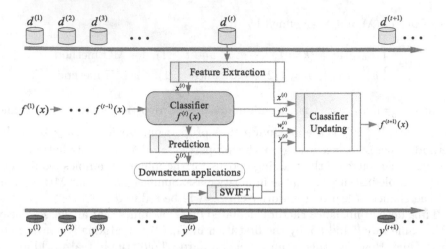

Fig. 1. Work flow of the proposed adaptive online learning framework

$\hat{y}^{(t)} = f^{(t)}(x^{(t)})$. For example, when $d^{(t)}$ is a newly published vulnerability, $\hat{y}^{(t)}$ is the predicted exploitation time, which can be used by decision-makers on cybersecurity-related resource allocation.

Classifier Updating. When the true label $y^{(t)}$ arrives, the classifier $f^{(t)}(\cdot)$ gets updated by retraining the classifier with the new labelled instance $(x^{(t)}, y^{(t)})$. Specifically, we design a strategy named SWIFT to improve the overall prediction performance. SWIFT is particularly designed for tackling the dynamic multiclass imbalanced problem by capturing the real-time multiclass imbalanced statuses and adjusting the class weight $w_c^{(t)}$ over time. More details on SWIFT will be presented in the following subsection. With the labelled sample $(x^{(t)}, y^{(t)})$ and the class weigh $w_c^{(t)}$, classifier $f^{(t)}(\cdot)$ can be retrained to $f^{(t+1)}(\cdot)$ for the usage of next time step.

The whole adaptive online learning framework is a loop on time step t. Within the loop, classifier $f^{(t)}(\cdot)$ adjusts its parameters adaptively according to the real time data stream. The in detailed implementation process of the whole adaptive online learning framework is shown in the pseudo-code of Algorithm 1.

3.2   Sliding Window Imbalance Factor Technique

The idea of SWIFT comes from a basic intuition that the multiclass imbalanced statuses are dynamically changed and samples come from current minority class should be assigned a larger class weight, because less samples are available to learn the patterns of the minority class.

Specifically, SWIFT consists of two steps at each time step t ($t \geq 1$), namely, updating the Sliding Window Imbalance Factor (SWIF) $w^{(t)}$ and updating the class weight $w_c^{(t)}$.

Algorithm 1. Adaptive Online Learning Framework

    Input: raw data stream $D = \{d^{(1)}, d^{(2)}, \cdots, d^{(t)}, \cdots\}$; corresponding labels $Y = \{y^{(1)}, y^{(2)}, \cdots, y^{(t)}, \cdots\}$; initialized classifier $f^{(1)}(\cdot)$; $w_{c[k]}^{(1)} = 1$.

    Output: predicted labels $\widehat{Y} = \{\hat{y}^{(1)}, \hat{y}^{(2)}, \cdots, \hat{y}^{(t)}, \cdots\}$; updated classifiers $F = \{f^{(2)}(\cdot), f^{(3)}(\cdot), \cdots, f^{(t+1)}(\cdot), \cdots\}$.

1: **for** each time step t **do**
2:    **if** instance $d^{(t)}$ arrives **then**
3:      extracting and selecting features $x^{(t)} \in \mathbb{R}^n$ from $d^{(t)}$
4:      predicting the exploitation time of $x^{(t)}$ by calculating $\hat{y}^{(t)} = f^{(t)}(x^{(t)})$
5:    **end if**
6:    **if** instance label $y^{(t)}$ arrives **then**
7:      **if** SWIFT==True **then**
8:        calculating Sliding Window Imbalance Factor by equation (4)
9:        calculating class weight $w_c^{(t)}$ by equation (5)
10:     **end if**
11:     updating classifier $f^{(t)}$ to $f^{(t+1)}$ based on $(x^{(t)}, y^{(t)})$ and $w_c^{(t)}$.
12:    **end if**
13: **end for**

Updating the Sliding Window Imbalance Factor. The SWIF of each class $w^{(t)} = [w_{[1]}^{(t)}, w_{[2]}^{(t)}, \cdots, w_{[N]}^{(t)}]$ is designed on the basis of (1) and (2), where $w_{[k]}^{(t)}$ ($k = \{1, 2, \cdots, N\}$) represents the imbalance factor of class $c_{[k]}$. Instead of calculating the overall imbalance factor so far like (1) or setting a fixed time decay factor θ for all classes and time steps like (2), we calculate the imbalance factor of class $c_{[k]}$ with (4).

$$
w_{[k]}^{(t)} = \begin{cases} \dfrac{(t-1)*w_{[k]}^{(t-1)} + [(x^{(t)}, c_{[k]})]}{t}, & (k = 1, 2, \cdots, N \text{ and } 1 \leq t < z) \\[3mm] \dfrac{z*w_{[k]}^{(t-1)} - [(x^{(t-z)}, c_{[k]})] + [(x^{(t)}, c_{[k]})]}{z}, & (k = 1, 2, \cdots, N \text{ and } t \geqslant z) \end{cases} \tag{4}
$$

where z ($z \geq N$) is the sliding window size and can be optimized by common hyper-parameter optimisation techniques, such as Grid Search, Random Search and Bayesian Optimisation. When $1 \leq t < z$, SWIF is identical with (1).

Updating Class Weight. Instead of training the classifier K times according to $w_{[k]}^{(t)}$ like the MO and MU methods, we define a class weight $w_c^{(t)}$ based on SWIF, where $w_c^{(t)} = [w_{c[1]}^{(t)}, w_{c[2]}^{(t)}, \cdots, w_{c[N]}^{(t)}]$ and $w_{c[k]}^{(t)}$ ($k = \{1, 2, \cdots, N\}$) is the current class weight for class $c[k]$, calculated as (5).

$$
w_{c[k]}^{(t)} = \frac{1}{N_z * w_{[k]}^{(t)}}, (w_{[k]}^{(t)} \neq 0, \ k = 1, 2, \cdots, N \text{ and } t \geq 1)) \tag{5}
$$

where N_z ($N_z \leq N$) is the number of unique classes located in the current sliding window z. Especially, when $w_{[k]}^{(t)} = 0$, no sample belonging to class $c_{[k]}$ is included in the current sliding window and therefore no need to calculate its class weight.

4   Exploitation Time Prediction

In this section, we apply the proposed adaptive online learning framework to predict the exploitation time of software vulnerabilities. Comparative studies between SWITF and previous works are also conducted in this section.

4.1   Dataset

The vulnerability data comes from the National Vulnerability Database (NVD)[1] maintained by the National Institute of Standards and Technology (NIST), U.S. Department of Commerce. A total number of 132,344 unique vulnerabilities are published between 1988 and 2020 in NVD. Exploitation data comes from ExploitDB[2], using the exploit publication dates as the exploitation dates. By integrating vulnerabilities in NVD and exploits in ExploitDB with the cveID (a unique vulnerability identifier), 23,302 vulnerabilities are identified as exploited vulnerabilities [30].

A vulnerability's exploitation time is the time interval between the publication date of vulnerability and the earliest publication date of its corresponding exploits in days. When a vulnerability is exploited after being published, its exploitation time is positive and the class label is marked as 'Pos'. Similarly, when a vulnerability is exploited before being published, the exploitation time is negative and the class label is marked as 'Neg'. When the vulnerability and its earliest exploit are published at the same day, the exploitation time equals to zero, and the corresponding class label is 'ZeroDay'. The total number of vulnerabilities in our collected dataset for classes 'Pos', 'Neg' and 'ZeroDay' are 3877(16.6%), 18113 (77.7%) and 1312 (5.7%) respectively. Figure 2 shows the exploitation time distributions of all collected 23,302 exploitable vulnerabilities.

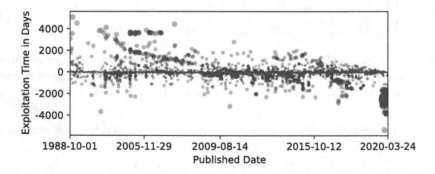

Fig. 2. Exploiltation time distributions of vulnerabilities from 1988 to 2020

To have an intuition on the multiclass imbalanced dynamics of the collected dataset, we draw the overall imbalance factor calculated by (1) and the SWIF calculated by (4) in Fig. 3. Although subplots (a) and (b) share the same trend, (a)

---

[1] https://nvd.nist.gov/vuln/data-feeds.

[2] https://www.exploit-db.com/.

is too smooth to reflect the latest imbalanced dynamics. For example, Fig. 3(b) shows that in recent years the ratio of 'Pos' vulnerabilities increased dramatically and sometimes the proportion is even over 0.6. However, due to the great influence of history data, the overall imbalance factor shown in (a) can only see a very smooth increase.

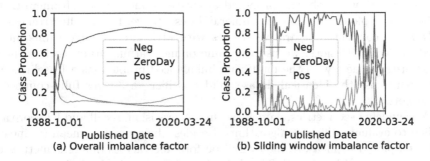

Fig. 3. Dynamical multiclass imbalanced statuses over time: (a) is calculated by Eq. (1); (b) is calculated by Eq. (4) when $z = 50$.

4.2   Feature Extraction and Selection

Recently developed machine learning and deep learning algorithms are powerful to extract features from unstructured raw data [25–27]. Previous studies usually extract features from CVSS metrics or vulnerability descriptions for vulnerability assessment or exploitability prediction [8,15,26]. In this paper, we extract features from both sides to maximize the representativeness of extracted features.

Specifically, the extracted feature attributes from CVSS metrics include 'referenceNum', 'baseScore', 'accessVector', 'accessComplexity', 'authentication', 'confidentialityImpact', 'integrityImpact', 'availabilityImpact', 'impactScore', 'exploitabilityScore', 'userInteractionRequired', 'severity', 'obtainUserPrivilege', 'obtainAllPrivilege', 'acInsufInfo' and 'obtainOtherPrivilege'. We adopt feature attributes from CVSS version V2.0, instead of version V3.0, because V2.0 is available for almost all published vulnerabilities, while V3.0 is only available for vulnerabilities published after 2015 in most cases.

Meanwhile, we employ a pre-trained Bidirectional Encoder Representations from Transformers (BERT) [6] model to extract sentence-level semantic features from vulnerability descriptions. The implementation and pre-trained BERT model are available at GitHub[3]. We use the token embedding of [CLS] at the last Transformer Layer of BERT model as the final feature representation of a vulnerability's description, which is a vector in a 768 dimensional feature space.

---

[3] https://github.com/google-research/bert.

Finally, we select 10 features from each side via Principal Component Analysis (PCA) and concatenate them into a 20-dimensional feature set for exploitation time prediction.

4.3 Comparative Study

In this section, we apply the proposed adaptive online learning framework to predict the exploitation time of vulnerabilities. To verify the effectiveness of SWIFT, we conduct comparative studies with MU and MO algorithms introduced in Sect. 2. Besides, we use the same online learning framework without any strategy to deal with the multiclass imbalanced problem as a Baseline. All experiments use the identical dataset and classifier—a 3-layer fully-connected neural network.

Four widely used metrics, namely, accuracy, precision, recall and F1 score are applied to evaluate different algorithms. Besides, the geometric mean (G-mean) is also employed to measure the overall performance across different metrics or classes, as adopted by [23,24]. The definition of G-mean is shown in (6),

$$G\text{-mean}(x_1, x_2, \cdots, x_n) = \sqrt[n]{x_1 \times x_2 \times \cdots \times x_n} \qquad (6)$$

where x_1, x_2, $\cdots$, x_n are the n elements to calculate the G-mean of them.

Performance Comparison on the Multi-majority Class—'Neg'. Figure 4 presents the four classification metrics of different algorithms on class 'Neg' over time. The overall fluctuation trends of all metrics are identical with the overall imbalance factor of 'Neg' shown in Fig. 3(a). In regards to accuracy and precision, SWIFT has a weak advantage compared with Baseline and is apparently better than MO and MU. However, MO achieves the best recall on class 'Neg', which is much better than the second level including Baseline and SWIFT. MU performs the worst on recall. All algorithms except MU achieve comparable performance on F1 score over time.

Table 1 shows the G-mean of different algorithms over the four metrics in Fig. 4. The best performance of each metric is emphasized in bold. SWIFT performs the best on accuracy, precision and F1 score, while MO achieves the best recall. SWIFT exceeds baseline on all four metrics and leads the Baseline by a large margin on three of them—accuracy, precision and F1 score. MO precedes Baseline only on precision and MU are worse than Baseline on all metrics.

Performance Comparison on the Multi-minority Class—'ZeroDay'. Figure 5 shows the performance comparison on class 'ZeroDay' over time. The performance fluctuation trends on all metrics are also consistent with the overall imbalance factor of 'ZeroDay' as shown in Fig. 3(a). In regards to accuracy, all algorithms except MU achieve comparatively good performance over time. MO achieves the best precision, following with SWIFT; both of them exceed Baseline in a large margin. MU performs the worst on accuracy and precision, while

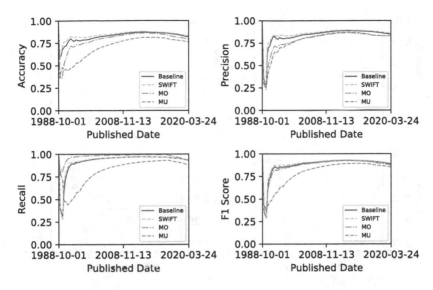

Fig. 4. Performance comparison on the multi-majority class—'Neg'.

Table 1. G-mean comparison on the multi-majority class—'Neg'.

Metrics	Baseline	SWIFT	MO	MU
Accuracy	0.8248	**0.8411**	0.7818	0.6973
Precision	0.8246	**0.8389**	0.7763	0.7676
Recall	0.9131	0.9172	**0.9740**	0.8033
F1 Score	0.8665	**0.8762**	0.8607	0.7815

MO performs the worst on recall and F1 score. SWIFT is apparently superior to Baseline, MU and MO on recall and F1 score.

Similarly, Table 2 summarizes the G-mean of different algorithms over the four metrics in Fig. 5. The best performance on each metric is marked in bold. For class 'ZeroDay', SWIFT wins the best accuracy at 0.8990. MO obtains the highest precision at 0.6252, which is much higher than the second place 0.5878 achieved by SWIFT. However, SWIFT achieves the best in both recall and F1 score, with a large advantage over Baseline, MU and MO.

Performance Comparison on the Multi-minority Class—'Pos'. Figure 6 shows the performance comparison on class 'Pos' over time. The performance fluctuation trends on all metrics are also consistent with the overall imbalance factor of class 'Pos' as shown in Fig. 3(a). All algorithms including MU achieve a comparably good performance on accuracy. In regards to precision, only SWIFT is always superior than the Baseline. MU is better than the Baseline only at the beginning years on precision. However, in recent years, Baseline increases faster than MU and eventually overtakes MU. In regards to recall and F1 score, only

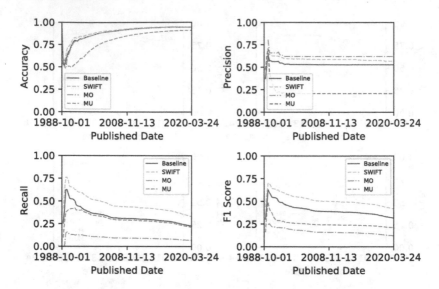

Fig. 5. Performance comparison on the multi-minority class—'ZeroDay'.

Table 2. G-mean comparison on the multi-minority class—'ZeroDay'.

Metrics	Baseline	SWIFT	MO	MU
Accuracy	0.8809	**0.8962**	0.8783	0.7836
Precision	0.5307	0.5878	**0.6252**	0.2202
Recall	0.3235	**0.4567**	0.09435	0.2947
F1 score	0.4001	**0.5123**	0.1637	0.2496

Table 3. G-mean comparison on the multi-minority class—'Pos'.

Metrics	Baseline	**SWIFT**	MO	MU
Accuracy	0.8716	**0.8863**	0.8833	0.8730
Precision	0.3088	**0.3999**	0.3246	0.1675
Recall	0.1436	**0.1550**	0.0593	0.0442
F1 Score	0.1939	**0.2215**	0.0971	0.0686

SWIFT has a weak advantage over the Baseline. MO and MU have a very poor performance on Recall and F1 Score.

Similarly, Table 3 presents the G-mean of different algorithms over the four metrics in Fig. 6. This time, SWIFT exceeds all other algorithms on all four metrics. The second place on accuracy and precision is MO; on recall and F1 score is Baseline. Although MO performs better than baseline on accuracy and precision, it has a very poor performance on recall and F1 score.

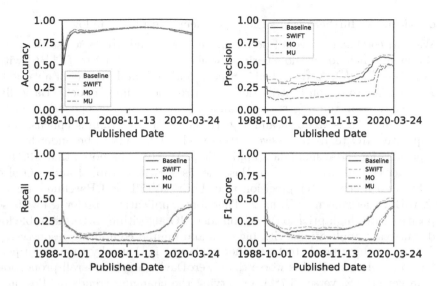

Fig. 6. Performance comparison on the multi-minority class—'Pos'.

G-Mean Comparison over All Classes. From Tables 1, 2 and 3, it is hard to sort the performance of those algorithms one by one. Because some algorithms may perform better than others on one class or one metric but at the same time worse on other classes or metrics. For example, SWIFT performs best on all metrics for class 'Pos' but it isn't the best algorithm for all metrics on class 'ZeroDay' and 'Neg', where MO performs best on precision of class 'ZeroDay' and on recall of class 'Neg'.

Therefore we further calculate the G-mean of different algorithms over all classes, as listed in the first 5 lines of Table 4. It shows that SWIFT achieves overwhelming advantages on all four metrics compared with other algorithms. Besides, an 'Overall' metric, the G-mean over the above four metrics, is used as the unique number to represent the final performance of each algorithm, as shown in the last line of Table 4. It shows that SWIFT performs the best at 0.6178 for online multiclass exploitation time prediction, excelling the second place Baseline at 0.5716 8.08%. While MO at 0.5117 is 10.47% lower than Baseline and MU at 0.4532 is even worse, with 20.71% lower than Baseline

Table 4. Comparison of G-mean over all classes

Metrics	Baseline	SWIFT	MO	MU
Accuracy	0.8591	**0.8746**	0.8478	0.7846
Precision	0.5547	**0.6089**	0.5754	0.3851
Recall	0.4601	**0.5097**	0.3759	0.3808
F1 score	0.4869	**0.5367**	0.3738	0.3666
Overall	0.5716	**0.6178**	0.5117	0.4532

Discussion. We further discuss the experimental results as follows.

(1) We noticed that the overall performance of MO and MU is worse than the Baseline, which adopts no algorithm when updating $f^{(t)}(\cdot)$ to $f^{(t+1)}(\cdot)$ with sample $(x^{(t)}, y^{(t)})$. As described in Sect. 2, both MO and MU adopt a variable K following a Poisson Distribution to control the training epochs. Basically, MO is an oversampling strategy allowing more training epochs on $(x^{(t)}, y^{(t)})$, while MU is an downsampling strategy with less training epochs. Consequently, MO tends to be over-fitting while MU tends to be under-fitting, especially for those small data applications. It should be noted that MO has a higher overall precision than Baseline, as shown in Table 4 and MO also achieves a much better precision than both SWIFT and Baseline on class 'ZeroDay', as shown in Table 2. Therefore, applications prefer a low false positive rate (high pricision) can consider MO algorithm. MU is suitable for downstream applications with big data and limited computing resources.

(2) The other thing needs to be further explained is that why the overall performance of class 'Pos' is worse than 'ZeroDay' when its overall proportion is larger (16.6% verses 5.7%). Observing the changing trends of 'Pos' and 'ZeroDay' shown in Fig. 3(a), we will find that class 'ZeroDay' accounts for a larger proportion at first and gradually goes down. Actually, in a relatively long period, 'ZeroDay' vulnerabilities account more than 'Pos'. In line with this, the performance of 'ZeroDay' decreases gradually in Fig. 5. While the performance of 'Pos' increases rapidly at the later stage as shown in Fig. 6. Of course, it should be acknowledged that classifiers have some time lags when adapting to the change of class imbalanced statuses in online learning mode. Therefore, it will take some time to make the performance of class 'Pos' overtaking class 'Zeorday'.

5   Conclusion

Software vulnerability exploitation analysis is of importance for vulnerability assessment and cybersecurity decision making. Existing studies all focus on predicting the exploitability of a vulnerability in an offline learning fashion. In this paper, we are the first to focus on a multiclass exploitation time prediction problem and tackle it as a multiclass imbalanced online learning problem.

To solve this practical problem, we propose an algorithm-agnostic adaptive online learning framework. Within this framework, we design a SWIFT algorithm to further improve the prediction performance. Experiments on real word vulnerability dataset from 1988 to 2020 have demonstrated the effectiveness of the proposed framework as well as the developed SWIFT algorithm.

In the future, we plan to improve the framework's efficiency by designing detective algorithms for multiclass imbalanced status changing and concept drift in data distributions.

Acknowledgment. The first author is partly supported by the Science and Technology Research Program of Chongqing Municipal Education Commission of China (Grant No. KJQN201901306)

References

1. Afzaliseresht, N., Miao, Y., Michalska, S., Liu, Q., Wang, H.: From logs to stories: human-centred data mining for cyber threat intelligence. IEEE Access **8**, 19089–19099 (2020)
2. Alazab, M., Tang, M.: Deep Learning Applications for Cyber Security. Springer, Switzerland (2019). https://doi.org/10.1007/978-3-030-13057-2
3. AlEroud, A., Karabatis, G.: A contextual anomaly detection approach to discover zero-day attacks. In: 2012 International Conference on Cyber Security, pp. 40–45. IEEE (2012)
4. Bozorgi, M., Saul, L.K., Savage, S., Voelker, G.M.: Beyond heuristics: learning to classify vulnerabilities and predict exploits. In: Proceedings of the 16th ACM SIGKDD International Conference on Knowledge Discovery and Data Mining, pp. 105–114. ACM (2010)
5. Cai, T., Li, J., Mian, A.S., Sellis, T., Yu, J.X., et al.: Target-aware holistic influence maximization in spatial social networks. IEEE Trans. Knowl. Data Eng. (2020)
6. Devlin, J., Chang, M.W., Lee, K., Toutanova, K.: Bert: Pre-training of deep bidirectional transformers for language understanding. arXiv preprint arXiv:1810.04805 (2018)
7. Eiram, C., Martin, B.: The cvssv2 shortcomings, faults, and failures formulation. In: Technical report, Forum of Incident Response and Security Teams (FIRST) (2013)
8. Han, Z., Li, X., Xing, Z., Liu, H., Feng, Z.: Learning to predict severity of software vulnerability using only vulnerability description. In: 2017 IEEE International Conference on Software Maintenance and Evolution (ICSME), pp. 125–136. IEEE (2017)
9. Li, J., Cai, T., Deng, K., Wang, X., Sellis, T., Xia, F.: Community-diversified influence maximization in social networks. Inf. Syst. **92**, 101522 (2020)
10. Li, M., Sun, X., Wang, H., Zhang, Y., Zhang, J.: Privacy-aware access control with trust management in web service. World Wide Web **14**(4), 407–430 (2011)
11. Liu, M., Zhang, X., Chen, Z., Wang, X., Yang, T.: Fast stochastic auc maximization with $o(1/n)$-convergence rate. In: International Conference on Machine Learning, pp. 3189–3197 (2018)
12. Rasool, R.U., Ashraf, U., Ahmed, K., Wang, H., Rafique, W., Anwar, Z.: Cyberpulse: a machine learning based link flooding attack mitigation system for software defined networks. IEEE Access **7**, 34885–34899 (2019)
13. Shen, Y., Zhang, T., Wang, Y., Wang, H., Jiang, X.: Microthings: a generic iot architecture for flexible data aggregation and scalable service cooperation. IEEE Commun. Mag. **55**(9), 86–93 (2017)
14. Tang, M., Alazab, M., Luo, Y.: Big data for cybersecurity: vulnerability disclosure trends and dependencies. IEEE Trans. Big Data **5**, 317–329 (2017)
15. Tavabi, N., Goyal, P., Almukaynizi, M., Shakarian, P., Lerman, K.: Darkembed: exploit prediction with neural language models. In: Thirty-Second AAAI Conference on Artificial Intelligence (2018)
16. Team, C.: Common vulnerability scoring system v3. 0: Specification document. First. org (2015)
17. Wang, B., Pineau, J.: Online bagging and boosting for imbalanced data streams. IEEE Trans. Knowl. Data Eng. **28**(12), 3353–3366 (2016)
18. Wang, H., Sun, L., Bertino, E.: Building access control policy model for privacy preserving and testing policy conflicting problems. J. Comput. Syst. Sci. **80**(8), 1493–1503 (2014)

19. Wang, H., Wang, Y., Taleb, T., Jiang, X.: Special issue on security and privacy in network computing. World Wide Web **23**(2), 951–957 (2020)
20. Wang, H., Yi, X., Bertino, E., Sun, L.: Protecting outsourced data in cloud computing through access management. Concurrency Comput. Pract. Exp. **28**(3), 600–615 (2016)
21. Wang, H., Zhang, Z., Taleb, T.: Special issue on security and privacy of iot. World Wide Web **21**(1), 1–6 (2018)
22. Wang, S., Minku, L.L., Yao, X.: A learning framework for online class imbalance learning. In: 2013 IEEE Symposium on Computational Intelligence and Ensemble Learning (CIEL), pp. 36–45. IEEE (2013)
23. Wang, S., Minku, L.L., Yao, X.: Dealing with multiple classes in online class imbalance learning. In: IJCAI, pp. 2118–2124 (2016)
24. Wang, S., Yao, X.: Multiclass imbalance problems: analysis and potential solutions. IEEE Trans. Syst. Man Cybern. Part B (Cybernetics) **42**(4), 1119–1130 (2012)
25. Wang, X., Wang, S., Xin, Y., Yang, Y., Li, J., Wang, X.: Distributed pregel-based provenance-aware regular path query processing on RDF knowledge graphs. In: World Wide Web, pp. 1–32 (2019)
26. Yang, Y., Guan, Z., Li, J., Huang, J., Zhao, W.: Interpretable and efficient heterogeneous graph convolutional network. arXiv preprint arXiv:2005.13183 (2020)
27. Yin, J., You, M., Cao, J., Wang, H., Tang, M.J., Ge, Y.-F.: Data-driven hierarchical neural network modeling for high-pressure feedwater heater group. In: Borovica-Gajic, R., Qi, J., Wang, W. (eds.) ADC 2020. LNCS, vol. 12008, pp. 225–233. Springer, Cham (2020). https://doi.org/10.1007/978-3-030-39469-1_19
28. Zhang, F., Wang, Y., Liu, S., Wang, H.: Decision-based evasion attacks on tree ensemble classifiers. In: World Wide Web, pp. 1–21 (2020)
29. Zhang, J., Li, H., Liu, X., Luo, Y., Chen, F., Wang, H., Chang, L.: On efficient and robust anonymization for privacy protection on massive streaming categorical information. IEEE Trans. Dependable Secure Comput. **14**(5), 507–520 (2015)
30. Zhang, J., Tao, X., Wang, H.: Outlier detection from large distributed databases. World Wide Web **17**(4), 539–568 (2014)

Recommender Systems

Double-Wing Mixture of Experts for Streaming Recommendations

Yan Zhao[1,2], Shoujin Wang[2], Yan Wang[2], Hongwei Liu[1(✉)],
and Weizhe Zhang[1,3]

[1] School of Computer Science and Technology, Harbin Institute of Technology,
Harbin, China
{yanzhao,liuhw,wzzhang}@hit.edu.cn
[2] Department of Computing, Macquarie University, Sydney, Australia
{shoujin.wang,yan.wang}@mq.edu.au
[3] Cyberspace Security Research Center, Peng Cheng Laboratory, Shenzhen, China

Abstract. Streaming Recommender Systems (SRSs) commonly train recommendation models on newly received data only to address *user preference drift*, i.e., the changing user preferences towards items. However, this practice overlooks the *long-term user preferences* embedded in historical data. More importantly, the common *heterogeneity* in data stream greatly reduces the accuracy of streaming recommendations. The reason is that different preferences (or characteristics) of different types of users (or items) cannot be well learned by a unified model. To address these two issues, we propose a Variational and Reservoir-enhanced Sampling based Double-Wing Mixture of Experts framework, called VRS-DWMoE, to improve the accuracy of streaming recommendations. In VRS-DWMoE, we first devise variational and reservoir-enhanced sampling to wisely complement new data with historical data, and thus address the user preference drift issue while capturing long-term user preferences. After that, we propose a Double-Wing Mixture of Experts (DWMoE) model to first effectively learn heterogeneous user preferences and item characteristics, and then make recommendations based on them. Specifically, DWMoE contains two Mixture of Experts (MoE, an effective ensemble learning model) to learn user preferences and item characteristics, respectively. Moreover, the multiple experts in each MoE learn the preferences (or characteristics) of different types of users (or items) where each expert specializes in one underlying type. Extensive experiments demonstrate that VRS-DWMoE consistently outperforms the state-of-the-art SRSs.

Keywords: Recommender system · Mixture of experts · Streaming recommendation

1 Introduction

Recommender Systems (RSs) have played an increasingly important role to assist users to make wise decisions. Nowadays, E-commerce platforms generate continuous data stream, e.g., continuous users' purchase records, at an unprecedented

Z. Huang et al. (Eds.): WISE 2020, LNCS 12343, pp. 269–284, 2020.
https://doi.org/10.1007/978-3-030-62008-0_19

speed, which poses new challenges for RSs. Conventional offline RSs train recommendation models with large-volume data periodically [7,26], and thus cannot process data stream in a real-time manner. To this end, *Streaming Recommender Systems* (SRSs) [23], which perform real-time recommendations based on the data stream, have emerged.

Various SRSs have been proposed through different ways. As an earlier attempt, researchers have constructed SRSs by adapting offline RSs to the streaming setting through training the recommendation models incrementally with new data, e.g., incremental collaborative filtering [12]. Later on, SRSs specifically devised for the streaming scenario have been proposed, e.g., Neural Memory Recommender Networks (NMRN) [20].

Despite many SRSs have been proposed, the following two challenges still need to be well addressed to improve the accuracy of streaming recommendations: **CH1:** how to address *user preference drift*, i.e., the user preferences towards items changing over time [17], while capturing *long-term user preferences*, and **CH2:** how to handle the *heterogeneity* of users and items, i.e., different types of users (or items) have different preferences (or characteristics). To address **CH1**, Wang et al. [20] and Wang et al. [23] have proposed a neural memory network based approach, i.e., NMRN, and a reservoir-based approach, i.e., Stream-centered Probabilistic Matrix Factorization (SPMF), respectively. However, NMRN has difficulties in capturing long-term user preferences as the preferences stored in memory might be overwritten frequently over the continuous data stream, while SPMF has limited capability to address user preference drift as it could not sufficiently learn from new data to capture the changing user preferences. Different from the first challenge, **CH2** has not been discussed in the literature of SRSs. Existing SRSs commonly utilize a unified model to learn user preferences and item characteristics for all users and items [4,6]. However, they cannot well deal with the intrinsic difference between user preferences and item characteristics. Moreover, the preferences (or characteristics) of different types of users (or items) cannot be well learned by a unified model, either.

Our Approach and Contributions. To address the above two challenges, we propose a novel Variational and Reservoir-enhanced Sampling based Double-Wing Mixture of Experts framework, called VRS-DWMoE, to improve the accuracy of streaming recommendations. Specifically, VRS-DWMoE contains two key components: 1) Variational and Reservoir-enhanced Sampling (VRS), which wisely samples historical data containing long-term user perferences from the reservoir, i.e., a set of representative historical data, with an adjustable sampling size to complement new data, and 2) Double-Wing Mixture of Experts (DWMoE), which first learns heterogeneous user preferences and item characteristics with the training data prepared by VRS, and then utilizes the learned preferences and characteristics to make recommendations. Specifically, DWMoE contains two elaborately devised Mixture of Experts (MoEs) to learn user preferences and item characteristics, respectively. Note that MoE is an effective ensemble learning model which wisely fuses the outputs of multiple experts, i.e., atomic models specializing in different types of input, for better learning performance [10]. Moreover, the multiple experts in each of the aforementioned two MoEs learn the preferences (or characteristics) of different types of users (or items) where each expert specializes in one underlying type.

The characteristics and contributions of our work are summarized as follows:

- In this paper, we propose a novel VRS-DWMoE framework, which consists of Variational and Reservoir-enhanced Sampling (VRS) and Double-Wing Mixture of Experts (DWMoE), for accurate streaming recommendations.
- To address **CH1**, we propose VRS to wisely complement new data with historical data while guaranteeing the proportion of new data. In this way, VRS not only captures long-term user preferences from the sampled historical data, but also effectively addresses user preference drift by highlighting the importance of new data.
- To address **CH2**, we propose DWMoE to first effectively learn heterogeneous user preferences and item characteristics, and then make recommendations with learned preferences and characteristics. Specifically, DWMoE not only learns user preferences and item characteristics with two elaborately devised MoEs, respectively, to deal with their intrinsic difference, but also allows each expert to specialize in one underlying type of users (or items) to more effectively learn the heterogeneous user preferences (or item characteristics).

2 Related Work

In this section, we first review streaming recommender systems, and then introduce mixture of experts, based on which we propose VRS-DWMoE.

2.1 Streaming Recommender Systems

The early SRSs enhance offline RSs with elaborately devised online update mechanisms for streaming recommendations. For example, Papagelis et al. [12] adapt user-based collaborative filtering to the streaming setting by incrementally updating the user-to-user similarities. Later on, several approaches have been proposed to adapt the matrix factorization to the streaming setting, including incremental stochastic gradient descent [19], randomized block coordinate descent [1], and fast alternating least square [6]. Moreover, the optimization methods for offline matrix factorization have also been applied to the streaming setting, including pair-wise personalized ranking [2] and Bayesian inference [15].

Recently, SRSs focusing on the challenges in streaming recommendations have emerged. To address user preference drift while capturing long-term user preferences, the neural memory network based approach, i.e., NMRN [20], and the reservoir-based approach, i.e., SPMF [23], have been proposed. Specifically, NMRN tries to capture user preferences by neural memory network. However, it has limited capability to capture long-term user preferences, as the long-term preferences stored in the memory might be overwritten frequently over the continuous data stream. In addition, SPMF maintains a reservoir and trains the recommendation model with both sampled historical data and sampled new data. However, SPMF has difficulties in addressing user preference drift, as it equally treats the historical data and new data when conducting the sampling process and thus overlooks the importance of new data. In addition, to avoid the limitations of a single model, OCFIF [25] ensembles multiple recommendation models to conduct streaming recommendations. However, it does not fully exploit the potential of these recommendation models, as it trains the recommendation models independently and selects only one model for recommendations.

Summary. More work is needed to address user preference drift while capturing long-term user preferences. In addition, studies on addressing the issue of user or item heterogeneity in streaming recommendations have not been reported.

2.2 Mixture of Experts

Mixture of Experts (MoE) [10] is an effective ensemble learning model which wisely fuses the results of multiple experts to achieve better learning performance. Specifically, MoE contains 1) multiple experts where each expert is an atomic model specializing in learning from a particular type of input data, 2) a gating network to calculate the gating weights, i.e., the expertise of each expert regarding the input, and 3) a fusion module to fuse the outputs of the experts with the gating weights. Its effectiveness has been verified in various areas [14], including offline recommendations [9]. However, these existing MoE based RSs commonly employ MoE to perform multi-task learning [9] or simply combine multiple independent RSs [18], rather than effectively learning the heterogeneous user preferences and item characteristics from the interactions. In addition, although MoE has achieved good performance regarding streaming data in multiple areas [11], its effectiveness in streaming recommendations has not been explored.

Summary. The potential of MoE has not been explored by existing SRSs. Although MoE based offline RSs have been proposed, they cannot well learn the heterogeneous user preferences and item characteristics. Moreover, conventional MoE based approaches only contain a single MoE, and thus can not be employed directly to well address the aforementioned challenge, i.e., CH2.

3   VRS-DWMoE Framework

We first formulate our research problem. After that, we propose Variational and Reservoir-enhanced Sampling based Double-Wing Mixture of Experts (VRS-DWMoE), and then introduce its two key components, i.e., VRS and DWMoE.

3.1   Problem Statement

In this section, we formulate the research problem of streaming recommendations with implicit interactions, e.g., the users' purchase records of items. Given the user set $\mathbf{U}$ and item set $\mathbf{V}$, we use $y_{u,v}$ to denote an interaction between user $u \in \mathbf{U}$ and item $v \in \mathbf{V}$. Then, the list of currently received interactions is denoted by $\mathbf{Y} = \{y_{u^1,v^1}^1, y_{u^2,v^2}^2, \dots, y_{u^k,v^k}^k, \dots\}$. Note that the interactions in $\mathbf{Y}$ are sorted based on their receiving time, e.g., y_{u^k,v^k}^k indicates the kth received interaction. In addition, as in the real-world data stream, the adjacent interactions, e.g., y_{u^k,v^k}^k and $y_{u^{k+1},v^{k+1}}^{k+1}$, may involve different users, i.e., user u^k is possibly different from user u^{k+1}. With the above information, given the target user u' and target item v', the task of SRSs can be formulated as $\hat{y}_{u',v'} = P(y_{u',v'}|\mathbf{Y})$, i.e., predicting the probability of an interaction between the target user and target item conditional on the currently received interactions.

3.2   Our Proposed VRS-DWMoE Framework

To simultaneously address user preference drift while capturing long-term user preferences and handle the heterogeneity of users and items, we propose VRS-DWMoE, which contains two key components: 1) Variational and Reservoir-enhanced Sampling (VRS), and 2) Double-Wing Mixture of Experts (DWMoE). Specifically, as shown in Fig. 1, VRS first complements the new data with sampled historical data while guaranteeing the proportion of new data in preparation for training. After that, with the training data prepared by VRS, DWMoE better learns heterogeneous user preferences and item characteristics with two elaborately devised Mixture of Experts (MoEs), i.e., Mixture of User Experts (MoUE) and Mixture of Item Experts (MoIE), respectively, and then makes recommendations based on the learned user preferences and item characteristics.

3.3   Variational and Reservoir-Enhanced Sampling

The continuous and infinite data stream makes it impractical for SRSs to train recommendation models with all the data. To this end, we propose VRS to prepare the data for training by wisely complementing new data with sampled historical data while guaranteeing the proportion of new data.

Specifically, following [8,23], we first maintain a reservoir to store a set of representative historical data. As newer data commonly reflect more recent user preferences, we put new data into the reservoir and discard the oldest data when the reservoir runs out of space. With this reservoir and new data, VRS generates the training data with two different strategies in two typical scenarios for streaming recommendations, i.e., the *underload* scenario, where the data receiving speed is lower than the data processing speed, and the *overload* scenario, where the data receiving speed is higher than the data processing speed, respectively. Note that the contribution of VRS mainly lies in the underload scenario, which is more common in the real world [3]. More details are presented below.

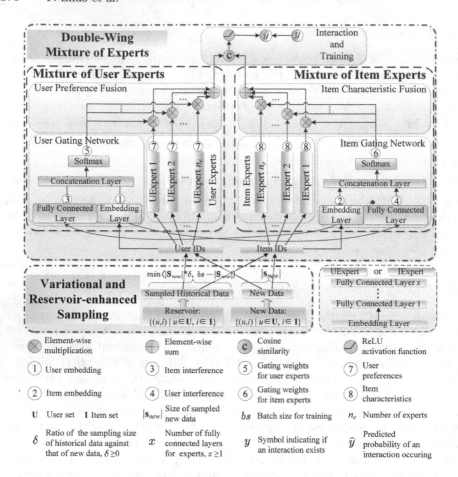

Fig. 1. Workflow of our proposed VRS-DWMoE.

Underload Scenario. In the underload scenario, VRS generates training data by first sampling representative historical data $\mathbf{S}_{his}$ from the reservoir with a variational sampling size, and then merging $\mathbf{S}_{his}$ and all the new data $\mathbf{N}$ (i.e., all the new data are sampled by VRS in the underload scenario). Specifically, with the training batch size bs and a predetermined parameter δ ($\delta \geq 0$) measuring the ratio of the sampling size $|\mathbf{S}_{his}|$ of reservoir against the size s_{new} of new data, the sampling size $|\mathbf{S}_{his}|$ of reservoir can be calculated as below,

$$|\mathbf{S}_{his}| = min(s_{new} * \delta, bs - s_{new}). \tag{1}$$

In this way, the sampling size of reservoir can be adjusted by δ based on the characteristics of data stream. For example, δ should be set to a small value, e.g., 0.1, for data stream where users' preferences change frequently to focus more on new data. Then, $\mathbf{S}_{his}$ is sampled from the reservoir based on their receiving time, i.e., more recent received interactions are assigned higher sampling probabilities.

Specifically, to reflect the importance of newer data, we employ a decay ratio λ_{res} ($\lambda_{res} > 1$) to assign higher sampling probabilities to newer data in the reservoir,

$$p_k = p_{k-1} * \lambda_{res}, \tag{2}$$

where p_i denotes the sampling probability of the ith received interaction. Assuming that the sampling probability of the earliest received interaction in the reservoir is p_1, we can get p_k by iteratively performing Eq. (2), i.e.,

$$p_k = p_1 * (\lambda_{res})^{k-1}. \tag{3}$$

Then, with Eq. (3), we can obtain the normalized sampling probabilities, taking the probability of the kth received interaction as an example,

$$P(k|\lambda_{res}, s_{res}) = \frac{p_k}{\sum_{i=1}^{s_{res}} p_i} = \frac{(\lambda_{res})^{k-1} * (1 - \lambda_{res})}{1 - (\lambda_{res})^{s_{res}}}, \tag{4}$$

where s_{res} denotes the size of the reservoir.

With the normalized sampling probabilities and the sampling size calculated in Eq. (1), VRS samples representative historical data $\mathbf{S}_{his}$ from the reservoir. Finally, the training data $\mathbf{T}$ is obtained by merging $\mathbf{S}_{his}$ and the new data $\mathbf{N}$.

Overload Scenario. In the overload scenario, VRS only samples $\mathbf{S}_{new}$ from new data to form the training data, i.e., $|\mathbf{S}_{his}|$ is set to 0, for effectively capturing the latest user preferences. The sampling probability of the new data can be calculated in a similar way as described by Eq. (2) to (4), i.e.,

$$P(k|\lambda_{new}, s_{new}) = \frac{p_k}{\sum_{i=1}^{s_{new}} p_i} = \frac{(\lambda_{new})^{k-1} * (1 - \lambda_{new})}{1 - (\lambda_{new})^{s_{new}}}, \tag{5}$$

where λ_{new} and s_{new} denote the decay ratio and size, respectively, of new data. With this sampling probability and the sampling size set to be the batch size bs, VRS samples $\mathbf{S}_{new}$ from the new data as the training data $\mathbf{T}$. Note that, in the case where the data receiving speed exactly equals to the data processing speed, VRS utilizes the entire new data to form the training data $\mathbf{T}$.

3.4   Double-Wing Mixture of Experts

With the training data $\mathbf{T}$ prepared by VRS, DWMoE first utilizes two MoEs, i.e., MoUE and MoIE, to learn user preferences and item characteristics, respectively, to deal with their intrinsic difference [21]. Moreover, each expert in MoUE (or MoIE) specializes in one underlying type of users (or items) for more effectively learning heterogeneous user preferences (or item characteristics). Then, DWMoE makes recommendations with the learned user preferences and item characteristics.

Specifically, MoUE and MoIE share the same structure with different parameters. This structure has three key parts: 1) multiple experts, 2) a gating network, and 3) a fusion module. Taking MoUE as an example, multiple experts first learn the user preferences in parallel. Then, the gating weights, which measure the expertise of each expert regarding each input user, are calculated by the gating network. After that, the fusion module calculates the unified preferences for each user by fusing the preferences learned by all experts with the gating weights. Note that we set the numbers (n_e) of experts in MoUE and MoIE the same in this paper, and will study the effect of different numbers of experts in MoUE and MoIE in the future work. More details are presented below.

Expert. The experts in MoUE and MoIE learn user preferences and item characteristics, respectively. Taking MoUE as an example, each expert first utilizes an embedding layer to learn the user embedding $\mathbf{p}_u^i$, where i denotes the index of the expert. Then, the user preferences are learned by the experts with x $(x \geq 1)$ fully connected layers, taking the user preferences $\mathbf{P}_u^i$ as an example,

$$\mathbf{P}_u^i = a_{i,x}^{MoUE}(\cdots a_{i,2}^{MoUE}(\mathbf{W}_{i,2}^{MoUE} a_{i,1}^{MoUE}(\mathbf{W}_{i,1}^{MoUE}\mathbf{p}_u^i + \mathbf{b}_{i,1}^{MoUE}) + \mathbf{b}_{i,2}^{MoUE})\cdots), \quad (6)$$

where a_*^*, $\mathbf{W}_*^*$, and $\mathbf{b}_*^*$ denote the activation function, weight matrix, and bias vector, respectively. For example, $a_{i,x}^{MoUE}$, $\mathbf{W}_{i,x}^{MoUE}$, and $\mathbf{b}_{i,x}^{MoUE}$ denote the activation function, weight matrix, and bias vector, respectively, in the xth layer for the ith expert in MoUE. Note that the symbols a_*^*, $\mathbf{W}_*^*$, and $\mathbf{b}_*^*$ are used in the rest of this paper with different superscripts and subscripts to introduce the fully connected layers. In a similar way as described by Eq. (6), item characteristics $\mathbf{Q}_v^j$ can be learned by the jth expert in MoIE with the item embedding $\mathbf{q}_v^j$,

$$\mathbf{Q}_v^j = a_{j,x}^{MoIE}(\cdots a_{j,2}^{MoIE}(\mathbf{W}_{j,2}^{MoIE} a_{j,1}^{MoIE}(\mathbf{W}_{j,1}^{MoIE}\mathbf{q}_v^j + \mathbf{b}_{j,1}^{MoIE}) + \mathbf{b}_{j,2}^{MoIE})\cdots). \quad (7)$$

Gating Network. The gating networks in MoUE and MoIE calculate the gating weights measuring the expertise scales of each expert in learning the preferences of input users and characteristics of input items, respectively. To achieve this goal, taking the gating network in MoUE as an example, we first employ an embedding layer and a fully connected layer to calculate user embedding $\mathbf{p}_u^{MoUE}$ and item interference $\mathbf{I}^{MoUE}$, respectively. Note that the item interference is employed to more accurately calculate the gating weights in MoUE by taking the items interacted with the corresponding users into consideration. Specifically, the item interference can be calculated with the item embedding $\mathbf{q}_v^{MoIE}$ in MoIE,

$$\mathbf{I}^{MoUE} = a_{gate}^{MoUE}(\mathbf{W}_{inter}^{MoUE}\mathbf{q}_v^{MoIE} + \mathbf{b}_{inter}^{MoUE}). \quad (8)$$

Then, the user embedding and item interference are concatenated,

$$\mathbf{c}^{MoUE} = \left[\mathbf{p}_u^{MoUE}; \mathbf{I}^{MoUE}\right]. \quad (9)$$

After that, $\mathbf{c}^{MoUE}$ is fed into a softmax layer to get the gating weights $\mathbf{g}^{MoUE}$,

$$\mathbf{g}^{MoUE} = softmax(\mathbf{W}_{soft}^{MoUE}\mathbf{c}^{MoUE} + \mathbf{b}_{soft}^{MoUE}). \quad (10)$$

Likewise, the gating weights $\mathbf{g}^{MoIE}$ for the experts in MoIE can be calculated in a similar way as described by Eq. (8) to (10).

Fusion Module. The user preferences (or item characteristics) learned by multiple experts in MoUE (or MoIE) are fused to the unified ones to fully utilize the expertise of all the experts. Specifically, with the above calculated user preferences, item characteristics, and their corresponding gating weights, the unified user preferences $\mathbf{P}_u^{uni}$ and unified item characteristics $\mathbf{Q}_v^{uni}$ can be calculated with the dot production, respectively,

$$\mathbf{P}_u^{uni} = [\mathbf{P}_u^1; \cdots; \mathbf{P}_u^{n_e}]^T \mathbf{g}^{MoUE}, \qquad (11)$$

$$\mathbf{Q}_v^{uni} = [\mathbf{Q}_v^1; \cdots; \mathbf{Q}_v^{n_e}]^T \mathbf{g}^{MoIE}, \qquad (12)$$

where n_e denotes the number of experts.

Interaction Module. To make recommendations based on unified user preferences and unified item characteristics, we first utilize cosine similarity to measure how the unified preferences match the corresponding unified characteristics,

$$cos\_sim_{<\mathbf{P}_u^{uni}, \mathbf{Q}_v^{uni}>} = cosine(\mathbf{P}_u^{uni}, \mathbf{Q}_v^{uni}) = \frac{(\mathbf{P}_u^{uni})^T \mathbf{Q}_v^{uni}}{\|\mathbf{P}_u^{uni}\|_2 \|\mathbf{Q}_v^{uni}\|_2}, \qquad (13)$$

and then we obtain the predicted probability $\hat{y}_{u,v}$ of the interaction between user u and item v by performing a nonlinear transformation of this cosine similarity,

$$\hat{y}_{u,v} = a_{out}^{predict}(\mathbf{W}_{out}^{predict} cos\_sim_{<\mathbf{P}_u^{uni}, \mathbf{Q}_v^{uni}>} + \mathbf{b}_{out}^{predict}). \qquad (14)$$

Optimization. To learn the parameters of our proposed DWMoE, we train the model by minimizing the following loss with stochastic gradient descent,

$$loss^{total} = loss^{acc} + \gamma(loss^{gate}), \qquad (15)$$

where $loss^{acc}$ is the loss for the recommendation accuracy, $loss^{gate}$ is used as the regularization term for gating weights to avoid local optimizations, and γ is the coefficient to adjust the importance of $loss^{gate}$.

Specifically, to measure the difference between the ground truth and the prediction, we employ the cross-entropy loss as below,

$$loss^{acc}(y_{u,v}, \hat{y}_{u,v}) = -(y_{u,v}log(\hat{y}_{u,v}) + (1 - y_{u,v})log(1 - \hat{y}_{u,v})), \qquad (16)$$

where $y_{u,v}$ is the label of the interaction between user u and item v, i.e., it is 1 if this interaction exists and 0 otherwise, and $\hat{y}_{u,v}$ denotes the predicted probability for this interaction. Moreover, we introduce $loss^{gate}$ to avoid the local optimization caused by the imbalanced utilization of experts, i.e., some experts receive large gating weights for most interactions while others always receive small gating weights. Specifically, we employ the standard derivations of gating weights in both MoUE and MoIE to form $loss^{gate}$,

$$loss^{gate} = loss_{MoUE}^{gate} + loss_{MoIE}^{gate}$$
$$= \sqrt{\frac{1}{n_e} \sum_{i=1}^{n_e} (\mathbf{g}_i^{MoUE} - \bar{\mathbf{g}}^{MoUE})^2} + \sqrt{\frac{1}{n_e} \sum_{j=1}^{n_e} (\mathbf{g}_j^{MoIE} - \bar{\mathbf{g}}^{MoIE})^2}, \qquad (17)$$

278 Y. Zhao et al.

where $\mathbf{g}_i^{MoUE}$ and $\mathbf{g}_j^{MoIE}$ denote the gating weight for the ith expert in MoUE and the gating weight for the jth expert in MoIE, respectively, and $\bar{\mathbf{g}}^{MoUE}$ and $\bar{\mathbf{g}}^{MoIE}$ denote the average of gating weights for MoUE and MoIE, respectively. Through minimizing $loss^{gate}$, DWMoE encourages MoUE and MoIE to more effectively utilize all their experts to learn the heterogeneous user preferences and item characteristics, respectively, and thus to increase the accuracy of streaming recommendations .

4 Experiments

In this section, we present the results of the extensive experiments we conducted which aim to answer the following four research questions:

RQ1. How does our proposed VRS-DWMoE perform when compared with the state-of-the-art approaches?

RQ2. How does our proposed DWMoE perform when compared with the existing recommendation models?

RQ3. How does our proposed VRS perform when compared with the existing sampling methods?

RQ4. How does the number of experts in VRS-DWMoE affect the recommendation accuracy?

Table 1. Statistics of three datasets used in our experiments.

Datasets	#Users	#Items	#Interactions	Sparsity
MovieLens	6400	3703	994169	95.81%
Netflix	5000	16073	1010588	98.74%
Yelp	25677	25815	731671	99.89%

The symbol # indicates the number, e.g., #Users indicates the number of users.

4.1 Experimental Settings

Datasets. In the experiments, we employ three widely-used real-world datasets [6,23], i.e., MovieLens (1M)[1], Netflix[2], and Yelp[3], to verify the effectiveness of our proposed VRS-DWMoE. Note that we extract the interactions of randomly selected 5000 users from the Netflix dataset for the experiments, as processing the original Netflix dataset, which contains more than 100 million interactions, is beyond our computational capacity. In addition, following the common practice [6,13], for each dataset, we retain the interactions from users

---

[1] https://grouplens.org/datasets/movielens/1m.
[2] https://www.kaggle.com/netflix-inc/netflix-prize-data.
[3] https://www.yelp.com/dataset/challenge.

who have more than 10 interactions to reduce data sparsity. The statistics of the tuned datasets are summarized in Table 1. Moreover, following [20,25], we transform the explicit ratings in all three datasets into the implicit ones, where it is 1 if an explicit rating exists and 0 otherwise, as this work focuses on the recommendations with implicit interactions.

Evaluation Policy. Following [6], we first sort the data in each of the three datasets by their receiving time, and then divide them into two parts, i.e., 1) the training set to simulate the historical data, and 2) the test set to simulate the upcoming data in the streaming scenario. Specifically, the data in the training set are used for incremental training while the data in the test set are first used for the test and then used for incremental training. We have set the proportion of training set to 85%, 90% and 95%, respectively, for evaluating the performance of our proposed VRS-DWMoE. Due to the space limit, we report the results in the case where the proportion of the training set is 90% only, as the work in [6] does, while the results in the other two cases are similar to the reported ones. Moreover, to verify the effectiveness of our proposed VRS-DWMoE in the underload scenario and overload scenario, we train the recommendation model with a fixed number n_p ($n_p = 256$ in this paper) of interactions each time and adjust the number n_r of interactions received in this training period to indicate different workload intensities. For the sake of simplicity, we use n_p and n_r to simulate the data processing speed s_p and data receiving speed s_r, respectively. In this way, the underload scenario and the overload scenario can be simulated by the cases where $s_p > s_r$ and $s_p < s_r$, respectively.

Evaluation Metrics. Following the common practice [6,23], we adopt the ranking-based evaluation strategy. Specifically, for each interaction between a target user and a target item, we first randomly sample 99 items which are not interacted with this user as negative items, and then rank the target item among these 100 items, i.e., the target one plus the 99 sampled ones. Finally, the recommendation accuracy is measured by two widely used metrics, i.e., Hit Ratio (**HR**) and Normalized Discounted Cumulative Gain (**NDCG**) [6,23].

Baselines. The following eight baselines are used for comparisons, including iBPR, iGMF, iMLP, iNeuMF, RCD, eAls, SPMF, and OCFIF.

- *Bayesian Personalized Ranking* (BPR) [13] is a representative personalized ranking method to optimize the matrix factorization. We adapt BPR to the streaming setting, named as **iBPR**, by training it with new data continuously via stochastic gradient descent.
- *Neural Matrix Factorization* (NeuMF) [5] is an advanced matrix factorization model, which combines two other recommendation models, i.e., *Generalized Matrix Factorization* (GMF) and *Multi-Layer Perceptron* (MLP), to achieve higher recommendation accuracy. We adapt these three recommendation models, i.e., NeuMF, GMF, and MLP, to the streaming setting, named as

iNeuMF, **iGMF**, and **iMLP**, respectively, by training the recommendation models with new data continuously via stochastic gradient descent.

- *Randomized block Coordinate Descent* (**RCD**) [1] and *Element-wise Alternating Least Squares* (**eAls**) [6] are two representative approaches for optimizing the matrix factorization models in the streaming setting. We enhance RCD and eAls with abilities of batch processing to increase their throughput for fair comparisons.
- *Stream-centered Probabilistic Matrix Factorization* (**SPMF**) [23] is a state-of-the-art SRS. SPMF is originally performed along with a time-consuming sampling method and does not perform well with our evaluation policy where sampling needs to be frequently performed. For a fair comparison, we employ our proposed VRS to prepare training data for SPMF.
- *Online Collaborative Filtering with Implicit Feedback* (**OCFIF**) [25] is the only reported SRS employing multiple models (i.e., matrix factorization) to avoid the limitations of a single model for higher recommendation accuracy.

In addition, we equip our proposed **VRS-DWMoE** with different numbers of experts (i.e., 2, 4, 6, and 8) for comparisons. For example, VRS-DWMoE_8 indicates VRS-DWMoE equipped with 8 experts for both MoUE and MoIE.

Parameter Setting. For a fair comparison, we initialize the baselines with parameters reported in their papers and optimize them for our settings. For our VRS-DWMoE, we empirically set the learning rate to 0.001, the batch size bs to 256, the loss coefficient γ to 0.01, and the volume of reservoir to 10000 interactions. Besides, we employ the widely used negative sampling technique [22, 24,25], where the reservoir is used to check if an interaction exists and the negative sampling size is set to four, to improve the learning performance. We also adopt L2 regularization and Adam optimizer to avoid overfitting and for the optimization purpose, respectively. Other parameters including δ, λ_{res}, and λ_{new} are adjusted via cross validation to achieve the best performance in different cases.

4.2 Performance Comparison and Analysis

Experiment 1: Comparison with Baselines (for RQ1 and RQ2)
Setting. To answer **RQ1** and **RQ2**, we compare our proposed VRS-DWMoE (the number n_e of experts is set to eight) with all eight baselines with a fixed data processing speed $s_p = 256$ and different data receiving speeds, where $s_r = 128$ and $s_r = 512$ indicate the underload scenario and overload scenario, respectively.

Result 1 (for RQ1). Table 2 shows the results of our proposed approach and eight baselines on all three datasets. In all the cases, VRS-DWMoE_8 delivers the highest recommendation accuracies (marked with **bold** font), and the improvement percentages of VRS-DWMoE_8 over the best-performing baselines

Table 2. Performance comparison with baselines.

Datasets	Metrics		HR@10			NDCG@10		
	Data receiving speeds (s_r)		128	256	512	128	256	512
MovieLens	Baselines	eAls	0.231	0.231	0.234	0.106	0.106	0.108
		RCD	0.287	0.297	0.278	0.139	0.145	0.138
		iBPR	0.303	0.303	0.279	0.147	0.147	0.134
		SPMF	0.460	0.453	0.440	0.251	0.246	0.238
		iGMF	0.525	0.529	0.477	0.295	0.297	0.265
		iMLP	0.538	0.539	0.488	0.304	0.303	0.272
		iNeuMF	<u>0.551</u>	<u>0.546</u>	<u>0.496</u>	<u>0.311</u>	<u>0.307</u>	<u>0.275</u>
		OCFIF	0.532	0.508	0.467	0.291	0.279	0.256
	Ours	VRS-DWMoE_8	**0.563**	**0.558**	**0.535**	**0.317**	**0.313**	**0.299**
	Improvement percentages[a]		2.20%	2.20%	7.90%	1.90%	2.00%	8.70%
Netflix	Baselines	eAls	0.395	0.389	0.362	0.211	0.207	0.192
		RCD	0.447	0.436	0.435	0.226	0.226	0.219
		iBPR	0.685	0.686	0.627	0.396	0.395	0.360
		SPMF	0.701	0.669	0.640	0.425	0.397	0.378
		iGMF	0.747	0.748	0.577	0.482	0.482	0.352
		iMLP	0.787	0.782	0.624	0.519	0.510	0.369
		iNeuMF	<u>0.801</u>	<u>0.798</u>	<u>0.711</u>	<u>0.531</u>	<u>0.529</u>	<u>0.430</u>
		OCFIF	0.745	0.734	0.606	0.457	0.453	0.357
	Ours	VRS-DWMoE_8	**0.821**	**0.814**	**0.790**	**0.553**	**0.548**	**0.515**
	Improvement percentages[a]		2.50%	2.00%	11.1%	4.10%	3.60%	19.8%
Yelp	Baselines	eAls	0.287	0.289	0.290	0.167	0.167	0.169
		RCD	0.454	0.452	0.447	0.260	0.257	0.259
		iBPR	0.307	0.295	0.188	0.180	0.172	0.108
		SPMF	0.197	0.192	0.184	0.104	0.100	0.097
		iGMF	0.499	0.470	0.396	0.294	0.276	0.228
		iMLP	<u>0.573</u>	<u>0.574</u>	<u>0.438</u>	<u>0.338</u>	<u>0.338</u>	0.246
		iNeuMF	0.566	0.570	0.435	0.331	0.334	<u>0.247</u>
		OCFIF	0.260	0.249	0.203	0.135	0.129	0.107
	Ours	VRS-DWMoE_8	**0.608**	**0.603**	**0.602**	**0.354**	**0.358**	**0.353**
	Improvement percentages[a]		6.10%	5.10%	37.4%	4.70%	5.90%	42.9%

[a] Improvement percentages over the best-performing baseline(s)

(marked with <u>underline</u>) are introduced in the last row for each dataset, ranging from 2.0% to 37.4% with an average of 8.5% in terms of HR@10, and ranging from 1.9% to 42.9% with an average of 10.4% in terms of NDCG@10.

The superiority of VRS-DWMoE can be explained in two aspects: 1) VRS addresses user preference drift while capturing long-term user preferences by wisely complementing new data with sampled historical data, and 2) DWMoE better learns the heterogeneous user preferences and item characteristics with two MoEs, where each expert specializes in one underlying type of users or items.

Result 2 (for RQ2). The superiority of our DWMoE is verified by the cases where $s_r = s_p$, i.e., $s_r = 256$, in Table 2. Specifically, in these cases, both our apporach and baselines utilize all the new data to train recommendation models, thus their recommendation accuracy only depends on their recommendation models. Therefore, the superiority of DWMoE is confirmed by the highest rec-

282   Y. Zhao et al.

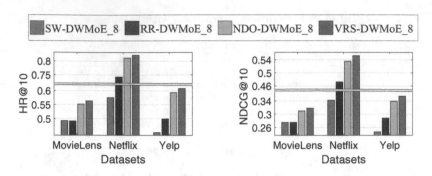

Fig. 2. Performance of VRS. VRS outperforms all the other sampling methods.

ommendation accuracy delivered by VRS-DWMoE in these cases. The reason for this superiority is that DWMoE not only learns heterogeneous user preferences and item characteristics with two dedicated MoEs, respectively, but also allows each of their experts to specialize in one underlying type of users or items.

Experiment 2: Performance of VRS (for RQ3)
Setting. To answer **RQ3**, we replace our proposed VRS with existing sampling methods, including New Data Only (NDO) [25], Reservoir-enhanced Random sampling (RR) [2], and Sliding Window (SW) [16], for comparisons. In this experiment, we report the results in the underload scenario only to save space while the results in the overload scenario are similar to the reported ones.

Result 3 (for RQ3). As Fig. 2 illustrates, our proposed VRS outperforms all the other sampling methods. The improvements of VRS over the best-performing baseline, i.e., NDO, range from 1.2% (on Netflix) to 2.0% (on Yelp) with an average of 1.9% in terms of HR@10, and range from 3.2% (on Netflix) to 4.3% (on Yelp) with an average of 3.4% in terms of NCDG@10. The effectiveness of VRS comes from wisely complementing new data with sampled historical data while guaranteeing the proportion of new data, and thus addressing user preference drift while capturing long-term user preferences.

Experiment 3: Impact of Number of Experts (for RQ4)
Setting. To answer **RQ4**, we compare the performance of VRS-DWMoE when equipped with different numbers (n_e) of experts, i.e., 2, 4, 6, and 8. In this experiment, we report the results in the overload scenario only to save space while the results in the underload are similar to the reported ones.

Result 4 (for RQ4). As Fig. 3 illustrates, our proposed VRS-DWMoE delivers higher recommendation accuracy when equipped with more experts. The improvements of VRS-DWMoE equipped with eight experts over that equipped with two experts range from 2.7% (on Netflix) to 6.5% (on Yelp) with an average

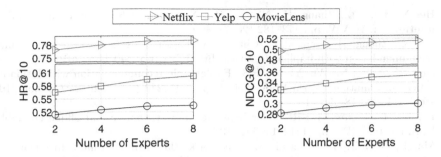

Fig. 3. Impact of the number (n_e) of experts. VRS-DWMoE delivers higher recommendation accuracy with more experts.

of 4.4% in terms of HR@10, and range from 4.0% (on Netflix) to 8.4% (on Yelp) with an average of 6.3% in terms of NCDG@10. The reason for the superiority of more experts is that more experts better complement one another with their expertise to more effectively learn user preferences and item characteristics.

5   Conclusions

In this paper, we have proposed a Variational and Reservoir-enhanced Sampling based Double-Wing Mixture of Experts framework (VRS-DWMoE) for accurate streaming recommendations. We first propose VRS to wisely complement new data with sampled historical data to address user preference drift while capturing long-term user preferences. After that, with these sampled data, DWMoE learns heterogeneous user preferences and item characteristics with two MoEs, i.e., MoUE and MoIE, respectively, and then makes recommendations with learned preferences and characteristics. The superiority of VRS-DWMoE has been verified by extensive experiments. In the future, we will wisely utilize different numbers of experts in MoUE and MoIE and study more effective reservoir maintenance strategy for higher accuracy of streaming recommendations.

Acknowledgements. This work was partially supported by Australian Research Council Discovery Projects DP180102378 and DP210101810.

References

1. Devooght, R., Kourtellis, N., Mantrach, A.: Dynamic matrix factorization with priors on unknown values. In: SIGKDD, pp. 189–198 (2015)
2. Diaz-Aviles, E., Drumond, L., Schmidt-Thieme, L., Nejdl, W.: Real-time top-n recommendation in social streams. In: RecSys, pp. 59–66 (2012)
3. Forbes Report. http://forbes.com/sites/benkepes/2015/06/03/30-of-servers-are-sitting-comatose-according-to-research/, Accessed 29 May 2020
4. Guo, L., Yin, H., Wang, Q., Chen, T., Zhou, A., Quoc Viet Hung, N.: Streaming session-based recommendation. In: SIGKDD, pp. 1569–1577 (2019)

5. He, X., Liao, L., Zhang, H., Nie, L., Hu, X., Chua, T.S.: Neural collaborative filtering. In: WWW, pp. 173–182 (2017)
6. He, X., Zhang, H., Kan, M.Y., Chua, T.S.: Fast matrix factorization for online recommendation with implicit feedback. In: SIGIR, pp. 549–558 (2016)
7. Hou, Y., Yang, N., Wu, Y., Yu, P.S.: Explainable recommendation with fusion of aspect information. World Wide Web 22(1), 221–240 (2018). https://doi.org/10.1007/s11280-018-0558-1
8. Lefakis, L., Fleuret, F.: Reservoir boosting: between online and offline ensemble learning. In: NIPS, pp. 1412–1420 (2013)
9. Ma, J., Zhao, Z., Yi, X., et al.: Modeling task relationships in multi-task learning with multi-gate mixture-of-experts. In: SIGKDD, pp. 1930–1939 (2018)
10. Masoudnia, S., Ebrahimpour, R.: Mixture of experts: a literature survey. Artif. Intell. Review 42(2), 275–293 (2012). https://doi.org/10.1007/s10462-012-9338-y
11. McKinnon, C.D., Schoellig, A.P.: Learning multimodal models for robot dynamics online with a mixture of gaussian process experts. In: ICRA, pp. 322–328 (2017)
12. Papagelis, M., Rousidis, I., Plexousakis, D., et al.: Incremental collaborative filtering for highly-scalable recommendation algorithms. In: ISMIS, pp. 553–561 (2005)
13. Rendle, S., Freudenthaler, C., Gantner, Z., Schmidt-Thieme, L.: BPR: bayesian personalized ranking from implicit feedback. In: UAI, pp. 452–461 (2009)
14. Shazeer, N., Mirhoseini, A., et al.: Outrageously large neural networks: the sparsely-gated mixture-of-experts layer. In: ICLR, pp. 1–19 (2017)
15. Silva, J.G., Carin, L.: Active learning for online bayesian matrix factorization. In: SIGKDD, pp. 325–333 (2012)
16. Soares, S.G., Araújo, R.: An on-line weighted ensemble of regressor models to handle concept drifts. Eng. Appl. Artif. Intell. 37, 392–406 (2015)
17. Song, D., Li, Z., Jiang, M., Qin, L., Liao, L.: A novel temporal and topic-aware recommender model. World Wide Web 22(5), 2105–2127 (2018). https://doi.org/10.1007/s11280-018-0595-9
18. Su, X., Greiner, R., Khoshgoftaar, T.M., Zhu, X.: Hybrid collaborative filtering algorithms using a mixture of experts. In: ICWI, pp. 645–649 (2007)
19. Vinagre, J., Jorge, A.M., Gama, J.: Fast incremental matrix factorization for recommendation with positive-only feedback. In: UMAP, pp. 459–470 (2014)
20. Wang, Q., Yin, H., Hu, Z., Lian, D., et al.: Neural memory streaming recommender networks with adversarial training. In: SIGKDD, pp. 2467–2475 (2018)
21. Wang, S., Cao, L.: Inferring implicit rules by learning explicit and hidden item dependency. IEEE Trans. Syst. Man Cybern. Syst. 50(3), 935–946 (2020)
22. Wang, S., Hu, L., Wang, Y., Sheng, Q.Z., Orgun, M., Cao, L.: Modeling multipurpose sessions for next-item recommendations via mixture-channel purpose routing networks. In: IJCAI, pp. 3771–3777 (2019)
23. Wang, W., Yin, H., Huang, Z., Wang, Q., Du, X., Nguyen, Q.V.H.: Streaming ranking based recommender systems. In: SIGIR, pp. 525–534 (2018)
24. Xu, Y., Zhu, Y., Shen, Y., Yu, J.: Leveraging app usage contexts for app recommendation: a neural approach. World Wide Web 22(6), 2721–2745 (2018). https://doi.org/10.1007/s11280-018-0543-8
25. Yin, J., et al.: Online collaborative filtering with implicit feedback. In: DASFAA, pp. 433–448 (2019)
26. Yu, Y., Gao, Y., Wang, H., Wang, R.: Joint user knowledge and matrix factorization for recommender systems. World Wide Web 21(4), 1141–1163 (2017). https://doi.org/10.1007/s11280-017-0476-7

Modelling Local and Global Dependencies for Next-Item Recommendations

Nan Wang, Shoujin Wang, Yan Wang$^{(\boxtimes)}$, Quan Z. Sheng, and Mehmet Orgun

Department of Computing, Macquarie University, Sydney, Australia
nan.wang12@students.mq.edu.au,
{shoujin.wang,yan.wang,michael.sheng,mehmet.orgun}@mq.edu.au

Abstract. Session-based recommender systems (SBRSs) aim at predicting the next item by modelling the complex dependencies within and across sessions. Most of the existing SBRSs make recommendations only based on *local dependencies* (i.e., the dependencies between items within a session), while ignoring *global dependencies* (i.e., the dependencies across multiple sessions), leading to information loss and thus reducing the recommendation accuracy. Moreover, they are usually not able to recommend cold-start items effectively due to their limited session information. To alleviate these shortcomings of SBRSs, we propose a novel *heterogeneous mixed graph learning (HMGL)* framework to effectively learn both local and global dependencies for next-item recommendations. The HMGL framework mainly contains a *heterogeneous mixed graph (HMG) construction* module and an *HMG learning* module. The HMG construction module map both the session information and the item attribute information into a unified graph to connect items within and across sessions. The HMG learning module learns a unified representation for each item by simultaneously modelling the local and global dependencies over the HMG. The learned representation is then used for next-item recommendations. Results of extensive experiments on real-world datasets show the superiority of HMGL framework over the start-of-the-art methods in terms of recommendation accuracy.

Keywords: Session-based recommendations · Heterogeneous mixed graph learning · Next-item recommendation · Graph neural network

1 Introduction

Recommender systems (RSs) have been playing an ever-increasingly important role in our daily lives to help users effectively and efficiently find items, services or contents that may be of their interests from a large amount of choices [14]. Conventional RSs, including content-based RSs and collaborative filtering RSs, usually make recommendations based on users' long-term and static preferences while ignoring their short-term and dynamic preference, which usually leads to the duplicated recommendations of similar items that cannot match users' changing preferences well [19]. To this end, *session-based recommender systems*

© Springer Nature Switzerland AG 2020
Z. Huang et al. (Eds.): WISE 2020, LNCS 12343, pp. 285–300, 2020.
https://doi.org/10.1007/978-3-030-62008-0_20

(SBRSs) were proposed to suggest the next item for a user given the purchased items in the current session [16]. Here a *session* refers to a shopping basket consisting of multiple purchased items in one transaction event [17,18].

In general, an SBRS recommends the next item by modelling the complex dependencies within and across sessions. *Markov chain based SBRSs* and *recurrent neural networks (RNN) based SBRSs* are two typical types of SBRSs. In particular, a *Markov chain based SBRS* [11] predicts the next item by modelling the transitions between any two adjacent items in a session. Therefore, it only captures the *first-order dependencies* (i.e., the dependency between any two adjacent items within a session) while ignoring the *high-order dependencies* (i.e., the cascaded dependencies across multiple items within a session) within a session and hence the recommendation accuracy may suffer. To model high-order dependencies within sessions, *RNN-based SBRSs* have been proposed. Particularly, Forsati et al. [12] took gated recurrent units (GRU) as the basic cells of an RNN to model the sequential dependencies among items with rigid order assumption within each session to predict the next item. However, these models only capture the single-way transitions from the starting item to the last one within a session, while neglecting complex transitions among distant items. To capture complex transitions among items within sessions, graph neural networks (GNN) have been employed in SBRSs to predict the next item and they have achieved superior performance [15]. However, three critical gaps still remain unresolved in existing GNN-based SBRSs:

Gap1: Existing GNN-based SBRSs model each session separately in a subgraph and thus can only capture the *local dependencies* (i.e., the dependencies between items within a session), failing to explicitly capture the *global dependencies* (i.e., the dependencies across multiple sessions). This leads to information loss and thus is harmful to the subsequent recommendations.

Gap2: Existing GNN-based SBRSs usually fail to recommend cold-start items (i.e., newly appearing items with few or even no session information like transaction records) as they are based on the session information only. Recommending cold-start items is necessary in real-world cases since new items are always coming successively for sale.

Gap3: Existing GNN-based SBRSs may easily be locally optimized to fit those minor frequent items and sessions, and thus cause the overfitting problem, which reduces the recommendation accuracy.

This study addresses the above three gaps by proposing a *heterogeneous mixed graph learning (HMGL) framework* to learn the complex local and global dependencies for next-item recommendations. To be specific, to address **Gap1**, we construct a *heterogeneous mixed graph (HMG)* to connect the items from all sessions by integrating the *item-item graph* built on session information (e.g., which items are purchased together in one session), and the *item-attribute graph* built on item attribute information (e.g., the brand and category of items) into a unified graph. In such a case, items from different sessions are easily connected by taking their shared attribute values as bridges. Here, a *mixed graph* means that there are both *directed edges* between sequential items within

Fig. 1. An example of HMG constructed on both session information and item attribute information. The arrowed lines indicate the sequential relations between items from each session, while the dotted lines mean the item-attribute relations.

sessions and *indirected edges* between items and their attribute values in the graph (see Fig. 1). Then, we propose an *HMG learning* module to learn a unified and informative representation for each item over HMG by modelling both the local and global dependencies in preparation for the subsequent next-item recommendations. To address **Gap2**, we incorporate item attributes into the HMG to effectively enhance the connections between clod-start and warm-start items through the shared attribute values. Therefore, it is much easier to identify and recommend those cold-start items that may be preferred by users. To address **Gap3**, we introduce a regularization term in the loss function by taking the overall graph structure as an additional constraint, which can prevent the local optimizations and over-fittings. The main contributions are summarized as follows:

- We construct a heterogeneous mixed graph (HMG) to encode both the session information and the item attribute information into one unified graph to connect all the items together. This enriches both local and global dependencies among items, especially for those clod-start ones.
- We propose an end-to-end HMG learning model to learn an informative representation for each item over the HMG by modelling both local and global dependencies to better prepare for the next-item recommendations.
- A new loss function with an additional regularization term is proposed for global optimization to improve the recommendation accuracy.

Extensive experiments have been conducted on two real-world transaction datasets to evaluate our proposed framework HMGL. The results show the superiority of HMGL over the state-of-the-art methods.

2   Related Work

In this section, we briefly review the representative SBRSs, which can be classified into: (1) conventional SBRSs, including Markov chain based SBRSs and factorization machine based SBRSs, and (2) deep learning based SBRSs, including RNN-based SBRSs and GNN-based SBRSs.

Conventional Session-Based Recommender Systems. Markov chain models and factorization machines are two conventional approaches for SBRSs. Given the prior items in a session, Markov chain based SBRSs adopt Markov chain models to model the transitions over these items in a sequence, for the prediction of the next item. However, due to the widely employed Markov property, Markov chain based SBRSs usually can only capture the first-order dependencies between two adjacent items while ignoring the high-order dependencies across multiple items. Factorization machines are adopted to factorize the observed transition from the current item to the next one into the latent representations of items, which are then used to estimate the unobserved transitions between items for predicting the next item [7]. As a combination of Markov chain models and factorization machines, Factorizing Personalized Markov Chain (FPMC) was built for SBRSs to take the advantages of both the models for better recommendations [11]. However, similar to Markov chain based SBRSs, factorization machine based SBRSs capture first-order dependencies only, in addition, they easily suffer from the data sparsity issue.

Deep Learning Based Session-Based Recommender Systems. With the capability of handling sequential data, recurrent neural network (RNN) is an intuitive choice to capture the sequential dependencies in SBRSs. The first RNN-based SBRSs is GRU4Rec [1], which employes gated recurrent unit (GRU) to capture the long-term dependencies within sessions. Later, Balazs, et al. [2] improved GRU4Rec by introducing a ranking loss function. To emphasize those items that are more relevant to the next-item, a self-attention model was combined with RNN for next-item recommendation. However, due to the employed rigid order assumption that any adjacent items within a session are highly sequentially dependent, RNN-based SBRSs may generate false dependencies [19].

In recent years, benefiting from the power of capturing the complex transitions among items, graph neural network (GNN) [21] has been employed to built more powerful SBRSs. The first GNN-based SBRS is SR-GNN [15], which first maps each session into a sub directed graph and then applies GNN on the graph to capture dependencies between items within each session. Later, Xu, et al. [3] proposed a graph contextualized self-attention model (GC-SAN) to incorporate an attention mechanism into GNN to learn the long-range dependencies within sessions for next-item recommendations. Qiu et al. [13] proposed a weighted attention graph layer and a readout function to learn embeddings of items while relaxing the order assumption over items for the next item recommendation. Yu et al. [5] proposed a novel target attentive graph neural network (TAGNN) to adaptively activate a user's different interest w.r.t. varied next-item for more accurate next-item recommendations. However, most of the existing GNN-based SBRSs [20] process each session separately, i.e., the GNN is employed on each subgraph successively to learn item representations for recommendations. In this way, they fail to explicitly capture the global dependencies based on multi-session and attribute information, which leads to information loss and thus is harmful to the subsequent recommendations. Moreover, these methods poor to recommend cold-start items due to their quite limited session information.

Only a few models have been proposed to capture the inter-session dependencies, like hierarchical RNN [10] and hierarchical attention [6] based SBRSs. However, they only capture the single-way sequential dependencies between sessions from the same user, totally different from the global dependencies in this paper. In practice, the dependencies across sessions are much more complex than the aforementioned sequential dependencies, since different sessions can be also widely connected via the shared items or shared item attribute values. This essentially motivates us to develop HMGL to learn the complex global dependencies across sessions to further improve the recommendation performance.

3   Problem Statement

We formulate the research problem in this section. Let $\mathcal{S} = \{s_1, s_2, ..., s_{|S|}\}$ denote the session set consisting of all the sessions in a dataset, where $|S|$ is the number of sessions in $\mathcal{S}$. $\mathcal{V}^I = \{v_1, v_2, ..., v_{|\mathcal{V}^I|}\}$ denotes the item set consisting of all the unique items from all sessions. $s_i = \{v_{i,1}, v_{i,2}, ..., v_{i,|s_i|}\}$ is a session consisting of sequentially interacted (e.g., purchased by an anonymous user) items. $F = \{f_1, f_2, ..., f_{|F|}\}$ is the set of attributes for all items and each attribute (e.g., category) $f_h = \{a_{h,1}, a_{h,2}, ..., a_{h,|f_h|}\}$ consists of multiple attribute values (e.g., food, beverage), denoted as $a_{h,b}(1 \leq b \leq |f_h|)$. All the attribute values constitute the attribute value set $\mathcal{V}^A = \{a_{h,1}, a_{h,2}, ..., a_{h,|f_h|}\}$ ($h \in \{1, 2, ..., |F|\}$). Usually, there are both categorical and numerical attributes for items. In this paper, we consider categorical attributes only since numerical attributes need to be handled different from the categorical ones. For each item $v_i \in \mathcal{V}^I$, all its attribute values form a set $\mathcal{A}_{v_i} = \{a_1, ..., a_h, ..., a_{|F|}\}$. For a target item v_l from a session s, all the items that occurred prior to v_l in s form the session context of v_l, denoted as $C_{v_l}^s = \{v_1, v_2, ..., v_{l-1}\}$, while all the attribute values of the items in $C_{v_l}^s$ form the corresponding attribute context $C_{v_l}^a = \{\mathcal{A}_{v_1}, \mathcal{A}_{v_2}, ..., \mathcal{A}_{v_{l-1}}\}$.

Given a context $C = [C^s, C^a]$ with $(l-1)$ precedent items associated with their attribute values, the task of our work is to recommend the lth item. Accordingly, our proposed HMGL learns the conditional probability distributions $P(v|C)$ for each candidate item $v(v \in \mathcal{V}^I)$ given the context C. Consequently, once all the model parameters have been learned, the next item to be recommended can be selected from the candidate items by maximizing $P(v|C)$.

4   Heterogeneous Mixed Graph Learning Framework

In this section, we present our proposed heterogeneous mixed graph learning (HMGL) framework. In particularly, the HMGL framework first constructs an HMG by mapping the session information and item attribute information into a unified graph and then jointly learns both local and global dependencies over the HMG for next-item recommendations. Accordingly, as shown in Fig. 2, the HMGL framework consists three components: the HMG construction module, the HMG learning module, the next-item prediction module. Specifically, the HMG learning model includes both the local dependency learning and the global dependency learning modules. In the following subsections, we first introduce some preliminaries and then introduce each component of the HMGL framework.

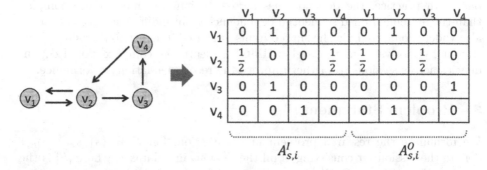

Fig. 2. The workflow of our proposed an HMGL framework. We integrate both session information and attribute information into an HMG. Then, the HMG learning module is devised to learn both local and global dependencies over HMG and export an informative representation for each item. Finally, we predict the conditional probability of each candidate item and recommend the item with maximum probability.

4.1 Preliminary

We define two types of graphs, i.e., heterogeneous mixed graph and underlying graph, which will be used in the following subsections.

Definition 1 *Heterogeneous Mixed Graph. A heterogeneous mixed graph (HMG) $\widetilde{\mathcal{G}} = \{\mathcal{V}, \mathcal{E}, \mathcal{D}\}$ is composed of a node set $\mathcal{V}$, an edge set $\mathcal{E}$, and a direction set $\mathcal{D}$. In addition, the numbers of both node types and edge types are larger than one, while there are both directed and undirected edges in the graph.*

Definition 2 *Underlying Graph. The underlying graph $\mathcal{G} = \{\mathcal{V}, \mathcal{E}\}$ of a given HMG $\widetilde{\mathcal{G}} = \{\mathcal{V}, \mathcal{E}, \mathcal{D}\}$ is an undirected graph extracted from the HMG by changing the directed edges into undirected ones. Hence, an underlying graph consists of the node set $\mathcal{V}$ and the edge set $\mathcal{E}$.*

4.2 Heterogeneous Mixed Graph Construction

We incorporate the item attribute information into SBRSs to enrich the connections between items for better next-item recommendations. In order to construct a unified HMG ($\widetilde{\mathcal{G}} = \{\mathcal{V}, \mathcal{E}, \mathcal{D}\}$), firstly, we model the items and their attribute values as two types of nodes in a graph, let $\mathcal{V} = \mathcal{V}^{\mathcal{I}} \cup \mathcal{V}^{\mathcal{A}}$ be the node set on the graph, where $\mathcal{V}^{\mathcal{I}}$ and $\mathcal{V}^{\mathcal{A}}$ are the node sets corresponding to items and item attribute values respectively, as defined in Sect. 3.

Secondly, we model the item-item relations within each session and the item-attribute value relations as two types of edges in HMG, namely, $\mathcal{E} = \mathcal{E}^{\mathcal{I}} \cup \mathcal{E}^{\mathcal{A}}$ constitute the edge set in HMG, where a directed edge $e_{i,j}^I \in \mathcal{E}^{\mathcal{I}}$ means item v_j occurs after item v_i in a given session, an undirected edge $e_{i,l}^A \in \mathcal{E}^{\mathcal{A}}$ means that item v_i has the attribute value a_l. As a result, an HMG connecting all items based on session information and attribute information is built. In this way, the shared attribute values of different items serve as bridges to connect items indirectly, since there is an edge between the shared attribute values and each of the items sharing these attribute values on the HMG. As a result, such an HMG not only enriches the connections between items, especially those items from different sessions, but also connects cold-start items and warm-start items.

4.3   Heterogeneous Mixed Graph Learning

As shown in Fig. 2, the HMG learning module contains two parts: *local dependency learning* and *global dependency learning*. First, a gated graph neural network (GGNN) is utilized for the local dependency learning by taking the initialized item representations as the input and output the local item representations which encode the local dependencies within sessions. Then, the local representations are imported into a path-based matrix factorization model to obtain the final item representations while further incorporating the global dependencies between items across sessions. Finally, the learned final representations encoding both local and global dependencies are fed into the prediction layer for next-item prediction. Next, we introduce the two parts one by one in detail.

Local Dependency Learning. In order to learn the local dependencies, we firstly extract a directed subgraph $\mathcal{G}_s = \{\mathcal{V}^{\mathcal{I}}{}_s, \mathcal{E}^{\mathcal{I}}{}_s\}$ based on a given session s from the HMG. Then, for the ith node $v_{s,i}$ in $\mathcal{G}_s$, we learn a latent representation $\mathbf{v}_{s,i}$ via gated graph neural networks (GGNN) [8]. GGNN updates the representation of each node through absorbing the information from other nodes in the same subgraph. Subsequently, the local dependencies embedded in each subgraph (session) are encoded into the representation of each node (item).

Firstly, we map each node $v \in \mathcal{V}^I$ into an unified low-dimension latent space to obtain the initial representation $\mathbf{v} \in \mathbb{R}^{1 \times d}$, where d denotes the dimension of the representation. Then for each node $v_{s,i} \in \mathcal{G}_s$, we employ GGNN to iteratively update its representation $\mathbf{v}_{s,i}$ by absorbing the information from other nodes in $\mathcal{G}_s$ to learn the local dependencies. In specific, in the tth iteration, we first extract the contextual information $\mathbf{a}_{s,i}^t$ from the neighborhoods of $v_{s,i}$ under the constraint matrix $\mathbf{A}^s$ based on $\mathcal{G}_s$:

$$\mathbf{a}_{s,i}^t = \mathbf{A}_{i,:}^s [\mathbf{v}_{s,1}^{t-1}, \mathbf{v}_{s,2}^{t-1}, ..., \mathbf{v}_{s,|s|}^{t-1}]^T \mathbf{H} + \mathbf{b}, \tag{1}$$

where $\mathbf{H} \in \mathbb{R}^{d \times d}$ is the weight matrix, $\mathbf{b}$ is the bias vector, $[\mathbf{v}_{s,1}^{t-1}, \mathbf{v}_{s,2}^{t-1}, ..., \mathbf{v}_{s,|s|}^{t-1}]$ is the list of hidden states (i.e., representations) of nodes in $\mathcal{G}_s$ at $(t-1)^{th}$ iteration.

The constraint matrix $\mathbf{A}^s = [\mathbf{A}^{s,I}, \mathbf{A}^{s,O}] \in \mathbb{R}^{|s| \times 2|s|}$ is the concatenation of two adjacency matrices $\mathbf{A}^{s,I}$ and $\mathbf{A}^{s,O}$, while $\mathbf{A}^s_{i,:}$ denotes the ith row of $\mathbf{A}^s$ and it corresponds to v_i. $\mathbf{A}^{s,I}$ and $\mathbf{A}^{s,O}$ represent weighted connections of incoming and outgoing edges in $\mathcal{G}_s$ respectively, which are calculated as occurrence times of the corresponding edge divided by the outdegree (the number of tail ends adjacent to a node) of the edge's starting node. In this way, the communications between nodes and different importance scales indicated by their frequencies of edges in the subgraph are captured.

Once the contextual information is extracted, we take $\mathbf{a}^t_{s,i}$ and the hidden state $\mathbf{v}^{t-1}_{s,i}$ of v_i in the $(t-1)^{th}$ iteration as the input to calculate the candidate hidden state $\widetilde{\mathbf{v}}^t_{s,i}$ of $v_{s,i}$ for the tth iteration as presented by Eq (4), where the reset gate vector $\mathbf{r}^t_{s,i}$ and the update gate vector $\mathbf{z}^t_{s,i}$ are calculated using Eqs (2) and (3) respectively:

$$\mathbf{z}^t_{s,i} = \sigma(\mathbf{W}_z a^t_{s,i} + \mathbf{U}_z \mathbf{v}^{t-1}_{s,i}), \tag{2}$$

$$\mathbf{r}^t_{s,i} = \sigma(\mathbf{W}_r a^t_{s,i} + \mathbf{U}_r \mathbf{v}^{t-1}_{s,i}), \tag{3}$$

$$\widetilde{\mathbf{v}}^t_{s,i} = tanh(\mathbf{W}_o a^t_{s,i} + \mathbf{U}_o(r^t_{s,i} \odot \mathbf{v}^t_{s,i})), \tag{4}$$

where $\mathbf{W}_z, \mathbf{W}_r, \mathbf{W}_o \in \mathbb{R}^{2d \times d}$ and $\mathbf{U}_z, \mathbf{U}_r, \mathbf{U}_o \in \mathbb{R}^{d \times d}$ are learnable weight matrices, $\sigma(\cdot)$ is the activation function and is specified as sigmoid function, $\odot$ denotes the element-wise multiplication operation.

Subsequently, the hidden state $\mathbf{v}^t_{s,i}$ of $v_{s,i}$ in the current t^{th} iteration can be determined by the update gate $\mathbf{z}^t_{s,i}$, the hidden state $\mathbf{v}^{t-1}_{s,i}$ of $v_{s,i}$ in the $(t-1)^{th}$ iteration and the candidate hidden state $\widetilde{\mathbf{v}}^t_{s,i}$:

$$\mathbf{v}^t_{s,i} = (1 - \mathbf{z}^t_{s,i}) \odot \mathbf{v}^{t-1}_{s,i} + \mathbf{z}^t_{s,i} \odot \widetilde{\mathbf{v}}^t_{s,i}. \tag{5}$$

In this way, the local dependencies between item $v_{s,i} \in s$ and other items in session s are encoded into the latent representation of $v_{s,i}$. Further, we can learn the representations of other items $v_{s,j} \in s(j \neq i)$ in the same way. After all the sessions in the dataset are processed in the same way, the local representations encoding local dependencies of all items are subsequently learned as $[\mathbf{v}^{local}_1, \mathbf{v}^{local}_2, ..., \mathbf{v}_{|\mathcal{V}^I|}{}^{local}]$.

Global Dependency Learning. In order to learn global dependencies, we firstly extract an underlying graph $\mathcal{G} = \{\mathcal{V}, \mathcal{E}\}$ from the whole HMG. Then two types of paths over $\mathcal{G}$ are defined to reveal two kinds of global inter-item dependencies. One type of path contain two adjacent item nodes on $\mathcal{G}$ to real their co-occurrence relations in the same sessions, denoted as "$v_i - v_j$"; the other type of path contains two item nodes plus their shared attribute value nodes, to reveal the indirect dependencies between items sharing the same attribute value, denoted as "$v_i - a_k - v_j$", where $v_i, v_j \in \mathcal{V}^I$ and $a_k \in \mathcal{V}^A$. Accordingly, the following two path matrices are built among all the item nodes: N^I with the entry $n^I_{i,j}$ to denote the number of the first type of paths between v_i and v_j, and

N^A with the entry $n_{i,j}^A$ to denote the number of the second type of paths between v_i and v_j. Inspired by [23] and [22], the number of paths between two nodes in a graph reflects the strength of dependency between them and can be estimated by the latent factors of them. Therefore, we factorize the path matrices into the latent factors of items to fine-tune the item representations to incorporate the global dependencies by taking the local representations of items as the input to initialize the latent vectors of items.

Hence, the two path matrices N^I and N^A are factorized jointly by estimating the values of entries in them respectively using the latent vectors of items below:

$$\hat{n}_{i,j}^I = f(\mathbf{p}^I, \mathbf{v}_i, \mathbf{v}_j) = \sum_{q=1}^{d} \mathbf{p}_q^I (\mathbf{v}_{i,q})^T \mathbf{v}_{j,q}, \tag{6}$$

$$\hat{n}_{i,j}^A = f(\mathbf{p}^A, \mathbf{v}_i, \mathbf{v}_j) = \sum_{q=1}^{d} \mathbf{p}_q^A (\mathbf{v}_{i,q})^T \mathbf{v}_{j,q}, \tag{7}$$

where $\mathbf{p}^I, \mathbf{p}^A \in \mathbb{R}^d$ are the latent vectors representing the types of paths, and d is the dimension of the latent space. $\mathbf{v}_{i,q}$ is the qth bit of the latent vector $\mathbf{v}_i$ of v_i. After the conduction of the factorization, the item representations are updated accordingly to incorporate the global dependencies. As a result, the informative final representation, e.g., $\mathbf{v}_i$, of each item, e.g., $v_i(v_i \in V^I)$, is achieved, which encodes both local and global dependencies for better next-item prediction.

4.4   Prediction and Optimization

Next-Item Prediction. Once the final representations $\mathbf{v}_1, \mathbf{v}_2, ..., \mathbf{v}_{|\mathcal{V}^I|}$ of all items are learned by HMGL framework, the prediction can be made by taking them as the input. Specifically, given a context $C = \{v_{c,1}, v_{c,2}, ..., v_{c,|C|}\}$, the representation $\mathbf{C}$ of C is obtained by integrating the representations of all items in it via a fully connected layer:

$$\mathbf{C} = [\mathbf{v}_{C,1}, \mathbf{v}_{C,2}, ..., \mathbf{v}_{C,|C|}]^T \mathbf{W}_C \tag{8}$$

where $\mathbf{W}_C$ is the weight matrix to be learned. Then, we feed the context representation $\mathbf{C}$ together with the latent representation of candidate item $\mathbf{v}_i$ into the output layer for the target item prediction. Specifically, a score that quantifies the relevance between the context and candidate item is computed as the inner product of the context representation $\mathbf{C}$ and the candidate item representation $\mathbf{v}_i$. Finally, we apply a softmax function to predict the conditional probability $\hat{y}_i$ for each candidate $v_i \in \mathcal{V}^I$:

$$\hat{y}_i = softmax((\mathbf{v}_i)^T \mathbf{C}).$$

Optimization. To learn the parameters of the proposed model, we utilize an end-to-end training scheme, the total objective is to minimize the following loss:

$$Loss^{Total} = Loss^{local} + \gamma(Loss^{Global}). \tag{9}$$

where γ is the coefficient to control the importance of the $Loss^{Global}$. The $Loss^{Global}$ also acts as regularization term for better optimization and prevent the over-fittings or local optimizations.

The local loss function $Loss^{local}$ is defined as the cross-entropy between the prediction $\hat{y}$ and the ground truth y, which can be written as:

$$Loss^{local} = -\sum_{i=1}^{m} y_i log(\hat{y}_i) + (1 - y_i)log(1 - \hat{y}_i) \tag{10}$$

where y is the label of each candidate item, its value is 1 when the candidate item is the true target item, and 0 otherwise.

The global loss function $Loss^{local}$ is defined as the root-mean-square error (RMSE), which is an estimator with respect to the true path matrices N^I, N^A and predicted matrix $\hat{N}^I, \hat{N}^A$, which is defined as the square root of the mean square error:

$$Loss^{Global} = Loss^I + Loss^A$$
$$= \frac{1}{|\mathcal{V}^I||\mathcal{V}^I|} \sum_{i,j \in \mathcal{V}^I} \sqrt{(\hat{n}_{i,j}^I - n_{i,j}^I)^2} + \sqrt{(\hat{n}_{i,j}^A - n_{i,j}^A)^2} \tag{11}$$

Finally, we use a mini-batch gradient descent to train the proposed HMGL model. Note that in session-based recommendation scenarios, most sessions are of relatively short lengths, SBRSs are easy to suffer from the over-fitting during the optimization. $Loss^{Global}$ is also acted as the penalty term in the total loss function, which can effectively prevent the local optimization and over-fitting.

5 Experiments

In this section, we introduce the datasets, evaluation metrics, comparison methods and parameter settings, and we evaluate the recommendation performance of our proposed HMGL framework by comparing it with the baseline methods.

5.1 Preparation

Data Preparation. The following two commonly used real-world transaction datasets are used for experiments:

- Tmall: released by IJCAI-15 competition, which records the pruchased items as well as their attribute information (i.e., category and brand) in each transaction on Tmall.com (Chinese version of Amazon.com).
- Dunnhumby: provides the household transactions of 2,500 households over 2 years (collected by the data science company Dunnhumby). In addition, the category information of each item is provided.

Firstly, a set of sessions is extracted from the original transaction data by putting all the items in one transaction together to form a session. Following a common practice [19], those sessions containing less than three items are removed since at least two items should be used to build an informative context and the addition one as the target item. The item information table contains the attribute values of each item occurred in the sessions. Secondly, the set of sessions is splitted into training set, test set and validation set respectively. We randomly select 70% as the training set, 20% as the test set, and the rest 10% for validation. Finally, to test the performance of our proposed model under different cold-start levels, part of the sessions in training set are removed to form the training sets with various cold-start levels. To be specific, we construct 3 different training sets with a drop rate of 0%, 40%, and 80%, respectively. Taking the one with the drop rate of 80% as an example, for each target item to be predicted in the testing set, 80% of all the sessions containing it in the training set are dropped. The statistic of the datasets are shown in Table 1.

Table 1. Statistics of experimental datasets.

Statistics	Tmall	Dunnhumby
#Sessions	125,111	173,913
#Items	26,251	24,897
#Item category	763	583
#Item brand	3,641	n.a
Avg. session length	5.21	8.09

Evaluation Metrics. We use the following widely used accuracy metrics to evaluate all the comparison approaches.

- Rec@k (Recall) is a metric to measure prediction accuracy. It represents the proportion of the correctly recommended items amongst the recommended top-k items. Here we choose $k \in \{10, 20\}$.
- Mrr@k (Mean Reciprocal Rank) is the average of reciprocal ranks of the true target items over all recommendation instances. Here we choose $k \in \{10, 20\}$.

Comparison Methods. To demonstrate the efficacy of our proposed HMGL framework, which extracts both global and local dependencies for next-item recommendations, we implemented two versions of our model: (1) full version of **HMGL** proposed in this work; and (2) **HMGL − L**, which only utilizes local dependencies. We take the representative methods for performance comparisons, which are built on representative frameworks including matrix factorization, Markov chains, recurrent neural networks (RNN), memory networks, convolutional neural networks (CNN) and graph neural networks (GNN).

- **POP** recommends the top-k frequent items in the training set and in the current session respectively.
- **BPR-MF** is the state-of-the-art method for non-sequential recommendations, which uses a pairwise ranking loss [11].
- **FPMC** is a classic model combining matrix factorization and first-order Markov Chain for next-basket recommendations [11]. Here it is utilized to factorize the transition matrix between any two items in the session data and thus predict the next-item based on the last item in the session.
- **iGRU4Rec-BPR** is the improved version of typical RNN-based SRBS, namely GRU4Rec, which uses GRU to model the sequences of purchased items in sessions for next-item recommendations. It takes Bayesian Personalized Ranking (BPR) as the loss function [2].
- **STAMP** is a novel short-term memory priority model to capture both the user's long-term preference from previous clicks and the current preference from the last click in a session for the next-item recommendations [9].
- **NextItNet** uses a convolutional generative network to model long range dependencies in sequences of items for next-item recommendations [4].
- **SR-GNN** is an SBRS model using gated graph neural networks to first generate latent representations of items in sessions and then use these representations for next-item recommendations [15].

Parameter Settings. We initialize all the baseline models with the parameter settings reported in their papers and then tune them on our datasets for best performance for fair comparison. For our model, all parameters are initialized using a Gaussian distribution with a mean of 0 and a standard deviation of 0.1. The sizes of item representations and hidden states in HMGL are set to 128. The mini-batch Adam optimizer is utilized to learn the model parameters, where the initial learning rate is set to 0.001, and the coefficient γ is set to 0.1 via cross validation on the specific datasets, the batch size is set to 100. We run 30 epoches to train our HMGL model for best performance.

5.2   Recommendation Accuracy Evaluation

Extensive experiments are conducted to answer the following questions:

- **Q1:** How does our approach perform compared with the state-of-the-art SBRSs in terms of recommendation accuracy in the warm-start situation?
- **Q2:** How does our approach perform compared with the state-of-the-art SBRSs in terms of recommendation accuracy in the cold-start situation?
- **Q3:** How does our full model HMGL which models both global and local dependencies perform compared with its simplified version HMGL-L which models local dependencies only?

Result 1 (for Q1): Comparison with Baselines Under Warm-Start Situation. We compare the recommendation accuracy of our HMGL model with

Table 2. Results of effectiveness experiments on two datasets.

Scenario	Model	Tmall				Dunnhumby			
		Rec@10	Rec@20	Mrr@10	Mrr@20	Rec@10	Rec@20	Mrr@10	Mrr@20
drop 0%	POP	0.0170	0.0310	0.0050	0.0060	0.0441	0.0666	0.0215	0.0229
	Item-KNN	0.1280	0.1490	0.0680	0.0720	0.0412	0.0596	0.0186	0.0200
	BPR-MF	0.1760	0.2150	0.0960	0.0968	0.0196	0.0337	0.0074	0.0081
	FPMC	0.2850	0.3070	0.2200	0.2500	0.0445	0.0749	0.0183	0.0223
	iGRU4Rec	0.3170	0.3410	0.2300	0.2360	0.0507	0.0731	0.0205	0.0220
	STAMP	0.3000	0.3090	0.2370	0.2490	0.0734	0.1124	0.0313	0.0354
	NextItNet	0.2490	0.2700	0.1710	0.1720	0.1204	0.1591	0.0604	0.0629
	SR-GNN	<u>0.3680</u>	<u>0.4100</u>	<u>0.2460</u>	<u>0.2490</u>	<u>0.1931</u>	<u>0.2527</u>	<u>0.0951</u>	<u>0.0992</u>
	HMGL-L	0.3681	0.4103	0.2462	0.2492	0.1935	0.2527	0.0952	0.0995
	HMGL	**0.3783**	**0.4185**	**0.2549**	**0.2577**	**0.2006**	**0.2627**	**0.0992**	**0.1035**
	Improvement(%)[1]	2.07	2.54	3.61	3.49	3.88	3.95	4.31	4.33
drop 40%	POP	0.0150	0.0270	0.0050	0.0060	0.0419	0.0641	0.0208	0.0224
	Item-KNN	0.0820	0.1210	0.0350	0.0380	0.0075	0.0108	0.0033	0.0034
	BPR-MF	0.1150	0.1380	0.0780	0.0810	0.0022	0.0339	0.0100	0.0112
	FPMC	0.1176	0.1205	0.0612	0.0661	0.0441	0.0666	0.0215	0.0229
	iGRU4Rec	0.1508	0.1533	0.0731	0.0852	0.0077	0.0128	0.0031	0.0034
	STAMP	0.1698	0.1770	0.0424	0.0555	0.0193	0.0238	0.0059	0.0076
	NextItNet	0.1418	0.1444	0.0752	0.0766	0.0859	0.1092	0.0468	0.0490
	SR-GNN	<u>0.1750</u>	<u>0.2120</u>	<u>0.0890</u>	<u>0.0923</u>	<u>0.1600</u>	<u>0.2140</u>	<u>0.0753</u>	<u>0.0790</u>
	HMGL-L	0.1751	0.2122	0.0893	0.0925	0.1603	0.2141	0.0754	0.0791
	HMGL	**0.1799**	**0.2174**	**0.09232**	**0.0953**	**0.1724**	**0.2284**	**0.0788**	**0.0825**
	Improvement(%)	2.80	2.54	3.73	3.25	7.75	6.72	4.46	4.43
drop 80%	POP	0.0080	0.0215	0.0050	0.0060	0.0291	0.0499	0.0172	0.0187
	Item-KNN	0.0510	0.0720	0.0300	0.0300	0.0052	0.0081	0.0022	0.0024
	DPR-MF	0.0300	0.0340	0.0350	0.0390	0.0215	0.0295	0.0080	0.0087
	FPMC	0.0333	0.0507	0.0328	0.0336	0.0065	0.0072	0.0027	0.0032
	iGRU4Rec	0.0408	0.0425	0.0420	0.0471	0.0075	0.0124	0.0236	0.0338
	STAMP	0.0951	0.1564	0.0689	0.0694	0.0087	0.0666	0.0041	0.0229
	NextItNet	0.0859	0.0938	0.0351	0.0413	0.0119	0.0145	0.0215	0.0313
	SR-GNN	<u>0.1337</u>	<u>0.1652</u>	<u>0.0659</u>	<u>0.0681</u>	<u>0.0652</u>	<u>0.0850</u>	<u>0.0342</u>	<u>0.0356</u>
	HMGL-L	0.1338	0.1653	0.0659	0.0683	0.0653	0.0851	0.0343	0.0357
	HMGL	**0.1395**	**0.1698**	**0.0694**	**0.0702**	**0.0702**	**0.0919**	**0.0359**	**0.0366**
	Improvement(%)	4.33	2.78	5.31	3.08	7.66	8.11	4.97	2.80

[1]Improvement achieved by HMGL over the best-performing compared methods

those of eight representative baselines in the warm-start situation. The results are shown in Table 2 (the drop 0% scenario). The performance of the first two methods PoP and Item-KNN is poor, since they make recommendations only based on the popularity or similarity of items, failing to effectively capture the dependencies between items in sessions. The performance of BPR-MF is a little better, but still poor. The main reason is that both datasets are extremely sparse and MF models easily suffer from sparse data. FPMC takes the advantages of both Markov Chain and factorization machines, and thus performs a slightly better than BPR-MF. But FPMC is a fist-order Markov Chain model, which can only learn the transitions over adjacent items while ignoring high-order dependencies. Benefiting from the capability of capturing complex relations of deep neural networks, iGRU4Rec, STAMP and NextItNet achieved better performance than the aforementioned models. For example, iGRU4Rec employs RNN built on gated recurrent unit (GRU) to model the sequential dependencies within sessions. STAMP and NextItNet employs attention mechanism and CNN respectively to model intra-session dependencies for next-item recommendations. By capturing the complex transitions among items, the performance of SR-GNN is better. However, it fails to explicitly capture the global dependencies, leading to information loss and thus is harmful to the subsequent recommendations.

By explicitly capturing both the local dependencies and the global dependencies, our HMGL framework achieves the best performance on both datasets. It outperforms the best-performing method SR-GNN with an average of 3.52%, ranging from 2.07% to 4.33%, in terms of Rec@10, Rec@20, Mrr@10 and Mrr@20.

Result 2 (for Q2): Comparison with Baselines Under Cold-Start Situations. In order to model the cold-start and warm-start scenario, we construct three different training sets with a drop rate of 0%, 40%, and 80%. It is clear that HMGL achieves higher accuracy than any of the eight baseline methods in these two cold-start scenarios, shown by the scenarios demoted as "drop 40%" and "drop 80%" in Table 2. Specifically, when dropping 40%, the average improvement percentage achieved by HMGL over the best-performing methods is 4.46%, range from 2.8% to 7.75%. When dropping 80%, the average improvement percentage is 4.88%, range from 2.8% to 8.11%. This verifies the effectiveness of of our proposed HMGL framework in handling the cold-start items.

Result 3 (for Q3): Global and Local Dependencies vs. Local Dependencies Only. To demonstrate the efficacy of the modeling of global dependencies, we compare the performance of HMGL with that of HMGL-L. As shown in Table 2, it is clear that HMGL achieves higher accuracy under all scenarios on both datasets than HMGL-L does. This justifies the necessity to explicitly and comprehensively model the global dependencies over the whole dataset for more accurate next-item recommendations.

6   Conclusions

In this paper, we propose an Heterogeneous Mixed Graph Learning (HMGL) framework for session-based recommendations. Firstly, we have constructed a heterogeneous mixed graph based on both session information and attribute information. Then we have designed an HMG learning model to learn a unified representation for each item by modelling both local and global dependencies. The learned representations are further used for next-item recommendations. Extensive experiments on two real-world datasets demonstrated the effectiveness of HMGL. As for future work, we plan to further utilize item attributes by combining them together with item IDs to build more informative item representations for better addressing the cold-start recommendation issue.

Acknowledgements. This work was supported by ARC Discovery Project DP180102378.

References

1. Hidasi, B., Karat-zoglou, A., et al.: Session based recommendations with recurrent neural networks. In: ICLR, pp. 1–10 (2016)
2. Hidasi, B., Karat-zoglou, A., et al.: Recurrent neural networks with top-k gains for session-based recommendations. In: CIKM, pp. 843–852 (2018)
3. Xu, C., Zhao, P., et al.: Graph contextualized self attention network for session-based recommendation. In: IJCAI, pp. 3940–3946 (2019)
4. Yuan, F., Karatzoglou, A., et al.: A simple convolutional generative network for next item recommendation. In: WSDM, pp. 582–590 (2019)
5. Yu, F., Zhu, Y., et al.: TAGNN: target attentive graph neural networks for session-based recommendation. In: SIGIR, pp. 1–5 (2020)
6. Ying, H., Zhuang, F., et al.: Sequential recommender system based on hierarchical attention networks. In: IJCAI, pp. 3926–3932 (2018)
7. Linden, G., Smith, B., York, J.: Amazon.com recommendations: item-to-item collaborative filtering. IEEE Internet Comput. **7**(1), 76–80 (2003)
8. Li, Y., Tarlow, D., et al.: Gated graph sequence neural networks. In: ICLR, pp. 1–14 (2015)
9. Liu, Q., Zeng, Y., et al.: STAMP: short-term attention/memory priority model for session-based recommendation. In: KDD, pp. 1831–1839 (2018)
10. Ruocco, M., Skrede, O.S.L., et al.: Inter-session modeling for session-based recommendation. In: 2nd Workshop on DLRS, pp. 24–31 (2017)
11. Rendle, S., Freudenthaler, C., et al.: Factorizing personalized Markov chains for next-basket recommendation. In: WWW, pp. 811–820 (2010)
12. Forsati, R., Meybodi, M.R., et al.: Web page personalization based on weighted association rules. In: ICECT, pp. 130–135 (2009)
13. Qiu, R., Li, J., et al.: Rethinking the item order in session-based recommendation with graph neural networks. In: CIKM, pp. 579–588 (2019)
14. Feng, S., Li, X., et al.: Personalized ranking metric embedding for next new POI recommendation. In: IJCAI, pp. 2069–2075 (2015)
15. Wu, S., Tang, Y., et al.: Session-based recommendation with graph neural networks. In: AAAI, pp. 346–353 (2018)
16. Wang, S., Hu, L., et al.: Modeling multi-purpose sessions for next-item recommendations via mixture-channel purpose routing networks. In: IJCAI, pp. 3771–3777 (2019)
17. Wang, S., Cao, L.: Inferring implicit rules by learning explicit and hidden item dependency. IEEE Trans. Syst. Man Cybern. Syst. **50**(3), 935–946 (2020)
18. Wang, S., Hu, L., et al.: Intention nets: psychology-inspired user choice behavior modeling for next-basket prediction. In: AAAI, pp. 6259–6266 (2020)
19. Wang, S., Hu, L., et al.: Attention-based transactional context embedding for next-item recommendation. In: AAAI, pp. 2532–2539 (2018)
20. Wang, W., Zhang, W., et al.: Beyond clicks: modeling multi-relational item graph for session-based target behavior prediction. In: WWW, pp. 3926–3932 (2020)

21. Li, X., Yin, H., et al.: Semi-supervised Clustering with Deep Metric Learning and Graph Embedding. World Wide Web **23**(2), 781–798 (2020). https://doi.org/10.1007/s11280-019-00723-8

22. Wang, Y., Feng, C., Chen, L., Yin, H., Guo, C., Chu, Y.: User identity linkage across social networks via linked heterogeneous network embedding. World Wide Web **22**(6), 2611–2632 (2018). https://doi.org/10.1007/s11280-018-0572-3

23. Wang, Z., Liu, H., et al.: Unified embedding model over heterogeneous information network for personalized recommendation. In: IJCAI, pp. 3813–3819 (2019)

Why-Not Questions & Explanations
for Collaborative Filtering

Maria Stratigi[1], Katerina Tzompanaki[2], and Kostas Stefanidis[1(✉)]

[1] Tampere University, Tampere, Finland
{maria.stratigi,konstantinos.stefanidis}@tuni.fi
[2] ETIS Lab, UMR 8051, CY Cergy Paris University, ENSEA, CNRS,
95302 Pontoise, France
aikaterini.tzompanaki@cyu.fr

Abstract. Throughout our digital lives, we are getting recommendations for about almost everything we do, buy or consume. However, it is often the case that recommenders cannot locate the best data items to suggest. To deal with this shortcoming, they provide explanations for the reasons specific items are suggested. In this work, we focus on explanations for items that do *not* appear in the recommendations they way we expect them to, expressed in why-not questions, to aid the system engineer improve the recommender. That is, instead of offering explanations on every item proposed by the system, we allow the developer give feedback about items that were not proposed. We consider here the most traditional category of recommenders, i.e., the collaborative filtering one, and propose ways for providing explanations for why-not questions. We provide a detailed taxonomy of why-not questions on recommenders, and model-specific explanations based on the inherent parameters of the recommender. Finally, we propose an algorithm for producing explanations for the proposed why-not questions.

Keywords: Explanations · Why-not questions · Recommendations · Collaborative filtering · Recommender systems

1 Introduction

Recommendations have been integrated in many of the services available to users in recent times. Although, recommenders try to accurately propose interesting items to users according to their preferences, it is often the case that they cannot locate the best data items to suggest. This can be due to many different reasons. One reason can be the cold start problem, where the system does not have enough information about a user to make accurate predictions. Another cause may be the over-specification on the part of the users. This means that a user has previously expressed a preference for a specific category and the system is unlikely to propose items that belong to a different category. Furthermore, often the systems can be misdirected due to ambiguous information on the users and

© Springer Nature Switzerland AG 2020
Z. Huang et al. (Eds.): WISE 2020, LNCS 12343, pp. 301–315, 2020.
https://doi.org/10.1007/978-3-030-62008-0_21

their preferences. Finally, as a system relies a lot on its hyper-parameters and thresholds, unlucky recommendations may be tight to the system's configuration.

The problem of explaining recommendations is a long-standing problem, which is most regularly approached by introducing explanations along with recommendations (e.g., [4,16,22]). This way, the user or the system's designer gets insights on why an item is suggested. The explanations can then vary on granularity or presentation format based on the final consumer, i.e., the final user of the recommender or the designer of the system. In this work, we expand on the concept of *post-hoc*, model-based explanations [23], i.e., explanations provided after the recommendations have been produced and based on the knowledge of the system, by exploiting the concept of why-not questions. These questions are not about why items were proposed, but why items *were not proposed*. We judge that this kind of questions are necessary for the system engineer, who needs to better understand the system and get hints on how to debug it. For example, assume a system that recommends products to users. If the engineer finds that the products of a specific company are never proposed to a user, he/she may need to understand why, and find the best way(s) to turn the situation around. This could be in the benefit of the diversity of recommendations proposed to the final user, or even for promotional campaigns of the specific company, who does not see their products proposed by the system. On the other side, asking a why-not question may not be such a straightforward task for a final user, who is totally unaware of the context or his/her preferences. However, it can still be applied in the case of a knowledgeable user, who is aware of the context of the recommendations. For instance, a female user of a career development site may wonder why she never gets suggestions for managerial positions. In this case, a why-not explanation would help the user gain trust on the system and promote its gender-fairness. In this work, we assume the system designer as the consumer of the explanations, and leave the case of the final user as a future work.

One could suggest that explaining why a certain item is not proposed is dual to explaining why all the recommended items are proposed. With the standard explanation method, a user has to go over the recommended list and understand the differences between the proposed items and the one(s) expected. This would be a time consuming, or even impossible, task, depending on the user's understanding of the data set and the recommendation model. For this reason, already existing systems that treat the 'why' aspect of explanations cannot trivially explain missing recommendations, especially without the user feedback in the form of a why-not question. By having the system answering why-not questions, this process is streamlined and not strictly dependent on the user knowledge.

In this paper, we consider the traditional paradigm of a user-based collaborative filtering recommendation system for providing explanations to why-not questions. First, we provide a detailed categorization of why-not questions characterized by three main properties: (i) the level of absenteeism that the why-not questions mention (absence or low position in the ranking of a result set), (ii) their granularity (referring to a single result or a set), and (iii) their dependency to existing recommended items. Note, that a why not question may belong to multiple classes. Second, we provide fine-grained and personalised model-based explanations targeted for *system engineers*. The explanations are not dependent on the context of the system (e.g., social, product, PoI recommendation). We distinguish explanations between the *general* ones, based on the general setting of the problem, and the *model-specific* explanations, based on the inherent parameters of the recommendation model. Fourth, we propose an algorithmic method for computing explanations for why-not questions in collaborative filtering. Finally, we conduct a preliminary experimental study that explores the explanations space and motivates their usage by a system designer.

2 Preliminaries and Related Work

For a general setting of our recommender, assume a set of data items I and a set of users U, where each user provides ratings for a subset of I. Specifically, a user $u \in U$ rates an item $i \in I$ with a score s. The subset of users that rated an item i is denoted by U_i, whereas the subset of items rated by a user u, is denoted by I_u. For every item i not rated by a user u, the recommender estimates a relevance score, $p(u, i)$. The items with a high relevance score for u will compose the recommendation list (called also recommended items) for the user.

The literature regarding how to estimate the relevance score of an item for a user is extensive. In this work, we will focus on *collaborative filtering*, a well established recommendation approach that recommends items that users with similar preferences like (e.g., [3,11,13]). Specifically, the collaborative filtering (CF) approach is based on the idea that people who agreed in their evaluation of certain items in the past are likely to agree again in the future. The steps of a CF algorithm to produce a list of recommendations for a user u are: (1) Find the most similar users $Peers_u$ with u by means of a similarity function $sim(u, u')$ between u and every other user u'. (2) Predict a relevance score p for each item not rated by u based on his/her similar users $Peers_u$. (3) Recommend a list R_u with the top-k items with the highest relevance score.

The first step of the CF algorithm is to compute similarities between the users. To measure the similarity $sim(u, u')$ between two users, we exploit their ratings that are available in the recommender. Several metrics appear in the related work for counting similarities between users based on ratings. We employ here the Pearson correlation measure [11], which is fast to compute and performs very well for the case of collaborative filtering. It directly calculates the correlation between two users with a score from -1 for entirely dissimilar users, to 1 for identical users. A user u' is considered similar to u if their similarity is above a threshold th and if they have rated more than $numI$ common items. We further refine the process, by keeping only the $numP$ users with the highest similarity scores. We name these users as *peers* of u, $Peers_u$. In the second step of the algorithm, we use the peers of u to predict a relevance score $p(u, i)$ for any item i that u has not yet rated. To this end, we use the *weighted sum of others ratings* [18]. We only recommend items to u if more than $numPI$ peers have rated them. In this way, we have a more robust understanding of the items' preference by the peers. Additionally, we do not have many *false positives*, by avoiding proposing an item only liked by one (or few) peer(s), while it is unknown to the rest. In the final step, we sort all the items we have predicted a score for and return the k items with the highest score in the list R_u. Furthermore, we denote by $pos_{R_u,i}$ the index of the item i in the list R_u.

Related Work on Explanations in CF. CF explanations are typically provided based on users implicit or explicit feedback (for a survey of explanations in recommenders, refer to [20]). For example, a direct solution is to first find a set of peers for the user in question and then produce a recommendation to this user. The explanation is that the user is similar to the peers, and the peers made good ratings on the recommended item [14]. [9] compares the effectiveness of different display styles for explanations in CF. Specifically, explanations can be displayed as an aggregated histogram of the ratings of the peers, or be displayed as the detailed ratings of the peers. Alternatively, explanations can be provided by telling the user that the recommended item is similar to other items the user liked before, where several highly rated items by the user are shown as explanations [15]. To study the usefulness of explanations in recommender systems, [19] developed a prototype system to study the effect of different types of explanations. In brief, this study shows that providing appropriate explanations can benefit the recommender system over specific goals, like transparency, persuasiveness, trustworthiness and satisfaction. More recently, there exist approaches, e.g., [2,6,21], for generating explanations with methods using matrix or tensor factorization, where the goal is to make latent factors more tangible. From a different perspective, [8] studies the problem of computing minimum subsets of user actions to change the top-ranked recommendations in a counterfactual setup. The concept of why-not questions is used also for probabilistic range queries in [5], either by modifying the original query or by modifying the why-not set. [7] offers a similar framework for why-not questions on reverse top-k queries. To the best of our knowledge, we are the first to define and study the problem of providing explanations based on why not questions in recommender systems.

3 Why-Not Questions

In this paper, we expand on the concept of explanations in recommender systems, by exploiting the concept of why-not questions. These questions are not about why an item is recommended but why an item is not recommended in the expected way. Instead of offering explanations on every item that is proposed by the system, we allow the user to give feedback in the form of questions. For instance, in a movies recommender if the user is not satisfied with the movies list provided by the system, he/she can ask questions like: *Why were there not any comedies recommended?* The system will answer with information based on the system characteristics and the data associated with these items. This paradigm is not yet explored in recommendation systems, while it has been recently explored in other contexts like in explaining query results in relational databases [1], in reverse skyline queries [10], and briefly in machine learning systems [12].

We propose to characterise why-not questions by three main properties: (i) their level of absenteeism, (ii) their granularity, and (iii) their dependency to existing recommended items. The first property is naturally derived from the notion of false negative results, i.e., the items that should have been returned (in a certain position) but are not. The second property goes one step beyond to express groupings of missing items (that can be regarded as false negatives). The third property, corresponds to the need of the system expert to express the fact that an item that is returned (true positive) and an item that is not (false negative) should be encountered together in a result set.

First, we examine why-not questions based on *absenteeism*. In this respect, we further distinguish between (i) *total* absenteeism, and (ii) *position* absenteeism. Question such as *Why not Titanic?* belong to the *total* category, since they are about items that do not appear in the recommendation list, without a specific requirement for the position on which they should appear. Questions that ask about the ranking of items, such as *Why not rank Titanic first?* belong to the *position* category. It is evident thus, that a position absenteeism why-not question can be applied on items that *are* recommended, but still not as highly as expected. Second, we review *Granularity*. Granularity describes the level of detail of the question that is asked, distinguishing between *atomic* cases and *group* cases. In more detail, the user is able to ask questions about specific items (atomic case), such as: *Why not Titanic?*, or about set of items that share a common characteristic (group case), such as: *Why not comedies?* Third, the *Dependency* property describes items that usually appear together in the answers, or should be returned in a specific order. Example questions are *Why there are not any comedies but there are dramas?* This kind of questions fall also in the case of group recommendations, when users expect to find groups of items together. We subsequently define why-not questions in a formal way.

Definition 1. *Let I be a set of items, u a specific user of a recommender system built on I, $R_u \subseteq I$ the set of recommended items for user u by a recommender system. A why-not question is a set of the form*

$$wn = \{(m, pos, d) \mid m \in I \text{ and } pos \in \{1, \ldots, |R_u|\} \text{ and } d \in R_u\}$$

Definition 1 is general enough to cover all the cases that we mentioned before, i.e., absenteeism (m and pos), granularity, and dependency (d). Even though not explicitly apparent, a *granularity* why-not question can be derived by expanding the group to the related items in I. For example, a why-not questions of the style *Why not comedies?* can be represented by $wn = \{(Big,,), (Zoolander,,)\}$, given that the system can find these two comedies in its database. Moreover, if the user wants to ask one specific type of a why-not question that does not involve all three parts m, pos, and d, then he/she can leave that part empty. For example, in the case of a total absenteeism why-not question of the style *Why not Titanic?*, the corresponding wn would be $\{(Titanic,,)\}$. In the next paragraphs we elaborate more on the different types of why-not questions that we consider and how they are expressed using Definition 1.

Next, we specify Definition 1 to express the different properties of why-not questions (absenteeism, granularity, dependency). As the *absenteeism* property is always apparent in a why-not question, we discuss both sub-categories of the *absenteeism* property (total and position), with respect to granularity and dependency. In order to keep the notation from becoming too cumbersome, we will only give the notation for set of items (the group subcategory of the granularity property). This does not affect the formalization of the why-not questions, since both granularity questions can be noted using a set format (an individual item belongs to a set that consists of just one item). For each case, we present an intuitive description, examples in the context of a movie recommendation system, and the formal expression corresponding to that case of why-not question.

- **Total Absenteeism:**
  - *Independent*: The user asks why some items do not exist in the recommendation list.
    Example-Atomic: *Why is there not Titanic?*
    Example-Group: *Why are there not any comedies?*
    Formally, an independent total absenteeism why-not question is:

    $$wn_{ti} = \{(m,,) \mid m \in I \setminus R_u\}$$

  - *Dependent*: The user asks why certain items do not exist while other (that usually appear together) exist.
    Example-Atomic: *Why is there not Titanic while there is Up?*
    Example-Group: *Why not any thrillers when there are action films?*
    Formally, a dependent total absenteeism why-not question has the form:

    $$wn_{td} = \{(m,,d) \mid m \in I \setminus R_u \text{ and } d \in R_u\}$$

- **Position Absenteeism:**
  - *Independent*: The user can question the ranking of a set.
    Example-Atomic: *Why is Titanic not ranked first?*
    Example-Group: *Why are comedies not in a higher ranking?*
    Formally, an independent position absenteeism why-not question is:

    $$wn_{pi} = \{(m,pos,) \mid m \in R_u \text{ and } pos_{R_u,m} < pos\}$$

  - *Dependent*: The user asks why certain items do not appear higher in the recommendation list than other recommended items.
    Example-Atomic: *Why not place Titanic before Up?*
    Example-Group: *Why not place comedies before dramas?*
    Formally, an independent position absenteeism why-not question is:

    $$wn_{pd} = \{(m,pos,d) \mid m \in I \text{ and } pos > pos_{R_u,d} \text{ and } d \in R_u\}$$

4   Why-Not Explanations

To answer a why-not question, we seek to provide meaningful explanations to the system designers. By meaningful, we mean the information that is adequate to help the designer understand why the items are not recommended in the expected way, and subsequently use this information in order to repair the system. For this reason, we split the input of the problem to distinct components that can explain - either individually or combined - the why-not question provided by the user. These components are: the input item set, the sets of all and of similar users given the user in question, the set of ranking scores, and the recommender system design (hyperparameters) itself. To accommodate the different sources of error, we define a multi-type structure, called an *explanation*, as follows:

Definition 2. *A why-not explanation for a why-not question on the recommendations of a user u is a set of parameters of the recommender system, responsible for the absence of the missing item(s) from the (specific positions of the) recommendation list.*

We distinguish between *general* explanations, which can appear in any recommender system, and *model-specific* explanations that are based on the inherent parameters of the CF recommendation model. We further describe the general and CF explanations in Sects. 4.1 and 4.2, respectively, while we provide an algorithm to compute them in Sect. 5. We accompany the discussion with examples of why-not questions and respective possible explanations, summarized in Table 1. For clarity, we include in this table a description in natural language for the question and the explanation. The descriptions of the explanations can be regarded as the output of a statistical analysis of the resulting explanations.

4.1   General Explanations

Users, in most cases, ask about an item that does not appear in the recommendation list. So, it is very likely that this item does not exist in the database of the system. The explanation, then, is straightforward; this item is not suggested because it is unknown to the system. Another explanation emerges from the number of returned top-k items. If that number is low, then the missing item may be further down the recommendation list. However, we do not consider that the selected k may be the problem if the item is found at an index greater than $2k$, to promote other potential (model-specific) explanations. Finally, the system may produce the same score for different items. To break the ties, it adopts a specific method, e.g., it will place first the first encountered item in the database. So, when a user poses a why-not question on an item that has been neglected due to the tie-breaking method, the system may designate the tie-breaking method as a culprit. E1-E3 in Table 1 are examples of such explanations.

4.2   CF Explanations

The concept behind CF is that the system suggests items to a user that his/her similar users have liked in the past. This makes all the possible explanations revolving around the user's peers. One scenario is that none of the peers have rated an item. In this case, the item is invisible to the system and cannot be suggested. A similar scenario is for that item to have never been rated before by any user. This again makes the item invisible to the system. Aside from these two explanations, an answer for a "Why not item A?" question is the combination of the results for the three following questions: (i) how many peers have rated it, (ii) what scores they have given it, and (iii) how similar they are to the user. If just one or two peers have rated an item, then the system ignores it, to avoid a false suggestion. If all or most of the peers have given a low score to an item, then the system, in turn, calculates a low score for it. Finally, the similarity that a peer shares with the user is also primordial. If a peer who has a high similarity with the user does not like an item, then this has an impact on that item's final predicted score. E4-E11 in Table 1 are examples of such explanations.

We represent the results of the three aforementioned questions "How many peers have rated it?", "What scores have they given it?" and "How similar are these peers to the user?" as a set of tuples of the form $(peer, score, similarity)$. Each tuple describes a peer who has rated the targeted item and consists of three values: (i) the peer's id, (ii) the score that he/she has given to the item, and (iii) the similarity shared between the peer and the user. If the set is empty then none of the peers have rated this item and we provide explanation E10 (Table 1). If the set consists of only one or two tuples (peers) then it corresponds to explanation E4. To produce the rest of the explanations, we combine into a user-friendly explanation the number of peers that have rated the item, the similarity that these peers share with the user and the scores they have given to the item.

When a user questions the item's ranking in the recommendation list, the system again checks the same information. The system answers questions like: "Why was not item A ranked higher?" by explaining the item's statistics: how many peers have rated the item, if they favored it and how similar these peers are to the user. This type of question is vague, in the sense that the user questions the general ranking of an item without comparing it to another item; the user issued an independent question. So the system treats it as if it was a total absenteeism question. For this reason, explanations E17-E19 are the same as for a total absenteeism why-not question. Another alternative for handling these types of questions is to transform them from independent to dependent by arbitrarily selecting an item in the list. We select this item according to the specifics of the question - higher or lower ranking. For example, a question like "Why was not item A ranked higher?" can be transformed into "Why not place item A before item B?". In this case, the system returns a more detailed explanation as shown in lines E20-E23 in Table 1.

Explanations to dependent why-not questions, such as "Why not item A but item B?" or "Why not place item A before item B?", are more complicated since they involve multiple items, some of which exist and others not, mixing explanations and why-not explanations. We explain the process for the first question. The second is answered in a similar way. First, we decompose the why-not question to two separate queries. The first is a part of the user's question:

Table 1. Examples of why-not questions and explanations in CF.

WN Question	Model	WN Explanation Description	Id
	General/I	Item A does not exists in the database	E1
Any why-not question	General/k	You asked for few items	E2
	General/Tie	Item had the same score as another item	E3
Why not suggest item A?	CF/numPI	Only x ($<$numPI) of your peers has rated this item	E4
	CF/{(peer, s, sim)}	x of your peers have given a low score to this item	E5
	CF/{(peer, s, sim)}	x of your most similar users have given a low score to A	E6
	CF/{(peer, s, sim)}	x peers like A, but y dislike it	E7
	CF/{(peer, s, sim)}	All of your peers have given the item a low score	E8
$wn = (A, ,)$	CF/numP	None of your most similar users have rated A, but x with a lower similarity have given it a high/low score	E9
	CF/Peers	None of your peers has rated this item	E10
	CF/S	No one has rated this item	E11
Why not suggest item A but suggest B?	CF/numPI	x of your peers have rated item B but only y ($<$numPI) has rated item A	E12
	CF/{(peer, s, sim)}	x of your peers like item B but dislike A	E13
	CF/{(peer, s, sim)}	x peers like item B and y dislike item A	E14
$wn = (A, , B)$	CF/numP	Your most similar peers have not rated A but have rated B	E15
	CF/Peers	Your peers have rated B but none of them have rated A	E16
Why is not item A ranked higher?	CF/{(peer, s, sim)}	x of your peers have given a low score to this item	E17
	CF/{(peer, s, sim)}	x of your most similar peers have given a low score to A	E18
$wn = \{(A, pos_{R_u}A - 1,)\}$	CF/{(peer, s, sim)}	x peers like A, but y dislike it	E19
Why is not item A higher than B?	CF/{(peer, s, sim)}	x of your peers like item B but dislike A.	E20
	CF/{(peer, s, sim)}	x peers like item B and y dislike A	E21
$wn =$	CF/numP	Your most similar peers have not rated A but have rated B	E22
$\{(A, pos_{R_u}B - 1, B)\}$	CF/Peers	Your peers have rated B but none of them have rated A	E23
Why not suggest comedies?	CF/Peers	None of your peers rated the same movie	E24
	CF/{(peer, s, sim)}	Your peers dislike comedies	E25
	CF/Peers	None of your peers has rated a comedy	E26
	CF/{(peer, s, sim)}	Only x of your peers like comedies	E27
$wn = \{(C_1, ,), ..., (C_n)\}$	CF/numP	Your most similar peers do not like comedies but x of your least similar do	E28

"Why not item A". The second query we make is "Why not item B"[1]. Intuitively, since the system has promoted item B to the user instead of A (either by not even suggesting A for *total* why-not questions or with a better ranking in the recommendation list for *position* why-not questions), the results of the questions "How many peers have rated it?", "What scores have they given it?" and "How similar are these peers to the user?" have higher values than the results for A. Then, we combine the answers of these two why-not questions. For example, see E12-E16 in Table 1. We choose to explain the existence of item B as a why-not explanation, because it allows us to combine the results of the two questions more effectively than if we used a generic explanation method, such as [9].

When the user formulates a *group* why-not question, for example "Why not more comedies?", the explanation that the system provides is a union of all the answers that it would have provided for individual items. For each item in the same category as the one the user asked about and a peer has preferred in the past, we formulate a why-not question. Since this can become very cumbersome for the user to consume, one can summarize the results into a user-friendly output. For instance, "Your peers like comedies, but they have not liked the

---

[1] An alternative here could be to employ a solution for explaining recommendations.

same one", meaning that the peers have shown some preference for comedies, but each peer has rated a different movie. This explains why none of them were suggested. To reduce the number of questions we issue to the system, we can take into consideration the items that the peers have shown a great preference for. These items are the most important for us, since they have the highest probability to be suggested. For example, in a system where the ratings are in the range from 1 to 5, we can only consider the items that have a rating higher than the average 2.5. We provide more explanations for a group why-not question in lines E24-E28 of Table 1.

Algorithm 1: *WNCF*

Input: item set I, user set U, user u, why-not question $\{(i,,)\}$, rating scores S, recommendation list R_u for user u, threshold $numPI$, threshold $numP$,peers of u *Peers*, relevance score function $p(u,i)$

Output: e, explanation

1   **if** $i \notin I$ **then**
2   e.add('I');
3   **else if** $\exists i' : p(u,i)=p(u,i')$ *and* $pos_{R_{i'}} \leq k$ **then**
4   e.add('Tie');
5   **else if** i *in the 2k first entries of expanded R* **then**
6   e.add('k');
7   **else if** i *has no ratings in S* **then**
8   e.add('S');
9   **else if** *at least one peer of u has rated i* **then**
10   **for** *peer* $\in$ *Peers* **do**
11   **if** *peer has rated i* **then**
12   e.add$((peer, s(peer,i),sim(u,peer)))$;

13   **if** *less than numPI most similar peers of u have rated i* **then**
14   e.add('numP');
15   **if** *less than numPI peers of u have rated i* **then**
16   e.add('numPI');
17 **else**
18   **for** u' *user in U* **do**
19   **if** u' *has rated i* **then**
20   e.add$((u', s(u',i),\text{-}))$;
21   e.add('Peers');
22 **return** e;

---

5   *WNCF* Algorithm

In this section, we introduce *WNCF* (standing for *Why-Not in Collaborative Filtering*), an algorithm for the computation of why-not explanations for the CF model (Algorithm 1). *WNCF* addresses total absenteeism for atomic granularity independent why-not questions. The extension of this algorithm to explain other types of why-not questions, as discussed in Sect. 4, is trivial for some cases, e.g., group independent why-not questions, but not for all. We postpone the extensions to future work.

As mentioned in Sect. 4, we first check if a general explanation can be provided; if not, we proceed with the model-specific explanations modeled in tuples representing the peers who have rated i, along with information on their rating on i and the similarity to user u. If such peers do not exist, we provide explanations based on other users who have rated i.

In more detail, *WNCF* receives as input the item set, user set, the ranking scores, and the threshold values $numPI$ and $numP$ of the CF system, as well as the why-not question for a user u. We also consider known the peers of u and the recommendation list calculated for u, as well as the relevance score function. Line 1 checks if the specific item exists in the database. If it does not, we return the explanation code I, to indicate that the source of error is the input data set. Line 3 checks if the item shares the same relevance score with another item that appears in the list. In this case, we return the explanation code Tie, to indicate that the source of error is the tie breaking method. Line 5 checks if the item appears between the kth and $2k$th entry. In this case, we return the explanation code k, to indicate the the k maybe too low. Line 7 checks if any user in the system has rated i. If none of them did, then we return the explanation code S, to indicate that there are not rating scores for i.

Lines 9–16 check the peers of the user. For every peer who has rated i, we report the score he/she has given, as well as the similarity he/she shares with u (Line 12). Then, we check the $numP$ most similar peers of the user (Line 13). If less than $numPI$ of them have rated the item, we return the code 'numP' to express that there are not enough most similar peers who have rated the item. Subsequently, we check the rest of the peers and if there were not at least $numPI$ peers who have rated i (Line 15), then we return the explanation code 'numPI'. This indicates that from all the peers of user u less than $numPI$ peers have rated this item. Finally, if none of the peers has rated this item (Line 17) we return the explanation code $Peers$ to indicate that there are no peers who have rated the item. We also return information about the users (non-peers) who have rated the item and their scores.

Overall, our rational for returning the extra information on the users (peers or not), their ratings for the item in question and their similarity to the user u for CF systems, is two-fold. First, we can compute statistics that can be easily consumed by the developers (in their raw format or as visualisations) and help them understand more about the setting. Second, we can use this information as input to a repair mechanism, which would propose changes to the system so as to make the missing item appear in the list.

6   Experiments

We study *WNCF* with respect to different parameters, namely the characteristics of the users for whom we pose the why-not question, and the popularity of the missing movie. Moreover, we perform an experiment that shows the next step that a developer can take, after he/she receives a *WNCF* explanation.

Experimental Setup. In the experiments we used the MovieLens 20 Million Ratings[2] that consists of 27.278 movies and 138.493 users. To study the behavior of the algorithm for different types of users we randomly selected 100 users that have rated a few items (45 to 55), called *Moderate Users*, and 100 users that have rated many items (145 to 155), called *Active Users*. To experiment with the characteristics of the movies that comprise our why-not questions we used movies of varying popularity. We randomly selected 4 sets of 100 movies that have 2K (least popular), 4K, 6K and 8K (most popular) ratings, respectively. We denote these sets as Movies2K, Movies4K, Movies6K, and Movies8K. We ensured that the movies selected are not in the recommendation lists of the users, in order to be able to run why-not questions with them. To find the peers of a user, we used the Pearson Correlation with a threshold of 0.8 (th), while to predict a score for an unrated item we utilized 100 ($numP$) peers. An item cannot be considered for addition in the recommendation least, unless 3 ($numPI$) or more peers of the user have given it a rating. We report to the user the top-10 movies with the highest predicted scores.

Explanations Study. First, we define a total absenteeism why-not question for each item in the varying popularity movie sets, for moderate and active users (Fig. 1(a) and (b) respectively). Then, we run *WNCF* for each user and why-not question, and we calculate the percentage of occurrences of each explanation depending on the different parameters as they appear in the different segments of Algorithm 1. The k explanation indicates that the item in question was further down in the list that was provided to the user. The $numP$ explanation means that we should augment the $numP$ threshold to be able to find enough most similar peers that have rated this specific item. The *Peers* explanation occurs when none of the peers of the user has rated an item. Finally, by *Tuples* we denote explanations comprised by information on the peers of the user, calculated in Algorithm 1 line 12, and when the conditions in lines 13 and 15 are false.

Let us further analyze the result of the experiment in Fig. 1. When we use the more popular movies, more of them are in the k range, almost all of them are rated by a peer of the user and most of them are rated by a top peer. This is most evident when we compare the results for Movies2K and Movies8K. For Movies2K, less than 20% of the movies could be explained by the information provided on the peers in the moderate case, while for the Movies8K more than 95% of explanations were composed by the peers. Additionally, in the Movies2K more than half of them were not rated by any of the user's peers, while in the

---

[2] https://grouplens.org/datasets/movielens.

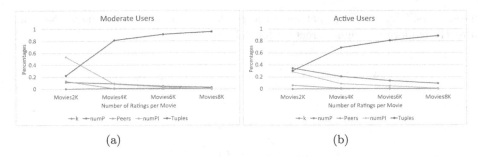

(a)   (b)

Fig. 1. Explanations for varying popularity of missing movies for (a) moderate, and (b) active users.

Movies8K this number has dropped down to almost zero. We can observe similar numbers for the Active Users for the Movies2K and Movies8K movie sets. This is a self-evident result since the more popular movies have a higher chance to be rated by the peers of a user. This is further corroborated by the *Tuples* values. For the Movies2K, it has low values, since the movies are not that popular. With each subsequent movies set, as the popularity of the movies rises, so does the the values of the *Tuples* variable. The most popular movies are more likely to have been voted by a top peer.

When comparing the results for the two sets of users, the Active Users have more explanations about the top peers not having rated an item ($numP$) than the Moderate Users. This is because the more ratings a user has given, the more similar other users he/she has. Since the number of peers we use is a constant variable and not a percentage, there is a higher chance the selected users have not rated the movies questioned. At the same time, we can see that the number of movies that were not rated by any of peers is lower than that of the Moderate Users. This is again because Active Users have a higher number of peers.

To demonstrate how the developer can proceed when he/she has acquired one explanation, we considered the case of *Peers* explanations. We took all the movies that were not rated by any peer (corresponding to *Peers* explanations), and we examined all the users in the system in order to find the new threshold needed in the similarity function, so as the recommender to be able to calculate a preference score for that item for the considered users. Then, we calculated the difference between the threshold we used originally (in our experiments, 0.8) and the new calculated threshold. Figures 2a and 2b show the results for the Moderate and Active users, respectively. In both experiments, we excluded the Movies8K set because the number of movies that were not rated by any of the peers is very small (less than 5 in both user sets). In both cases the adjustments needed in the similarity threshold are small. The average values (denoted with x in the figures) is below 0.04. While the Active Users have more outliers (dots in the figures) their ranges are similar to those in the Moderate user set. Finally, the median value (line inside the box) is comparable for both user sets across all movie sets. Thus, we see that with the provided explanation the developer can directly explore the right direction for debugging his system.

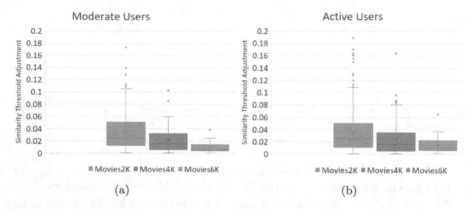

Fig. 2. The similarity threshold (*th*) adjustment needed for the recommender to be able to calculate a preference score for the missing items corresponding to a *Peers* explanation for (a) moderate, and (b) active users.

7   Summary

In this work, we pay special attention on transparency provided via explanations in recommender systems. We exploit the concept of why-not questions, allowing the user to give feedback in the form of questions about why items are not proposed in the expected way. We consider the collaborative filtering approach, and propose ways for providing explanations for why-not questions. We provide a detailed taxonomy of why-not questions with respect to three main properties: (i) the level of absenteeism that the why-not questions mentions (absence or low position in the ranking of a result set), (ii) their granularity (referring to a single result or a group), and (iii) their dependency to existing recommended items. An explanation for a why-not question is meant to inform the user about the possible sources of error linked to the why-not question. We distinguish explanations between general ones, i.e., explanations that are independent to the recommendation model used, and model-specific ones, based on the inherent parameters of CF. Finally, we provide an algorithm for computing why-not explanations in CF systems. Clearly, there are many directions for future work, including proposing explanations for content-based, hybrid and sequential [17] recommendation models, as well as efficient algorithms and implementations in specific contexts. Furthermore, we target the automatic refinement of the recommendations computed for the users, by exploiting the defined explanations.

8 Questions & Explanations for Collaborative Filtering   315

References

1. Bidoit, N., Herschel, M., Tzompanaki, K.: Immutably answering why-not questions for equivalent conjunctive queries. In: TaPP (2014)
2. Borges, R., Stefanidis, K.: On measuring popularity bias in collaborative filtering data. In: EDBT/ICDT Workshops (2020)
3. Breese, J.S., Heckerman, D., Kadie, C.M.: Empirical analysis of predictive algorithms for collaborative filtering. In: UAI (1998)
4. Chang, S., Harper, F.M., Terveen, L.G.: Crowd-based personalized natural language explanations for recommendations. In: RecSys (2016)
5. Chen, L., Gao, Y., Wang, K., Jensen, C.S., Chen, G.: Answering why-not questions on metric probabilistic range queries. In: ICDE (2016)
6. Chen, X., Qin, Z., Zhang, Y., Xu, T.: Learning to rank features for recommendation over multiple categories. In: ACM SIGIR (2016)
7. Gao, Y., Liu, Q., Chen, G., Zheng, B., Zhou, L.: Answering why-not questions on reverse top-k queries. Proc. VLDB Endow. 8(7), 738–749 (2015)
8. Ghazimatin, A., Balalau, O.D., Roy, R.S., Weikum, G.: PRINCE: provider-side interpretability with counterfactual explanations in recommender systems. In: WSDM (2020)
9. Herlocker, J.L., Konstan, J.A., Riedl, J.: Explaining collaborative filtering recommendations. In: CSCW (2000)
10. Islam, M.S., Zhou, R., Liu, C.: On answering why-not questions in reverse skyline queries. In: ICDE (2013)
11. Konstan, J.A., Miller, B.N., Maltz, D.A., Herlocker, J.L., Gordon, L.R., Riedl, J.: Grouplens: applying collaborative filtering to usenet news. Commun. ACM 40(3), 77–87 (1997)
12. Lim, B.Y., Dey, A.K., Avrahami, D.: Why and why not explanations improve the intelligibility of context-aware intelligent systems. In: CHI (2009)
13. Ntoutsi, E., Stefanidis, K., Rausch, K., Kriegel, H.: Strength lies in differences: Diversifying friends for recommendations through subspace clustering. In: CIKM (2014)
14. Resnick, P., Iacovou, N., Suchak, M., Bergstrom, P., Riedl, J.: Grouplens: an open architecture for collaborative filtering of netnews. In: CSCW (1994)
15. Sarwar, B.M., Karypis, G., Konstan, J.A., Riedl, J.: Item-based collaborative filtering recommendation algorithms. In: WWW (2001)
16. Stefanidis, K., Ntoutsi, E., Petropoulos, M., Nørvåg, K., Kriegel, H.: A framework for modeling, computing and presenting time-aware recommendations. Trans. Large-Scale Data- Knowl.-Centered Syst. 10, 146–172 (2013)
17. Stratigi, M., Nummenmaa, J., Pitoura, E., Stefanidis, K.: Fair sequential group recommendations. In: SAC (2020)
18. Su, X., Khoshgoftaar, T.M.: A survey of collaborative filtering techniques. In: Advances in Artificial Intelligence (2009)
19. Tintarev, N.: Explanations of recommendations. In: RecSys (2007)
20. Tintarev, N., Masthoff, J.: A survey of explanations in recommender systems. In: ICDE (2007)
21. Wang, N., Wang, H., Jia, Y., Yin, Y.: Explainable recommendation via multi-task learning in opinionated text data. In: ACM SIGIR (2018)
22. Yu, C., Lakshmanan, L.V.S., Amer-Yahia, S.: Recommendation diversification using explanations. In: ICDE (2009)
23. Zhang, Y., Chen, X.: Explainable recommendation: a survey and new perspectives. Found. Trends Inf. Retrieval 14(1), 1–101 (2020)

Learning from Multiple Graphs of Student and Book Interactions for Campus Book Recommendation

Qiaomei Zhang, Yanmin Zhu$^{(\boxtimes)}$, Tianzi Zang, and Jiadi Yu

Shanghai Jiao Tong University, Shanghai 200240, China
{qmayzhang,yzhu,zangtianzi,jiadiyu}@sjtu.edu.cn

Abstract. The development of university digitalization has generated a large amount of student data, such as academic performance records, book loaning records, etc. This provides a data foundation for many student-related tasks, such as book recommendations, academic performance predictions. Campus book recommendation is very important, which can not only help students to find books they might be interested in but also improve the utilization of book resources in campus libraries. Existing work proved the effectiveness of incorporating pairwise relationships of users and items, i.e., social network and item dependency graph. However, most of them model item dependency based on co-occurrence and ignore occurrence order between items which is crucial in the campus book recommendation. In this paper, we propose to construct multiple graphs including a student-student graph, a student-book interaction graph, and a book-book graph to simultaneously capture student correlation, student-book interaction, and book dependency. We design a graph-based convolution network that contains three modules with different convolution operations to learn representations for students and books from the three graphs, and a prediction module to make recommendations. Experiments on the real-world dataset demonstrate the effectiveness of the proposed method.

Keywords: Campus book recommendation · Collaborative filtering · Multiple student and book interaction graphs · Graph convolution

1 Introduction

With the advance and wide appliance of information and communication technologies [20,27], in universities, various student behaviors can be easily recorded, such as book loaning, library visiting, and academic performance. These records are very valuable for providing a data foundation for student-related tasks such as book recommendations, academic performance predictions, etc [6,15,16].

In this paper, we focus on the book recommendation task, which is very meaningful for both students and university management. As the number of books increases in campus libraries, students are often overwhelmed by a large

Z. Huang et al. (Eds.): WISE 2020, LNCS 12343, pp. 316–330, 2020.
https://doi.org/10.1007/978-3-030-62008-0_22

number of book resources, which leads to two problems. First, it's becoming more difficult for students to find the books they might be interested in. Second, the utilization of abundant resources in libraries is very low. Some books may never be checked out, for example, 75% of the books in the Changsha University of Science and Technology Library have never been used [25]. By applying the recommendation system technique to analyze book-loan histories of students, we can establish a book recommendation system for students, which can solve the two problems.

The difference between campus book recommendation and general recommendation is that book loaning behaviors of students have a stronger dependency. According to statistics, over 40% of book pairs in our book loaning dataset have a unidirectional borrowing dependency probability no less than 0.8. That is, for a book pair (A, B) that often borrowed by the same users, 80% of the situation, A is borrowed before B. While the percentage for movie dependency is only 11% (we take the MovieLens as an example). This is consistent with our common sense as courses that students take have a dependency. For example, for a student who has borrowed books related to advanced mathematics and probability theory, books related to advanced mathematics often appear in front of books related to probability theory, because advanced mathematics is a prerequisite course of probability theory.

Collaborative filtering (CF) [12] is the most popular recommendation method and can be used for book recommendation. One of the most successful methods of performing CF is matrix factorization (MF) [14]. MF parametrizes users and items as latent vectors(also termed as embeddings) in the same space and adopts inner product between user embeddings and item embeddings as the estimated preference scores. More recent works [10,11,22] introduce deep neural network to boost the performance of traditional MF methods. However, as [19] mentioned, those methods only use the interactions to define objective functions while lacking explicit incorporation in the embeddings.

To explicitly encode interactions into embeddings, recent efforts [2,5,13,21] convert the history interactions between users and items into graphs, and employ graph convolution network(GCN) on graphs to capture the interactions. GC-MC [2] employs one convolutional layer on a user-item interaction graph to exploit the direct connections between users and items. NGCF [19] applies multiple convolution layers on the user-item bipartite graph to capture high-order connectivity. PinSage [26] applies GCN on an item-item graph to explore the similarities between items. A recent work [18] further proves the effectiveness of mining information from multiple graphs, i.e., a user-user graph, a user-item graph, and an item-item graph.

Despite the effectiveness of those GCN-based methods, there are still some limitations. First, those methods construct the user-user graph or the item-item graph from the user-item interactions to measure similarities between users or items. On one hand, those similarities can be implicitly captured in the user-item interaction graph, i.e., the 2-hop neighbors of users in the interaction graph measure user similarities. On the other hand, the user-user graph fails to model cold users. Besides, most of these models construct the item-item graph as an undirected co-occurrence graph that ignores occur order dependency between two items, while this dependency is more common in campus book recommendations. Finally, the user-item interaction graph is heterogeneous and unweighted, those models [2,19,26] ignore the intrinsic difference between two node types and aggregate information from neighbors in the same way. What's worse, they don't explicitly explore the influence strength of different neighbors.

In this paper, we propose a novel graph convolution network-based campus book recommendation model named MGCN to simultaneously learn from multiple graphs including a student-student graph, a student-book bipartite graph, and a book-book dependency graph, with three key innovations:

- *Capturing the intrinsic difference of different node types and influence weights of different neighbors in the bipartite user-item graph.* We apply different convolution operations to aggregate information from neighborhoods for users and items, respectively. To capture the influence of neighbors, we employ graph attention network to dynamically decide influence weights.
- *Modeling the co-occurrence and sequence dependencies between books simultaneously.* We construct the book-book graph as a directed graph that captures those two kinds of dependency at the same time. We also design a novel graph convolution operation for the directed graph.
- *Incorporating auxiliary information of students into proposed MGCN model to measure the correlation between students and boost the performance.* The recommendation performance will decrease due to high data sparsity of the book loaning dataset. Thus, we also construct a student-student graph from auxiliary information of students and employ graph convolution operation on the graph to capture student correlations.

We conduct extensive experiments on a real-world campus book loaning dataset. The results show that 1) our proposed approach outperforms other baseline methods with regard to different metrics, and 2) the use of multiple graphs are effective for improving the recommendation performance.

This paper is organized as follows. We first introduce the definition of the research problem in this paper and the three graphs in Sect. 2. In Sect. 3, we introduce the proposed model with detailed descriptions of the modules we designed for multiple graphs. Extensive experiments are given in Sect. 4 to evaluate the effectiveness of the proposed model. In Sect. 5, we give a brief review of related works followed by the conclusion and future work in Sect. 6.

2   Preliminary

2.1   Notations and Definition

Definition 1 (Interaction Sequence). For a student u_i, his/her book loaning records (i.e., $< u_i, b_j, d >$) are sorted by borrowing dates and form a sequence. $< u_i, b_j, d >$ denotes a student u_i borrowed a book b_j at date d.

Definition 2 (Library Visiting Sequence). For a student u_i, his/her library visiting records (i.e., $< u_i, l, t >$) are sorted by visiting time and form a sequence. $< u_i, l, t >$ denotes a student u_i visit the library l at time t.

Definition 3 (Student-book Interaction Graph). The student-book interaction graph $G^I = (V^U \cup V^B, E^I)$ is a heterogeneous graph that represents the interactions between students and books. V^U and V^B denote the student set and book set, respectively, and $|V^U| = N$ and $|V^B| = M$. $E^I \subset V^U \times V^B$ is the set of interaction edges between students and books.

Definition 4 (Student-student Graph). The student-student graph $G^U = (V^U, E^U)$ is designed to explore student correlations. $E^U \subset V^U \times V^U$ is the set of edges between students. The weight of e_{u_i,u_j} in E^U is defined as the similarity between u_i and u_j which takes the student profiles and library visiting behaviors into consideration. Its formulation is defined as follows.

$$e_{u_i,u_j} = \alpha e^p_{u_i,u_j} + (1-\alpha)e^l_{u_i,u_j}, \tag{1}$$

where $e^p_{u_i,u_j}$ is the similarity weight of student profiles, $e^l_{u_i,u_j}$ reflects the closeness of relationship between u_i and u_j mined from their library visiting behaviors. α is a hyperparameter to balance the two weights. $e^p_{u_i,u_j} = 1$ if the two students have the same profiles, otherwise $e^p_{u_i,u_j} = 0$.

$$e^l_{u_i,u_j} = \frac{f(u_i, u_j)}{f(u_i, u_i) + f(u_j, u_j) - f(u_i, u_j)}, \tag{2}$$

where $f(u_i, u_j)$ denotes the number of times that u_i and u_j visit the library within the predefined time interval Δt. For each library visit pair $\{(u_i, t_i), (u_j, t_j)\}$ of the two students, if $|t_i - t_j| < \Delta t$, then value of $f(u_i, u_j)$ will plus 1. This similarity measurement is consistent with Jaccard similarity coefficient by computing the intersection as the co-occurrence times of two students.

Definition 5 (Book-book Graph). The book-book graph $G^B = (V^B, E^B)$ is a directed graph designed to retain the dependency between books. $E^B \subset V^B \times V^B$ is the set of edges between books. There are two kinds of book dependencies, one is the co-occurrence dependency that are often borrowed together, the other is the sequential dependency that are often borrowed one after another. The directed book-book graph can simultaneously capture these two dependencies.

For each book loaning pair $\{(b_i, d_i), (b_j, d_j)\}$ of a student, if $d_i - d_j < \Delta d$, it is regarded as a co-occurrence dependency, and there exists one edge from b_i to b_j as well as one edge from b_j to b_i. Otherwise it's treated as sequence dependency, and there only exists one edge from b_i to b_j.

2.2 Problem Formulation

In our work, the interactions between students and books are converted into the interaction graph G^I and edges indicate students' performance over items. Thus the recommendation task can be redefined as predicting the existence of edge between vertex u_i and b_j given the three graphs G^I, G^U, G^B, i.e., $p(e_{u_i,b_j}|G^I, G^U, G^B)$.

3 Proposed Method

We now present the proposed model named MGCN, and the overview is illustrated in Fig. 1 (we take the inference of student u_2's preference score over book b_1 as an example). There are four modules in this model: the WGCN module applies the weighted graph convolution operation on the student-student graph to learn representations for students based on their relationship with others; the AHGCN module is proposed to capture the connectivity with high-order neighbors of different node types (i.e., student nodes and book nodes) contained in heterogeneous student-book interaction graph, and learn representations for students and books; the BGCN module employs bidirectional graph convolution operation on the directed book-book graph to aggregate information from both downstream neighbor nodes and upstream neighbor nodes; finally, the embeddings learned from the first three modules are used to make predictions.

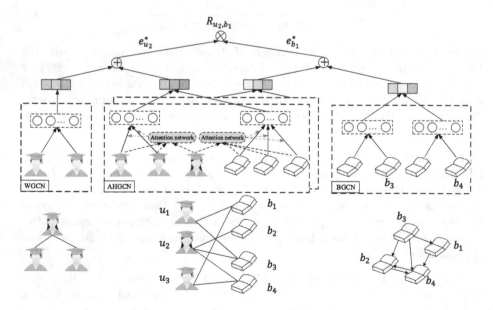

Fig. 1. Overview of the proposed MGCN model.

3.1 WGCN Module for Student Correlation Modeling

Following prior work [4,10], we use randomly initialized embedding vectors $p_{u_i} \in \mathbb{R}^d$ and $q_{b_j} \in \mathbb{R}^d$ to respectively represent the student u_i and the book b_j, where d represents the embedding size. p_{u_i} and q_{b_j} act as initial embeddings.

The WGCN module for student correlation modeling is shown in the bottom left in Fig. 1. It applies the weighted graph convolution operation on the student-student graph $G^U = (V^U, E^U)$ to aggregate information from similar students and generate a representation vector $e_{u_i}^U$ for each student.

For a vertex u_i in the graph, the general graph convolution operation is defined as follows.

$$e_{u_i}^U = \sigma(\sum_{u_j \in N^U(u_i)} \gamma_{u_i,u_j} W^U p_{u_j} + b^U), \tag{3}$$

where $N^U(u_i)$ denotes the direct neighbors of u_i in the student-student graph. $\gamma_{u_i,u_j} = A_{u_i,u_j}^U / D_{u_i,u_i}^U$ denotes the normalized edge weights of neighbor student u_j and u_i in the student-student graph, A^U denotes the adjacency matrix of the student-student graph, D^U is a diagonal matrix, $D_{u_i,u_i}^U = \sum_j A_{u_i,u_j}^U$.

In this module, we get a new representation $e_{u_i}^U$ for student u_i that capture student correlation.

3.2 AHGCN Module for Interaction Modeling

The AHGCN module employs attention-guided graph convolution operation on the heterogeneous student-book interaction graph to capture the influence of high-order neighbors with different node types, and learn representations for students and books that capture the student-book interaction.

In this paper, for each vertex, we consider the influence of the two kind vertexes with the nearest distance, i.e., neighbors within 2-hop. Although we can consider higher-order neighbors, we believe that higher-order neighbors have a weak impact on the target node, which is easy to introduce noise and consume much time. Besides, the general GCN operates on the whole graph and vertexes of different types (i.e, students and books) share the same parameters, which ignores the heterogeneity of the student-book interaction graph. In this paper, we employ two convolution operations for student vertexes and book vertexes, respectively.

For each student vertex u_i, we aggregate the information from the 1-st hop neighbor books and the 2-nd hop neighbor students to get an additional representation for u_i. To reduce parameters, AHGCN module shares the initial embeddings p_{u_i} and q_{b_j}. The convolution operation for vertex u_i is defined as follows.

$$e_{u_i}^I = \sigma(\sum_{b_j \in N_1^I(u_i)} W_1^I q_{b_j} + \sum_{u_k \in N_2^I(u_i)} W_2^I p_{u_k} + b_1^I), \tag{4}$$

where $N_1^I(u_i)$ and $N_2^I(u_i)$ denote the 1-st hop neighbors and the 2-nd hop neighbors of u_i in the student-book interaction graph. W_1^I denote the weight matrix

322   Q. Zhang et al.

of the convolution operation of first-hop neighbors, as the embedding space of u_i and 1st-hop neighbors (i.e., books) are different, W_1^I also acts as a mapping function to transform the book-space embeddings into the student-space. W_2^I is the weight matrix of the second convolution operation, it controls the influence strength of all 2nd-hop neighbors as 2nd-hop neighbors should have smaller influence compared to 1st-hop neighbors, and b_1^I is the bias vector.

However, in this convolution operation, the neighbors in the same order share the same parameters and contribute equally to the target vertex which violate the facts that different neighbors have different influence strength. And the edge weights of interaction graph is 1 and we can't use edge weights to distinguish the influence strength of different neighbors. Therefore, we employ the attention mechanism to discriminate the importance of different neighbors automatically, as defined below.

$$e_{u_i}^I = \sigma(\sum_{b_j \in N_1^I(u_i)} \alpha_{u_i,b_j} W_1^I q_{b_j} + \sum_{u_k \in N_2^I(u_i)} \beta_{u_i,u_k} W_2^I p_{u_k} + b_1^I), \qquad (5)$$

where α_{u_i,b_j} and β_{u_i,u_k} are the attention score for first hop neighbor b_j and second hop neighbor u_k of vertex u_i. Here we parameterize the attention score for neighbor b_j by:

$$h_{u_i} = tanh(W_1^a p_{u_i} + b_1^a)$$
$$\alpha_{u_i,b_j} = \frac{exp(h_{u_i}{}^T h_{b_j})}{\sum_{b_l \in N_1^I(u_i)} exp(h_{u_i}{}^T h_{b_l})}, \qquad (6)$$

where W_1^a and b_1^a are trainable parameters of the attention layer.

The attention score for neighbor u_k is computed as below.

$$\beta_{u_i,u_k} = \frac{exp(p_{u_i}{}^T p_{u_k})}{\sum_{u_l \in N_2^I(u_i)} exp(p_{u_i}{}^T p_{u_l})} \qquad (7)$$

The process of learning representation $e_{b_j}^I$ for book vertex b_j is similar to that of $e_{u_i}^I$. For the sake of space, we omit the specific process and just give the formulation.

$$e_{b_j}^I = \sigma(\sum_{u_l \in N_1^I(b_j)} \lambda_{b_j,u_l} W_3^I p_{u_l} + \sum_{u_k \in N_2^I(b_j)} \theta_{b_j,b_k} W_4^I q_{b_k} + b_2^I), \qquad (8)$$

where $N_1^I(b_j)$ and $N_2^I(b_j)$ denote the 1-st hop neighbors and the 2-nd hop neighbors of b_j in the student-book interaction graph. λ_{b_j,u_l} and θ_{b_j,b_k} are the attention score for first hop neighbor u_l and second hop neighbor b_k of vertex b_j. W_3^I and W_4^I are the weight matrixes for convolution, and b_2^I is bias vector.

In this module, we get new representations $e_{u_i}^I$ and $e_{b_j}^I$ for u_i and b_j, respectively.

3.3   BGCN Module for Book Dependency Modeling

The BGCN module employs a bidirectional graph convolution operation on the directed book-book graph, which takes both downstream neighbor nodes and upstream neighbor nodes into consideration, and learn representations for books that encode the dependency between books. As the book-book graph is directed, each vertex can act as a starting point, an ending point, or both. In this paper, we initialize two embedding vectors $e_{b_j}^{B-}$ and $e_{b_j}^{B+}$ for each vertex, where $e_{b_j}^{B-}$ and $e_{b_j}^{B+}$ denote the embedding vectors for vertex b_j when it act as a starting point and an ending point, respectively.

Given the directed book-book graph $G^B = (V^B, E^B)$, let A_B denote the adjacency matrix, $D_O = diag(A_B \mathbf{1})$ is the out-degree diagonal matrix and $\mathbf{1} \in \mathbb{R}^N$ denotes the all-ones vector. $P^O = D_O^{-1} A_B$ denotes a transaction matrix, and each row b_j in P, i.e., $P_{b_j,:}^O$, denotes the likelihood of diffusion from vertex b_j to other vertexes within one hop. Similarly, we can define a transaction matrix $P^I = D_I^{-1} A_B^T$, where each row b_j in P^I, i.e., $P_{b_j,:}^I$, denotes the likelihood of diffusion from other vertexes to vertex b_j within one hop. $D_I = diag(A_B^T \mathbf{1})$.

For a vertex b_j, the bidirectional graph convolution operation is defined as follows.

$$e_{b_j}^B = \sigma \Big(\sum_{b_i \in N_{b_j}^{B-}} P_{b_j,b_i}^O W_1^B e_{b_i}^{B-} + \sum_{b_k \in N_{b_j}^{B+}} P_{b_j,b_k}^I W_2^B e_{b_k}^{B+} + b^B \Big), \qquad (9)$$

where $N_{b_j}^{B-}$ and $N_{b_j}^{B-}$ denote the downstream and upstream neighbors, respectively. P_{b_j,b_i}^O denotes the likelihood of transaction from b_j to b_i, and P_{b_j,b_k}^O denotes the likelihood of transaction from b_k to b_j. W_1^B and W_2^B are the weight matrixes for convolution, and b^B is bias vector.

In this module, we generate a new representation $c_{b_j}^B$ for book b_j that encodes the dependency between books.

3.4   Prediction Module

In the prediction module, we utilize the representation learned from the first three modules to get the prediction. In this work, we adopt the inner product as the interaction function. Formally, student u_i's preference score towards book b_j $\hat{R}_{u_i,b_j}$ is computed as follows.

$$\begin{aligned} \hat{R}_{u_i,b_j} &= e_{u_i}^* {}^T e_{b_j}^* \\ e_{u_i}^* &= e_{u_i}^U \oplus e_{u_i}^I \oplus p_{u_i} \\ e_{b_j}^* &= e_{b_j}^B \oplus e_{b_j}^I \oplus q_{b_j}, \end{aligned} \qquad (10)$$

where $\oplus$ represents element-wise addition and $e_{u_i}^U, e_{u_i}^I, e_{b_j}^I, e_{b_j}^B$ are representations for students and books learned from first three modules. We also include initialized embeddings to cover general matrix factorization model (i.e., if we remove the first three modules, the model becomes a pure matrix factorization model)

3.5 Model Learning

To estimate model parameters, we optimize the BPR loss, which is widely used in many recommender systems [3,9]. The core assumption of BPR loss is that the observed student-book pairs are more reflective of student preference, which should have higher scores than unobserved ones. Formally, the loss function for recommendation is defined as follows.

$$Loss = \sum_{(u_i,b_j,b_k)\in O} -ln\ \sigma(\hat{R}_{u_i,b_j} - \hat{R}_{u_i,b_k}) + \lambda\|\theta\|_2^2, \tag{11}$$

where $O = \{(u_i, b_j, b_k)|(u_i, b_j) \in R^+, (u_i, b_k) \in R^-\}$ is the training set, R^+ is the set of observed student-book pairs, R^- is the set of negative pairs sampled from unobserved ones. $\sigma\,(.)$ is the sigmoid function. θ denotes all trainable parameters, λ is the hyper parameter to control the strength of L_2 regularization.

4 Experiment

In this section, we conduct experiments on one real-world book datasets from the university library to evaluate the effectiveness of the proposed method. We aim to answer the following research questions:

RQ1 How does our approach perform compared to state-of-the-art recommendation methods?

RQ2 Is the utilization of multiple graphs useful for improving recommendation performance?

RQ3 How do the hyperparameters(embedding size, message dropout ratio) affect results?

In the following, we will first present the experimental settings, followed by answering the above research questions.

4.1 Experimental Settings

Dataset. We conducted experiments on a real world book loaning dataset of university T. The original data is highly sparse, making it difficult to evaluate the collaborative filtering algorithm. As such, following previous work [10], we filter the data and retained only students and books with more than 5 interactions. This results in a subset of the dataset that contains 6,619 students, 9,791 books and 102,932 interactions.

Evaluation Metrics. To evaluate the recommendation performance, we adopt the widely used *leave-one-out* evaluation [1,17]. For each student, we hold out the lastest interaction for testing and the remaining for training. For each test item, we randomly sample 99 unobserved items as negative samples. the test item is ranked among the 100 items. To evaluate the result, we adopt two metrics named *Hit Ratio* (HR) and *Normalized Discounted Cumulative Gain* (NDCG) which are the same as [10]. HR measures whether the test item appears in the top-k list, while NDCG accounts for hit position by assigning higher scores to top ranks. We use HR@k and NDCG@k to evaluate the top-k items and in our experiments we set k to 5 and 10.

Baselines. We compare our method with the following methods:

- **Itempop.** Items are ranked by their popularity which is calculated by the number of interactions. This is a non-personalized method to benchmark the recommendation performance.
- **BPR** [17]. This method optimizes a standard matrix factorization method with BPR loss. It's a highly competitive baseline for item recommendation.
- **NeuMF** [10]. It's a state-of-the-art neural CF model that models linear and nonlinear feature interactions through MF and MLP together .
- **GC-MC** [2]. This model employs a graph convolution autoencoder on bipartite interaction graph to capture direct interactions.
- **PinSage** [26]. This model applies GraphSAGE [7] on item-item graph to encode similarities between items. We employ two graph convolution layers as suggested in [26].
- **NGCF** [19]. It's a graph neural network-based CF model that utilizes graph neural network to encode the high-order connectivity into embeddings. Especially, we employ three graph convolution layers which report best performance in [19].
- **Multi-GCCF** [18]. This model employs GCN on multiple graphs to simulatenously model relationships of users and/or items. Our work is different from this work in graph construction and convolution operations. We employ two graph convolution layers in the bipartite graph as suggested in [18].

Parameter Settings. We implement our method based on Tensorflow. Codes and data after desensitization are available at https://github.com/qmayzhang/MGCN. The embedding size is fixed to 64 for all models. All baselines are trained by optimizing the BPR loss in Eq. 11 with Adagrad optimizer. For each positive instance, we sampled four negative instances. The batch size is 1024 and the learning rate is 0.002. We also adopt message dropout technique to avoid overfitting and the ratio is set to 0.1.

4.2   Overall Performance Comparison (RQ1)

We first compare the overall recommendation performance of all methods. Table 1 reports the results. We have the following key observations:

- Our MGCN model consistently outperforms all baselines on the dataset w.r.t different evaluation metrics. The results indicate the effectiveness of MGCN that learns from multiple graphs of student and book interactions to simultaneously explore student correlation, book dependency and student-book interaction and make recommendations.

Table 1. Overall performance comparison.

	K=5		K=10	
	HR	NDCG	HR	NDCG
Itempop	0.1039	0.0677	0.1736	0.0900
BPR	0.2218	0.1481	0.3288	0.1825
NeuMF	0.2231	0.1450	0.3256	0.1821
GC-MC	0.2197	0.1434	0.3360	0.1807
PinSage	0.2380	0.1592	0.3505	0.1954
NGCF	<u>0.2489</u>	0.1654	<u>0.3653</u>	<u>0.2090</u>
Multi-GCCF	0.2466	<u>0.1663</u>	0.3590	0.2022
MGCN	**0.2626**	**0.1797**	**0.3866**	**0.2195**

– The three GCN-based models including PinSage, NGCF, and Multi-GCCF
  achieve great improvement compared to other baselines. We due this improvement to the effectiveness of capturing high-order connectivity in graph structure. Compared to Multi-GCCF, NGCF performs better although it uses only
  the bipartite graph. This may have two reasons, first, NGCF exploits higher-
  order neighborhoods, i.e., 3-hop neighbors while Multi-GCCF only exploits 2-
  hop neighbors. Second, as we described in Sect. 1, the two additional user-user
  graph and item-item graph in Multi-GCCF are constructed from user-item
  interaction graph, which can be implicitly captured by the 2-hop neighbors
  in the interaction graph.

4.3 Effects of Multiple Graphs (RQ2)

To demonstrate the effectiveness of multiple graphs, we conduct experiments
using variants of the proposed MGCN model with different graphs. The results
are shown in Table 2. The variant model MGCN_U only exploits the student
correlation graph, and we compare the performance of student-student graph
with different construction methods. MGCN_B only uses the undirected book
dependency graph, while MGCN_BD only uses the directed graph. MGCN_U&B
uses the two graphs simultaneously. The MGCN_I utilizes the interaction graph
to capture high-order connectivity contained in the student-book interaction.
The MGCN exploits three graphs concurrently. We can see that:

– MGCN model that learns from all three graphs achieves the best performance
  on the dataset. This demonstrates the effectiveness of exploring student correlation, book dependency and high-order connectivity with multiple graphs
  of student and book interactions.
– Other varients also achieve better performance compared to BPR. And the
  varient model MGCN_U&BD achieves comparable perfomance, as it is trained
  using the bpr loss which implicitly explores the direct-connected student-book
  interaction.

Table 2. Effects of Multiple Graphs.

	K=5		K=10	
	HR	NDCG	HR	NDCG
BPR	0.2303	0.1600	0.3448	0.1938
MGCN_U(behavior)	0.2455	0.1686	0.3535	0.2024
MGCN_U(behavior+profile)	0.2473	0.1673	0.3626	0.2041
MGCN_B	0.2430	0.1670	0.3544	0.2029
MGCN_BD	0.2478	0.1700	0.3575	0.2050
MGCN_U& BD	0.2603	0.1781	0.3763	0.2151
MGCN_I	0.2467	0.1684	0.3593	0.2043
MGCN	**0.2626**	**0.1797**	**0.3866**	**0.2195**

- Compared to MGCN_B, MGCN_BD achieves limited improvement. This may because the sequential dependency introduces noisy dependency related to non-professional books such as novels.

4.4 Effects of Hyperparameters (RQ3)

We conduct experiments to explore the effects of hyperparameters, i.e., the embedding size and the message dropout ratio.

Effect of Embedding Size. Figure 2 shows the performance w.r.t. different embedding sizes on the dataset. We can see that:

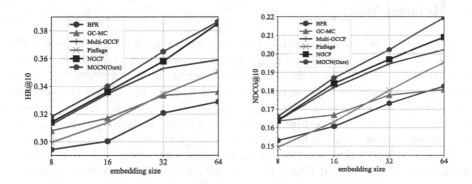

Fig. 2. Effect of embedding size

- MGCN model achieves the best performance under all embedding size on the whole, and with the increase of embedding size, the performance of recommendation also improved a lot compared to baselines. Even with a smaller

embedding size of 32, the proposed model MGCN can outperform most baselines except NGCF under most evaluation metrics. This demonstrates the
high expressiveness of the representations we learned from the three modules
with multiple graphs of student and book interactions.
– Multi-GCCF and NGCF achieve competitive performance. This demonstrates
the effectiveness of exploring high-order connectivity in student-book interactions and the pairwise relationship between students and books.

Effect of Dropout. Following prior work [19], we employ the message dropout
technique to prevent MGCN from overfitting. Figure 3 shows the effect of the
message dropout ratio against different evaluation metrics. We can see that
MGCN achieves the best performance with a dropout ratio of 0.1.

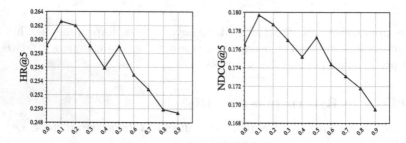

Fig. 3. Effect of Dropout Ratio

5   Related Work

In this section, we review existing recommendation work of model-based collaborative filtering methods and graph-based methods and highlight the difference. Model-based CF methods parameterize user/items with vectorized embeddings and reconstruct user-item interactions with interaction functions based on
embeddings. As a representative work, Matrix Factorization (MF) represents
each user/item as an embedding vector and adopts the inner product as the
interaction function [14]. While recent works exploit deep learning techniques
to model the interaction function, so as to capture the nonlinear feature interactions between users and items. For example, NeuMF [10] adopt deep neural
networks to model the interaction function. Despite the effectiveness of these
methods, it's argued that the embedding functions are not sufficient as they lack
explicitly encoding of the collaborative signal contained in the interaction data.

In recommendation systems, the user-item interactions can naturally be represented as bipartite graphs. Early work [8,24] adopt label propagation on graphs
to capture CF effects. With the advancement of graph neural networks (GNN),
recent works [2,19,23] devote much effort to exploring GNN. GC-MC [2] regards

the recommendation as a matrix completion task and employs a graph convolutional layer as an autoencoder. PinSage [26] applies multiple convolution layers on the item-item graph that measures similarities between items. NGCF [19] adopts multiple graph convolution layers on the user-item interaction graph to explicitly encode high-order connectivity into embeddings. GraphRec [5] employs one attentive graph convolution layer on both the interaction graph and social network to distinguish the weights of different neighbors. However, these methods don't explicitly explore the influence strength of neighbors and the high-order connectivity simultaneously.

6   Conclusion and Future Work

In this work, we construct multiple graphs of student and book interaction, and design a graph-based neural network with three modules to learn from these graphs for the campus book recommendation. Experiments on the real-world dataset show the effectiveness of our method. This work demonstrates the importance of exploring student correlation and book (to be more general, items) dependency. In the future, we'll further explore more complex and reasonable methods to construct graphs to enhance the recommendation performance.

Acknowledgement. This research is supported in part by the 2030 National Key AI Program of China 2018AAA0100503 (2018AAA0100500), National Science Foundation of China (No. 61772341, No. 61472254, No. 61772338 and No. 61672240), Shanghai Municipal Science and Technology Commission (No. 18511103002, No. 19510760500, and No. 19511101500), the Program for China Top Young Talents, the Program for Changjiang Young Scholars in University of China, the Innovation and Entrepreneurship Foundation for oversea high-level talents of Shenzhen (No. KQJSCX20180329191021388), the Program for Shanghai Top Young Talents, Shanghai Engineering Research Center of Digital Education Equipment, and SJTU Global Strategic Partnership Fund (2019 SJTU-HKUST).

References

1. Bayer, I., He, X., Kanagal, B., Rendle, S.: A generic coordinate descent framework for learning from implicit feedback. In: WWW, pp. 1341–1350 (2017)
2. Berg, R.V.D., Kipf, T.N., Welling, M.: Graph convolutional matrix completion (2017). arXiv preprint arXiv:1706.02263
3. Chen, J., Zhang, H., He, X., Nie, L., Liu, W., Chua, T.: Attentive collaborative filtering: multimedia recommendation with item- and component-level attention. In: SIGIR, pp. 335–344 (2017)
4. Cheng, W., Shen, Y., Zhu, Y., Huang, L.: DELF: a dual-embedding based deep latent factor model for recommendation. In: IJCAI, pp. 3329–3335 (2018)
5. Fan, W., et al.: Graph neural networks for social recommendation. In: WWW, pp. 417–426. ACM (2019)
6. Goodall, D., Pattern, D.: Academic library non/low use and undergraduate student achievement: a preliminary report of research in progress. Libr. Manag. 32(3), 159–170 (2011)

7. Hamilton, W., Ying, Z., Leskovec, J.: Inductive representation learning on large graphs. In: Advances in Neural Information Processing Systems, pp. 1024–1034 (2017)
8. He, X., Gao, M., Kan, M.Y., Wang, D.: Birank: towards ranking on bipartite graphs. TKDE **29**(1), 57–71 (2016)
9. He, X., He, Z., Du, X., Chua, T.: Adversarial personalized ranking for recommendation. In: SIGIR, pp. 355–364 (2018)
10. He, X., Liao, L., Zhang, H., Nie, L., Hu, X., Chua, T.S.: Neural collaborative filtering. In: WWW, pp. 173–182. IW3C2 (2017)
11. Hu, G., Zhang, Y., Yang, Q.: Conet: collaborative cross networks for cross-domain recommendation. In: CIKM, pp. 667–676. ACM (2018)
12. Hu, Y., Koren, Y., Volinsky, C.: Collaborative filtering for implicit feedback datasets. In: ICDM, pp. 263–272. IEEE (2008)
13. Kipf, T.N., Welling, M.: Semi-supervised classification with graph convolutional networks. In: ICLR (2017)
14. Koren, Y., Bell, R., Volinsky, C.: Matrix factorization techniques for recommender systems. Computer **8**, 30–37 (2009)
15. Lian, D., Ye, Y., Zhu, W., Liu, Q., Xie, X., Xiong, H.: Mutual reinforcement of academic performance prediction and library book recommendation. In: ICDM, pp. 1023–1028. IEEE (2016)
16. Mezick, E.M.: Return on investment: libraries and student retention. J. Acad. Librarianship **33**(5), 561–566 (2007)
17. Rendle, S., Freudenthaler, C., Gantner, Z., Schmidt-Thieme, L.: BPR: bayesian personalized ranking from implicit feedback. In: UAI, pp. 452–461 (2009)
18. Sun, J., et al.: Multi-graph convolution collaborative filtering. In: ICDM, pp. 1306–1311. IEEE (2019)
19. Wang, X., He, X., Wang, M., Feng, F., Chua, T.: Neural graph collaborative filtering. In: SIGIR, pp. 165–174 (2019)
20. Wu, Z., Li, J., Yu, J., Zhu, Y., Xue, G., Li, M.: L3: sensing driving conditions for vehicle lane-level localization on highways. In: INFOCOM, pp. 1–9. IEEE (2016)
21. Xu, Y., Zhu, Y., Shen, Y., Yu, J.: Learning shared vertex representation in heterogeneous graphs with convolutional networks for recommendation. In: IJCAI, pp. 4620–4626 (2019)
22. Xue, H.J., Dai, X., Zhang, J., Huang, S., Chen, J.: Deep matrix factorization models for recommender systems. In: IJCAI, pp. 3203–3209 (2017)
23. Yang, C., Bai, L., Zhang, C., Yuan, Q., Han, J.: Bridging collaborative filtering and semi-supervised learning: a neural approach for POI recommendation. In: SIGKDD, pp. 1245–1254. ACM (2017)
24. Yang, J.H., Chen, C.M., Wang, C.J., Tsai, M.F.: Hop-rec: high-order proximity for implicit recommendation. In: RecSys, pp. 140–144 (2018)
25. Yang, X., Zeng, H., Huang, Y.: Artmap-based data mining approach and its application to library book recommendation. In: IUCE, pp. 26–29. IEEE (2009)
26. Ying, R., He, R., Chen, K., Eksombatchai, P., Hamilton, W.L., Leskovec, J.: Graph convolutional neural networks for web-scale recommender systems. In: SIGKDD, pp. 974–983 (2018)
27. Yu, J., et al.: Sensing human-screen interaction for energy-efficient frame rate adaptation on smartphones. IEEE Trans. Mob. Comput. **14**(8), 1698–1711 (2015)

A Clustering-Based Collaborative Filtering Recommendation Algorithm via Deep Learning User Side Information

Chonghao Zhao[1], Xiaoyu Shi[2(✉)], Mingsheng Shang[2], and Yiqiu Fang[1]

[1] School of Computer Science and Technology, Chongqing University
of Posts and Telecommunications, Chongqing 400714, China
`S180231916@stu.squpt.edu.cn, fangyq@squpt.edu.cn`
[2] Chongqing Key Laboratory of Big Data and Intelligent Computing, Chongqing Institute of
Green and Intelligent Technology, Chinese Academy of Sciences, Chongqing 400714, China
`{xiaoyushi,msshang}@cigit.ac.cn`

Abstract. Collaborative filtering (CF) is a widely used recommendation approach that relies on user-item ratings. However, the natural sparsity of user-item
ratings can be problematic in many domains, limiting the ability to produce accurate and effective recommendations. Moreover, in some CF approaches only rating
information is used to represent users and items, which can lead to a lack of recommendation explained. In this paper, we present a novel deep CF-based recommendation model, which co-learns users' abundant attributes. To better understanding
the user's preference, we explore user deeper and unseen factors on the user-item
ratings and user's side information by adopting the AutoEncode network. After
that, we conduct the k-mean algorithm with extracted deep user factors to classify
users. Then the user-side CF algorithm is employed to produce the recommendation list based on the classification results, for alleviating recommendation speed.
Finally, we conduct lots of experiments on real-world datasets. Compared with
state-of-the-art methods, the results show that the proposed method has a significant improvement in recommendation performance, in terms of recommendation
accuracy and diversity. Furthermore, it also enjoys high effectiveness, and the
approach is useful when it comes to assigning intuitive meanings to improve the
explainability of recommender systems.

Keywords: Collaborative filtering algorithm · Recommendation system ·
K-means++ · Clustering

1 Introduction

With the rapid development of the Internet and mobile technologies, the recommender
system has become an essential part of e-business applications, which can help people to
find the potential interesting information and services [1]. Collaborative Filtering (CF)
[2, 3] is the most popular approach in RS and has received a great deal of attention in
industry, such as Amazon, Netflix, Taobao, and so on. Generally, most CF approaches

© Springer Nature Switzerland AG 2020
Z. Huang et al. (Eds.): WISE 2020, LNCS 12343, pp. 331–342, 2020.
https://doi.org/10.1007/978-3-030-62008-0_23

rely on user-item ratings that predict the users' preferences based on the users or the items having similar ratings. In this kind of RS, a typical matrix of user-item ratings is exploited to compute similarities between users (user-based) or items (item-based), then make a prediction based on the computed user/item similarities.

However, the UCF algorithm also has several limitations such as low scalability when dealing with large amounts of data [4], and the problem of a cold start. Further, the traditional CF algorithm needs to compute the similarities of the increasing number of users to all other users, and it requires higher computation efficiency. It is a significant challenge to improve computation speed for an online recommender system. Also, the number of users and items is vast; however, most users just rate a small part of items, so the data used to calculate similarities between users and items is sparse. Finally, it comes to the condition that the recommendation results may not be satisfactory.

In fact, many social media are obtaining user side information when they register. It is an effective way to deal with the cold start problem. Therefore, the combination of item interaction information and user side information can get a better recommendation effect theoretically.

In recent years, deep learning has made breakthrough progress in image processing [5], natural language processing [6], and speech recognition [7]. Meanwhile, deep learning has a subversive effect on the recommendation system, which brings more opportunities to improve the recommendation performance. However, neither of them has modeled the product's comments and user's auxiliary information simultaneously.

In this paper, we present a novel collaborative filtering (CF) method for a top-N recommendation named Autoencoder and K-means++ in Collaborative Filtering (AK-UCF) AK-UCF generalizes several previously mentioned clustering technology, deep learning technology, and user attribute information. But its structure is much more flexible. For instance, it is easy to incorporate nonlinearities into the model to achieves better top-N recommendation results. We compare the performance of AK-UCF with other collaborative filtering methods in different data sets. Experimental results show that AK-UCF consistently outperforms other recommended algorithms by a significant margin on a number of common evaluation metrics.

Our contributions can be summarized as follows:

- In the cluster stage, we use deep learning technology to reduce the dimension of the scoring matrix and improve the problem of the sparse scoring matrix.
- We use the clustering data for collaborative filtering recommendation and reduce the time consumption of collaborative filtering recommendation.
- We use the user side information to improve the cold start problem in the recommendation system and collaborate to produce user portraits.

The remainder of this paper is organized as follows. Related works to our contributions are presented in Sect. 2. The implementation details of our AK-UCF method are shown in Sect. 3. The datasets are described and the performance is analyzed via experiments in Sect. 4. Finally, we summarize our paper with some concluding remarks in Sect. 5.

2 Related Works

We describe the current situation of methods Auto-Encoder dimensionality reduction algorithm, K-means++ clustering algorithm, and user side information used in AK-UCF.

2.1 Auto-Encoder

Literature [8] proposes to use AutoEncoder to extract the compressed representation of users and projects in the scoring matrix. As a deep feature of users and projects, the extracted features are used for scoring prediction. Experiments prove that the number of RMSE indicators is better than traditional models such as collaborative filtering. On the other hand, literature uses an automatic encoder that does not extract the deep features of the user. It can be considered to use a stack-type noise reduction encoder, so that deep feature vectors can be obtained and the recommendation quality can be improved.

The CDAE model [9] takes the row of the user-item evaluation matrix as input, obtains the hidden representation of the user through a layer of neural network coding, and restores the user's interaction behavior through a layer of neural network. Unlike the simplest Autorec model, the CDAE model incorporates user-specific considerations when coding for hidden representations, with more semantics. In order to make the model more robust, the CDAE model performs noise processing on the input features, either by dropout or by adding Gaussian noise. A common shortcoming of both approaches is that they do not take into account user side information.

2.2 K-Means++

There has been diverse research to enhance recommendation accuracy by means of clustering methods. In [10], CF and content-based filtering methods were conducted by finding similar users and items, respectively, via clustering, and then a personalized recommendation to the target user was made. As a result, improved performance on the precision, recall, and F1 score was shown. Similarly, as in [10], communities (or groups) were discovered in [11] before the application of matrix factorization to each community. In [12], social activeness and dynamic interest features were exploited to find similar communities by item grouping, where items are clustered into several groups using cosine similarity. As a result of grouping, the K most similar users based on the similarity measure were selected for recommendation. The performance of user-based CF with several clustering algorithms including K-means++, self-organizing maps (SOM), and fuzzy C-Means (FCM) clustering methods was shown in [13]. It was shown that K-means++ user-based CF has the best performance in comparison with user-based CF based on the FCM and SOM clustering methods.

2.3 User Side Information

In this paper, we mainly consider some basic information, like users' age, gender, and occupation. Further, consider the deep statistical information of users and items, user-item rating matrix. We directly integrate the side information and the rating information of users in the deep neural network. Through combining the two parts, we jointly build the recommendation model to fully characterize the interaction between users and items.

3 Methodology

For exploiting the full advantage of the available user side information, we propose a clustering-based collaborative filtering recommendation algorithm via deep learning user side information, which uses the Auto-Encoder and K-means++ in User-based Collaboration Filter algorithm. The model consists of two-stage. As shown in Fig. 1.

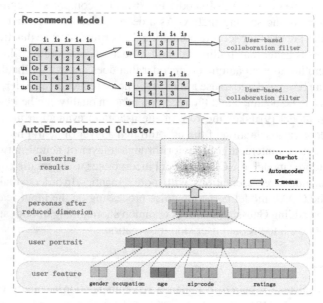

Fig. 1. The architecture of AK-UCF model

3.1 Auto-Encoder-Based Clustering

In the clustering stage, we use the user's characteristic information to add user category information to the user. We believe that the interests of users with the same attributes are also similar. The user feature information used in our model includes the user's gender, age, occupation, geographic location. According to experience, we divide the user feature information into corresponding number categories. For example, we divide gender data into two categories.

The categories are independent of each other, so we use one-hot encoding to encode this information. For example, the one-hot encoding value for male gender is [1, 0], and the encoding value for female gender is [0, 1]. Finally, a user feature information code $f_u = [g_u, o_u, a_u, z_u]$ with d_f-dimension is obtained. Where g_u, o_u, a_u, z_u represent the One-Hot encoding of gender, age, occupation, and geographic location respectively.

Another important feature information must be considered when clustering users: user-score information. List user rating information into the matrix as Table 1.

Use the rating vector for all items of user u_i to represent the rating features r_u: [$r_{ui,v1}$, $r_{ui,v2}$, ..., $r_{ui,vn}$]. Where $r \in \{0, 1, ..., 5\}$ donate users' ratings for movies. AE is used to

Table 1. Information of user-item rating

	v_1	v_2	...	v_n
u_1	$r_{u1,v1}$	$r_{u1,v2}$	...	$r_{u1,vn}$
u_2	$r_{u2,v1}$	$r_{u2,v2}$	...	$r_{u2,vn}$
...	...	...	...	...

reduce the high dimensional sparse vector r_u. The sample data r_u of the AE is encoded by the encoder function f to obtain the coding feature $r_u^{(n)}$, and r_u and $r_u^{(n)}$ satisfy the following Eq. 1:

$$
\begin{aligned}
r_u^{(1)} &= f_{\theta_1}(r_u) = s(W_1 r_u + b_1) \\
r_u^{(2)} &= f_{\theta_2}(r_u^{(1)}) = s(W_2 r_u^{(1)} + b_2) \\
&\quad \cdots \\
r_u^{(n)} &= f_{\theta_n}(r_u^{(n-1)}) = s(W_n r_u^{(n-1)} + b_n)
\end{aligned}
\tag{1}
$$

where: s is a neural network excitation function, generally using a nonlinear function such as sigmoid Function; $\theta = \{W, b\}$ is a set of parameters. Then pass the following Eq. 2:

$$
\begin{aligned}
\hat{r}_u^{(n)} &= g_{\theta_n}(r_u^{(n)}) = s(W_n' r_u^{(n)} + b_n') \\
\hat{r}_u^{(n-1)} &= g_{\theta_{(n-1)}}(\hat{r}_u^{(n)}) = s(W_{(n-1)}' r_u^{(n-1)} + b_{(n-1)}') \\
&\quad \cdots \\
\hat{r}_u &= g_{\theta_1}(\hat{r}_u^{(1)}) = s(W_1' \hat{r}_u^{(1)} + b_1')
\end{aligned}
\tag{2}
$$

Converting the coded d_r-dimension vector $r_u^{(n)}$ into a reconstructed representation of the original input r_u, Eq. 3 is the optimization goal of AE.

$$
L = \left\| r_u - \hat{r}_u \right\|^2
\tag{3}
$$

By continuously correcting the parameters θ, the average reconstruction error L is minimized, and the obtained $r_u^{(n)}$ can be considered to retain most of the information of the original sample, the equivalent feature of the sample r_u.

Add the rating features $r_u^{(n)}$ to the feature information code to get personas $p_u = [f_u, r_u^{(n)}]$ with d_p-dimension. Where $d_p = d_f + dr$. The dimension of the user profile may be too high to use K-means++ clustering algorithm. Then use Auto-encoder again to reduce the dimension of the user profile from d_p-dimension to d_u-dimension.

In order to solve the time-consuming problem of the collaborative filtering algorithm, we cluster the obtained user portraits. Since the category of each user is not known in advance, we use an unsupervised K-means++ clustering algorithm. Each user portrait is taken as a cluster sample, and a user portrait vector is randomly selected as the first cluster center. K-means++ algorithm is used to calculate K categories of users.

The K value is related to specific datasets and usually determined by approaches based on Silhouette Coefficient [14] or Elbow method [15]. In our work, we will adjust the K value to meet our specific requirements. It has been shown that the initial clustering centers should be selected uniformly to get a good clustering result [16]. Thus, we will use the K-means++ algorithm given by D. Arthur to determine the initial cluster center. The K-means++ algorithm is shown in Algorithm 1.

Algorithm 1: The K-means++ algorithm

Input: The amount of cluster K.
Output: k initial cluster center $c_1, c_2, ..., c_k$.
1: Choose a user randomly as the first initial cluster center c_1.
2: **for all** i = 2:k **do**
3:   Calculate the shortest distance $D(u)$ between each user and all current cluster centers.
4:   Sample every user $u \in U$ with probability $P(u)$ as the next cluster center c_i.

$$P(u) = \frac{D(u)^2}{\sum_{u \in U} D(u)^2}$$

5: **end for**
6: **for** c_i changes **do**
7:   For each user in the set, calculate the distance from each center, and classify this user as the nearest center.

8:   For each category c_i, recalculate its cluster center $c_i = \frac{1}{|c_i|} \sum_{x \in c_i} x$ (which is

the center of mass of all samples belonging to the class).
9: **end for**

3.2   Recommended Model

UCF algorithm mainly includes two steps as follows:

Calculate Interest Similarity Between Users. Given user u, v which belongs to the same cluster as user u, $N(u)$ represents a collection of items where u has had positive feedback, and $N(v)$ represents a collection of items where v has had positive feedback. To calculate the similarity between users based on the user-item rating matrix, we use cosine similarity to calculate the similarity, which is expressed as follows Eq. 4:

$$w_{uv} = \frac{|N(u) \cap N(v)|}{\sqrt{|N(u)||N(v)|}} \tag{4}$$

In the traditional UCF recommendation system, the user similarity needs to be calculated for any two users, so the time complexity is $O(|U|*|U|)$, which is very time-consuming when the number of users is large. Our AK-UCF improves this method. It calculates user similarity in the same category instead of calculating user similarity in all users. For example, K-means++ clustering algorithm divides users into K clusters on average in the preprocessing stage, and the time complexity of calculating user similarity in each cluster is $O((|U|/K)^2)$. So the overall time complexity is $O(|U|^2/K)$. AK-UCF is of great significance in solving the time-consuming problem in the recommendation stage of the traditional UCF recommendation system.

Find the Items that the Target User May Like from the Items that the Users with Similar Interests have Interacted With. After obtaining the interest similarity between users, our AK-UCF will recommend K items that the user has the most similar interests to the user. The Eq. 5 measures the interest of user u in the AK-UCF algorithm for items.

$$p(u, i) = \sum_{v \in S(u,K) \cap N(i)} w_{uv} r_{vi} \tag{5}$$

where $S(u, K)$ donates the K users who are most similar in interest to user u, $N(i)$ is the set of users who have interacted with item i, w_{uv} is the similarity of interest between user u and user v, r_{vi} represents user v's interest in item i, because it uses implicit feedback data of a single behavior, so all $r_{vi}=1$. In the top-N recommendation algorithm, we first rank the user u's interest in the items and then recommend the first n items that user u have not interacted with to the user u.

In the method of solving the cold start problem, we calculate the distance between the feature embedding provided by the new user u and K clustering centers and find the nearest clustering center which is the category c_u of the new user u. Then the top n items with the highest popularity that users in cluster c_u have ever interacted with are recommended to user u. The popularity of item i (P_i) is calculated as Eq. 6:

$$P_i = \sum_{v \in c_u} I_{N(v)}(i) \tag{6}$$

where $I_{N(v)}(i)$ is the indicator function. When $i \in N(v)$, the value of $I_{N(v)}(i)$ is 1, else the value of $I_{N(v)}(i)$ is 0.

4   Experimental Results and Analysis

4.1   Datasets

Movielens is a rating-based movie recommendation system, created by the GroupLens research team of the University of Minnesota which is specifically used to study recommendation technology. This article uses the Movielens_100k dataset and Movielens_1m dataset to verify the performance of the algorithm. These datasets are recognized as the main datasets for evaluating recommended algorithms. The information on these two datasets is shown in Table 2:

Table 2. Statistics of dataset

Dataset	Number of users	Number of movies	Number of ratings	Rating range	Least number of one user interacted with	Sparsity
Movielens_100k	943	1,682	100,000	1–5	20	93.69%
Movielens_1m	6,040	3,900	1,000,209	1–5	20	95.75%

4.2 Measurement

We compare our AK-UCF with the baseline models through the evaluation indicators Precision, Coverage, MAP, and running time which are commonly used in the recommendation system.

Precision. Precision is a measure widely used in the field of information retrieval and statistical classification to evaluate the quality of results. At present, it is widely used in the evaluation of the TOP-N recommendation system. Precision is the ratio of the number of correct items recommended and the number of all recommended items. The quality of the recommended results can be judged by the Precision.

Coverage. Coverage describes the ability of a recommendation system to explore the long tail of items. Coverage has different definitions of methods. The simplest definition is the proportion of items that the recommendation system can recommend to the total set of items. It is an important index to measure the novelty of recommendation results.

Mean Average Precision (MAP). In Eq. 7, Average Precision (AP) is a ranked precision metric that gives larger credit to correctly recommended items in top ranks. AP@N is defined as the average of precisions computed at all positions with an adopted item, namely.

$$AP@N = \frac{\sum_{k=1}^{N} \text{Pr}\, ecision@k \times rel(k)}{\min\{N, |C_{adopted}|\}} \tag{7}$$

where Precision(k) is the precision at cut-off k in the top-N list $C_{N,\,rec}$, and rel(k) is an indicator function equaling 1 if the item at rank k is adopted, otherwise zero. Finally, Mean Average Precision(MAP@N) is defined as the mean of the AP scores for all users.

4.3 Analysis of Experimental Results

To test our proposed method is efficient, three algorithms are compared. The first compared algorithm is traditional UCF. The second compared algorithm is User-based Collaborative Filtering with K-means++ clustering (K-UCF). And the third compared algorithm is CDAE as we mentioned in Sect. 2.

For every user from the input set their ratings were divided into training (80%) and testing parts (20%). All the models have some parameters to tune. In our experiment, The dimension of rating features r_u was reduce from 3952-dimension to 80-dimension. The dimension of the user profile p_u from 209-dimension to 20-dimension. Individual users are divided into four clusters in K-UCF and AK-UCF.

The impact of the number of movies recommended on recommendation quality: Firstly, this paper will set the number of users to be 20, set the number of movies recommended to be the independent variables, and set precision and MAP to be the dependent variable. Figure 2 compares Precision results on UCF, K-UCF, CDAE, and our AK-UCF. And Fig. 3 compares MAP results on UCF, K-UCF, CDAE, and our AK-UCF.

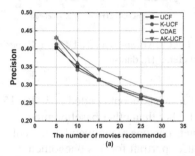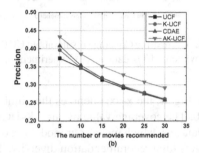

Fig. 2. Comparing Precision results about UCF, K-UCF, CDAE, and our AK-UCF. (a) Experiment on Movielens_100k dataset. (b) Experiment on Movielens_1m dataset.

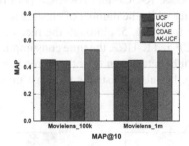

Fig. 3. Comparing MAP results on UCF, K-UCF, CDAE, and our AK-UCF.

In Fig. 2, the x-axis denotes the number of movies recommended is 5, 10, 15, 20, 25, 30. Obviously, the Precision of the other three algorithms is lower than our AK-UCF, which means our AK-UCF gains better performance than UCF, K-UCF, and CDAE. Besides, compared to the results in the UCF, the Precision increased by 9% in the Movielens_100k dataset and 16% in the Movielens_1m dataset. In Fig. 3, the x-axis denotes experiments on the UCF, K-UCF, CDAE, and our AK-UCF algorithms in Movielens_100k and Movielens_1m. The MAP results are obtained by recommending 10 movies to each user. The MAP value of our AK-UCF is higher than other baseline models in the two datasets, which means our model is more accurate and sensitive to the order of recommendation at the same time.

The impact of the number of movies recommended on recommendation diversity: Our results of the diversity of recommendation systems through Coverage are represented in Fig. 4.

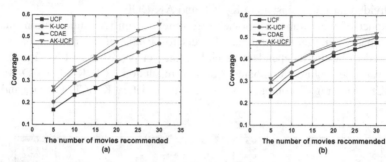

Fig. 4. Comparing Coverage results about UCF, K-UCF, CDAE, and our AK-UCF. (a) Experiment on Movielens_100k dataset. (b) Experiment on Movielens_1m dataset.

In Fig. 4, the x-axis denotes the number of movies recommended is 5, 10, 15, 20, 25, 30. The coverage value of our AK-UCF is higher than other baseline models in the two datasets, which means our model not only gets persona recommendation quality but also gets high recommendation diversity. The user portrait for this phenomenon is that products with low popularity become more popular in a small range of user clusters.

Figure 5 shows the difference in the running time of distinct recommendation models in the recommendation process. K-UCF and AK-UCF both run collaborative filtering algorithms in the same cluster while the number of clusters K is equal to 4, so the results are similar. As we can see from Fig. 5, although the recommended quality of CDAE algorithms is better than UCF and K-UCF, the time consumption is the most. In comparison, our algorithm is not only better than the other three baseline algorithms in terms of recommendation quality, but also consumes the least time in the recommendation process.

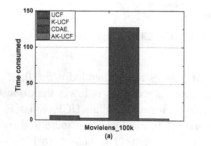
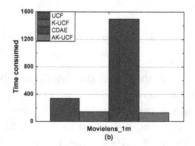

Fig. 5. Comparing running time results on UCF, K-UCF, CDAE, and our AK-UCF.

5 Conclusion

A multi-task recommendation system based on collaborative filtering is proposed in this paper. Firstly, user category information is derived from the user's auxiliary information and user-item rating matrix in the data preprocessing stage. In the recommendation stage, we only calculate the most similar users among the same cluster users. Compared with the traditional collaborative filtering algorithm, K-means++ clustering only recommendation algorithm, deep learning only algorithm, our algorithm has a great improvement in recommendation quality and time consumption in the recommendation process.

The future work includes tuning the algorithm based on an online experiment and trying other clustering algorithms such as Mixture-of-Gaussian clustering to optimize the collaborative filtering algorithm.

Acknowledgments. This work was supported by the Chongqing Research Program of Technology Innovation and Application under grants cstc2019jscx-zdztzxX0019, in part by Chongqing Natural Science Foundation under grants cstc2018jcyjAX0047, and Youth Innovation Promotion Association CAS, No. 2017393.

References

1. Adomavicius, G., Tuzhilin, A.: Toward the next generation of recommender systems. IEEE Trans. Knowl. Data Eng. **17**(6), 734–749 (2005)
2. Konstan, J.A., Riedl, J.: Recommender systems: from algorithms to user experience. User Model. User-Adap. Inter. **22**(1–2), 101–123 (2012)
3. Su, X., Khoshgoftaar, T.M.: A survey of collaborative filtering techniques. Adv. Artif. Intell. **2009**, 12 (2009)
4. Sarwar, B.M., Karypis, G., Konstan, J., Riedl, J.: Recommender systems for large-scale e-commerce: scalable neighborhood formation using clustering. In: International Conference on Computer and Information Technology. IEEE, Dhaka (2002)
5. Gu, F., Zhang, H., Wang, C.: A two-component deep learning network for SAR image denoising. IEEE Access **8**, 17792–17803 (2020)
6. Deep Learning for Natural Language Parsing: S. Jaf, C. Calder. IEEE Access **7**, 131363–131373 (2019)
7. Khalil, R.A., Jones, E., Babar, M.I., Jan, T., Zafar, M.H., Alhussain, T..: Speech emotion recognition using deep learning techniques: a review. IEEE Access **7**, 117327–117345 (2019)
8. Sedhain, S., Menon, A.K., Sanner, S., Xie, L.: AutoRec: autoencoders meet collaborative filtering. In: WWW 2015 Companion Proceedings of the 24th International Conference on World Wide Web, pp. 111–112 (2015)
9. Wu, Y., DuBois, C., Zheng, A.X., Ester, M.: Collaborative denoising auto-encoders for top-N recommender systems. In: WSDM 2016 Proceedings of the Ninth ACM International Conference on Web Search and Data Mining (2016)
10. Huang, C.-L. Yeh, P.-H., Lin, C.-W., Wu, D.-C.: Utilizing user tag-based interests in recommender systems for social resource sharing websites. Knowl. Based Syst. **56**, 86–96 (2014
11. Yin, B., Yang, Y., Liu, W.: Exploring social activeness and dynamic interest in a community-based recommender system. In: Proceedings of the 23rd International Conference World Wide Web, Seoul, Korea, pp. 771–776 (2014)

12. Guerraoui, R., Kermarrec, A.-M., Patra, R, Taziki, M.: D2P: distance-based differential privacy in recommenders. VLDB **8**(8), 862–873 (2015)
13. Koohi, H., Kiani, K.: User based collaborative filtering using fuzzy c-means. Measurement **91**, 134–139 (2016)
14. Rousseeuw, P.J., Silhouettes: a graphical aid to the interpretation and validation of cluster analysis. J. Comput. Appl. Math. **20**(1), 53–65 (1987)
15. Ketchen, D.J., Shook, C.L.: The application of cluster analysis in strategic management research: an analysis and critique. Strat. Manage. J. **17**(6), 441–458 (1996)
16. Arthur, D., Vassilvitskii, S..: k-means++: the advantages of careful seeding. In: Proceedings of the Eighteenth Annual ACM-SIAM Symposium on Discrete Algorithms, Society for Industrial and Applied Mathematics, pp. 1027–1035 (2007)

Path-Based Academic Paper Recommendation

Shengjun Hua[1], Wei Chen[1,2], Zhixu Li[1], Pengpeng Zhao[1], and Lei Zhao[1(✉)]

[1] School of Computer Science and Technology, Soochow University, Suzhou, China
20184227008@stu.suda.edu.cn,
{robertchen,zhixuli,ppzhao,zhaol}@suda.edu.cn
[2] Institute of Artificial Intelligence, Soochow University, Suzhou, China

Abstract. The overabundance of academic papers makes it difficult for researchers to find relevant or interested papers. To address the problem, existing studies have developed many approaches to recommend academic papers effectively. Most of them mainly utilize content-based filtering or citation analysis to measure similarity or relatedness of two papers and recommend relevant papers to the given query. However, these recommended papers are usually discrete from each other, i.e., the relationship between recommended papers are omitted, which disables researchers from having an sight into the time-oriented development of the topic they are interested in. To overcome the drawbacks of existing work, we propose a novel academic paper recommendation method called PAPR (Path-based Academic Paper Recommendation). Our method aims to recommend an ordered path of relevant papers, which are of great benefit in helping researchers understand the development of a specific topic. During process, we take both content and network structure into account to learn the representation of a paper. Next, the similarity between papers are measured based on the representation. The experimental results based on real data show that the proposed method outperforms the state-of-art methods.

Keywords: Paper recommendation · Representation learning · Citation relations.

1 Introduction

Due to the overwhelming amount of academic papers published every year, researchers need to spend much time on searching relevant or interested papers, i.e., recommending academic papers becomes a challenging task. To address the problem, many digital libraries and recommender systems have been developed to recommend relevant literature according to the provided keywords or users' profiles. Although search engines accelerate search process, finding satisfactory scientific articles is still time-consuming and inconvenient.

A number of techniques have been proposed to recommend academic papers. Content-based filtering (CBF) [11] views content of document as keywords to

© Springer Nature Switzerland AG 2020
Z. Huang et al. (Eds.): WISE 2020, LNCS 12343, pp. 343–356, 2020.
https://doi.org/10.1007/978-3-030-62008-0_24

measure similarity between papers. Collaborative filtering (CF) is a classical recommendation method which recommends items to users based on others who have similar preferences [13]. Citation-based analysis uses citation relationships to calculate relatedness among papers [16]. Hybrid methods have been proposed to improve the recommendation results, which combine two or more recommendation techniques to get better performance [6].

The techniques mentioned above are mostly used to recommend academic papers that are similar to a given paper in some respects. As shown in Fig. 1(a), given a paper u_1 in the field of graph embedding, the recommended papers are u_6, u_2, u_4, and u_8. In these recommended papers: u_6 surveys the study of dimensionality reduction, which has the same background knowledge with u_1; papers u_2 and u_4 are topically related to u_1 and focus on addressing problem of network representation; paper u_8 provides a new fundamental theory which can be transferred to deal with the research problem in u_1. Existing recommendation methods rank papers based on relevance and return a set of discrete papers relevant to the given query. However, for researchers who concentrate on learning existing literature about the topic of the given paper, they want to know how the topic develops and evolves, and understand what it originates from. Conventional academic paper recommendation ignores the relationships of recommended papers and can not provide effective results to help researchers trace the development of the topic.

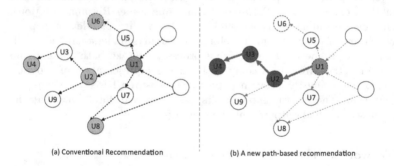

(a) Conventional Recommendation   (b) A new path-based recommendation

Fig. 1. Comparison of conventional recommendation and path-based recommendation.

To address above-mentioned problem, we propose a novel method called PAPR (Path-based Academic Paper Recommendation), which considers time-oriented development of the topic of the given paper, and recommends a sequence of academic papers. Figure 1(b) illustrates how our method works. Our method takes a paper u_1 as input and produces a 3-hop path u_2, u_3, u_4 from the paper. The proposed method concentrates on creating a deep and consecutive paper path. Nodes on the path are papers highly relevant and important, which represent the evolution and development of the paper's topic. Addressing this task will be beneficial to researchers who start to investigate a new topic.

In this paper, we take both content and network information into account to obtain better performance on paper recommendation. We learn the textual vector for each paper based on the semantic information (e.g., title and abstract) and utilize the network information to get relatedness of papers. Then, we combine semantic similarity and network relatedness to obtain the united similarity between papers. To get the path of recommendation, we use beam search [18] to get papers which have high relevance with all papers in the existing path.

The main contributions of this paper can be summarized as follows:

- We introduce a novel academic paper recommendation method to trace the origin and development of the topic of the given paper, which is beneficial for people to have an sight into the development of a research topic.
- We propose a novel strategy to measure the relevance between two papers and generate a consecutive paper path for recommendation, where both text semantic information and network structural information are considered.
- We conduct extensive experiments on two real-world datasets DBLP and ACM, and the results show that our proposed approach performs better than all compared methods.

The remaining sections of this paper are as follows. Section 2 presents the concept of scientific network and the problem definition. The overview of the path-based paper recommendation is introduced in Sect. 3. Section 4 describes the experimental setup and discusses the experimental results. Section 5 presents related work on recommending research papers. Section 6 concludes the paper.

2 Problem Definition

In this section, we introduce the concept of scientific network and then formally define the problem. The notations we will use are summarized in Table 1.

Table 1. Notations and explanations.

Notation	Explanation
G	Academic Network
V	Vertices in the graph G
E	Edges in the graph G
T_V	A set of object types
T_E	A set of relation types
Φ	Mapping function to embed vertices
N_i	Neighbors cited by node v_i

Definition 1. *(Academic Network). An academic network is defined as a graph* $G = (V, E)$, *where* $V = \{v_1, ..., v_n\}$ *represents n vertices and E is the set of edges in G. Each node v and each link e are associated with their mapping functions* $\psi(v) : V \to T_V$ *and* $\varphi(e) : E \to T_E$, *respectively.* T_V *and* T_E *denote the sets of predefined objects and relation types. A mapping function* $\Phi : V \to \mathbb{R}^d$ ($d \ll |V|$) *is used to learn the latent representation of papers.*

Example 1. As shown in Fig. 2(a). we construct an academic graph. It consists of multiple types of objects(Author, Paper, Venue) and relations(publish relation between papers and venues, written relation between papers and authors).

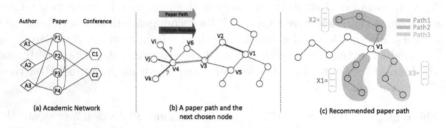

(a) Academic Network   (b) A paper path and the next chosen node   (c) Recommended paper path

Fig. 2. Result of a path-based paper recommendation.

Definition 2. *(Academic Paper Path). An academic paper path is defined as a path in the form of* $v_{i_1} \to v_{i_2} \to ... \to v_{i_k}$, *which describes a chain from new papers to old papers. For adjacent nodes* $v_{i_{k-1}}$ *and* v_{i_k} *in the path, there is a citation relation between them.*

Example 2. From Fig. 2(b), we can observe there is a paper path T which can be denoted as $T = v_1 \to v_2 \to v_3 \to v_4$. To extend the path, papers cited by v_4 will be considered and one of them will be chosen as next node according to the similarity between cited papers and T.

Definition 3. *(Path Recommendation Probability). Given a path* $T = v_{i_1} \to v_{i_2} \to ... \to v_{i_k}$, *which denotes a path of recommended papers, the probability of recommending the paper path is defined as the similarity between the representation of the given paper and that of the recommended path.*

Example 3. According to Fig. 2(c), there are three paper paths named *path1*, *path2*, *path3*. Then, we can return the path having the maximum similarity with v_1, after computing the similarity between v_1 and *path1*, *path2*, *path3* separately.

Problem Formalization. Given a graph G and a query paper v_{i_1}, the Path-based Academic Paper Recommendation is to select a path T defined on the graph G, which has the maximum path recommendation probability.

3   Proposed Approach

To generate an ordered path for paper recommendation, we define the relatedness measurement of papers and the construction of path recommendation. We give an overview of our recommendation framework in Fig. 3.

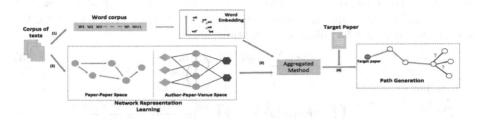

Fig. 3. Overview of recommendation framework.

3.1   Text Learning

The semantic relatedness of papers is measured mainly based on textual information. We use words from metadata (e.g., title, abstract) to build texts of documents. An effective method to compute semantic similarity of two papers is to generate word embeddings and combine them to form paper embeddings. Specifically, let c_i, c_j denote the vector representation of two papers P_i, P_j and the semantic relatedness $F_s(P_i, P_j)$ is defined by cosine function:

$$F_s(P_i, P_j) = cos(c_i, c_j) = \frac{c_i \cdot c_j}{\parallel c_i \parallel \parallel c_j \parallel} \tag{1}$$

where c_i, c_j are weighted average vector representation of words in P_i, P_j respectively. For the vector representation c_i of paper P_i, let $w = (w_1, ..., w_p)$ denote word vectors of unique words in paper and we have

$$\hat{c_i} = \sum_{k=1}^{p} \text{tfidf}(w_k) \cdot w_k \tag{2}$$

where TF-IDF is used to measure importance of word vectors and we use L2-normalization to get final text-level representation, $c_i = \hat{c_i}/ \parallel \hat{c_i} \parallel$.

3.2   Network Learning

The academic network is fundamentally a multi-relational heterogeneous graph where edges indicate many relationships and contain different semantic relatedness. In our model, we consider paper-paper network and author-paper-venue network and adopt two different methods to obtain relatedness between papers.

Paper-Paper Space. The paper-paper relationship constructs a homogeneous network for the academic graph. To evaluate the latent interactions between nodes, we use random walk based sampling strategy to learn latent representation of nodes. In the generation of random walks, we use two parameters p and q proposed by [5] to bias our random walks towards local neighborhood or tend to move further away. After generating a random walk, we define a window size k and $N_s(u)$ denotes neighborhood for node u in the slide window. Let $\phi_1 : V \rightarrow \mathbb{R}^d$ be the mapping function from nodes to feature representations. We now try to maximize the likelihood function

$$\max \sum_{u \in V} log Pr(N_s(u)|u) \tag{3}$$

and we have

$$Pr(N_s(u)|u) = \prod_{n_i \in N_s(u)} Pr(n_i|u) = \prod_{n_i \in N_s(u)} \frac{\exp(\phi_1(n_i) \cdot \phi_1(u))}{\sum_{v \in V} \exp(\phi_1(v) \cdot \phi_1(u))} \tag{4}$$

Finally, we optimize above equation using stochastic gradient ascent and skip-gram architecture to get node representation. Then we can get the similarity of two nodes v_i, v_j in paper-paper space:

$$f_c(v_i, v_j) = cos(\phi_1(v_i), \phi_1(v_j)) = \frac{\phi_1(v_i) \cdot \phi_1(v_j)}{\| \phi_1(v_i) \| \| \phi_1(v_j) \|} \tag{5}$$

Author-Paper-Venue Space. The paper-paper relation can not discover relatedness information between papers without citation relationship. To get a better relatedness evaluation, we also consider other relationships. Specifically, we use two most common and effective meta-path schemes, which are "author-paper-author"(APA) and "author-paper-venue-paper-author"(APVPA). We use meta-path-based random walk strategy [4] to incorporate different types of nodes into skip-gram and learn effective node representations. Given a node v, we maximize the probability of having the heterogeneous context $N_t(v), t \in T_V$:

$$argmax_\theta \sum_{v \in V} \sum_{t \in T_V} \sum_{c_t \in N_t(v)} \log p(c_t|v; \theta) \tag{6}$$

where $N_t(v)$ denotes v's neighborhood with the tth type of nodes. Let $\phi_2 : V \rightarrow \mathbb{R}^d$ be the mapping function for nodes and $p(c_t|v; \theta)$ is commonly defined as a softmax function:

$$p(c_t|v; \theta) = \frac{\exp^{\phi_2(c_t) \cdot \phi_2(v)}}{\sum_{u \in V} \exp^{\phi_2(u) \cdot \phi_2(v)}} \tag{7}$$

To achieve efficient optimization, we also use negative sampling for network learning. After learning network representation, the similarity of two nodes v_i, v_j in author-paper-venue space can be defined as:

$$f_a(v_i, v_j) = cos(\phi_2(v_i), \phi_2(v_j)) = \frac{\phi_2(v_i) \cdot \phi_2(v_j)}{\| \phi_2(v_i) \| \| \phi_2(v_j) \|} \tag{8}$$

Network-Based Relatedness. To make full use of network information, the final network embedding for a paper consists of paper-paper and author-paper-venue space. Formally, we formulate the relatedness of two paper v_i and v_j:

$$F_n(v_i, v_j) = \alpha f_a(v_i, v_j) + (1 - \alpha)f_c(v_i, v_j) \tag{9}$$

where $\alpha \in [0, 1]$ is to adjust the weights of two parts.

3.3   Aggregated Method

The united relatedness measurement includes semantic similarity and network closeness. We can get the relatedness $Aggr(P_i, P_j)$ between papers P_i and P_j:

$$Aggr(P_i, P_j) = \lambda F_s(P_i, P_j) + (1 - \lambda)F_n(P_i, P_j) \tag{10}$$

where $\lambda \in [0, 1]$ trades off the weight of F_s against F_n.

Time Decay. To get a path which can illustrate the development of the field of study, it is necessary to avoid overemphasizing old important articles. In this paper, we use a time-based decay parameter γ to reduce influence of older articles. The final relatedness $Sim(P_i, P_j)$ can be defined as:

$$Sim(P_i, P_j) = Aggr(P_i, P_j) \cdot \exp^{-\gamma \cdot |Time(P_i) - Time(P_j)|} \tag{11}$$

where γ is positive, $Time(P_i)$ and $Time(P_j)$ denote papers' published years.

3.4   Path Generation

To get an ordered path, we define a strategy to extend the path from a given node. We propose a two-stage beam search component for path generation, as shown in Fig. 4.

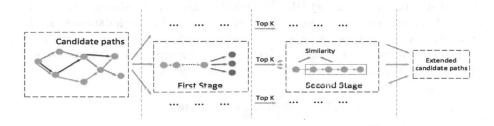

Fig. 4. Two stage path generation.

First Stage. In the first stage, we improve cohesiveness among nodes in the generated path. Specifically, given a path T in the candidate paths, which is denoted as $(u_1, u_2, ..., u_l)$, we need to choose the next node to extend the path. The similarity of candidate node $\widehat{u}$ and path T can be defined as:

$$F_h(T, \widehat{u}) = \frac{1}{l} \sum_{i=1}^{l} Sim(u_i, \widehat{u}) \qquad (12)$$

To get better results, we use beam search to get a set of candidate nodes $CanSet = \{u_{l+1}^{(1)}, ..., u_{l+1}^{(k)}\}$, which has top-$k$ high similarity with the path. We extend the given path T to k new paths. For all candidate paths, we do the same operation and get the new set of all extended candidate paths.

Second Stage. In the second stage, we try to reduce the size of the set of extended candidate paths. To keep topic similarity between the generated paths and the original paper, we define the similarity between the path $T = (u_1, u_2, ..., u_l)$ and the original node u_1 as follows:

$$F_h(T, u_1) = \frac{1}{l-1} \sum_{i=2}^{l} Sim(u_i, u_1) \qquad (13)$$

We choose the most similar k candidate paths as the i-step generated paths. We do two-stage path-generation repeatedly to get the final generated paths and rank these paths in order of decreasing similarity for paper recommendation.

4   Experiment

In this section, we introduce the datasets and experimental setup, as well as the evaluation metrics and baseline methods. We also present and analyze the results of our experiments.

4.1   Data Preparation

Dataset. The experiments are conducted on two different datasets: the DBLP dataset and the ACM dataset [19]. The information of title, abstract, keywords, authors, publication venue and publication year is used in both datasets. Statistics of the two constructed heterogeneous bibliographic networks are summarized in Table 2. We preprocess texts of each paper, remove stop words and words appearing less than 10 times and then stem each word.

Table 2. Statistics of two datasets.

Dataset	Papers	Authors	Venues	Terms	Relationships
DBLP	1,782,700	2,052,414	18,936	100,000	9,590,600
ACM	2,385,057	2,004,398	269,467	61,618	12,048,682

Evaluation Settings. To conduct the experiments, we invite 10 experts to help on evaluating the quality of recommended results. We randomly choose 200 papers in DBLP and ACM as test set respectively. For each given paper, 3 experts annotate ordered paths and we choose the overlap results as ground truth papers. All meta-data (e.g., title and abstract) can be available for annotators.

4.2   Experimental Settings

Baseline Methods. Several widely deployed paper recommendation approaches were implemented. We compared the recommendation results of the following methods in academic network:

LDA: LDA [1] is a celebrated generative model for text documents that learns representations for documents as distributions over word topics.

LSI: LSI [3] uses singular value decomposition on the BOW representation to arrive at a semantic feature space.

PathSim: PathSim [17] is a meta path-based method to search similar papers in heterogeneous information networks. It considers different linkage paths to study similarity among the same type of objects in academic networks.

HeteSim: HeteSim [15] is a path-constrained method to measure the relatedness of heterogeneous objects in heterogeneous networks.

PageRank: PageRank [10] is a method to derive an object's importance based on authority propagation in the heterogeneous bibliographic network. It tends to rank papers based on citation analysis in citation networks.

Deepwalk: Deepwalk [12] is a network-only representation learning method. Deepwalk takes random walk paths from network as sentences and nodes as words to learn the node representations by applying the Skip-Gram algorithm.

Node2vec: Node2vec [5] proposes an improvement to the random walk phase of DeepWalk and combines DFS-like and BFS-like neighborhood exploration.

PAPR: Our proposed method combines text and network structure information to learn representations for nodes. To evaluate the benefit brought by time decay introduced in Sect. 3.3 and two-stage path-generation introduced in Sect. 3.4, we design two variations of PAPR. **PAPR1** only uses two-stage path-generation and **PAPR2** only uses time decay.

Parameter Settings. In the section of text learning, we adopt 100 dimension, 10 window size, skip-gram model and 5 negative samples for word2vec. In the section of network learning, we set 128 dimension, 10 window size as the basic parameters. The number of walks per node is 20 and the walk length is 30. In all methods, biased parameters are also fine-tuned to report the best performance.

Table 3. Recommendation performance comparisons on DBLP and ACM datasets in terms of Precision and Recall.

Dataset	DBLP						ACM					
	P@3	P@5	P@8	R@3	R@5	R@8	P@3	P@5	P@8	R@3	R@5	R@8
LDA	0.3686	0.2725	0.1873	0.1382	0.1538	0.1586	0.2923	0.2128	0.1726	0.1182	0.1287	0.1578
LSI	0.3813	0.2796	0.2043	0.1419	0.1564	0.1715	0.3769	0.2615	0.1794	0.1516	0.1575	0.1625
PathSim	0.3729	0.2599	0.1761	0.1402	0.1469	0.1497	0.3684	0.2519	0.1716	0.1476	0.1513	0.1540
HeteSim	0.3665	0.2556	0.1760	0.1383	0.1450	0.1505	0.3549	0.2529	0.1857	0.1195	0.1273	0.1415
PageRank	0.3841	0.2848	0.2133	0.1410	0.1568	0.1761	0.4211	0.2895	0.1934	0.1654	0.1700	0.1705
Deepwalk	0.4067	0.2966	0.2071	0.1526	0.1677	0.1761	0.3730	0.2538	0.1692	0.1509	0.1541	0.1542
Node2vec	0.4258	0.3220	0.2222	0.1589	0.1807	0.1872	0.4576	0.3179	0.2119	0.1837	0.1911	0.1931
PAPR1	0.4892	0.3714	0.2762	0.1807	0.2060	0.2308	0.5423	0.4153	0.3042	0.2187	0.2507	0.2756
PAPR2	0.5084	0.3771	0.2777	0.1895	0.2109	0.2330	0.4346	0.3179	0.2154	0.1758	0.1927	0.1957
PAPR	**0.5443**	**0.4218**	**0.3325**	**0.2023**	**0.2345**	**0.2691**	**0.6231**	**0.4948**	**0.3709**	**0.2521**	**0.3001**	**0.3356**

Evaluation Metrics. We employ Precision and Recall at position M (P@M and R@M) as the evaluation metrics. Precision@M is defined as the fraction of ground truth papers contained by the M-length recommended path and Recall@M is defined as the fraction of ground truth papers that appear in the M-length recommended path contained by the whole ground truth. For each given paper P_i, we have:

$$P@M = \frac{|N(M;P_i)|}{|N_p(P_i)|} \qquad R@M = \frac{|N(M;P_i)|}{|N_r(P_i)|} \tag{14}$$

where $N(M;P_i)$ is the set of ground truth papers in the M-length recommended path, $N_p(P_i)$ is the set of papers in the recommended path and $N_r(P_i)$ is the set of papers in the ground truth path.

4.3 Result Analysis

Evaluation results on different methods are shown in Table 3. We compare the proposed method with different baselines using Precision and Recall. Table 3 summarizes the comparison results on both DBLP and ACM datasets. It can be easily observed that our proposed method outperforms other methods and gains a performance improvement of more than 20% over the best competitive algorithm. In general, the method LDA gets the lowest precision on ACM dataset, which shows that the similarity of papers can not be evaluated only based on content information. PathSim and HeteSim have the similar performance. However, they don't perform well on DBLP dataset because they heavily depend on hand-engineered meta paths. To analyze the effect of time-decay component and two-stage path generation component in our method, we also compare the proposed method with its variation PAPR1 and PAPR2. We can observe that PAPR gains a performance improvement of more than 10% over PAPR1 and more than 6% over PAPR2 on precision metrics, since time-decay strategy and two-stage path-generation play important roles for better recommendation.

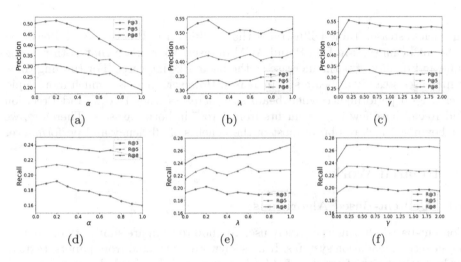

Fig. 5. Precision and recall with varying α, λ and γ for PAPR.

4.4   Parameter Study

In this section, we study the impact of parameters α, λ and γ on dataset DBLP. α is proposed to balance paper-paper space and author-paper-venue space. Observed from Fig. 5(a)(d), PAPR achieves the best performance when $\alpha = 0.2$. This is because, citation relation has a leading role and two meta-path relations (APA, APVPA) make a great supplement for relatedness measurement. λ trades off the weight of text semantic and network structure. We set $\alpha = 0.2$ to evaluate influence of text similarity and network similarity. From Fig. 5(b)(e), we find that the lower λ can get better results of Precision@3 and Recall@3. Meanwhile, the higher λ gets better results of Precision@8 and Recall@8. It means that short path depends more on network structure and long path depends more on text semantics. γ is used to reduce the influence of older papers. Figure 5(c)(f) show the influence of time decay. We can observe that the precision and recall increase quickly at first and then decline slowly, which means time decay is helpful to get a better time-oriented path but excessive decay has no better effects.

To achieve the trade-off between running time and the quality of the paths, we also perform experiments over average running time of different beam sizes

Table 4. Performance comparision over running time of different beam sizes.

Beam size	DBLP							ACM						
	Time(s)	P@3	P@5	P@8	R@3	R@5	R@8	Time(s)	P@3	P@5	P@8	R@3	R@5	R@8
1	0.323	0.510	0.395	0.296	0.172	0.209	0.237	0.410	0.518	0.439	0.356	0.229	0.267	0.311
3	0.662	0.526	0.413	0.328	0.181	0.222	0.256	0.695	0.579	0.473	0.365	0.234	0.277	0.328
5	0.935	0.545	0.421	0.333	0.202	0.234	0.269	1.020	0.623	0.495	0.371	0.252	0.300	0.335
7	1.194	0.551	0.424	0.335	0.213	0.235	0.273	1.371	0.617	0.494	0.373	0.249	0.300	0.336

on each testcase. Table 4 illustrates the results w.r.t. Precision and Recall on different beam sizes in DBLP and ACM dataset. We can see that both running time and quality of paths increase as the beam size increases. For running time, it increases steadily as beam size increases. However, there is a much more rapid increase on performance when beam size increases from 1 to 5, and Precision and Recall have few improvements from 5 to 7 in both datasets. Therefore, we set beam size as 5 to obtain most of the benefits of efficiency and performance.

5 Related Work

5.1 Content-Based Algorithms

Content-based filtering is a widely used method to compare similarity of items in paper recommendation systems. It uses keywords extracted from papers' texts to evaluate similarity of articles. [21] feeds documents' titles and abstracts into TF-IDF model and learns probabilistic model to evaluate relatedness of documents. [9] develops a paper recommendation system which constructs users' profiles based on candidate papers' titles. However, these methods mostly use Bag-of-Words model, which has difficulty in finding conceptually similar work [2].

5.2 Collaborative Filtering

Collaborative filtering is a popular and widely used method in recommender systems. However, collaborative filtering cannot generate accurate recommendations without sufficient initial ratings from users. In order to alleviate the cold start problem, [8] proposes a context-based collaborative filtering, which incorporates co-occurrence relations into rating matrix. [14] combines the matrix factorization with the topic modeling to achieve a better performance. Nevertheless, collaborative filtering needs much computing time and offline data processing.

5.3 Network-Based Algorithms

Network-based paper recommendations concentrate on analyzing citation network to understand the relationship between scholarly papers. Two main methods in citation analysis are co-citation analysis and bibliographic coupling. However, these methods can not address complex relationships in networks [22].

[20] proposes a method based on random walk with restart(RWR) to measure vertex-to-vertex relevance. [7] considers a heterogeneous network to recommend similar papers. Nevertheless, these works mainly extract hand-engineered structural information to calculate paper similarity, which is inflexible and time-consuming.

6 Conclusion

In this paper, we propose a Path-based Academic Paper Recommendation, namely PAPR, to perform academic paper recommendation. Specifically, to acquire an ordered path of relevant scholarly papers, we design a flexible and expressive model, which takes both text semantics and network structure into account. To facilitate evaluation of similarity between papers, textual attributes and network structure are embedded in vector space. The experiments based on real datasets show that our model outperforms other methods.

Acknowledgment. This work was supported by the Major Program of the Natural Science Foundation of Jiangsu Higher Education Institutions of China under Grant No. 19KJA610002 and 19KJB520050, and the National Natural Science Foundation of China under Grant No. 61902270, a project funded by the Priority Academic Program Development of Jiangsu Higher Education Institutions.

References

1. Blei, D.M., Ng, A.Y., Jordan, M.I.: Latent Dirichlet allocation. J. Mach. Learn. Res. **3**, 993–1022 (2003)
2. Chen, T.T., Lee, M.: Research paper recommender systems on big scholarly data. In: Yoshida, K., Lee, M. (eds.) PKAW 2018. LNCS (LNAI), vol. 11016, pp. 251–260. Springer, Cham (2018). https://doi.org/10.1007/978-3-319-97289-3_20
3. Deerwester, S., Dumais, S.T., Furnas, G.W., Landauer, T.K., Harshman, R.: Indexing by latent semantic analysis. J. Am. Soc. Inform. Sci. **41**(6), 391–407 (1990)
4. Dong, Y., Chawla, N.V., Swami, A.: metapath2vec: scalable representation learning for heterogeneous networks. In: Proceedings of the 23rd ACM SIGKDD International Conference on Knowledge Discovery and Data Mining, pp. 135–144 (2017)
5. Grover, A., Leskovec, J.: node2vec: scalable feature learning for networks. In: Proceedings of the 22nd ACM SIGKDD International Conference on Knowledge Discovery and Data Mining, pp. 855–864 (2016)
6. Hammou, B.A., Lahcen, A.A., Mouline, S.: APRA: an approximate parallel recommendation algorithm for big data. Knowl.-Based Syst. **157**, 10–19 (2018)
7. Lao, N., Cohen, W.W.: Relational retrieval using a combination of path-constrained random walks. Mach. Learn. **81**(1), 53–67 (2010)
8. Liu, H., Kong, X., Bai, X., Wang, W., Bekele, T.M., Xia, F.: Context-based collaborative filtering for citation recommendation. IEEE Access **3**, 1695–1703 (2015)
9. Nascimento, C., Laender, A.H., da Silva, A.S., Gonçalves, M.A.: A source independent framework for research paper recommendation. In: Proceedings of the 11th Annual International ACM/IEEE Joint Conference on Digital Libraries, pp. 297–306 (2011)

10. Page, L., Brin, S., Motwani, R., Winograd, T.: The PageRank citation ranking: bringing order to the web. Proceedings of the WWW Conference, vol. 1998, pp. 161–172 (1999)
11. Pazzani, M.J., Billsus, D.: Content-based recommendation systems. In: Brusilovsky, P., Kobsa, A., Nejdl, W. (eds.) The Adaptive Web. LNCS, vol. 4321, pp. 325–341. Springer, Heidelberg (2007). https://doi.org/10.1007/978-3-540-72079-9_10
12. Perozzi, B., Al-Rfou, R., Skiena, S.: DeepWalk: online learning of social representations. In: Proceedings of the 20th ACM SIGKDD International Conference on Knowledge Discovery and Data Mining, pp. 701–710 (2014)
13. Schafer, J.B., Frankowski, D., Herlocker, J., Sen, S.: Collaborative filtering recommender systems. In: Brusilovsky, P., Kobsa, A., Nejdl, W. (eds.) The Adaptive Web. LNCS, vol. 4321, pp. 291–324. Springer, Heidelberg (2007). https://doi.org/10.1007/978-3-540-72079-9_9
14. Shan, H., Banerjee, A.: Generalized probabilistic matrix factorizations for collaborative filtering. In: 2010 IEEE International Conference on Data Mining, pp. 1025–1030. IEEE (2010)
15. Shi, C., Kong, X., Huang, Y., Philip, S.Y., Wu, B.: HeteSim: a general framework for relevance measure in heterogeneous networks. IEEE Trans. Knowl. Data Eng. **26**(10), 2479–2492 (2014)
16. Small, H.: Co-citation in the scientific literature: a new measure of the relationship between two documents. J. Am. Soc. Inform. Sci. **24**(4), 265–269 (1973)
17. Sun, Y., Han, J., Yan, X., Yu, P.S., Wu, T.: PathSim: meta path-based top-K similarity search in heterogeneous information networks, vol. 4, pp. 992–1003. CiteSeer (2011)
18. Sutskever, I., Vinyals, O., Le, Q.V.: Sequence to sequence learning with neural networks. In: Advances in Neural Information Processing Systems, pp. 3104–3112 (2014)
19. Tang, J., Zhang, J., Yao, L., Li, J., Zhang, L., Su, Z.: ArnetMiner: extraction and mining of academic social networks. In: Proceedings of the 14th ACM SIGKDD International Conference on Knowledge Discovery and Data Mining, pp. 990–998 (2008)
20. Tian, G., Jing, L.: Recommending scientific articles using bi-relational graph-based iterative RWR. In: Proceedings of the 7th ACM Conference on Recommender Systems, pp. 399–402 (2013)
21. Wang, C., Blei, D.M.: Collaborative topic modeling for recommending scientific articles. In: Proceedings of the 17th ACM SIGKDD International Conference on Knowledge Discovery and Data Mining, pp. 448–456 (2011)
22. Zhou, D., et al.: Learning multiple graphs for document recommendations. In: Proceedings of the 17th International Conference on World Wide Web, pp. 141–150 (2008)

Database System and Workflow

A Slice-Based Method to Speed Up Join View Maintenance for Transactions

Huichao Duan, Huiqi Hu[✉], Xuan Zhou, and Aoying Zhou

East China Normal University, Shanghai, China
stevenduan@stu.ecnu.edu.cn, {hqhu,xzhou,ayzhou}@dase.ecnu.edu.cn

Abstract. With the increments of data volumes and user numbers, big data applications require higher transaction throughput but lower query latency for database systems. The materialized view accelerates analytical queries by trading space for query efficiency. Nevertheless, it has to be updated under transactional workloads to obtain up-to-second results. Unfortunately, the cost of view maintenance is expensive, which requires examining its maintenance strategies carefully. In this paper, we redesign the view maintenance strategy from the transaction perspective. Compared with conventional methods that compute the modifications of different operations separately, we implement a slice-based method that maintains the updates of several base tables with join relations in one transaction as the increments of a slice. Then we optimize the view maintenance process based on the slices such as avoiding invalid expression evaluation and base table access. We conduct experiments in PostgreSQL under CH-benCHmark. Experiments show that our method can increase transaction throughput by 17%–121%, reduce query latency by 30%–85%, and achieve 1.9× higher query throughput than those of conventional methods.

Keywords: Incremental view maintenance · Materialized views · Join view

1 Introduction

Multi-table join efficiency is an important factor that restricts the performance of complex queries. To accelerate multi-table join queries, the materialized view is a powerful tool that significantly facilitates query by trading space cost for execution time. To get an up-to-second query result, materialized views must be updated along with the base tables (or base relations), which is known as view maintenance. Any update falling on the base tables will cause the changes of join views. To reduce the expensive cost of updates, a typical approach is to leverage incremental view maintenance (IVM). IVM builds on the observation that computing only updates to view contents induced by minor changes of base tables is more efficient than recomputing the entire view from the base tables [4]. Unfortunately, it is still non-trivial since obtaining the incremental results requires to join computation cross multiple base tables.

© Springer Nature Switzerland AG 2020
Z. Huang et al. (Eds.): WISE 2020, LNCS 12343, pp. 359–375, 2020.
https://doi.org/10.1007/978-3-030-62008-0_25

It is important to make view maintenance efficient. Conventional IVMs are based on the changes of a base table or a DML operation, i.e., they mainly discuss the impact of an INSERT/DELETE/UPDATE operation on the view when it occurs to a base table. In general, conventional IVMs are divided into two categories according to the timing of view maintenance: (i) synchronous IVM inserts records into the base table and calculates view increments in the same transaction. Each change to the base table triggers an additional view computation. (ii) Asynchronous IVM, which calculates the view increments before it is queried, sometimes consider multiple computations as a batch. As shown in [2,16,20], asynchronous IVM can merge operations which perform insertions, deletions, or updates on the same record. For the DML operations on different base tables, it is still difficult to collapse them together to provide optimization.

In many situations, database systems rely on transactional workload to manipulate tables, where a transaction contains a number of DML operations, which can update a number of base tables together. And these base tables may have multi-join relationships within an analytical query. Recall that conventional views only consider the updates on separate base tables, so the view maintenance has to reconstruct the associations via the increments of each individual base relation. In this paper, we attempt to calculate the increment of a transaction to the entire view at once. Facing the fact that multiple base tables are co-updated in the transaction, we propose a slice-based method to speed up view maintenance. Base tables co-updated in the transaction are considered as a slice. When we look at IVM in terms of the slice, we compute the increments caused by multiple DML operations together instead of computing them separately. More importantly, we can use the transaction information to facilitate the computation of the slice's increment, which significantly speeds up the overall view maintenance. The slice-based method can be seamlessly into the current standard asynchronous IVM process without maintaining any materialized information. To summary, the contributions of the paper are listed as follows:

(1) We propose a slice-based method to speed up view maintenance. Co-updated base tables in the transaction are formed as a slice. We directly calculate the increment of the slice to the view. This method is lightweight with powerful optimization capability.
(2) We propose two optimization methods named *expression optimization* and *table access optimization* to accelerate the computation of the slice's increment. By exploiting the information embedded in the transaction, these two optimization methods reduce the overhead of view maintenance by reducing computation cost and base table accesses.
(3) We implement the proposed method on PostgreSQL. The speed-up method is integrated into the existing asynchronous IVM method. Experiments on CH-benCHmark show it achieves much better performance than current methods in terms of both transaction throughput and query latency.

The rest of the paper is organized as follows. Related works are outlined in Sect. 2. The sliced based method is introduced in Sect. 3. Section 4 and Sect. 5

describe the expression optimization and table access optimization respectively. Section 6 presents a comprehensive experimental study. The paper is concluded in Sect. 7.

2   Related Works

When a sequence of insertions, deletions, and updates are conducted to base tables, the view needs to be recalculated according to its definition. In IVM, it calculates the difference (increments) rather than recomputing the entire view. IVM has been studied for decades [2,9–12]. [15] proves the correctness through the relational algebra. [4] comprehensively review the view maintenance methods appearing in the latest researches. To the best of our knowledge, there are no research works whichever studied maintaining views from the transactions.

Synchronized IVM. In early works, synchronized IVMs [6,9] are the main methods in database systems. A view is maintained immediately upon any base table modification, as part of the transaction making that modification. The view contents under this maintenance strategy are always consistent with the base tables. While queries benefit from the immediate availability of up-to-date views, the overhead of view maintenance will greatly obstruct the transaction processing. Thus, for heavy transactional workloads, synchronized IVM is expensive. [9] presents an incremental evaluation algorithm to compute changes to materialized views, which is compared with our method in the experiment as an implementation of the synchronized IVM. View stacking which is known as views on views splits the view into multiple slices to respond to queries, which is also made as a prototype in SQL Server [6]. Our strategy of dividing slices as transaction granularity and view stacking are similar to the divide and conquer strategy. The difference is that we can use the semantic logic and runtime information in the transaction to speed up the maintenance of the slice itself.

Asynchronous IVM. Asynchronous view maintenance strategy can reduce transaction latency and improves transaction performance. The view is allowed to be inconsistent with the base table and to be synchronized as necessary when queried. Asynchronous IVM methods are being accepted in modern databases. [20] presents a novel way to lazily maintain materialized views in SQL Server that relieves update overhead. [10] assumes the base tables have primary keys and performs IVM by computing ID-based diff sets that compactly identify the to-be-modified tuples through their IDs. This paper has expanded more general optimizable scenarios in the transactional workload based on [10].

Table Access Optimization. For the data warehouse, accessing base tables from the data source (e.g., database) usually results in higher query delay and cost. To avoid the table access, there are a series of studies [8,16] on self-maintainable views, which uses its materialized contents and the delta info without base tables. [8] exploits primary key constraints to speed up IVM. However, self-maintainability is a rather strong property; making views self-maintainable for all possible base table modifications may require maintaining a lot of auxiliary data and may be expensive [4]. Using auxiliary data to reduce the frequency of base table accesses is another option. [13] quickly determines whether the update transaction is relevant to the view by maintaining a hash-based data structure which summarizes the values of the join columns. Similarly, maintaining top-k [19] info and Min/Max [18] can also help to reduce the table access frequency. DBToaster [11] depends on runtime in-memory structure to keep query results, which stores multiple copies of data and relies entirely on memory resources. In general, the above methods either have to maintain auxiliary data or fall into the limited specific scenarios, while our modular approach can be universally applied to transactional workload without redundant data and other constraints.

3   IVM Based on Slices

Given the transaction that contains several operations, we no longer compute the increment of each base table separately. Instead, we attempt to calculate them together as a slice. To this end, first we give the standard process of IVM and show how slice can speed up the process in Sect. 3.1. Next we describe how to generate slices based on the view schema and the transaction in Sect. 3.2. The algorithm of computing increments of the slice is introduced in Sect. 3.3.

3.1   Increments of the View

We follow the standard process [20] to compute the increments of the view. $V = T_1 \bowtie T_2 \cdots T_n$ is a view of multi-table join, where T_i is a base table. The updates of relation T are represented as increments, denoted by ΔT [15]. It captures two kinds of updates of T, new records inserted and old records deleted. In this paper, we also claim the functionality of operation "$\oplus$", which defines over records of two relations. In specific, $\oplus$ means fusing the records of two relations. Thus for $T \oplus \Delta T = T'$, it means T gets a new state T' by merging increments ΔT [3]. If each table (T_i) in this view schema has an increment (denoted by ΔT_i), then the criterion to compute ΔV is described in Eq. 1.

$$
\begin{aligned}
\Delta V = {} & \Delta T_1 \bowtie T_2 \bowtie \cdots \bowtie T_n \\
& \oplus (T_1 \oplus \Delta T_1) \bowtie \Delta T_2 \bowtie T_3 \bowtie \cdots \bowtie T_n \\
& \oplus \cdots \\
& \oplus (T_1 \oplus \Delta T_1) \bowtie \cdots \bowtie (T_{i-2} \oplus \Delta T_{i-2}) \bowtie \Delta T_{i-1} \bowtie T_i \bowtie \cdots \bowtie T_n \\
& \oplus (T_1 \oplus \Delta T_1) \bowtie \cdots \bowtie (T_{i-1} \oplus \Delta T_{i-1}) \bowtie \Delta T_i \bowtie T_{i+1} \bowtie \cdots \bowtie T_n \\
& \oplus \cdots \\
& \oplus (T_1 \oplus \Delta T_1) \bowtie \cdots \bowtie (T_{j-1} \oplus \Delta T_{j-1}) \bowtie \Delta T_j \bowtie T_{j+1} \bowtie \cdots \bowtie T_n \\
& \oplus (T_1 \oplus \Delta T_1) \bowtie \cdots \bowtie (T_j \oplus \Delta T_j) \bowtie \Delta T_{j+1} \bowtie T_{j+2} \bowtie \cdots \bowtie T_n \\
& \oplus \cdots \\
& \oplus (T_1 \oplus \Delta T_1) \bowtie \cdots \bowtie (T_{n-1} \oplus \Delta T_{n-1}) \bowtie \Delta T_n .
\end{aligned} \tag{1}
$$

In each row of this formula, the increment of a base table is used to join with other base tables or base tables augmenting their increments. To a certain extent, each row in this equation (e.g., the i-th row) can be seen as the increments contributed by the modification on a base table (e.g., ΔT_i) to the entire view. All these increments are then fused to form the ΔV.

Our key insight is that by consolidating several rows of computation, the overall cost of the entire view can be reduced. We can build up a *slice* that is composed of several base tables with join relationships. Assuming that the increments contributed from ΔT_i to ΔT_j (i.e., the underlined part of Eq. 1) can be merged into one *slice*, then the calculation of ΔV becomes:

$$
\begin{aligned}
\Delta V = {}& (T_1 \oplus \Delta T_1) \bowtie \cdots \bowtie (T_{i-1} \oplus \Delta T_{i-1}) \bowtie \Delta T_i \bowtie T_{i+1} \bowtie \cdots \bowtie T_n \\
& \oplus \cdots \\
& \oplus (T_1 \oplus \Delta T_1) \bowtie \cdots \bowtie (T_{i-2} \oplus \Delta T_{i-2}) \bowtie \Delta T_{i-1} \bowtie T_i \bowtie \cdots \bowtie T_n \\
& \oplus \underline{(T_1 \oplus \Delta T_1) \bowtie \cdots \bowtie (T_{i-1} \oplus \Delta T_{i-1}) \bowtie \Delta S \bowtie T_{j+1} \bowtie \cdots \bowtie T_n} \quad (2) \\
& \oplus (T_1 \oplus \Delta T_1) \bowtie \cdots \bowtie (T_j \oplus \Delta T_j) \bowtie \Delta T_{j+1} \bowtie T_{j+2} \bowtie \cdots \bowtie T_n \\
& \oplus \cdots \\
& \oplus (T_1 \oplus \Delta T_1) \bowtie \cdots \bowtie (T_{n-1} \oplus \Delta T_{n-1}) \bowtie \Delta T_n.
\end{aligned}
$$

where ΔS equals:

$$
\begin{aligned}
\Delta S = {}& \Delta T_i \bowtie T_{i+1} \bowtie \cdots \bowtie T_j \\
& \oplus \cdots \\
& \oplus (T_i \oplus \Delta T_i) \bowtie \cdots \bowtie (T_{k-1} \oplus \Delta T_{k-1}) \bowtie \Delta T_k \bowtie T_{k+1} \bowtie \cdots \bowtie T_j \quad (3) \\
& \oplus \cdots \\
& \oplus (T_i \oplus \Delta T_i) \bowtie \cdots \bowtie (T_{j-1} \oplus \Delta T_{j-1}) \bowtie \Delta T_j.
\end{aligned}
$$

This indicates that we can unite the increments on these base tables within a slice into an increment of a slice, and directly compute its increment to the entire view. Equation 1 totally has $n*(n-1)$ joins. By observing Eq. 2, assuming that ΔS can be figured out in a very efficient way, then the remaining number of joins will be reduced to $n*(n-1-j+i)$. For example, for a common three-table join instance with modifications falling on each table, if a transaction updates two base tables together, then the total number join can be reduced from 6 to 3, which shows a powerful optimization opportunity. It is also worth noting that this method can be seamlessly embedded into the standard process when facing transactional workloads. A transaction can update a number of base tables at the same time, thus we attempt to exploit transaction information to optimize the computation of ΔS based on the straightforward way described in Eq. 3.

3.2   Slice Definition

We bunch up the base tables that are co-updated in a transaction into slices. Given the transaction and the schema, we first identify the dynamic tables.

Definition 1 (Dynamic Table). *A dynamic table has at least one column modified within the transaction, including columns updated in the* UPDATE *statements (i.e., columns after syntax "SET") and all columns involved in the* INSERT/DELETE *statements.*

A dynamic table indicates that the transaction inserting, deleting, or updating its columns. A transaction may modify several dynamic tables at once. Meanwhile, these dynamic tables may also have join relationships in the view schema. We identify these co-updated dynamic tables with join relations and merge them into a slice.

Definition 2 (Slice). *A slice denoted by S consists of dynamic tables, which are updated by the transaction and has join relations in the view schema.*

BEGIN TRANSACTION *Neworder*	CREAT VIEW V_1 AS
SELECT c_last FROM customer WHERE c_id=c₃;	SELECT c_last, o_cnt, ol_amount
INSERT INTO order VALUES (o₃,c₃,9);	FROM customer c, ⌐order o,¬⌐orderline ol¬
INSERT INTO orderline VALUES (ol₄,o₃,87),	WHERE c.c_id = o.c_id AND o.o_id = ol.o_id
(ol₅,o₃,66);	⌐Dynamic Table¬ Slice
COMMIT TRANSACTION *Neworder*	

Fig. 1. A running example of view & *Neworder* workload

Example 1. We take the CH-benCHmark[1] as an example in our next description. A main transactional load of CH-benCHmark is *Neworder*. *Neworder* transaction includes four SQLs, where three tables ORDER, NEW-ORDER, and ORDERLINE are modified while CUSTOMER only appears in a SELECT statement. As shown in Fig. 1, we define a view based on the CH-benCHmark. The view selects from CUSTOMER, ORDER, ORDERLINE and contains two join conditions. According to our definition, we can find out: (i) Dynamic tables. Our method classifies ORDER and ORDERLINE into dynamic tables as they all have an insert statement. (ii) Slice. Dynamic tables ORDER and ORDERLINE are co-updated by *Neworder*. Meanwhile, ORDER and ORDERLINE have the join relation in the view definition. So they are compacted into one slice.

3.3 Computing ΔS

Following Eq. 3, we describe the basic algorithm of computing ΔS for each slice S. For an arbitrary slice S, it contains a number of base tables T_k ($i \leq k \leq j$). Based on the definition, these tables have join relationships and are always co-updated. A transaction will produce its increments ΔT_k for each T_k. If the defined join order is $T_i \bowtie T_{i+1} \cdots \bowtie T_j$, then ΔS can be computed through

---

[1] CH-benCHmark [5,14] is a well-known hybrid workload which combines the transactional workload TPC-C with the analytical workload TPC-H.

Algorithm 1. For each ΔT_k, we need to calculate the increment to ΔS brought by it (line 2). In each iteration, we join the base tables one by one starting from ΔT_k. Meanwhile, we use ΔT to store the intermediate result (line 3). It obtains the results following the order: first it computes the increments by joining ΔT_k with T_{k-1} as $\Delta T = \Delta T_k \bowtie (\Delta T_{k-1} \oplus T_{k-1})$. Next, we join it with T_{k-2} as $\Delta T = \Delta T \bowtie (\Delta T_{k-2} \oplus T_{k-2})$. It keeps using the increments ΔT to join with all the remaining tables to obtain the result, i.e., $\Delta T_{k-3} \oplus T_{k-3}, \cdots, \Delta T_i \oplus T_i$ (line 3–6). Next it joins ΔT with $T_{k+1}, T_{k+2} \cdots, T_j$ (line 7–9). Finally, it obtains the increment yielded by ΔT_k and the result is fused into ΔS (line 10).

Next we describe how to optimize the basic algorithm. By observing the transaction, the algorithm can be accelerated when certain conditions are met.

4 Expression Optimization

Expression Simplification. Our first method to speed up computing ΔS is to simplify its expression. Consider the join between T_k and T_{k+1} within the slice. Following Eq. 3, ΔT_k and ΔT_{k+1} produce two increments to ΔS:

$$\Delta S^k = (T_i \oplus \Delta T_i) \bowtie \cdots \bowtie (T_{k-1} \oplus \Delta T_{k-1}) \bowtie \underline{\Delta T_k \bowtie T_{k+1}} \bowtie \cdots \bowtie T_j,$$

$$\Delta S^{k+1} = (T_i \oplus \Delta T_i) \bowtie \cdots \bowtie \underline{(T_k \oplus \Delta T_k)} \bowtie \Delta T_{k+1} \bowtie T_{k+2} \bowtie \cdots \bowtie T_j.$$

We can observe if $(T_k \oplus \Delta T_k) \bowtie \Delta T_{k+1} = \Delta T_k \bowtie \Delta T_{k+1}$, then ΔS^{k+1} becomes:

$$\Delta S^{k+1} = (T_i \oplus \Delta T_i) \bowtie \cdots \bowtie (T_{k-1} \oplus \Delta T_{k-1}) \bowtie \underline{\Delta T_k \bowtie \Delta T_{k+1}} \bowtie \cdots \bowtie T_j.$$

Similarly, if $\Delta T_k \bowtie T_{k+1} = \Delta T_k \bowtie \Delta T_{k+1}$, then ΔS^k becomes:

$$\Delta S^k = (T_i \oplus \Delta T_i) \bowtie \cdots \bowtie (T_{k-1} \oplus \Delta T_{k-1}) \bowtie \underline{\Delta T_k \bowtie \Delta T_{k+1}} \bowtie \cdots \bowtie T_j.$$

Algorithm 1: Computation of ΔS

Input: $\{T_i, \cdots T_j\}$; $\{\Delta T_i, \cdots \Delta T_j\}$
Output: ΔS

```
1  ΔS ← ∅;
2  for k←i to j do
3  │    ΔT ← ΔT_k;
4  │    for m ← k − 1 to i do
5  │    │    ΔT ← ΔT ⋈ (ΔT_m ⊎ T_m) ;
6  │    end
7  │    for m ← k + 1 to j do
8  │    │    ΔT ← ΔT ⋈ T_m ;
9  │    end
10 │    ΔS ← ΔS ⊕ ΔT;
11 end
12 return ΔS ;
```

This makes $\Delta S^k = \Delta S^{k+1}$ when $\Delta T_k \bowtie T_{k+1} = \Delta T_k \bowtie \Delta T_{k+1}$ and $(T_k \oplus \Delta T_k) \bowtie \Delta T_{k+1} = \Delta T_k \bowtie \Delta T_{k+1}$ happen. As a result, we only need to get the increments brought by ΔT_k and ΔT_{k+1} in one calculation, instead of computing them separately. We can check and execute this optimization for each join which significantly reduces the computation cost. In practice, this is a very powerful enhancement to the computation of ΔS. If the slice contains a three-table join $T_1 \bowtie T_2 \bowtie T_3$, which is most common in many real-world queries, then the basic algorithm contains $\Delta T_1 \bowtie T_2 \bowtie T_3$, $(T_1 \oplus \Delta T_1) \bowtie \Delta T_2 \bowtie T_3$ and $(T_1 \oplus \Delta T_1) \bowtie (T_2 \oplus \Delta T_2) \bowtie \Delta T_3$, 6 joins computed in total. The main cost is that they need to access the base table 6 times through IO operations. If we can simplify the expression, e.g.., the join between T_1 and T_2, then we only need to compute $\Delta T_1 \bowtie \Delta T_2 \bowtie T_3$ and $(T_1 \oplus \Delta T_1) \bowtie (T_2 \oplus \Delta T_2) \bowtie \Delta T_3$. As $\Delta T_1 \bowtie \Delta T_2$ just use the increments caused by the transaction in memory, it contains 3 joins to access base tables. The maintenance cost is almost halved after the simplification.

A Sufficient Condition. The key insight is to use the information provided by the transaction to determine whether we can meet the requirements $\Delta T_k \bowtie T_{k+1} = \Delta T_{k+1} \bowtie (T_k \oplus \Delta T_k) = \Delta T_k \bowtie \Delta T_{k+1}$. For $\Delta T_{k+1} \bowtie (T_k \oplus \Delta T_k)$, a sufficient condition to simplify it to $\Delta T_{k+1} \bowtie \Delta T_k$ is that all records of T_k participating in join with ΔT_{k+1} are already contained in ΔT_k, where ΔT_k is the increment by the transaction to base table T_k. Formally represented with relational algebra, suppose the join key (column) between T_k and T_{k+1} is denoted by JK, then we have

$$\sigma_{JK \text{ IN } \pi_{JK}\Delta T_{k+1}} T_k \subseteq \Delta T_k \Rightarrow \Delta T_{k+1} \bowtie (T_k \oplus \Delta T_k) = \Delta T_{k+1} \bowtie \Delta T_k.$$

$\sigma_{JK \text{ IN } \pi_{JK}\Delta T_{k+1}} T_k$ selects records from T_k which satisfies the predicate (i.e., "JK IN $\pi_{JK}\Delta T_{k+1}$") over the join column. Similarly, $\Delta T_{k+1} \bowtie (T_k \oplus \Delta T_k)$ can be simplified to $\Delta T_{k+1} \bowtie \Delta T_k$, when $\sigma_{JK \text{ IN } \pi_{JK}\Delta T_{k+1}} T_k \subseteq \Delta T_k$. When both of them are satisfied, then $\Delta S^k = \Delta S^{k+1}$.

Table 1. Expression optimization conditions

Case 1	UPDATE T_k SET... = ... WHERE [JK =/BETWEEN/IN *;]	/	SELECT... FROM T_k WHERE [JK =/BETWEEN/IN *]; UPDATE T_k SET... = ...
Case 2	DELETE FROM T_k WHERE [JK =/BETWEEN/IN *;]	/	SELECT... FROM T_k WHERE [JK =/BETWEEN/IN *]; DELETE FROM T_k ...;
Case 3	INSERT INTO T_{k+1} JK,... VALUES(*,...)	&	INSERT INTO T_k ...,JK,... VALUES(..., *,...)

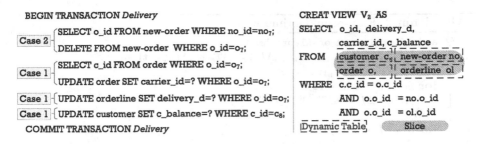

Fig. 2. A running example of view & *Delivery* workload

Condition Detection. Next, we use $\sigma_{JK \text{ IN } \pi_{JK} \Delta T_{k+1}} T_k$ to describe how to quickly determine whether this condition is met. Despite that we can direct obtain ΔT_k when executing the transaction, the straightforward way is still non-trivial because it has to access the base table T_k to select the records which satisfy the predict. Thanks to the information embedded in the transaction, it offers the potential to reach conclusion without accessing any base table. For an arbitrary table T_k, we make inspection based on Table 1.

We make a deduction based on the logic of the transaction. As listed in Table 1, there are three situations. For cases 1 and 2, the records in ΔT_k can be obtained through an inline WHERE condition in the UPDATE/DELETE/SELECT-INTO statement, or through the previous SELECT statement within the same transaction. Therefore, we can check whether the select predicates over T_k (i.e., "JK IN $\pi_{JK} \Delta T_{k+1}$") are included in this WHERE condition (predicates contained in brackets [ ] in Table 1).

Example 2. For example, the optimizable condition is satisfied if there is one predicate on JK in the WHERE condition. As shown in Fig. 2, we give the *Delivery* transaction of CH-benCHmark. According to our definition of the slice, the four tables on view V_2 belong to the same slice. For the join ($order \bowtie_{o\_id=ol.o\_id} orderline$), the transaction has a DML "UPDATE FROM $orderline$ WHERE o_id (JK) $= o_7$", all records satisfying $o\_id = o_7$ in $orderline$ are recorded in $\Delta orderline$ (case 1 of Table 1), i.e., $\sigma_{ol.o\_id \text{ IN } \pi_{o\_id} \Delta order} orderline \subseteq \Delta orderline$. So it satisfies the constraint.

Generally, database schema form the join relationships via the primary-foreign relation, which means join key is also the primary key or foreign key of the base tables. Based on the entity integrity and referential integrity constraints [17], we validate the condition as case 3 illustrated in Table 1. For ease of presentation, we denote that T_k has the foreign key which references the primary key of T_{k+1}. For insert statements, if a transaction inserts a record into T_{k+1}, the related records with corresponding JK (JK is also the primary key of T_{k+1}) in T_k will never exist before (entity integrity). Thus the to-be-inserted records of T_k must be contained in the transaction, the condition can be detected, i.e., $\sigma_{JK \text{ IN } \pi_{JK} \Delta T_{k+1}} T_k = \emptyset \Rightarrow \sigma_{JK \text{ IN } \pi_{JK} \Delta T_{k+1}} T_k \subseteq \Delta T_k$.

Algorithm 2: Expression Optimization

```
    // add after line 2 of Algorithm 1
```
1 if $\Delta T_k \bowtie T_{k+1} = \Delta T_{k+1} \bowtie (T_k \oplus \Delta T_k) = \Delta T_k \bowtie \Delta T_{k+1}$. **then**
2 | $\Delta T \leftarrow \Delta T_k \bowtie \Delta T_{k+1}$;
3 | **for** $m \leftarrow k - 1$ *to* i **do**
4 | | $\Delta T \leftarrow \Delta T \bowtie (\Delta T_m \oplus T_m)$;
5 | **end**
6 | **for** $m \leftarrow k + 2$ *to* j **do**
7 | | $\Delta T \leftarrow \Delta T \bowtie T_m$;
8 | **end**
9 | $k = k + 1$;
10 | $\Delta S \leftarrow \Delta S \oplus \Delta T$;
11 **end**
12 **return** ΔS ;

The algorithm to leverage expression optimization to speed up computing ΔS is described in Algorithm 2. For each join, we can detect whether the optimization condition satisfies (line 1). If ΔT_k and ΔT_{k+1} satisfy the condition, then the computation incurred by ΔT_{k+1} can be omitted (line 2).

5 Table Access Optimization

According to Algorithm 1, for each increment of dynamic table ΔT_k, the most cost consuming computation is $\Delta T \bowtie (\Delta T_m \oplus T_m)$ and $\Delta T \bowtie T_m$ (line 5 and 8), where ΔT starts with ΔT_k and becomes the intermediate increments for the next join. Each join computation requires an access to the base table T_m on the disk with expensive IO cost. This table access can be avoided in two situations: (i) There are no records in the disk that can be produced the join result, which has been studied in existing work [12]. (ii) The records to be accessed have already been acquired in previous operations. Thus we can directly join the obtained records instead of the disk access. This is the main optimization in this Section.

First, the optimization conditions detected in Sect. 4 confirm to this optimization. Recall that if we observe $\sigma_{JK \text{ IN } \pi_{JK} \Delta T_{k+1}} T_k \subseteq \Delta T_k$, we can compute $\Delta T_{k+1} \bowtie (T_k \oplus \Delta T_k)$ as $\Delta T_{k+1} \bowtie \Delta T_k$, which eliminates the table access to T_k. This condition can be applied to any $\Delta T \bowtie (\Delta T_m \oplus T_m)$ and $\Delta T \bowtie T_m$, when $\sigma_{JK \text{ IN } \pi_{JK} \Delta T} T_m \subseteq \Delta T_m$ satisfies.

Next we extend the above condition. During the execution of the transaction, besides ΔT_m, other records serving the transaction logic can also be obtained as runtime information. For T_m, we denote all records obtained during transaction execution as *runtime table* RT_m. We can leave out accessing base table T_m when all its records involved in the join with ΔT have been obtained in the runtime information, i.e., $\sigma_{JK \text{ IN } \pi_{JK} \Delta T} T_m \subseteq RT_m$. Thus we can simplify $\Delta T \bowtie (\Delta T_m \oplus T_m)$ or $\Delta T \bowtie T_m$ to $\Delta T \bowtie RT_m$.

Next we analyze how to detect scenarios that can be optimized. Since ΔT_m must be a subset of RT_m, the optimization conditions in Table 1 are still available. A transaction may contain more runtime information than ΔT_m, thus we extend Table 1 with the case 4 illustrated in Table 2. The records in ΔT_k can be obtained through the WHERE condition in the SELECT statement. Thus we can check whether the predicates over T_k (i.e., "JK IN $\pi_{JK}\Delta T_{k+1}$") are included in this WHERE condition.

Table 2. Table access optimization condition

Case 4	SELECT... FROM T_k WHERE [JK =/BETWEEN/IN $*$];

Example 3. As shown in Fig. 1, the transaction needs to access *customer* to obtain *c_last* with the predicate $c\_id = c_3$ (JK), which does not affect the view content but can be used as runtime information to assist the view calculation.

To leverage the table access optimization, we can avoid the table access cost (lines 5 and 8 of Algorithm 1 and lines 4 and 7 of Algorithm 2).

6 Experiment

6.1 Experiment Setup

We implement our method by extending the PostgreSQL database [1]. All experiments are conducted on a Linux server which has two Intel(R) Xeon(R) E5-2630 v4@2.2 GHZ processors (each with 10 physical cores), 160 GB main memory and an HP Smart Array P420i RAID controller with a battery-backed cache.

Benchmark. We use the CH-benCHmark [5,14] with scaling factor $w = 10,000$. The OLTP-Bench [7] tool is leveraged to carry out the hybrid workloads. We load all twelve tables of CH-benCHmark.

Due to that the performance of views varies from slices, dynamic tables and base tables with the optimization of the slice increment calculation, we select four analytical queries (SQL-3, SQL-5, SQL-12, SQL-18) to construct views (V_3, V_5, V_{12}, and V_{18} respectively), which inspires us to focus on the performance of our optimization strategy in various settings. We utilize all five transactional workloads of CH-benCHmark with default proportions to modify those tables.

Comparing Methods. We realize synchronous IVM and asynchronous IVM. For synchronous IVM (Sync), we refer to [9] to implement a synchronous IVM mechanism. We execute the created stored procedures to update the view directly. These stored procedures are bound to the execution of the transaction. For asynchronous IVM, we use the latest ID-based strategy (IDs) through procedures and delta tables, which are utilized to compute and store the delta records based on the to-be-modified tuples with their IDs [10]. Our slice-based method (S-MV) is integrated into the IDs. When the transaction contains the modifications of several tables, S-MV is used to speed up the calculation of the increments of the entire slice. While the transaction only includes the updates of a single table, S-MV maintains the increments of each base table individually, which degenerates to IDs due to the failure to meet the optimization conditions.

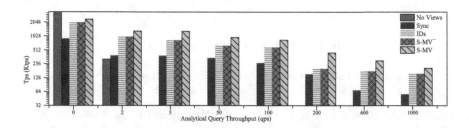

Fig. 3. Impact of view maintenance on transaction processing

6.2 Effect on Transaction Processing

We first measure the impact of view maintenance on transaction processing for different methods. When verifying the performance without optimizations, in order not to affect the transaction throughput which is limited by the conflicts between transactions, S-MV⁻ forces each dynamic table to correspond to only one slice. Therefore neither expression optimization nor table access optimization can take effect. We use four views (V_3, V_5, V_{12}, and V_{18}) with the same frequency to generate the analytical load, under which we evaluate the maximum transaction throughput. We plot the trends under different throughput of queries in Fig. 3.

At the Qps of 0, we measure the impact on transaction processing without the impact of queries. The method that does not maintain views (NO Views) achieves 3, 212 Tps, which is the maximum throughput that the PostgreSQL can achieve for CH-benCHmark workload under current hardware and its isolation level (read committed). As the throughput of queries rises, the transaction throughput declines. Among the other four methods, we can observe the following trends:

(1) Sync always has the lowest performance since the entire view maintenance cost is incurred during transaction execution.
(2) The Tps of IDs is higher than that of Sync. This is because IDs only needs to record the modifications of each base table during the transaction execution phase with a relatively small amount of computation. The rest of the heavy work is done periodically or when querying views.
(3) Without optimization, each slice of S-MV⁻ only records the increments of one single dynamic table as IDs does, which leads to the similar performance. It also indicates that the performance of S-MV⁻ tends to an ID-based asynchronous maintenance method (IDs) in the worst case.

(4) S-MV always has the highest performance. There are two factors: (i) Thanks to the optimizations of expression and table access, the experiment meets the optimization conditions, so S-MV consumes the lowest cost compared with other methods during transaction execution. (ii) S-MV compacts the updates of several base tables (as a slice) into one joined results (ΔS) without extra table access which can save IO cost. While IDs needs to store the updates to individual delta tables. In comparison, when the qps is 0, the performance of IDs drops by 38.5% (2,006 Tps), while S-MV has the least impact on throughput with only 27.6% decline (2,358 Tps).

6.3 Effect of Transaction Processing on View Query

Our next experiment studies the impact of transactions on view queries under different maintenance strategies.

Both V_5 and V_{12} contain only one slice, the difference is that V_5 also includes five additional base tables joined to the slice. Thus V_{12} can be used to reflect the maintenance performance of the slice. In Fig. 4(a)–(d), we fix the load of the analytical query in 10/100 Qps and increase the Tps.

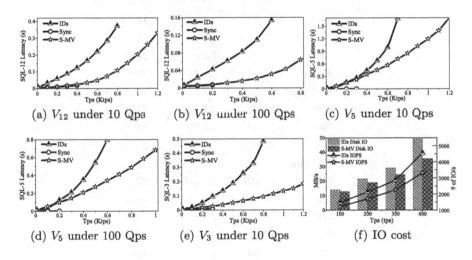

(a) V_{12} under 10 Qps   (b) V_{12} under 100 Qps   (c) V_5 under 10 Qps

(d) V_5 under 100 Qps   (e) V_3 under 10 Qps   (f) IO cost

Fig. 4. Analytical query latency

First, compared with IDs, S-MV shows advantages in query latency, which proves that our two optimizations are effective. Since IDs has to join all the increments stored in each individual delta table. While S-MV only needs a few calculations to complete the join between the computed slice and the rest of the base tables. Second, as shown in Fig. 4(a), at 600 Tps, the latency of S-MV (0.048s) is 78.3% lower than that of IDs (0.221s). While at 600 Tps shown in Fig. 4(c), the latency of S-MV (0.57s) is 42.7% lower than that of IDs (0.995s). Compare with V_5, the latency gap between S-MV and IDs of V_{12} continues to grow with the increment of Tps. Because V_{12} is built on ORDER and ORDERLINE, both of which are maintained in one slice, and their join result has been maintained every time the transaction is completed. When querying the view V_{12}, S-MV only requires to merge the slice into the view table, but V_5 needs to join the slice with the other five base tables. Third, as the transaction load continues to rise, the superiority of S-MV becomes prominent. By comparing Fig. 4(b) and 4(a), (d) and (c), we can find that the higher Qps, the more obvious advantage. Because when resources are limited, S-MV can play a better role in reducing computing and IO cost.

The view V_3 on SQL-3 has two slices: S_1 consists of tables ORDER and ORDER-LINE, and S_2 consists of table NEW-ORDER. Therefore, V_3 can be used to reflect the maintenance performance of the multiple slices. Corresponding to Fig. 4(e), we fix the analytic queries at 10 Qps and adjust transactions throughput to measure I/O cost under S-MV and IDs as shown in Fig. 4(f).

Table 3. Resource overhead

| Workload | | Method | Resource overhead | | |
Qps	Tps		CPU (%)	IO (MB/s)	# of IOPS
0	800	NO Views	**11.26**	**17.92**	**1604**
		S-MV	15.29	25.42	2183
		IDs	22.93	33.62	2870
		Sync	68.13	38.91	3289
2	300	NO Views	51.65	23.27	2027
		S-MV	**7.91**	**22.29**	**2136**
		IDs	12.52	25.45	2392
		Sync	24.05	27.74	2611
50	300	NO Views	–	–	–
		S-MV	**11.65**	**23.56**	**2243**
		IDs	17.62	28.99	2671
		Sync	29.24	29.03	2882
200	200	S-MV	**17.52**	**49.06**	**4522**
		IDs	28.05	91.3	8306
		Sync	–	–	–

As the Tps rises, the increments that need to be recorded in the transaction processing are increased. Meanwhile, the increments that need to be appended to the view are increased for each query interval. First, the gap of IOPS and the total amount of disk I/O between S-MV and IDs becomes larger. At 100 Tps, the total disk I/O of S-MV (12.9 MB/s) is 91.5% of IDs (14.1 MB/s), and the IOPS of S-MV (1, 215) is 82.1% of IDs (1, 480). At 600 Tps, the total disk I/O of S-MV (35.4 MB/s) is 72.1% of IDs (49.1 MB/s), and the IOPS of

S-MV $(3,320 \text{ IO/s})$ is 72.9% of IDs $(4,557 \text{ IO/s})$. Within the execution of the transaction, S-MV can directly capture ΔS_2 (ΔNEW-ORDER), and get ΔS_1 by calculating ΔORDER $\bowtie$ ΔORDERLINE. Thus S-MV only needs to insert increments into two slices (ΔS_1 and ΔS_2). However, IDs needs to insert increments into three delta tables (ΔORDER, ΔORDERLINE, ΔNEW-ORDER), which means more index access and IO cost. Note that since the number of rows of joined ΔORDER and ΔORDERLINE is equal to that of ΔORDERLINE, storing the joined rows (ΔS_1) directly will reduce the IO cost.

Meanwhile, S-MV can take advantage of table access optimization, which can significantly reduce the IO and computation overhead. In general, the more incremental records processed during the same time interval, the more obvious the advantages of S-MV.

6.4 View Maintenance Cost

From phenomenon to the essence, we choose several loads to measure resource overhead. In Table 3, we list CPU and I/O consumptions under different loads. It can be found that: (1) The overhead of S-MV is always lower than that of IDs. Because S-MV reduces overall resource consumption through the two optimizations. (2) At 0 Qps and 800 Tps, resources are used for transaction processing and view maintenance (except NO Views). CPU and I/O consumption are positively correlated, so under the same hardware resources, the maximum transaction throughput achieved by various methods is inversely proportional to resource consumption, which is also confirmed when Qps is 0 in Fig. 3. This conclusion also applies to other records in Table 3. (3) Without the help of views, it demands to access all the base tables and perform more join calculations. When Qps is 2 and Tps is 300, the CPU utilization rate of NO Views reached 51.65%, which is the highest compared with other methods. Analytical queries under NO Views consume a lot of system resources. So the Tps of NO Views is the lowest when Qps at 2 compared with Fig. 3. (4) In addition, because Sync needs to perform more inefficient view maintenance synchronously in transaction execution, it requires longer locking time and higher conflict. As a result, Sync's IOPS differs slightly from other methods, but its CPU consumption varies greatly.

6.5 Overall Performance

We pay attention to both query throughput and latency to measure the overall performance. In Fig. 5(a), we test the peak query throughput of four selected queries under varying Tps. In Fig. 5(b), we fix Qps to 10 to measure the latency of the four queries under varying Tps. From the two figures, we can observe:

First, as shown in Fig. 5(a), without views, the query performance is terrible. Since large amounts of system resources are required to answer analytical queries, as the transactional load increases, resource competition becomes more intense, which leading to a rapid decline in analytical throughput.

Second, Sync has the highest analytical throughput when Tps is below 20. Since Sync can get the query results directly without any additional calculations.

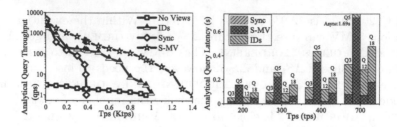

Fig. 5. Overall performance comparison

However, the calculation of executing transactions requires a lot of resources, causing the analytical throughput to drop rapidly as the throughput of transactions increases. Therefore, in Fig. 5(b), `Sync` has the lowest latency when Tps is below 300, but `Sync`'s Tps can hardly reach 400.

Next, compared with `Sync`, `IDs` has better performance when the transaction throughput increases. In the execution of transactions, lower maintenance cost has less impact on transaction processing, which enables `IDs` to achieve a higher query throughput. When recording the delta records, the ID-based strategy avoids redundancy and saves IO and storage overhead. Meanwhile, `IDs` asynchronously calculates cumulative increments in the query interval, which makes the batch processing possible.

Last, based on the `IDs` strategy, S-MV has the best performance by treating all runtime information in a transaction as a whole with two optimizations. In Fig. 5(a), S-MV has the highest query throughput as the transaction throughput increases. When Tps is $0.6k$, S-MV can reach 181 Qps, while `IDs` only reaches 62 Qps. S-MV can not only take advantage of `IDs` but also benefit from the optimization of expression and table access, which achieves the lowest resource consumption. This benefit becomes more evident as the load increases. As shown in Fig. 5(b), compared to `IDs`, the proportion of S-MV's latency is getting smaller and smaller, which means that the advantage of S-MV is expanded.

7   Conclusion

In this paper, we propose a slice-based method to speed up join view maintenance for transactional workloads. Tables co-updated by the transaction are divided into the slices, which are treated as the basic units for maintenance. Efficient calculation methods and optimizations are developed based on slices, which reduce the overall IO and computation cost of view maintenance. Compared with conventional methods, our slice-based method achieves better performances in terms of both transaction throughput and query latency.

Acknowledgements. This is work is partially supported by National Key R&D Program of China (2018YFB1003303), National Science Foundation of China under grant number 61772202, Youth Program of National Science Foundation of China under grant number 61702189.

References

1. Postgresql. https://www.postgresql.org/
2. Agrawal, P., Silberstein, A., Cooper, B.F., Srivastava, U., Ramakrishnan, R.: Asynchronous view maintenance for VLSD databases. In: SIGMOD, pp. 179–192 (2009)
3. Chavan, A., Deshpande, A.: DEX: query execution in a delta-based storage system. In: SIGMOD, pp. 171–186 (2017)
4. Chirkova, R., Yang, J.: Materialized views. Found. Trends Databases 4(4), 295–405 (2012)
5. Cole, R.L., Funke, F., et al.: The mixed workload CH-benCHmark. In: DBTest, p. 8 (2011)
6. DeHaan, D., Larson, P., Zhou, J.: Stacked indexed views in Microsoft SQL server. In: SIGMOD, pp. 179–190 (2005)
7. Difallah, D.E., Pavlo, A., et al.: OLTP-Bench: an extensible testbed for benchmarking relational databases. PVLDB 7(4), 277–288 (2013)
8. Gupta, A., Jagadish, H.V., Singh Mumick, I.: Data integration using self-maintainable views. In: Apers, P., Bouzeghoub, M., Gardarin, G. (eds.) EDBT 1996. LNCS, vol. 1057, pp. 140–144. Springer, Heidelberg (1996). https://doi.org/10.1007/BFb0014149
9. Gupta, A., Mumick, I.S., Subrahmanian, V.S.: Maintaining views incrementally. In: SIGMOD, pp. 157–166 (1993)
10. Katsis, Y., Ong, K.W., Papakonstantinou, Y., Zhao, K.K.: Utilizing IDs to accelerate incremental view maintenance. In: SIGMOD, pp. 1985–2000 (2015)
11. Koch, C., et al.: DBToaster: higher-order delta processing for dynamic, frequently fresh views. VLDB J. 1–26 (2014). https://doi.org/10.1007/s00778-013-0348-4
12. Larson, P., Zhou, J.: Efficient maintenance of materialized outer-join views. In: ICDE, pp. 56–65 (2007)
13. Luo, G., Yu, P.S.: Content-based filtering for efficient online materialized view maintenance. In: CIKM, pp. 163–172 (2008)
14. Psaroudakis, I., et al.: Scaling up mixed workloads: a battle of data freshness, flexibility, and scheduling. In: Nambiar, R., Poess, M. (eds.) TPCTC 2014. LNCS, vol. 8904, pp. 97–112. Springer, Cham (2015). https://doi.org/10.1007/978-3-319-15350-6_7
15. Qian, X., Wiederhold, G.: Incremental recomputation of active relational expressions. IEEE Trans. Knowl. Data Eng. 3(3), 337–341 (1991)
16. Quass, D., Widom, J.: On-line warehouse view maintenance. In: SIGMOD, pp. 393–404 (1997)
17. Türker, C., Gertz, M.: Semantic integrity support in SQL: 1999 and commercial (object-) relational database management systems. VLDBJ 10(4), 241–269 (2001). https://doi.org/10.1007/s007780100050
18. Xu, M., Ezeife, C.I.: Maintaining horizontally partitioned warehouse views. In: Data Warehousing and Knowledge Discovery, pp. 126–133 (2000)
19. Yi, K., Yu, H., Yang, J., Xia, G., Chen, Y.: Efficient maintenance of materialized top-k views. In: ICDE, pp. 189–200 (2003)
20. Zhou, J., Larson, P., Elmongui, H.G.: Lazy maintenance of materialized views. In: VLDB, pp. 231–242 (2007)

A Chunk-Based Hash Table Caching Method for In-Memory Hash Joins

Xing Wei, Huiqi Hu$^{(\boxtimes)}$, Xuan Zhou, and Aoying Zhou

East China Normal University, Shanghai, China
simba_wei@stu.ecnu.edu.cn, {hqhu,xzhou,ayzhou}@dase.ecnu.edu.cn

Abstract. In-memory query processing can be accelerated by caching intermediate query results. Among various types of intermediate results, hash tables used by hash join are ideal objects for caching, as they can benefit a wide range of queries. In this paper, we introduce a fine-grained hash table caching method to benefit the hash-join operator. Our insight is that the fine-grained management of cached hash tables at the granularity of chunks can achieve optimal caching efficiency. As hash chunks can be reused more effectively, we further propose a cache-enabled hash join operator to adapt the cache under chunk granularity and accelerate in-memory hash join execution. Furthermore, we also consider concurrent accesses to cached hash chunks and present the scheduling strategy to improve throughput and consider response time simultaneously. To the best of our knowledge, our work is the first one that studies the delicate management of intermediate result caching for the concurrent workload. We integrated our caching method into a prototype to evaluate its performance. Experiments show that it can achieve significant performance improvement over brute force caching methods.

Keywords: Cache management · Concurrent sharing · Hash join

1 Introduction

Caching is a classical approach to speedup traditional database operations. It is based on the basic intuition of trading space for time [4,7]. For in-memory databases, caching is still helpful [4]. In particular, existing works [2,13] have shown that caching internal data structures of query operators can accelerate in-memory query processing remarkably. However, there are a large number of operators involved with various data structures in a database system. Each operator can be confronted with complex predicates. Moreover, a database commonly needs to deal with concurrent workloads, which may compete for caches. They all make the management of an intermediate-result cache become a challenge.

Choosing the right internal data structures for caching is a crucial issue. Intuitively, the data structure must play a decisive role during an operator's execution. In this paper, we focus on the hash table cache. Meanwhile, we also believe that the techniques also can be used for other operators. The granularity

© Springer Nature Switzerland AG 2020
Z. Huang et al. (Eds.): WISE 2020, LNCS 12343, pp. 376–389, 2020.
https://doi.org/10.1007/978-3-030-62008-0_26

of cached contents is another critical issue. Database queries are usually faced with complex predicates, such that the hash table built by one join operation could not be necessarily reusable for another join operation, due to the difference in their predicates. To achieve good cache utilization, we need to perform caching at a fine granularity–while a complete hash table is unlikely to be reused, after broken into small pieces, the pieces can be more easily reused. Meanwhile, fine-grained caching method is beneficial to concurrent query processing. Concurrent workload could lead to intensive contention on the cached data structures [10]. If such contention is not dealt with effectively, cache may become a performance bottleneck rather than an accelerator.

Existing approaches for intermediate-result caching do not offer satisfactory solutions to either the granularity issue or the concurrency issue. The most recent work [2] on hash table caching employ a coarse-grained method for cache updating–to add uncovered data into a cached hash table, it replaces the hash table completely with an updated one. This results in the lower cache utilization in the coarse granularity and the deficiencies in handling the concurrent workload. In this paper, we tackle the granularity and concurrency issues. As mentioned earlier, we focus on the hash-join operator. Through the design of a fine-grained hash table cache, we illustrate how to design a delicate caching method for concurrent in-memory hash-join execution.

Our caching method regards a hash table as a union of *hash chunks*. Each chunk is a sub-hash-table of a certain key range. The granularity of hash chunks allows us to achieve high cache utilization, good memory efficiency and efficiently concurrent scheduling. At the same time, it avoids unnecessary overheads incurred by overly high granularity. To utilize the cache, we extend the conventional hash join operator to the cache-enabled hash join operator (*CHJO*), which involves additional cache lookup and update operations in the execution path of hash join. We implemented our cache in a prototype called *Simba*. Extensive experiments were conducted to evaluate their practicality. In summary, we made the following contributions in this paper:

1. We propose a chunk-based caching method for hash-join operator. To promote cache utilization, the method manages cached hash tables at the granularity of chunk. We show that the cache not only accelerates the building phase but also benefits the probing phase of a hash-join operation. (Sect. 3)
2. For the first time, we consider concurrent cache accesses to cached hash tables. We propose the cache scheduling strategy to reduce the contention of reusing cached hash tables and also consider the response time of each hash-join operator under the concurrent workload. (Sect. 4)
3. We validate our chunk-based caching method through an extensive experimental study. Results show that it achieves a significant performance improvement (about 60%) under the concurrent workload. (Sect. 5)

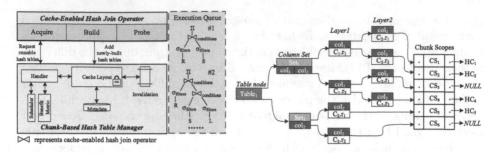

Fig. 1. The Chunk-based method   **Fig. 2.** The division tree structure

2   Related Work

Recycling Intermediates. Recycling aims to make database query engines profit from reusing partial previous query intermediates, which is able to continuously adapt to the changes in the workloads [13]. Prior recycling works [3,6,13], such as Recycler Graph [13], are devoted to answer what intermediates from the current query should be considered for further queries. Later work [2] further explore which type of intermediates could be more worthy of being recycled. Internal data structures (e.g., hash tables) have been proved to be a type of lighter intermediates for in-memory environment [2]. But no matter for predicating recyclable intermediates or choosing the lighter internal data structures, these works are lacked of a fine-grained management for the recycled intermediates. Considering that a query is usually involved with several predicates, upcoming queries could only reuse a small part of intermediates from previous queries. Meanwhile, the overlapping among complex predicates can cause recycled intermediates overlap each other so that the memory space is squandered.

Concurrent Sharing. Another area of related work is to study how to exploit sharing opportunities of cached intermediates for concurrent multiple queries. For multi-query optimization (MQO) [5], all concurrent queries in a batch are assembled into a globally optimal execution plan in order to share each other's intermediates during the execution of a gloabl plan [5]. Identifying sharing possibilities of mutiple queries within a batch and exploiting them by reusing other's intermediates is a main part of optimizations. In practice, the MQO strategy is only adaptive for offline analytical queries, which will sacrifice the response time to exploit the sharing opportunities. Similar to the MQO, the work-sharing studies [9,11,12] also process concurrently multiple queries at a time to share the overlapped work among multiple queries. No matter for MQO or work-sharing strategy, batching is an indispensable preprocessing which makes query response time become the Achilles' heel of these two strategies.

3   Chunk-Based Method

In this section, we introduce the details of our chunk-based caching method.

3.1   Overview

The core of our caching method is depicted in Fig. 1. Its design aims to reuse fine-grained hash tables (chunks) to accelerate hash-join execution. To utilize cached hash tables, a new hash-join operator named *cache-enabled hash join operator* (CHJO) is introduced, to replace the conventional hash-join operator.

CHJO: The cache-enabled hash join operator (CHJO) is derived from the conventional hash join operator. A typical hash join execution includes two phases called *building* and *probing* phases. The building phase firstly retrieves tuples from the building relation to construct a hash table, then the probing phase takes tuples from the probing relation to seek matched tuples in the hash table. While the CHJO augments hash join by adding an *acquiring* phase to find properly reusable hash chunks before the building and probing phases.

CHTM: The *chunk-based hash table manager* (CHTM) provides hash chunks for the CHJO and manages those cached hash chunks. It consists of five main components. (1) The *cache layout* manages reusable hash chunks in the cache and indexes chunk accesses. (2) The *handler* provides interfaces (*look-up* and *insert*) for CHJO to interact with the cache. (3) The *validation* is responsible for invalidating hash chunks that have been updated and evicting cold hash chunks, according to a given elimination policy. (4) The *benefit metric* is used to evaluate the benefit and cost of reusing hash chunks. (5) The *scheduler* provides a scheduling algorithm among concurrent CHJOs.

3.2   Chunk-Based Hash Table Manager

We first define the concepts of hash chunks and then introduce how to efficiently manage hash chunks in CHTM.

Hash Chunk. Let T be a table with n columns, we can pick up any m columns to construct a *dimensional column set*, which is denoted as $D = \{C_1, \ldots, C_m\}$ $(0 < m \le n)$. For each column, we can divide its domain into disjointed sub-ranges with uniform size. Suppose $C_i.R$ is the range of the column C_i. Then, $C_i.R$ can be divided into several sub-ranges $\{C_i.r_1, \ldots C_i.r_j, \ldots\}$, where $C_i.r_j$ is the j-th sub-range of $C_i.R$. If we select a sub-range from each column in the dimensional column set, then we can obtain a chunk scope like $CS_{1,2,\ldots,m} = \{C_1.r_1 \land C_2.r_2 \land \cdots \land C_m.r_m\}$. Suppose $Num(C_i)$ represents the sub-ranges number of column (C_i), the scope of table T can be divided into $\prod_{i=1}^{m} Num(C_i)$ chunk scopes. For a specific hash join, the columns associated with scan and join predicates will be considered as dimensional columns and construct hash scopes. Meanwhile, the hash table within a chunk scope will be called as hash chunk.

Cache Layout. The hash chunks in CHTM are organized in a tree structure (i.e., division tree) which is built according to the division of chunk scope. As is shown in Fig. 2, a table scope can be divided according to different groups of dimensional columns. E.g., the column set $Set_1(col_1, col_2)$ with two layers means that the table scope division is based two dimensional columns col_1 and col_2. Subsequently, each layer further makes the range of its dimensional column divided into several sub-ranges. To pick up a sub-range from first layer to the last layer, these sub-ranges can assemble a chunk scope.

Handler. The CHTM also provides the handler as an interface for the CHJO. It supports *lookup* operation for processing the requested scope from CHJO and *insert* operation for adding the newly-built hash chunks into the CHTM.
(1) Lookup: A lookup operation needs to choose a candidate set of cached hash chunks that can be totally or partially reused for current join execution. Based on the division tree structure, lookup operation firstly locate a target dimensional column set whose columns exactly match the dimensional columns of request scope. Then, lookup operation efficiently traverses division tree to select all proper chunk scopes (intersected with the request scope) as candidates.
(2) Insert: The insert operation is responsible for adding newly-built hash chunks (from CHJO) into the CHTM. The newly-built hash chunks are built according to the constraints of their target chunk scopes. Subsequently, the insert operation also needs to link the newly-built hash chunks to their traget chunk scopes.

Benefit Metric. This component provides a metric for evaluating the benefit of a candidate hash chunk. During the acquiring phase of hash join, the CHJO picks out the proper reusable hash chunks from the candidate set (by using above lookup operation) based on the benefit metric.

Invalidation. Based on the statistics on metadata, the organization of hash chunks is integrated with a MRU policy, which periodically picks out the hash chunks and evicts them from CHTM. When CHTM has no available memory for upcoming hash chunks, the evict process will be triggered.

3.3   Cache-Enabled Hash Join Operator

The cache-enabled hash join operator (CHJO) is designed to efficiently utilize the cached hash chunks. The procedure of CHJO (shown in Algorithm 1) is consist of three phases: acquiring phase, building phase and probing phase.

Acquiring Phase. As the first step of CHJO, the acquiring phase assembles the columns in hash table's predicates and their conditions (from join and scan) into a request scope and sends the request scope to the CHTM. By leveraging the lookup operation in the handler, it can obtain a set of candidate hash chunks (line 1 in Algorithm 1) whose dimensional columns exactly match the request

---

Algorithm 1: Cache-Enabled Hash Join

Input: Requested Scope $\mathcal{R}$ From Predicates;
```
// acquiring phase
```
1 $C \leftarrow$ handler.$Lookup(\mathcal{R})$; `// C is the candidate set`
2 **foreach** $hc \in C$ and hc is partial covered **do**
3   **if** $benefit(hc) < 0$ **then**
4   remove hc from C;

```
// building phase
```
5 $\Phi \leftarrow$ find uncovered scopes by comparing C and $\mathcal{R}$;
6 **foreach** tuple $t \in \Phi$ **do**
7   **if** $t \in$ a hollow scope **then**
8   build the hash chunk with t;
9   **else**
10   build the normal hash table with t;
11 add newly-built hash chunks into CHTM;
```
// probing phase
```
12 **foreach** tuple $t \in$ ProbeRelation **do**
13    $hc \leftarrow$ prejudge t;
14   **while** bucket $\leftarrow$ hc.hasNextHashBucket (t) **do**
15   **if** bucket.key $==$ t.key **then**
16   add the matching to result ;

---

scope. Then, the acquiring phase further selects the hash chunks that deserve to be reused based on the benefit metric (lines 2−4). Based on the comparison between request scope and table chunk scopes (shown in Fig. 3), we can identify the selected hash chunks into four types:

(*i*) *hollow chunks.* Its chunk scope falls entirely within the request scope, but its hash chunk is not cached in the cache layout yet (e.g., CS_2 and CS_6). Their hash chunks will be built and cached in the building phase.

(*ii*) *partial hollow chunks.* Its chunk scope partially falls within the request scope, but its chunk is not cached yet (e.g., CS_7). For the partial hollow chunk, we only create a hash table for the intersected part instead of building a whole hash chunk for them, i.e. the built hash table will not be cached. The reason is that the hash chunk is regarded as a basic unit for caching management.

(*iii*) *covered chunks.* Its chunk scope falls entirely within the request scope, and the hash chunk is cached already (e.g., CS_3 and CS_5). Its hash chunks can be directly reused for the join task.

(*iv*) *partial covered chunks.* Its chunk scope partially falls within the request scope, and hash chunk is cached already (e.g., CS_1, CS_4, CS_8 and CS_9). However, whether reusing them has a tradeoff. It is because that the redundant parts can introduce additional costs of filtering out unnecessary tuples during the probing phase. Thus, the benefit and costs should be evaluated by the metric component with a cost model. The specific evaluation is as follows:

Let hc denote a hash chunk, $|hc|$ be the number of tuples within the hash chunk and $|reusable\ part|$ indicate the number of tuples that can be reused for building. Thus, the reusing gains $(Gain(hc))$ saved from reusing partial hash chunk can be conducted based on equation 1, where $Cost_b(tuple)$ represents the CPU cycles of building one tuple. Furthermore, the overhead costs $(Cost(hc))$ are the spent CPU cycles on probing the unnecessary tuples according to equation 2, where $Cost_p(tuple)$ represents the CPU cycles on probing one tuple. All in all, the benefit $(Benefit(hc))$ is shown in equation 3 which compares the gains and costs. If the reusing gains exceed its costs, reusing this hash chunk can benefit the building phase. Otherwise, the partial covered chunks should be evicted from the candidate set (line 4).

$$Gain(hc) = Cost_b(tuple) \times |reusable\ part| \qquad (1)$$
$$Cost(hc) = Cost_p(tuple) \times (|hc| - |reusable\ part|) \qquad (2)$$
$$Benefit(hc) = Gain(hc) - Cost(hc) \qquad (3)$$

Building Phase. When encountering the second step, the candidates with cached hash chunks can be directly reused without any doubt. However, the building phase still needs to figure out the uncovered scopes (line 5) and build hash tables for them. If some of them belong to hollow chunk, these hash tables will be cached as hash chunks (lines 6 to 8 & 11). While belonging to the partial hollow chunks, the hash tables will not be cached as hash chunks (line 9 to 10).

Probing Phase. As the third step, the probing phase can also benefit from fine-grained hash chunks. Naturally, the probing phase can select a matched hash chunk in advance for a tuple from probe relation (line 13). When comparing to a whole hash table, a matched hash chunk can help the tuple avoid massive redundant tuple comparisons (line 14 to 16). It is because a hash chunk could contain less items with a same hash key than an entire hash table.

3.4 Chunk Granularity

Note that we have yet to define an appropriate chunk scope number (i.e., chunk granularity) for a table. The smaller sub-range means more chunk scopes which bring better flexibility for reusing. But, when chunk scope number exceeds TLB entry number, the frequent accesses to chunks could cause massive TLB missing which degrades overall performance. Thus, the chunk scope number for each table should be tuned to equal to TLB entry number.

4 Caching Scheduling

In this section, we describe how to schedule the cached hash chunks for concurrent hash-join tasks over the multi-core environment.

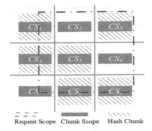

Fig. 3. Table Chunk scopes.

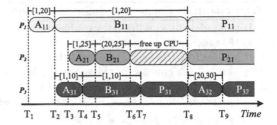

Fig. 4. Concurrent scheduling

4.1   Concurrent Reusing

Concurrency Control. Suppose each processor (thread) runs a hash-join task at a time, then the scheduler should ensure the correct concurrent access to those cached hash chunks. Intuitively, the concurrency control should be straightforwardly guaranteed by read/write locks on hash chunks. When multiple join tasks proceed to build hash chunks for the same hollow chunks simultaneously, all of them have to add write locks on these hollow chunks before entering the building phase. After the building phase, the newly-built hash chunks will be linked to the hollow chunk scopes and the write locks will be switched into the read locks for probing phase. While for covered or partial covered chunks, which are already cached, the join tasks must add read locks on acquired hash chunks. The reusing for partial hollow chunks does not refer to any lock. All the read locks will be released when the probing phase is finished.

Sharing Hash Chunks. Based on above concurrency cntrol policy, the covered chunks and partial covered chunks exhibit excellent sharing potential that their cached hash chunks can be directly reused by multiple join tasks without any conflict. But for a hollow hash chunk that be accessed by two (or more) join tasks simultaneously, there are two execution ways for concurrent join tasks, which are named as *waiting-for-sharing* and *running-instant* respectively. As is show in Fig. 4, there are several successive join tasks. The previous one on $\mathcal{P}_1$ will add write lock on the acquired hollow hash chunk, which avoids the latter simultaneously building hash chunks within $[1, 20]$. Meanwhile, the latter tasks on $\mathcal{P}_2$ and $\mathcal{P}_3$ have two options. *(i)* When taking *waiting-for-sharing*, the latter task within $[1, 25]$ on $\mathcal{P}_2$ will choose to build the hash table within $(20, 25]$ ($B_{21} : T_5 \quad T_6$) which has no intersection with previous task, and then free up CPU resources ($T_6 - T_8$) to wait to reuse hash chunk that being built by $\mathcal{P}_1$. *(ii)* When taking *running-instant*, the latter task within $[1, 10]$ on $\mathcal{P}_3$ will directly ignore the opportunity of sharing the hash chunk which is being built by $\mathcal{P}_1$ and build its owned hash table within $[1, 10]$ immediately ($B_{32} : T_4 - T_7$). For $\mathcal{P}_2$ with *waiting-for-sharing*, it saves the CPU resources of building hash chunks within $[1, 20]$ which could benefit the throughput. While for $\mathcal{P}_3$ with *running-*

Algorithm 2: Cache Scheduling Algorithm

Input: globally conditional variable cv
// Pseudo-code for case as $\mathcal{P}_1$
1 if $unique\_lock\ lck(S_{request}.mutex)$ then
2 | build hash chunks for the requested hollow chunks;
3 | switch the unique lock into shared lock;
4 | cv.notify_all() // wake up blocked threads

  // Pseudo-code for case as $\mathcal{P}_2$
5 if $T_{wait} + T_{unshared} \leq T_{run}$ then
6 | // waiting-for-sharing
     | build hash tables for unshared hollow chunks;
7 | while $!\ shared\_lock\ lck(S_{request}.mutex)$ do
8 | | cv.wait(lck); // block the thread
9 else
    | // running-instant
10 | building own hash tables instantly;
11 execute the probing phase;
12 release the shared lock on chunks;

---

instant, it can quickly finish building phase at T_7 moment which is faster than *waiting-for-sharing* at T_8 moment.

To this end, the scheduler is integrated with a cost model to evaluate these two execution way from the perspective of response time. For two concurrent hash-join tasks running on $\mathcal{P}_1$ and $\mathcal{P}_2$ respectively, we suppose the scheduler find that the requested scopes of $\mathcal{P}_2$ could cover or overlap the being-built scopes of hollow chunks running on $\mathcal{P}_1$. If the cost model estimates the time of $\mathcal{P}_2$ waiting for $\mathcal{P}_1$ exceeds the time of direct building execution, $\mathcal{P}_2$ will choose the *running-instant*. Otherwise, $\mathcal{P}_2$ will choose the *waiting-for-sharing* to reuse the being-built hash chunk by $\mathcal{P}_1$.

4.2 Scheduling Algorithm

Formally, a scheduling algorithm (shown in Algorithm 2) is designed to leverage the released CPU resources to maximize throughput while guaranteeing response time. In Algorithm 2, $\mathcal{P}_1$ and $\mathcal{P}_2$ are running two join tasks respectively which could request overlapped hash chunks. If $\mathcal{P}_1$ occupies the requested hollow chunks, it will first place a exclusive lock (i.e., write lock) on the acquired hollow chunks (line 1), then build the hash chunks (line 2), nextly switch the exclusive lock into a shared lock (i.e., read lock, line 3), finally notify other threads that these hash chunk have been built (line 4). Suppose $\mathcal{P}_2$ occupies the requested hollow chunks behinds $\mathcal{P}_1$. If $\mathcal{P}_2$ chooses *waiting-for-sharing* based on above cost model (line 5), it will try to add a shared lock on requested chunks (line 7). If the requested chunks are still being built by $\mathcal{P}_1$, $\mathcal{P}_2$ will be blocked and free up CPU resources for tasks (line 8). Once $\mathcal{P}_2$ is awake up by the notification from $\mathcal{P}_1$, it will directly reuse the acquired hash chunks for the following probing phase. If the $\mathcal{P}_2$ chooses to *running-instant*, it will construct its own hash chunks instantly (line 10), and the chunks will not be cached and shared by other tasks (line 11).

5   Experimental Evaluation

5.1   Methodology

Systems. We implement an in-memory prototype supporting the concurrent execution of multiple queries. Under the concurrent scenario, each hash-join task is assigned with a specified thread. Besides the chunk-based caching method (CBCM), we also provide two baseline methods: simple hash join (SHJ) and simple reusing hash join (SRHJ). The SHJ is a conventional hash join which is derived from the widely-used hash join implementation. The SRHJ is a variant of SHJ, which adopts the idea of recycling intermediates [13]. The SRHJ supports reusing cached hash tables in a coarse manner and takes a batching strategy to process multiple queries [2]. We allocate 20 MB cache space for CBCM and SRHJ. To provide approximate parameters ($Cost_b$ and $Cost_p$) for the cost model under different execution environments, we embedded a small piece of codes to obtain the accurate values in our prototype.

Dateset and Workload. We employ the table schemas and template queries defined in SSB [1] to generate our dataset and workloads for huge tables since it models a realistic scenario. *(i) Dataset:* It contains five tables (totally around 12 GB): *date* (225 KB), *part* (2490 MB), *customer* (2770 MB), *supplier* (3310 MB) and *lineorder* (5670 MB). The lineorder is a fact table and the rest of four tables are all dimensional tables which have at least two dimensional columns. *(ii) Workload:* To generate the hash-join queries, we convert four types of queries into four templates by substituting the range predicate in the query. In default, the selectivity S of each predicate is set as 30%. In the concurrent scenario, four template queries are randomly assigned to each working thread.

Hardware Configuration All experimental results are conducted on a machine with two Intel(R) Xeon(R) Silver 4110 CPU @ 2.10 GHz processors (total 16 physical cores) based on NUMA and 150 GB RAM running CentOS 6.5. The cache sizes of the processors are 32 KB data+32 KB instruction L1 cache, 256 KB L2 cache and 25 MB L3 cache. The number of TLB entries in each physical CPU core is 64 under the small-page mechanism.

5.2   Single-Threaded Scenario

In this section, we firstly analyze the benefits of our chunk-based from different perspectives under the single-threaded scenario.

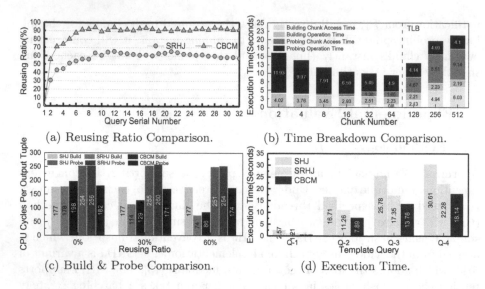

(a) Reusing Ratio Comparison.   (b) Time Breakdown Comparison.

(c) Build & Probe Comparison.   (d) Execution Time.

Fig. 5. Single-threaded scenario.

Resuing Ratio. The reusing ratio for a hash-join task indicates how many percents of the tuples (in the cache) that can be provided from the requested hash chunks. We take the first template query $Q - 1$, to execute 32 times under 30% selectivity. One significant thing to note from Fig. 5(a) is that the CBCM always has a higher reusing ratio than the SRHJ at any rime. The reusing ratio of CBCM is around 90%, along with continuous query execution. While the SRHJ is around 50%. The reason is that the CBCM takes a finer granularity than SRHJ, which can pick up more suitable hash chunks for reusing.

Time Breakdown. Nextly, we will try to take a hash-join task between table part and lineorder to analyze the efficiency of cache-enabled hash join operator under the default 60% reusing ratio. We break the execution time of CBCM into four parts: building chunk access time, building operation time, probing chunk access time and probing operation time. Results are shown in Fig. 5(b). Note that there is an acquiring phase to find candidate hash chunks by invoking the *lookup* operation of handler. However, we observe the time cost is rather small, thus we omit it in the result. We notice that the building/probing chunk access time is highly relevant with the chunk number. When the chunk number is less than TLB enty number (64), we observe that the chunk access only occupies a small part of total execution time. But when chunk number (e.g., 512) exceeds TLB entry number, the chunk access will occupy the majority (70.7%) of execution time due to the frequent TLB missing.

Build and Probe. Subsequently, we take above hash-join task to compare SHJ, SRHJ and CHJO from the perspectives of build and probe under the optimal chunk number (64). In Fig. 5(c), we vary the reusing ratio from 0% to 60% to analyze its impacts on build and probe operations. We observe build operaton (including chunk access) of CBCM and SRHJ can cost less CPU cycles than SHJ under the 30% and 60% reusing ratios. The reason is that CBCM and SRHJ both can reuse the cached hash chunks to accelerate the build operation. We also find that CBCM's build operation costs a little more CPU cycles than SRHJ due to the costs of chunk access. But at the 0% reusing ratio, CBCM costs 198 CPT for build operation, which is little larger than the SHJ (177 CPT) and SRHJ (178 CPT). This is because that CHJO needs little more CPU cycles to build hash chunks than common hash table construction. Another observation is that the probe operation (including chunk access) of CBCM can always work better than SHJ and SRHJ. It is because that CBCM can probe a fine-grained hash chunk so that it can avoid many unnecessary tuple comparisons.

Execution Time. To confirm our above analysis, we try to take SHJ, SRHJ and our CBCM to respectively execute four template queries for 32 times and record each query's average time in Fig. 5(d). We observe that CBCM always has the less execution time than SHJ and SRHJ. E.g., CBCM costs 18.14 s to execute $Q - 4$, which is less than 30.61 s (SHJ) and 22.28 s (SRHJ). Based on the results, we can draw a conclusion that our CBCM is really superior to the SHJ and SRHJ from the perspective of query execution time.

5.3   Concurrent Scenario

To evaluate the concurrent performance of our CBCM, we run a series of experiments under the multi-threaded scenario. The number of concurrent threads should be less than (or equal to) the number of physical cores (16).

Caching Scheduling Strategy. From now on, we have listed four strategies: *running-instant, waiting-for-sharing, batching-processing* and our *caching scheduling*. We implement these four strategies in our CBCM and compare their performance from the perspectives of throughput and response time under various thread number. In Fig. 6(a), we observe that *running-instant* obviously has lower throughput than other three strategies under various thread number. The reason for *running-instant* is due to ignore and wastes the sharing opportunity of being-built hash table. While for the *waiting-for-sharing* and *batching-processing*, they wait for sharing opportunity or batch queries into a global execution to seek better throughput. Since our caching scheduling strategy also considers the waiting time, sometimes it will abandon sharing opportunity to guarantee response time. Thus, the throughput of our caching strategy could fall a little behind *waiting-for-sharing* and *batching-processing*. In Fig. 6(b), our caching scheduling totally outperforms *waiting-for-sharing* and *batching-processing*, achieves similar response time to *running-instant*. It is because that caching scheduling does not trade response time for throughput.

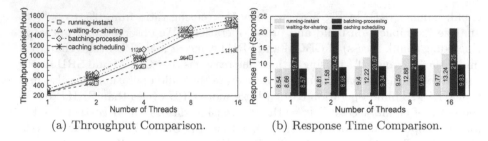

(a) Throughput Comparison.   (b) Response Time Comparison.

Fig. 6. Performance Comparison among Scheduling Strategies.

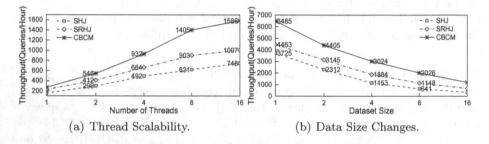

(a) Thread Scalability.   (b) Data Size Changes.

Fig. 7. Performance under different thread number and data sizes.

Thread and Data Size. To prove the superiority of our method, we vary the thread number (from 1 to 16) to take an overall performance comparison between CBCM and other two methods from the perspectives of throughput and average response time. Figure 7(a) shows our CBCM is significantly superior to SHJ and SRHJ. It is because SHJ does not consider resuing previous queries' hash tables and SRHJ takes a coarse cache management which has a low cache utilization and cause a lot of lock contentions. To treat the dataset size as a factor of overall performance, we compare CBCM to other two baseline methods under different dataset sizes. We vary the dataset size from 1 GB to 16 GB and run these concurrent queries using 16 threads. In Fig. 7(b), the comparison results show that the CBCM is also superior to others even under different dataset sizes.

6 Conclusion

In this paper, we propose a chunk-based method to delicately manage the cached hash tables for concurrent hash-join tasks. Under the chunk-based granularity, our method can obtain the higher cache utilization to better accelerate the building phase during the hash-join execution. Meanwhile, the fine-grained hash table can also benefit the probing phase by avoiding filtering out unnecessary tuples. For the concurrent workload, we also design a cache scheduling strategy to exploit the sharing opportunities on building costs. We evaluate our method in a prototype and the results show it is significantly superior to existing methods.

Acknowledgements. This is work is partially supported by National Key R&D Program of China (2018YFB1003303), National Science Foundation of China under grant number 61772202, Youth Program of National Science Foundation of China under grant number 61702189, and ECNU Academic Innovation Promotion Program for Excellent Doctoral Students (YBNLTS2019-020).

References

1. Pat O'Neil, X.C., O'Neil, B.: The star schema benchmark. https://www.cs.umb.edu/~poneil/StarSchemaB.PDF
2. Dursun, K., Binnig, C., Çetintemel, U., Kraska, T.: Revisiting reuse in main memory database systems. In: SIGMOD, pp. 1275–1289 (2017)
3. Ivanova, M., Kersten, M.L., Nes, N.J., Goncalves, R.: An architecture for recycling intermediates in a column-store. In: SIGMOD, pp. 309–320 (2009)
4. Tan, K., Goh, S., Ooi, B.C.: Cache-on-demand: recycling with certainty. In: ICDE, pp. 633–640 (2001)
5. Roy, P., Seshadri, S., Sudarshan, S., Bhobe, S.: Efficient and extensible algorithms for multi query optimization. In: SIGMOD, pp. 249–260 (2000)
6. Chen, C., Roussopoulos, N.: The implementation and performance evaluation of the ADMS query optimizer: integrating query result caching and matching. In: EDBT, pp. 323–336 (1994)
7. Shim, J., Scheuermann, P., Vingralek, R.: Dynamic caching of query results for decision support systems. In: SSDBM, pp. 254–263 (1999)
8. Gupta, H., Mumick, I.S.: Selection of views to materialize in a data warehouse. IEEE Trans. Knowl. Data Eng. **17**(1), 24–43 (2005)
9. Johnson, R.: To share or not to share. In: VLDB, pp. 351–362 (2017)
10. Feliu, J., Petit, S., Sahuquillo, J., Duato, J.: Cache-hierarchy contention-aware scheduling in CMPs. IEEE Trans. Parallel Distrib. Syst. **25**(3), 581–590 (2014)
11. Psaroudakis, I., Athanassoulis, M., Ailamaki, A.: Sharing data and work across concurrent analytical queries. PVLDB **6**(9), 637–648 (2013)
12. Candea, G., Polyzotis, N., Vingralek, R.: A scalable, predictable join operator for highly concurrent data warehouses. PVLDB **2**(1), 277–288 (2009)
13. Nagel, F., Boncz, P.A., Viglas, S.: Recycling in pipelined query evaluation. In: ICDE 2013, 8–12 April, Brisbane, Australia, pp. 338–349 (2013)

An Asynchronous View Maintenance Approach Based on IO Sharing

Huichao Duan, Huiqi Hu$^{(\boxtimes)}$, Weining Qian, and Aoying Zhou

East China Normal University, Shanghai, China
{hqhu,wnqian,ayzhou}@dase.ecnu.edu.cn, stevenduan@stu.ecnu.edu.cn

Abstract. The materialized view is considered as a panacea that significantly facilitates query by trading space cost for execution time. However, it also involves the expensive cost of maintenance which trades away the latency. Disk IO cost is an important factor restricting view maintenance performance. To solve this problem, we decouple a complete procedure of view maintenance into a number of tasks and execute them asynchronously. We put off the executions of those tasks until necessary data access from disk can be completed by subsequent transactions or queries. The execution of view maintenance tasks and transactions or queries can share disk IO, which fundamentally reduces IO resource consumption and view maintenance overhead. We accomplish the complete architecture of view maintenance to support IO sharing. To maximize the advantages of sharing IO, the job of view maintenance is split into fine-grained tasks. An effective cost evaluator is also proposed to select proper tasks to be executed for subsequent transactions or queries. We implement our method on the PostgreSQL. Experiments on the CH-benCHmark show it achieves better performance than existing methods.

Keywords: Materialized views · View maintenance · IO sharing

1 Introduction

Hybrid online transaction and analytical processing (HTAP) is becoming a core requirement for database services. With the increasing number of end-users, applications have put forward higher performance needs for online transaction processing (OLTP). Meanwhile, the rapid development of real-time business intelligence demands analytical applications to provide lower response time for queries on the latest data (OLAP). Despite its importance of requirement, there is no gold-standard solution established for HTAP at present. Some existing works attempt to integrate both row-stores and column-stores into a database system, applying different storage strategies for different workloads [9,18]. Another practical solution leverages various query optimization techniques on conventional OLTP systems with row-stores. Among them, materialized views can significantly facilitate query by reducing execution time. Many database systems, such as PostgreSQL, have realized the functionality. Materialized views are queries whose results are stored and maintained in order to

© Springer Nature Switzerland AG 2020
Z. Huang et al. (Eds.): WISE 2020, LNCS 12343, pp. 390–403, 2020.
https://doi.org/10.1007/978-3-030-62008-0_27

facilitate access to the data in their underlying base tables [3]. It brings the low query latency while correspondingly leads to the view refresh cost which is paid to catch up with the updates on the base table. Many studies have tried to reduce the view maintenance cost to expand its application area. Typical optimization methods include batch execution of maintenance tasks [1,19], efficiently representing updates to base tables [8,10], or reusing the results of common expressions [15]. However, the maintenance cost still can be further optimized.

In this paper, we attempt to share the IO resource utilized in the transaction and query executions with view maintenance processes, which fundamentally reduces the maintenance cost. Disk IO is an essential factor which restricts view maintenance performance. We target at optimizing the complicated join view maintenance, where any base table modification will cause access to the other base tables for the calculation of the changes to the entire view. When view updates are completed synchronously with base table modifications (also known as synchronous view maintenance), accesses to all base tables inevitably need to be conducted at once. While for asynchronous view maintenance, maintenance tasks can be completed at any time before querying the view, where disk IO operations can be postponed for further optimizations. We optimize asynchronous view maintenance by sharing the IO between view maintenance and transaction/query processing. We temporarily freeze the view maintenance task until subsequent DMLs[1] can complete the necessary data access to the disk. Therefore, these maintenance tasks can be a free ride without consuming the already-obtained IO resources, which essentially saves view maintenance cost.

We accomplish an entire view maintenance solution to support IO sharing for the join view. To maximize the advantages of sharing IO, we split the job of view maintenance sufficiently. Compared with asynchronous view maintenance, which only postpones the entire maintenance job, we divide the maintenance job into multiple maintenance tasks that fall on those participated base tables. Therefore, IO can be shared each time the base table is accessed to complete more view maintenance tasks. In brief, we construct a view graph to maintain the join relationships between base tables. Transactions will bring changes to views, thus a transaction execution in our framework not only produces accessing records for a base table but also generates related maintenance tasks according to its joined tables. As a result, the follow-up transactions or queries can complete those tasks without additional IO cost. We ensure all the necessary tasks are generated so that the view increments can be correctly calculated. Since subsequent DMLs can share IO with previous tasks at the execution time, we also build a cost evaluator to decide whether to invoke the task for a DML.

In summary, the contributions of the paper are listed as follows:

(1) We propose to optimize asynchronous view maintenance by sharing disk IO. The core methodology is to put off the view maintenance plan until its

---

[1] A data manipulation language (DML) is a computer programming language used to *SELECT*, *INSERT*, *UPDATE*, and *DELETE* data in a database.

necessary data access to disk can be completed by the subsequent DMLs, which effectively reduces the critical cost of view maintenance.

(2) We accomplish the complete architecture of view maintenance to support IO sharing. To maximize the advantages of sharing IO, the job of view maintenance is split into fine-grained tasks that fall on base tables. As a result, those maintenance tasks can be performed along with regular transaction or query executions. We also design an effective cost evaluator to choose proper tasks for subsequent DMLs.

(3) We implement the method on PostgreSQL. Experiments on the popular CH-benCHmark show the method achieves much better performance than existing methods.

The rest of the paper is organized as follows. Section 2 introduces related works. Section 3 depicts the ideas of IO sharing. Section 4 describes the process of task maintenance. Section 5 discusses the cost evaluator for task selection. Section 6 shows the experiment results. Section 7 concludes the paper.

2   Related Works

Asynchronous View Maintenance. Compared with conventional maintenance approaches which usually update the view synchronously with the base table transaction, asynchronous view maintenance can reduce transaction latency and improve processing performance. The view is allowed to be inconsistent with the base table and brought up to date as necessary when queried [1, 19]. [19] presents the calculation workflow and correctness proof of the asynchronous maintenance view in the database. Asynchronous view maintenance provides a basic optimization opportunity: it reduces redundant computation by consolidating update operations (especially operations performing on the same data) accumulated for a period of time [8, 14, 19].

Current Maintenance Optimizations. To reduce the expensive cost of view updates, a typical approach is to leverage incremental view maintenance (IVM). IVM builds on the observation that computing only updates to view contents induced by minor changes of base tables is more efficient than recomputing the entire view from the base tables [3]. IVMs are based on the changes of a base table, i.e., they mainly discuss the impact of an INSERT/DELETE/UPDATE operation on the view when it occurs to a base table. A batch of existing works has proposed IVM methods to optimize the maintenance cost. [3, 6] survey the main optimizations of IVM. We conclude those most widely used strategies:

Exploiting Key/Foreign-Key Constraints [8, 10]. These works make use of the uniqueness of primary key and the fact that the foreign key is constrained by the primary key to speed up IVM by reducing its access to base tables.

Using Auxiliary Data Information [7, 16, 17]. The motivation is materializing appropriate auxiliary data such as derivation count [7], Min/Max value [16], and top-k value [17] to reduce the computation cost of IVM. [7] keep track of the derivation (reference) count for each row in the view. The derivation count can reduce the computation cost of operations such as projection and aggregation. For instance, the deletion is not likely to alter a project view for a distinct operation when its derivation count of each row exceeds zero. For those deletions, the job of view maintenance is avoided. Similarly, maintaining top-k [17] info and Min/Max [16] can also help to reduce the IO costs of accessing base tables.

Leverage a Dedicated Data Structure to Speed up IVM Computation [2,5,12,14]. At the cost of maintaining a hash-based data structure, [12] quickly determines whether the update transaction is relevant to the view. DBToaster-based methods [2,14] depend on runtime in-memory structure to maintain the auxiliary data structure. From the perspective of the storage structure of the materialized view itself, [5] organizes the view into d-dimensional data cube to speed up selection and aggregation.

Multi-query Optimization [11,13,15]. The main idea is to reduce the calculation of common substructure. [13] applies its previous work on multi-query optimization [15] to exploit common subexpressions among maintenance expressions. So that the result of the common expression is evaluated only once. [11] optimizes the access times of the base table, which proposes the delta propagation strategy to share common intermediate results among views.

The above works are distinct with our IO sharing strategy, which takes advantage of the IO accesses already incurred by update transactions and queries to reduce the cost of IVM. To the best of our knowledge, we are the first to introduce sharing IO into IVM. Our method is independent of any assumptions and orthogonal to those current optimizations.

3   Overview of Sharing IO

Both transactions and queries need to be executed immediately to respond to users, this inspires us to seek the most straightforward method to maintain views asynchronously: to complete the IO of view maintenance by directly attaching it to the transaction and query execution. As a job of view maintenance contains many IO operations, it needs to be further divided into more fine-grained tasks, where each task performs IO on a separate base table. As a result, we can determine a number of view maintenance tasks and transactions/queries that share IO, by observing whether they access the same base table. Next, we use a simple example to give a whole picture of the idea of sharing IO.

Suppose we create a view $\mathcal{V}$, which is defined over the base tables $\mathcal{C}$ with the schema $\mathcal{C}$ ($c\_id, c\_name, c\_balance$), and $\mathcal{O}$ with the schema $\mathcal{O}(o\_id, c\_id, o\_cnt)$. The view $\mathcal{V}$ is formalized as

$$\mathcal{V} = \pi_{c\_name, o\_cnt} \left(\mathcal{C} \bowtie_{\mathcal{C}.c\_id = \mathcal{O}.c\_id} \mathcal{O} \right).$$

As shown in Fig. 1, we give some DMLs $\{T_1, T_2, T_3, Q_1, Q_2\}$ over a period of time to sketch the maintenance mechanism. The IO that must be paid to complete the DMLs is called *essential IO*. The IO that can be completed together with the

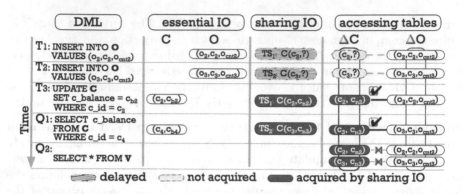

Fig. 1. A example of IO sharing

essential IO for the view maintenance task is called *sharing IO*. Here we assume that base tables are stored in the form of B+ trees, so the records with adjacent primary keys are also adjacent in their physical storage. The projection which is necessary to compute view increments on each base table is stored in respective *accessing tables*. Note that the records stored in the accessing table include all the records required to calculate the view increment, which are the modifications of the base table or the records stored in the base table prepared to be joined with other modifications of base tables to generate the view increments.

Suppose transactions T_1 and T_2 insert two rows into $\mathcal{O}$ which cause V to be updated. To complete the view updates, tuples in $\mathcal{C}$ (with values c_id equals c_2 and c_3) should also be retrieved. Instead of synchronously accessing $\mathcal{C}$, we temporarily freeze two view maintenance tasks and record these tasks (denoted by TS) : TS_1 which reads the records with c_id of c_2 in $\mathcal{C}$, TS_2 which reads the records with c_id of c_3 in $\mathcal{C}$. At this point, two view maintenance tasks are maintained asynchronously. In the subsequent execution of the transaction T_3, it requires IO to access the page where c_2 is located. Take a free ride to execute T_3, the view maintenance task TS_1 can be completed incidentally by sharing the IO. Next, when Q_1 reads the record with c_id of c_4, we found this record is physically stored next to the record with c_id of c_3 which the task TS_2 needs to access. Thus Q_1 can complete TS_2 with only a few extra IO cost. Eventually, the maintenance of the view can be done asynchronously by sharing IO. Query Q_2 can directly get the latest results of the view by joining all the accessing tables without additional IO cost.

For a view V defined over multiple base tables whose increments are called ΔV, the accessing table ΔT_i is the projection results of ΔV on the base table T_i which consists of *accessing records*. Simply speaking, to compute ΔV, we need to obtain all the required records on related base table T_i ($1 \leq i \leq n$). This includes (i) the modified records brought by the update transaction on T_i; (ii) the records stored in T_i which are prepared to be joined with modified records of other base tables. Those accessing records are separately obtained and stored into ΔT_i ($1 \leq i \leq n$). Once we obtain all the accessing records of base tables, we can easily compute ΔV following the standard formula: $\Delta V = \Delta T_i \bowtie \Delta T_{i+1} \bowtie \cdots \bowtie \Delta T_{n-1} \bowtie \Delta T_n$ [19]. Thus the purpose of IO sharing is to reduce IO operation to obtain accessing records for each base table. After fusing views and ΔV, we can get the latest view content.

4   Task Management

We divide the view maintenance job into multiple tasks that fall on different base tables, so each fine-grained *maintenance task* can benefit from the IO access. In this section, we analyze how to generate view maintenance tasks based on the view schema and how to perform these tasks to generate accessing records.

Task Memo. View maintenance tasks are generated according to the update transaction, which formally describes a partial job of view maintenance. All the tasks are stored in task memo.

View Graph. We construct a view graph $G = (T, E)$: vertices $T_i \in T$ in the graph are base tables, and edges $E_j \in E$ between them are join relationships. To guide the view explainer to generate tasks in the correct order, the view graph is constructed as a DAG (directed acyclic graph). A DAG determines a topological order of computing the increment of a join view. We denote $G(T_i)$ as the DAG which takes table T_i as the first updated base table.

Task Generation. Next we define the task. A task (denoted by TS) is formally described as a triple $\{T_{id}, \Phi, G(T_i)\}$:

- T_{id}: the base table ID to be accessed for the task.
- Φ: required set of tuples in base table T_{id} for maintenance. In practice, Φ is expressed as the predicates over on the columns of a base table.
- $G(T_i)$: the DAG utilized to generate the task.

A task is defined on a specific base table T_{id}, which access the target tuples in Φ. In the task memo, a base table can attach multiple tasks. The task generation process includes three stages. The three stages are as follows:

(I) Running an update transaction. An update transaction performing on a base table triggers the related views to be updated. Suppose we update a base table T_i, we first choose the corresponding DAG $G(T_i)$ which takes T_i as its initial base table to be updated. Next we generate a number of tasks based on the $G(T_i)$. The (accessing) records of base tables that join with T_i should also be retrieved. Thus, we generate a task and put it into the task memo for each of these base tables. Apparently, different transactions make use of different DAGs.

(II) Running a sharing IO task. By sharing IO, tasks maintained in the task memo can be picked and executed along with subsequent DMLs. To run a transaction or a query on the base table, it will first access the task memos of these base tables to select tasks that are worth sharing. It is worth noting that executing a task will also generate subsequent tasks of other base tables. This is because we calculate accessing records of different tables with a chaining manner following the DAG structure.

(III) The final stage. The final stage completes all the maintenance tasks. Before querying the view, it executes all the unfinished tasks without opportunities to share IO. Similar to the above process of task execution, running those unfinished tasks may also produce follow up tasks. We run all the tasks until no more new tasks are produced.

Formally, the critical point is to generate a task is to establish its $\{T_{id}, \Phi, G(T_i)\}$. In specific, given an update transaction acts on the table T_i, then $G(T_i)$ is established as the DAG for the following tasks. Next according to the topological ordering of the DAG, for each ending endpoint T_j of T_i ($T_j \in T$, $\langle T_i, T_j \rangle \in E$): the task should be constructed, where T_{id} of the task is exactly T_j. Φ of the task depends on the predicates defined in the view schema and the join relationship between T_i and T_j. As we are aware of Φ of the update transaction performing on T_i, we can compute Φ for T_j. Similarly, if a task acts on table T_j, then it will first look up its DAG ($G(T_i)$), then tasks should be constructed for every ending points T_k of T_j in $G(T_i)$.

Accessing Record Computation. Accessing records are retrieved and stored in the accessing tables. Accessing records are generated in two ways: (i) when an update transaction on table T_i is executed, it produces accessing records for ΔT_i; (ii) and when a task of $\{T_i, \Phi, G(T_x)\}$ (T_x is the first table) is executed, (no matter it is performed by sharing IO or in the final stage), it also produces accessing records within the scope of Φ. By sharing IO operations, we reduce the cost to generate those accessing records.

Example 1. We also give a working example in Fig. 2. As shown in Fig. 2(A), we assume a view $\mathcal{V}$ is defined on the base tables T_1, T_2, and T_3:

$$\mathcal{V} = \pi_{T_1.c_4,\ T_2.c_4,\ T_3.c_4}(T_1 \bowtie_{T_1.c_2 = T_2.c_1} T_2 \bowtie_{T_1.c_3 = T_3.c_1\ \&\ T_2.c_2 = T_3.c_2} T_3).$$

Correspondingly, a DAG $G(T_1)$ taking T_1 as the initial table is expressed as Fig. 2(B). When performing an insert transaction on T_1, the maintenance process mainly includes three stages: (i) The view explainer first constructs the tasks and calculates their Φ according to $G(T_1)$. Based on the record $\langle 1, 2, 3, 4 \rangle$ and the join relationship $T_1.c_2 = T_2.c_1$, we can obtain that Φ of the task for T_2 is $c_1 = 2$. Thus, the task can be represented as $\{T_2, \langle c_1 = 2 \rangle, G(T_1)\}$ and stored into T_2's task memo. Meanwhile, the accessing record $\langle 1, 2, 3, 4 \rangle$ can be directly generated and inserted into ΔT_1. (ii) Subsequently, when performing the query on the T_2, the view explainer attaches the task $\{T_2, \langle c_1 = 2 \rangle, G(T_1)\}$ to the query. So that the task and the query can share the IO result, where the record $\langle 2, 3, 4, 5 \rangle$ is accessed and later inserted into ΔT_2. Then, we can build the task for T_3 based on $G(T_1)$ and the join relationship $T_2.c_3 = T_3.c_1$, where Φ is $c_1 = 4$. (iii) Until then, the remaining task $\{T_3, \langle c_1 = 4 \rangle, G(T_1)\}$ has not been completed yet. Finally, when querying the view, this unfinished task is enforced to keep the view up to date, which gets the accessing record $\langle 4, 5, 6, 7 \rangle$. Then we fuse all the accessing records in the accessing tables and merge them with the view table.

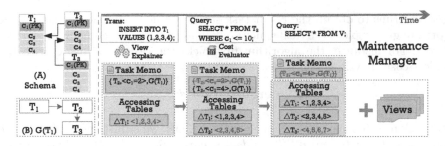

Fig. 2. Sample workflow

5   Task Selection

Our vision is to benefit view maintenance by making full use of every disk IO. However, running DMLs along with tasks increases the latency. Thus, we need to ensure its controllability. The cost evaluator focus on whether to attach the tasks to a DML. It measures the impact of tasks on DML according to the controllability parameter τ, and uses it as a threshold to decide whether tasks match DML or not. For ease of presentation, we use d to describe a DML operation. In specific, τ is treated as a proportion of cost growth, which means that after invoking the tasks into the execution of d, the cost growth is less than τ. For a DML to be executed, the problem is to make maximum use of its IO by sharing a maximum task set within the constraints of cost growth τ.

Next, we discuss the cost model to measure cost growth. The cost model mainly evaluates the increase of IO cost when a task is invoked into a DML operation for execution. The most accurate IO cost is counting the number of pages they retrieve, which is infeasible to obtain without performing the physical data access. Therefore we turn to compare their access ranges. Intuitively, if a DML and a task have the same access range, then the task is a good candidate for the DML. For simplicity, we use TS_{min} and TS_{max} to represent the access range of the task TS. And we use d_{min} and d_{max} to denote the access range of the DML operation d. Therefore the essential range of the DML d is $\langle d_{min}, d_{max} \rangle$. After adding the task into the DML, the access range extends to $\langle min(TS_{min}, d_{min}), max(TS_{max}, d_{max}) \rangle$. By comparing the two scopes, we can estimate the cost growth of invoking TS into DML d for execution as:

$$\Delta cost(d, TS) = \frac{max(TS_{max}, d_{max}) - min(TS_{min}, d_{min})}{d_{max} - d_{min}}.$$

Similarly, when a DML invokes multiple tasks, suppose that these tasks are represented by a task set $\mathcal{C} = \{TS^1, TS^2, \cdots TS^m\}$. Then the cost growth of invoking $\mathcal{C}$ into d for execution is:

$$\Delta cost(d, \mathcal{C}) = \frac{max(TS^1_{max}, \cdots TS^m_{max}, d_{max}) - min(TS^1_{min}, \cdots TS^m_{min}, d_{min})}{d_{max} - d_{min}}.$$

To invoke tasks as many as possible under the constraint of the cost growth τ, we leverage a greedy algorithm. Given the DML operation d, the greedy algorithm runs to obtain the result task set $\mathcal{C}$ as follows. In each iteration, the algorithm greedy picks the task which contributes the least increment on current IO cost. For instance, the algorithm first picks the task TS^* into $\mathcal{C}$ with the smallest cost growth, which is:

$$TS^* = \arg\min_{TS} \Delta cost(d, TS).$$

After picking TS^*, the access range expands. Thus, we adjust the current range from $\langle d_{min}, d_{max} \rangle$ to $\langle min(TS^*_{min}, d_{min}), max(TS^*_{max}, d_{max}) \rangle$. Then we pick the task with the smallest cost growth (i.e. $\Delta cost(d, \mathcal{C})$) for the current range into $\mathcal{C}$ in the next round of iteration. Iteratively, we pick tasks into $\mathcal{C}$ until the cost growth $\Delta cost(d, \mathcal{C})$ exceeds τ.

6   Experiment

6.1   Experiment Setup

Setup. Effectiveness and cost of our approach are evaluated in an open-source database PostgreSQL through the experiments conducted on a Linux server, which has two Intel Xeon Silver-4110@2.10 GHz processors (each with 8 physical cores), 256 GB main memory. We use RAID5 with flash-based write cache (FBWC) which has high performance on I/O accesses.

Benchmarks. In our experiment, all queries are executed on CH-benCHmark [4] with scaling factor w = 100,000. We load all twelve tables and define two views based on the common subqueries of all SQLs in CH-benCHmark:

$$\mathcal{V}_1 = \pi_{o\_id, o\_c\_id, o\_entry\_d, o\_carrier\_id, ol\_amount, ol\_delivery\_d}$$
$$(Order \bowtie_{o\_id=ol\_o\_id \ \& \ o\_d\_id=ol\_d\_id \ \& \ o\_w\_id=ol\_w\_id} Orderline),$$

$$\mathcal{V}_2 = \pi_{c\_id, c\_last, c\_state, o\_id, o\_entry\_d, o\_carrier\_id, ol\_amount}$$
$$(Customer \bowtie_{c\_id=o\_c\_id \ \& \ c\_d\_id=o\_d\_id \ \& \ c\_w\_id=o\_w\_id} Order$$
$$\bowtie_{o\_id=ol\_o\_id \ \& \ o\_d\_id=ol\_d\_id \ \& \ o\_w\_id=ol\_w\_id} Orderline).$$

$\mathcal{V}_1$ extracts from SQL 3, 5, 7, 8, 9, 10, 12, 18, and 21. And $\mathcal{V}_2$ extracts from SQL 3, 5, 7, 8, and 10. We utilize all five update transactions of CH-benCHmark with default proportions to modify the tables.

Comparing Methods. We compare our IO sharing IVM method (S-IO) with synchronous IVM and asynchronous IVM. In practice, S-IO set the controllability parameter τ to 108% by default. For synchronous IVM (Sync), we refer to [7] to implement a synchronous IVM mechanism, which updates the view directly by executing the created stored procedures when executing the transaction. For asynchronous IVM, we implement the latest ID-based strategy (IDs) with stored procedures and Δ-scripts according to [8]. In specific, we identify the to-be-modified tuples with their IDs into delta tables through the stored procedures and fuse those records with the view by executing Δ-scripts.

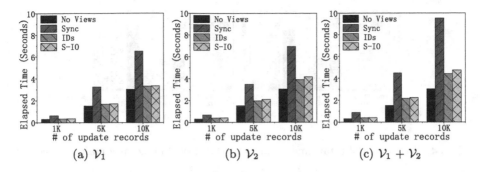

Fig. 3. Update transaction performance

6.2 Transaction Processing Performance

Our first experiment demonstrates the performance of transaction processing when maintaining views through sharing IO. We execute update transactions and maintain the corresponding views. For example, *Payment* transaction randomly modifies the user balance in *Customer* table, which results in the updates of view $\mathcal{V}_2$; *Neworder* transaction inserts new rows into *Order* table and *Orderline* table, which affects both view $\mathcal{V}_1$ and $\mathcal{V}_2$.

We compared the elapsed time of executing transactions for Sync, IDs, and S-IO. Figure 3(a) and Fig. 3(b) show three scenarios in which we update 1K, 5K, and 10K records when maintaining views $\mathcal{V}_1$ and $\mathcal{V}_2$. The following features can be observed: (i) Not maintaining the view (NO Views) does not affect transaction processing. On the contrary, Sync has the highest cost due to the additional view maintenance process accompanying the transaction execution. (ii) As the number of updated records reaches 10K in Fig. 3(a), the performance of Sync is 6.5 s, 2.16 times of that of NO Views which is 3.01 s, while the performance of IDs is 3.32 s, which is only 10% higher than that of NO Views. Compared with Sync which significantly increases the response time, IDs is hardly affected. (iii) Also, S-IO has exhibited obvious advantages in elapsed time. With the updated records of 10K, the performance of S-IO is 3.35 s increased by 1% compared with IDs (3.32 s), which benefits from our cost model and introduces minor overhead. At the same time, compared with IDs, the impact of sharing IO on the latency is also controlled within 107%, which shows that the cost evaluation and the configuration of the controllability parameter τ are effective. (iv) The gap of the elapsed time between S-IO and IDs in $\mathcal{V}_2$ is larger than that of $\mathcal{V}_1$. This is because view $\mathcal{V}_1$ is mainly affected by the *Neworder* transaction, where the ranges that view maintenance tasks need to access are already included in the transactions, that is, view $\mathcal{V}_1$ can be maintained directly through the IO sharing.

As shown in Fig. 3(c), the more views are maintained, the more advantages the approaches based on asynchronous maintenance is. Since the view maintenance is no longer need to be completed within the execution of the transaction, IDs and S-IO can significantly reduce the response time. Under asyn-

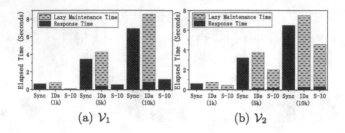

Fig. 4. Breakdown of total time spent on updates (including view maintenance)

chronous based maintenance, the performance depends only on the cost of the asynchronous incremental calculation. Thus, we next analyze the cost of view maintenance.

6.3   View Maintenance Cost

Even if view maintenance is performed in the background, maintenance overhead still affects other tasks of the database since the overall system resources are limited. We need to ensure that the maintenance cost of S-IO is lower than those of Sync and IDs methods.

We measure the total time required to complete the maintenance process for each view, which includes the response time of the update transaction and the lazy maintenance time of the actual asynchronous calculation. Figure 4 shows the total time spent on updates under the same two experiments in the previous section. It can be found that: (i) Although IDs reduces the response time, its total maintenance time is higher than that of Sync. This is because even if the update task is postponed, the overall execution cost of the asynchronous maintenance process does not decrease since it introduces additional overhead. (ii) However, with lower response time through asynchronous maintenance strategy, the sharing IO method reduces the disk IO cost which fundamentally leads to the decrease of total maintenance time. As shown in Fig. 4(b), with the updated records of 5K, the total elapsed time of S-IO is 2.04 s, which is only 54% than that of IDs (3.79 s) and only 64% than that of Sync (3.21 s). Compared to $\mathcal{V}_2$, the lazy maintenance time of S-IO in $\mathcal{V}_1$ is too small to be visible in Fig. 4(a).

Overhead for Asynchronous Based Maintenance. Compared with synchronous maintenance, asynchronous based maintenance has a few extra steps. In general, the cost is ultimately reflected in the CPU and IO load. Thus, we further verify the resource overhead of different methods from the perspective of CPU and IO.

As shown in Fig. 5, when there are only transactional loads, Sync consumes the most resources, with the most massive CPU and IO overhead, because synchronous view maintenance tasks increase the transaction response time. With the addition of view queries, the resource overhead of NO Views becomes the highest, because using base tables to respond to queries consumes a lot of IO

Workload		Method	Resource Overhead		
Qps	Tps		CPU% (%)	IO (MB/s)	# IOPS
0	700	NO Views	10.25	16.81	1548
		IDs	20.82	31.46	2722
		S-IO	26.12	33.42	2895
		Sync	65.32	39.85	3395
0.5	300	NO Views	46.25	28.37	2672
		IDs	12.32	25.12	2436
		S-IO	13.92	**23.39**	**2277**
		Sync	26.13	28.14	2628
2	500	NO Views			
		IDs	36.62	38.05	3536
		S-IO	**23.25**	**28.81**	**2582**
		Sync	-	-	-

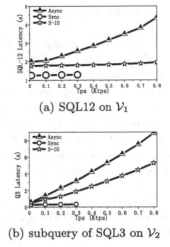

(a) SQL12 on $\mathcal{V}_1$

(b) subquery of SQL3 on $\mathcal{V}_2$

Fig. 5. IO cost   **Fig. 6.** View query performance

and CPU resources. But when qps is 2 and tps is 500, the method of NO Views and Sync are no longer competent. Compared with IDs, when there is no analytical load, S-IO needs to share IO, which introduces more overhead. But as the analytical load increases, the advantages of S-IO gradually appear, which reduces the total IO bandwidth and IOPS to save the maintenance cost.

6.4   Query Performance

In this experiment, we continuously apply five transactions to the views and compare the view response times of different methods. We execute SQL12 and SQL3 every 10 seconds to measure $\mathcal{V}_1$ and $\mathcal{V}_2$ respectively.

As shown in Fig. 6, for Sync, the update transaction cannot reach 350 tps due to the impact of completing the view update within the transaction. Under both S-IO and IDs, as the update load increases, continuously update transactions generate a large number of unfinished maintenance tasks, which results in the declining query latency. For S-IO, subsequent transactions can help complete the previously generated view maintenance tasks, so at the time of querying the view, the total incomplete maintenance tasks are less than those of IDs without optimizations which can minimize the impact of view maintenance on the query latency. In Fig. 6(a), when tps at 600, the latency of S-IO (1.87 s) is 53.6% lower than that of IDs (3.49 s). Compared to Fig. 6(b), the latency of S-IO (3.71 s) is 59.1% lower than that of IDs (6.38 s) with 600 tps. The advantage of S-IO on SQL12 is more obvious than that on SQL3 because the view tasks on the *Order* table and *Orderline* table are more easily to be completed with the continuous transaction executions through IO sharing as described in Sect. 6.2. In general, the more IO that view maintenance can share, the more query latency benefits.

7 Conclusion

In this paper, we study view maintenance for the join view. The main idea is to put off the executions of view maintenance tasks until necessary data access from disk can be completed by subsequent transactions or queries. Thus the execution of view maintenance tasks and transactions or queries can share disk IO. To support view maintenance with IO sharing, we design a complete solution to generate and maintain tasks. An effective cost evaluator is proposed to select proper tasks for subsequent transactions or queries. The experiments are conducted on the open-source database PostgreSQL. Results on the CH-benCHmark show it achieves much better performance than existing methods.

Acknowledgements. This is work is partially supported by National Key R&D Program of China (2018YFB1003300), Youth Program of National Science Foundation of China under grant number 61702189. The corresponding author is Huiqi Hu.

References

1. Agrawal, P., Silberstein, A., Cooper, B.F., et al.: Asynchronous view maintenance for VLSD databases. In: SIGMOD, pp. 179–192. ACM (2009)
2. Ahmad, Y., Kennedy, O., Koch, C., Nikolic, M.: DBToaster: higher-order delta processing for dynamic, frequently fresh views. PVLDB **5**(10), 968–979 (2012)
3. Chirkova, R., Yang, J., et al.: Materialized views. Found. Trends Databases **4**(4), 295–405 (2012)
4. Cole, R.L., Funke, F., et al.: The mixed workload CH-benCHmark. In: DBTest, p. 8 (2011)
5. Gray, J., Chaudhuri, S., et al.: Data cube: a relational aggregation operator generalizing group-by, cross-tab, and sub totals. Data Min. Knowl. Discov. **1**(1), 29–53 (1997)
6. Gupta, A., Mumick, I.: Materialized Views: Techniques, Implementations, and Applications (1999)
7. Gupta, A., Mumick, I.S., Subrahmanian, V.S.: Maintaining views incrementally. In: SIGMOD, pp. 157–166 (1993)
8. Katsis, Y., Ong, K.W., Papakonstantinou, Y., et al.: Utilizing IDs to accelerate incremental view maintenance. In: SIGMOD, pp. 1985–2000. ACM (2015)
9. Kemper, A., Neumann, T.: HyPer: a hybrid OLTP&OLAP main memory database system based on virtual memory snapshots. In: ICDE, pp. 195–206. IEEE (2011)
10. Larson, P., Zhou, J.: Efficient maintenance of materialized outer-join views. In: ICDE, pp. 56–65 (2007)
11. Lee, K.Y., Kim, M.H.: Optimizing the incremental maintenance of multiple join views. In: DOLAP, pp. 107–113. ACM (2005)
12. Luo, G., Yu, P.S.: Content-based filtering for efficient online materialized view maintenance. In: CIKM, pp. 163–172 (2008)
13. Mistry, H., Roy, P., Sudarshan, S., Ramamritham, K.: Materialized view selection and maintenance using multi-query optimization. In: SIGMOD, pp. 307–318 (2001)
14. Nikolic, M., Dashti, M., Koch, C.: How to win a hot dog eating contest: Distributed incremental view maintenance with batch updates. In: SIGMOD, pp. 511–526. ACM (2016)

15. Roy, P., Seshadri, S., Sudarshan, S., Bhobe, S.: Efficient and extensible algorithms for multi query optimization. In: SIGMOD, pp. 249–260 (2000)
16. Xu, M., Ezeife, C.I.: Maintaining horizontally partitioned warehouse views. In: Data Warehousing and Knowledge Discovery, pp. 126–133 (2000)
17. Yi, K., Yu, H., Yang, J., Xia, G., Chen, Y.: Efficient maintenance of materialized top-k views. In: ICDE, pp. 189–200 (2003)
18. Zhan, C., Su, M., et al.: AnalyticDB: real-time olap database system at Alibaba cloud. Proc. VLDB Endowment **12**(12), 2059–2070 (2019)
19. Zhou, J., Larson, P., Elmongui, H.G.: Lazy maintenance of materialized views. In: PVLDB, pp. 231–242 (2007)

AQapprox: Aggregation Queries Approximation with Distribution-Aware Online Sampling

Han Wu[1], Xiaoling Wang[1,2], and Xingjian Lu[1(✉)]

[1] Shanghai Key Laboratory of Trustworthy Computing,
East China Normal University, Shanghai, China
51184501059@stu.ecnu.edu.cn, xlwang@sei.ecnu.edu.cn, xjlu@cs.ecnu.edu.cn
[2] Shanghai Institute of Intelligent Science and Technology, Tongji University,
Shanghai, China

Abstract. Approximate query processing (AQP) is an effective way to provide approximate results for SQL queries, which relaxing accuracy in exchange for higher processing speed. In sampling-based AQP techniques, random sampling works well for uniformly distributed data but performs poorly on skewed data. To address this problem, we propose a distribution-aware approximation framework called AQapprox (aggregation queries approximation), to approximate queries more efficiently and accurately by extending Sapprox. We construct a probabilistic Map, which records the occurrences of sub-datasets on categorical columns and related statistics on numerical columns at each segment of the whole dataset. When a query arrives, AQapprox will combine Map and adaptively use different sampling methods based on the distribution. Experimental results on both real and synthetic datasets show that AQapprox can achieve a speedup by up to $5.9\times$ for skewed data, $64\times$ for uniform data over Sapprox, and has higher accuracy on multi-column queries.

Keywords: AQP · Distribution-aware approximation · Probabilistic map

1 Introduction

As data volumes are growing exponentially, in recent years, approximate query processing (AQP) has been proposed to provide approximate results for SQL queries by relaxing accuracy in exchange for higher query processing speed. AQP includes two categories: offline AQP and online AQP. Offline AQP techniques use generated offline samples or data synopses to answer queries, such as BlinkDB [3] and BAQ [12], etc. Online AQP techniques answer queries based on online sampling results, such as ApproxHadoop [8], Sapprox [18] and Quickr [11], etc.

Sampling and error estimation are two necessary phases for online aggregation. How to prepare representative samples is essential to approximation systems. In this paper, we define "the subset of data relevant to a set of attribute

© Springer Nature Switzerland AG 2020
Z. Huang et al. (Eds.): WISE 2020, LNCS 12343, pp. 404–416, 2020.
https://doi.org/10.1007/978-3-030-62008-0_28

values" as sub-dataset, for example, ("England", "Linux") is a value for (Country, OS), records containing ("England", "Linux") form a sub-dataset, for convenience, the unique value ("England", "Linux") represents a sub-dataset. Take the following two simple aggregation queries Q1, Q2 as examples.

Table 1. Example table T

ID	market	product	sales
1	FR	Books	1
...	FR	Books	1
60	FR	Books	1
61	FR	Toys	10
...	FR	Toys	10
100	FR	Toys	10
101	US	Videos	1
...	US	Videos	1
198	US	Videos	1
199	US	Books	100
200	US	Books	100

Q1
SELECT COUNT(*)
FROM T
WHERE product= 'Books';

Q2
SELECT market, SUM(sales)
FROM T
GROUP BY market;

Our goal is to prepare a good sample for this "Books" sub-dataset in Q1 and a representative sample for numerical column "sales" in Q2. Traditional sampling methods (Bernoulli sampling [7], Stratified sampling [16], etc.) require a full scan of the whole dataset, they are too expensive to be employed. A recent online sampling-based system, ApproxHadoop [8] works well on uniformly distributed data but performs poorly on skewed data. Like Q1 on Table 1, the "Books" sub-dataset is stored in a skewed distribution, ApproxHadoop may sample a large portion of the whole dataset while obtaining a small amount of sampled data belonging to "Books" sub-dataset. Aiming at the problem of skewed storage distribution of sub-datasets, Sapprox [18] was proposed to improve the I/O efficiency and poor estimation accuracy. However, as the number of columns in queried sub-dataset increases, the average error will increase, and its execution time is short for skewed sub-datasets but long for the uniform. Sapprox only supports queries on sub-datasets, it does not support the approximation for arbitrary aggregation queries (queries like Q2 without sub-datasets and having GROUP BY clauses, etc.).

Like Q2 on Table 1, suppose a sample of 20 rows are drawn uniformly from T. For the second group (market = "US"), there are two cases with high chance. Case I: when exactly one of rows 199 and 200 is in the sample, the estimation is (US, (9 * 1 + 1 * 100)/0.1). Case II: when neither of rows 199 and 200 is taken, the estimated answer is (US, (10 * 1)/0.1). These estimation results deviate far from the ground truth (US, 298), this shows that uniform sampling can produce large errors, stratified sampling in [3,16] and small group sampling in [4] cannot help here, they use different sampling rates for groups of different sizes, but the two groups in answer to Q2 have the same size. ApproxHadoop also performs particularly poorly. For the problem of skewed data on a numerical column, Sample+Seek [5] presents a measure-biased sampling method, but it is sampling in row-level, the cost is expensive.

To solve the above problems, we propose a distribution-aware approximation framework called AQapprox, to approximate aggregation queries more efficiently and accurately. For any relation table, we construct an offline probabilistic Map to record the occurrences of sub-datasets on categorical columns and related statistics on each single numerical column at each segment (segment refers to a list of continuously stored records) of the entire dataset, then non-parametric tests are used to check whether the occurrences of sub-datasets or all segments' coefficients of variation (CVs) are uniform or not, and p-values (test results) are also stored in Map. When a query arrives, AQapprox parses it first and then combines Map to perform sampling. If the sub-datasets or numerical columns to be queried are distributed uniformly, we apply segment-based reservoir sampling. Else if skewed, we apply cluster sampling with unequal probability, each segment is assigned with an inclusion probability proportional to the sub-dataset's occurrence or CV value on the segment. AQapprox can both handle queries on skewed storage distribution of sub-datasets and skewed values on numerical columns.

In summary, we make the following contributions:

- We propose a distribution-aware approximation framework AQapprox for online queries, it combines Map and adaptively uses different sampling methods based on the distribution, and can provide estimation with high precision quickly.
- We provide a guide on segment partition and a heuristic rule to determine which QCSs (query column sets) information needs to be involved in Map.
- We compare AQapprox with Sapprox and other baselines, experimental results demonstrate that AQapprox is more efficient and accurate than others.

The remainder of this paper is organized as follows. In Sect. 2, we introduce related work on online AQP techniques. Section 3 overviews the architecture and specific implementations of AQapprox for queries. The experimental study is reported in Sect. 4. Conclusion and future work are in Sect. 5.

2   Related Work

Online AQP usually uses online aggregation method. Common sampling techniques are random uniform sampling, stratified sampling etc., they require at least one full scan of the underlying dataset.

ApproxHadoop [8] applies multi-stage sampling theory to MapReduce, its sampling and error estimation methods work well on uniformly distributed data. Unfortunately, in many real situations, sub-datasets and values on numerical columns are usually distributed in a skewed form. For a sub-dataset, a small portion of the whole dataset may contain most records of it, while other portions have few records belonging to the sub-dataset.

Some works do consider the storage distribution skewness. Pansare, N. et al. [15] designs an operational model for online aggregation in MapReduce, it correlates the processing time of each block with the aggregated value, if a block contains more relevant data, it will need more time to process. Gan, Y. et al. [6] presents a keep-order approach to reduce biases, which can arise when estimating aggregations over a skewed dataset. However, these works do not address the sampling inefficiency problem. Sapprox [18] builds a SegMap to capture the occurrences of each unique value on a single commonly used column at each segment, and the occurrences of sub-datasets on multiple columns are estimated using the independence/dependence conditional probability formulas. However, as the number of columns in queried sub-dataset increases, the average error will increase, and it needs to know whether there is a dependent or independent relationship between multiple columns in advance. In AQapprox, we record more information in Map to provide more accurate results for multi-column queries quickly, and support more types of queries.

Sample+Seek [5] presents a measure-biased sampling method to solve queries on skewed values of a numerical column, each row is picked into the sample with probability proportional to its value. But it is sampling in row-level and requires to traverse the entire dataset every time sampling, the cost is expensive. In AQapprox, we sample in segment-level and assign each segment with different sampling probability.

3   AQapprox Design

3.1   Definitions

Data Model. Given a relation table T with n columns, we denote its column set as $\mathbb{C}$, and $\mathcal{C}$ represents a column, which is either a categorical column with a limited number of distinct values or a numerical column with real numbers. To apply cluster sampling, we define a cluster as a segment, hence a segment is the sampling unit. The number of records in each segment is the same and can be an arbitrary integer. In Sect. 3.6, we will give a practical guide on how to set the optimal segment partition. And we note the total number of segments as N.

Query Model. AQapprox can support any SQL aggregation query, here we focus on the following query.
SELECT $\mathcal{C}_1^s$, $\mathcal{C}_2^s$, $\cdots$, $\mathcal{C}_{|\mathbb{C}^s|}^s$, **AGG**($\mathcal{C}_1^a$), **AGG**($\mathcal{C}_2^a$), $\cdots$, **AGG**($\mathcal{C}_{|\mathbb{C}^a|}^a$)
FROM T
WHERE ($\mathcal{C}_1^w$ **VP** v_1) **OP** ($\mathcal{C}_2^w$ **VP** v_2) $\cdots$ **OP** $\cdots$ ($\mathcal{C}_{|\mathbb{C}^w|}^w$ **VP** $v_{|\mathbb{C}^w|}$)
GROUP BY $\mathcal{C}_1^g$, $\mathcal{C}_2^g$, $\cdots$, $\mathcal{C}_{|\mathbb{C}^g|}^g$;

where $\mathbb{C}^s = \left\{ \mathcal{C}_1^s, \mathcal{C}_2^s, \cdots, \mathcal{C}_{|\mathbb{C}^s|}^s \right\}$ denotes a set of selected columns, and $\mathbb{C}^a = \left\{ \mathcal{C}_1^a, \mathcal{C}_2^a, \cdots, \mathcal{C}_{|\mathbb{C}^a|}^a \right\}$ denotes a set of aggregation columns, AGG can be COUNT, SUM, AVG, MAX, MIN, STDDEV, VAR, PERCENTILE, etc.

$\mathbb{C}^w$ is a set of selection conditions (VP denotes a value comparison operation in $\{<, \leq, >, \geq, =, \neq\}$, v_i denotes a value, OP is either OR or AND), and $\mathbb{C}^g = \left\{ \mathcal{C}_1^g, \mathcal{C}_2^g, \cdots, \mathcal{C}_{|\mathbb{C}^g|}^g \right\}$ is a set of grouping columns.

Definition (Query Column Set). Given a query q_i, the query column set (QCS) of q_i is $\pi_i = \mathbb{C}^s \cup \mathbb{C}^a \cup \mathbb{C}^w \cup \mathbb{C}^g$.

3.2   Workflow

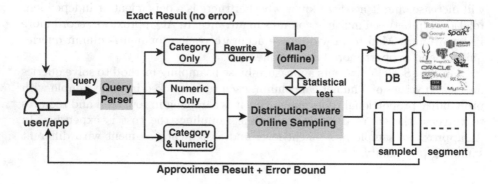

Fig. 1. AQapprox architecture

Figure 1 shows the overall architecture of AQapprox. At runtime, when a user issues a SQL query, query parser parses it first, if the QCS contains categorical columns only, then the query will be rewritten on the Map, and exact result will be returned directly. Else AQapprox uses distribution-aware online sampling based on the results of statistical non-parametric tests on Map information to sample segments, returns an approximate result and error bound to the user.

3.3   Construction of Map

Map is a data structure used to store statistics. Before creating Map, we must determine to build Map information on which QCSs of categorical columns.

Determination of QCSs. We use the following heuristics. **Case I:** If there is no historical query workload, we assume that there are total f categorical columns, when f is very large, the number of QCSs (2^f) and the total number of sub-datasets is exponential. It is impossible to construct and store all QCSs' Map information. Our goal is to select m categorical columns, and the value of m should be guaranteed to cover at least $b\%$ (set b as 90, 95, etc.) of all possible QCSs, that is $\frac{C_f^1 + C_f^2 + ... + C_f^m}{2^f} \geq b\%$. **Case II:** If we can obtain historical query workload, a query has a QCS q_i with some (unknown) probability p_i, that is, QCSs are drawn from a Multinomial $(p_1, p_2, \cdots)$ distribution. The best estimate of p_i is simply the frequency of queries with QCS q_i in past queries. In practice, we can calculate the frequency of each QCS q_i from historical workloads, sort these frequencies in descending order, and take the top y QCSs to make that $y/2^f \geq b\%$. For both cases, if a new query has a QCS which is not covered by the Map, AQapprox uses segment-based reservoir sampling [7].

Creation of Map. For categorical columns, Map records the occurrence of a QCS's sub-dataset on each segment, then p_value of Kolmogorov-Smirnov test [9] on these occurrences is stored in Map, which used to check the uniformity and skewness, finally Map stores these statistics in a quintuple form, see Fig. 2. For each numerical column, Map captures the maximum, minimum and CV on each segment $i(i = 1, \cdots, N)$, and p_value of Feltz and Miller test [14] on these CVs is also stored, which used to test is there a significant difference between CVs and reflect the fluctuation of data values, Map stores them in a quintuple form.

Example of Map Structure. Figure 2 gives an example of Map structure.

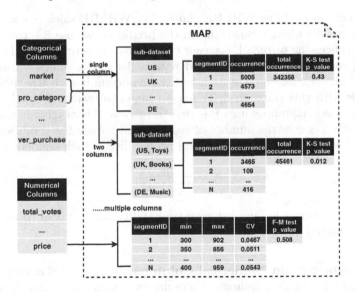

Fig. 2. Example of map structure

3.4 Online Sampling for Queries

After being parsed by the parser, a query q_i will be classified into the following three scenarios according to QCS. In this paper, we assume that in the cases discussed next, the QCS of q_i exists in the Map.

Categorical Columns Only. If the QCS of q_i only contains categorical columns, AGG can only be COUNT, it does not make sense for SUM, AVG, etc. aggregations on categorical columns [12]. For any column $\mathcal{C}$, AQapprox rewrites q_i by replacing each COUNT($\mathcal{C}$) with SUM(occurrence) on Map. What's more, if q_i has a group-by clause, AQapprox rewrites it by replacing each COUNT($\mathcal{C}$) with total_occurrence and returns an exact answer quickly without sampling.

Numerical Columns Only. If the QCS of q_i only contains numerical columns, Case I: query q_i without WHERE clause, AQapprox can provide accurate results for MAX and MIN query, as other aggregations (SUM, AVG, etc.), AQapprox combines with the results of F-M test stored in Map, if $p\_value \geq \alpha$ (α is a significance level), it shows that values on this column have little variation and fluctuation, AQapprox will apply segment_based reservoir sampling to sample desired number of segments for query. Else if $p\_value < \alpha$, it can be concluded that values on this column are skewed to some extent, AQapprox applies cluster sampling with unequal probability, the inclusion probability of each segment i:

$$\pi_i = \frac{CV(seg\_i)}{\sum_{j=1}^{N} CV(seg\_j)} \tag{1}$$

Case II: query q_i contains WHERE clause, i.e. WHERE sales $\geq$ low and sales $\leq$ high. To handle a range constraint in the predicate, we use B+ tree index. In offline pre-processing work, a B+ tree is built for each commonly used numerical column $\mathcal{C}$, each leaf node is associated with a list of $\mathcal{C}$'s values within certain range, and for each $\mathcal{C}$'s value at each leaf, we associate it with IDs (pointers) of the records with this value on $\mathcal{C}$. When a range constraint $low \leq \mathcal{C} \leq high$ is in the predicate, AQapprox uses B+ tree to retrieve IDs of records with $low \leq t_{\mathcal{C}} \leq high$, then counts the number of records that satisfy the condition in each segment i, denote as η_i. So the inclusion probability of each segment i is:

$$\pi_i = \frac{\eta_i}{\sum_{j=1}^{N} \eta_j} \tag{2}$$

That is, each segment is sampled with probability proportional to the number of records selected in the segment by the query.

Both Categorical and Numerical Columns. If the QCS of q_i contains both categorical and numerical columns, there are two main cases. Case I: only categorical columns in WHERE clause. When the query is do aggregate on numerical columns and the sub-dataset in WHERE clause belongs to categorical columns, like the following example Q3:

SELECT SUM(price) FROM T WHERE product_category='Books';

In AQapprox, we use a sampling ratio of $p\%$, note that this ratio is the percentage of data sampled for each sub-dataset, while not the percentage of the whole dataset. Our goal is to sample enough records for this sub-dataset to satisfy the sampling ratio. Combining Map, if the K-S test $p\_value \geq \alpha$, it shows the sub-dataset is distributed uniformly on the whole dataset, AQapprox uses segment-based reservoir sampling. Otherwise, it can be concluded that the sub-dataset is distributed skewed to some extent, AQapprox applies cluster sampling with unequal probability, segments containing more records belonging to the queried sub-dataset will have a higher probability to be sampled. Suppose a sub-dataset is distributed over N segments, each segment i contains M_i records belonging to the queried sub-dataset, where M_i is referred to as the occurrences of a sub-dataset in segment i. We calculate the inclusion probability of each segment i as:

$$\pi_i = \frac{M_i}{\sum_{j=1}^{N} M_j} \tag{3}$$

Case II: both categorical and numerical columns in WHERE clause. The process of AQapprox to handle this type of query is similar to Case I.

3.5   Approximation and Error Estimation

If AQapprox applies segment-based reservoir sampling, the error bounds can be computed using statistics [13]. Now we focus on the approximate results and error estimation of cluster sampling. Take Q3 as an example, approximation for other aggregation functions and other types of QCSs are similar.

Suppose a table T contains R records, and the population is partitioned into N segments, each segment i contains M_i records belonging to the queried sub-dataset, each record in segment i has an associated value v_{ij}. The ground sum is $\tau = \sum_{i=1}^{N} \sum_{j=1}^{M_i} v_{ij}$. To approximate τ, we need to sample a list of n segments based on the inclusion probability π_i. Combined with one stage cluster sampling [13] with unequal probability, the approximation can be estimated as:

$$\hat{\tau} = \frac{1}{n} \sum_{i=1}^{n} (\tau_i/\pi_i) \pm \epsilon \tag{4}$$

the error bound ϵ is defined as:

$$\epsilon = t_{n-1,1-\alpha/2} \sqrt{\hat{V}(\hat{\tau})}, \hat{V}(\hat{\tau}) = \frac{1}{n} \frac{1}{n-1} \sum_{i=1}^{n} (\frac{\tau_i}{\pi_i} - \hat{\tau})^2 \tag{5}$$

where $t_{n-1,1-\alpha/2}$ is the value of Student t-distribution with $n-1$ degrees of freedom at the confidence α, s_τ^2 is the variance of τ_i from each sampled segment.

3.6   The Optimal Segment Partition

In this section, we provide a practical guide on how to set the optimal segment size. The cost of a sample design with m segments is:

$$Cost(sample) = mc_1 + mSc_2 \tag{6}$$

S represents segment size, c_1 is the cost of one segment seek time and storage cost of storing one segment information in Map, c_2 is the cost of reading one record in a segment. With a fixed variance $\hat{V}(\hat{\tau})$, our goal is to minimize the cost. To derive the optimal S from Eq. 6, we incorporate Kish's formula on cluster sampling [17]:

$$\hat{V}(\hat{\tau}) = \hat{V}_{srs}(\hat{\tau}) \times deff, \ deff = 1 + (S-1)\rho \tag{7}$$

where $\hat{V}_{srs}(\hat{\tau})$ represents the variance using simple random sampling, $deff$ is the design effect of clustering sampling, ρ denotes the homogeneity of records in a segment. Then we can get the following formulas:

$$Cost(sample) = \frac{c_1 R^2 s^2 (1 + Sc_2/c_1)[1 + (S-1)\rho]}{S \times \hat{V}(\hat{\tau})} \tag{8}$$

R is the total number of records, s^2 is the variance of values in all the records. By minimizing $Cost(sample)$ in Eq. 8, the optimal segment size is:

$$S = \sqrt{\frac{c_1}{c_2} \frac{1-\rho}{\rho}} \tag{9}$$

4   Experimental Evaluation

4.1   Experimental Setup

Hardware. We ran experiments on an Ubuntu 18.04 Server with 8 Intel i7-9700K cores, 64 GB RAM and 1 TB SSD, PYTHON 3.6, PostgreSQL 12.0.

Datasets. We evaluate AQapprox on three datasets: Amazon review dataset [1], LINEITEM table from TPC-H benchmark [2] and a synthetic dataset used to experiment with skewed numerical columns. Table 2 gives the details.

Table 2. Datasets

Dataset	Size(kb)	Number of records	Avg record size(kb)	Total columns	S
Amazon	2960163	4,000,000	0.74	15	800
TPC-H	613808	5,000,000	0.12	16	1000
Synthetic	406954	5,000,000	0.08	4	1000

Baselines. For queries that are supported by Sapprox, we compare AQapprox with Sapprox (as [18] had verified that Sapprox was better than other sampling schemes, we did not evaluate them here). Else we compare AQapprox with PostgreSQL System Tablesampling [10] and Stratified Sampling [16].

Evaluation Metric: $RelativeError = \frac{|approximation\ answer - precise\ answer|}{precise\ answer}$.
Sampling ratio $p = 10\%$. We ran each method 10 times and reported the average.

4.2 Experimental Results

Storage Cost Comparison. The coverage of (past queries' or all possible) QCSs reaches to 93.75% for both Amazon and TPC-H.

Table 3. Storage cost ratio

Dataset	Method	
	Sapprox	AQapprox
Amazon review	0.026%	1.580%
TPC-H	0.152%	0.787%

As shown in Table 3, the storage cost of AQapprox is larger than Sapprox, but its proportion of total size is still relatively small, which is acceptable.

Categorical Columns Only. Due to space constraints, we only show the results on Amazon dataset, results on TPC-H are similar. The running query is:
SELECT COUNT(*) FROM Amazon WHERE product='Books';

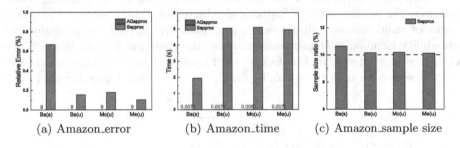

(a) Amazon_error   (b) Amazon_time   (c) Amazon_sample size

Fig. 3. Approximation accuracy and efficiency of AQapprox and Sapprox on Amazon dataset. Bs, Mc, Me: Books, Music, Mobile. (u, s): (uniform, skewed).

As shown in Fig. 3, whether it is a uniformly distributed or skewed sub-dataset, AQapprox is very fast and error-free when facing such type of queries.

Both Categorical and Numerical Columns. Due to space constraints, only the results of TPC-H are shown here. We also experiment with other forms of queries, the results are similar.
SELECT AVG(extendedprice) FROM LINEITEM WHERE shipmode='MAIL';

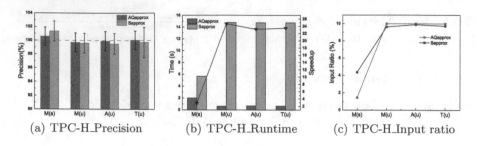

(a) TPC-H_Precision   (b) TPC-H_Runtime   (c) TPC-H_Input ratio

Fig. 4. Approximation accuracy, efficiency and input ratio on TPC-H

In Fig. 4(a), all approximation results are normalized to the precise results indicated by the 100% guideline, the error bars show the 95% confidence intervals. For sub-dataset queries on a single column, Map information established by AQapprox and Sapprox are the same, so the accuracy is similar. In Fig. 4(b), AQapprox achieves a speedup by up to 3.5× for skewed sub-dataset and 25.1× for the uniform over Sapprox, because AQapprox uses different methods to sample uniform and skewed sub-datasets. The speedup for skewed sub-dataset can also be explained by the actual low input ratio in Fig. 4(c).

Experiments on Multi-column Sub-datasets. We evaluate the following queries:
SELECT AVG(extendedprice) FROM LINEITEM WHERE shipmode='MAIL';
(*WHERE shipmode='MAIL' and returnflag='N';*)
(*WHERE shipmode='MAIL' and returnflag='N' and linestatus='O';*)

Figure 5(a) reports that as the number of query columns in WHERE clause increases, the relative error of Sapprox also increases, this is because the occurrences of multi-column sub-datasets in Sapprox are derived using conditional probability formula, as the columns increase, the error will become large. While AQapprox is no affected and performs well. Figure 5(b) shows AQapprox achieves 24.6× speedup for uniform and 5.9× for skewed sub-dataset over Sapprox.

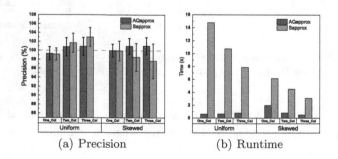

(a) Precision   (b) Runtime

Fig. 5. Approximation accuracy and efficiency for multi-columns on TPC-H.

Numerical Columns Only. We compare AQapprox with other baselines.
SELECT SUM(sales) FROM synthetic_table;

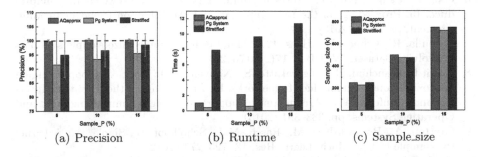

     (a) Precision   (b) Runtime   (c) Sample_size

Fig. 6. Approximation accuracy and efficiency on synthetic dataset

As results in Fig. 6(a), the relative error of AQapprox is much smaller than others, and the confidence intervals produced by baselines are extremely wide, which are unacceptable. In Fig. 6(b), the runtime of AQapprox is much shorter than Stratified Sampling, and slightly longer than PostgreSQL System Table-sampling, but AQapprox achieves a good trade-off between time and accuracy.

5   Conclusion and Future Work

In this paper, we propose AQapprox to enable both efficient and accurate approximations for aggregation queries. It employs a probabilistic Map to store statistics about the original dataset, and combines Map to perform distribution-aware online sampling. Experimental results show that AQapprox is more accurate and fast than baselines. In the future work, we will optimize the selection of QCSs to be created with Map information.

Acknowledgments. This work was supported by NSFC grants (No. 61532021 and 61972155), Shanghai Knowledge Service Platform Project (No. ZF1213).

References

1. Amazon review data. http://jmcauley.ucsd.edu/data/amazon/
2. Tpc-h benchmark. http://www.tpc.org/tpch/
3. Agarwal, S., Mozafari, B., Panda, A., Milner, H., Madden, S., Stoica, I.: BlinkDB: queries with bounded errors and bounded response times on very large data. In: Proceedings of the 8th ACM European Conference on Computer Systems, pp. 29–42 (2013)
4. Chaudhuri, S., Ding, B., Kandula, S.: Approximate query processing: no silver bullet. In: Proceedings of the 2017 ACM International Conference on Management of Data, pp. 511–519 (2017)

5. Ding, B., Huang, S., Chaudhuri, S., Chakrabarti, K., Wang, C.: Sample+seek: approximating aggregates with distribution precision guarantee. In: Proceedings of the 2016 International Conference on Management of Data, pp. 679–694 (2016)
6. Gan, Y., Meng, X., Shi, Y.: Processing online aggregation on skewed data in mapreduce. In: Proceedings of the Fifth International Workshop on Cloud Data Management, pp. 3–10 (2013)
7. Gemulla, R., Lehner, W., Haas, P.J.: Maintaining bounded-size sample synopses of evolving datasets. VLDB J. **17**(2), 173–201 (2008)
8. Goiri, I., Bianchini, R., Nagarakatte, S., Nguyen, T.D.: ApproxHadoop: bringing approximations to mapreduce frameworks. In: Proceedings of the Twentieth International Conference on Architectural Support for Programming Languages and Operating Systems, pp. 383–397 (2015)
9. Gretton, A., Borgwardt, K.M., Rasch, M.J., Schölkopf, B., Smola, A.: A kernel two-sample test. J. Mach. Learn. Res. 13, 723–773 (2012)
10. Haas, P.J., König, C.: A bi-level bernoulli scheme for database sampling. In: Proceedings of the 2004 ACM SIGMOD International Conference on Management of Data, pp. 275–286 (2004)
11. Kandula, S., et al.: Quickr: lazily approximating complex AdHoc queries in bigdata clusters. In: Proceedings of the 2016 International Conference on Management of Data, pp. 631–646 (2016)
12. Li, K., Zhang, Y., Li, G., Tao, W., Yan, Y.: Bounded approximate query processing. IEEE Trans. Knowl. Data Eng. **12**, 2262–2276 (2019). https://doi.org/10.1109/TKDE.2018.2877362
13. Lohr, S.L.: Sampling: Design and Analysis. Nelson Education (2009)
14. Marwick, B., Krishnamoorthy, K.: cvequality: tests for the equality of coefficients of variation from multiple groups. R software package version, vol. 1, p. 3 (2018)
15. Pansare, N., Borkar, V.R., Jermaine, C., Condie, T.: Online aggregation for large mapreduce jobs. Proc. VLDB Endowment **4**(11), 1135–1145 (2011)
16. Park, Y., Mozafari, B., Sorenson, J., Wang, J.: VerdictDB: universalizing approximate query processing. In: Proceedings of the 2018 International Conference on Management of Data, pp. 1461–1476 (2018)
17. Wiegand, H.: Kish, l.: Survey Sampling. Wiley, New York (1965). ix + 643 s., 31 abb., 56 tab., preis 83 s. **10**(1), 88–89 (2010)
18. Zhang, X., Wang, J., Yin, J.: Sapprox: enabling efficient and accurate approximations on sub-datasets with distribution-aware online sampling. Proc. VLDB Endowment **10**(3), 109–120 (2016)

LogRank+: A Novel Approach to Support Business Process Event Log Sampling

Cong Liu[1], Yulong Pei[2], Qingtian Zeng[3(✉)], Hua Duan[3(✉)], and Feng Zhang[3]

[1] Shandong University of Technology, Zibo, China
liucongchina@163.com
[2] Eindhoven University of Technology, Eindhoven, The Netherlands
y.pei.1@tue.nl
[3] Shandong University of Science and Technology, Qingdao, China
qtzeng@163.com, hduan59@163.com, zhangfengsdkd@163.com

Abstract. Massive amounts of business process event logs are collected and stored by modern information systems. Numerous process discovery approaches have been proposed to extract descriptive process models from such event logs in the past decades. To improve process discovery efficiency, event log sampling techniques are proposed. A sample log is a delicately selected subset of the original log that requires less computational cost. However, existing sampling techniques have difficulties, e.g., low efficiency, in handling large-scale event logs. To tackle this challenge, we propose a novel ranking-based event log sampling approach, denoted as *LogRank+*, to support efficient sampling. In addition, we introduce a framework to evaluate the effectiveness of different sampling techniques by quantifying the sampling efficiency and the quality of sample logs. The proposed sampling approach has been implemented in the open-source process mining toolkit ProM. Experimental evaluation with both synthetic and real-life event logs demonstrates that the proposed sampling approach provides an effective solution to improve event log sampling efficiency as well as ensuring high quality of the obtained sample logs from a process discovery perspective.

Keywords: Event logs · Efficient sampling · Process discovery · Effectiveness evaluation

1 Introduction

Process mining [2, 14, 25] aims at extracting process-oriented insights from business process event logs that are readily available from modern information systems. Process discovery, as one of the most fundamental tasks of process mining, allows to uncover descriptive process models from event logs. Various process discovery approaches, e.g., *Alpha Miner* [3], *Heuristic Miner* [23], and *Inductive Miner* [12], that take as input an event log and produce a process model have been proposed in the past two decades. However, existing process discovery approaches are unable to handle properly or may cause low efficiency when facing large-scale event logs.

© Springer Nature Switzerland AG 2020
Z. Huang et al. (Eds.): WISE 2020, LNCS 12343, pp. 417–430, 2020.
https://doi.org/10.1007/978-3-030-62008-0_29

To improve discovery efficiency, one effective strategy is to re-implement existing discovery approaches using MapReduce to make them scalable to large-scale event logs. *Evermann* presents the MapReduce implementations of the *Alpha Miner* and *Heuristic Miner* in [11]. However, the re-implementation process is extremely time-consuming and requires developers to have extensive knowledge on the underlying discovery approach. Moreover, re-implementation techniques are specially tailored for specific approach and cannot be generalized. Rather than re-implementing existing discovery approaches, event log sampling techniques provide an alternative mean to improve discovery efficiency. Considering for example the *LogRank*-based sampling technique in [18] and [19]. It implements a graph-based ranking model to extract a sample log by taking an arbitrary event log as input. The sample log is much smaller and can be processed more efficiently than the original log. Although sampling techniques facilitate efficient process discovery, the sampling itself is sometimes time-consuming when handling large-scale event logs. To tackle this challenge, this paper proposes a novel ranking-based event log sampling approach, denoted as $LogRank^+$, to support efficient log sampling. In addition, we introduce a framework to evaluate the effectiveness of different sampling techniques from a process discovery perspective.

The rest of this paper is organized as follows. Section 2 presents a brief review of the related work. Section 3 defines some preliminaries. Section 4 presents the research questions and introduces an overview of our approach. Section 5 introduces the $LogRank^+$-based sampling technique and a framework to evaluate the effectiveness of a sampling technique. Section 6 presents tool support. Section 7 presents experimental evaluation. Finally, Sect. 8 concludes the paper.

2   Related Work

Process mining aims to discover, monitor and improve real business processes by extracting knowledge from event logs [2]. As one of the most challenging process mining tasks, *process discovery* has received a lot of attention in the past years. In general, existing process discovery approaches can be categorized into two types. One type discovers a process model that can guarantee 100% fitness against the input log, i.e., all traces from the input log can be replayed by the discovered model. *Inductive Miner* is one typical approach [12] of this type. The other type discovers process models that do not provide 100% fitness guarantee, e.g., *Heuristic Miner* [23]. These approaches typically consider traces not included in the discovered process model as exceptional behavior or noise. Therefore, they are excluded during discovery.

With the growing availability of event logs from current information systems, large-scale event logs have posed new performance challenges for existing process discovery approaches. The main reason is that most discovery approaches are no longer feasible to process an entire large data set using a single machine, due to the hardware limitations such as I/O and memory. In such scenarios, the discovery of process models from large-scale event logs has to resort to current distributed platforms. Considering for example the well-known MapReduce framework [8] has been used to implement the existing process discovery algorithms, e.g., [11]. Their implementations reply on constructing log abstractions such as directly-follows graphs required by current discovery approaches, and the computation progress of which is done in parallel using one or several MapReduce jobs. Although these approaches have shown that they can efficiently speed up standalone algorithms in the presence of large-scale event logs, their designs still follow a conventional way, i.e., applying computation to all traces in a log.

Rather than re-implementing existing discovery approaches, event log sampling techniques provide an alternative mean to handle large-scale event logs. For example, the *LogRank*-based sampling technique in [18] and [19] is capable of sampling a large-scale event log to a smaller size that can be efficiently processes by existing discovery techniques. In addition, this work also provides a solution on evaluating the quality of the sample log against of the original one. Although the *LogRank*-based sampling technique accelerates the discovery process, the sampling itself is sometimes time-consuming when facing large-scale event logs. To handle this problem, this paper proposes a novel ranking-based log sampling approach, denoted as *LogRank$^+$*, to support efficient log sampling.

3   Preliminaries

Let S be a set and $\emptyset$ be the empty set. $|S|$ denotes the number of elements in S. $+$ and $-$ represent the union and difference of two sets. $\mathbf{B}(S)$ is the set of all multisets over set S. $f \in X \to Y$ is a function. A sequence over S of length n is a function $\sigma : \{1, 2, ..., n\} \to S$. If $\sigma(1) = a_1, \sigma(2) = a_2, ...\sigma(n) = a_n$, we write $\sigma = \langle a_1, a_2, ...a_n \rangle$. The set of all finite sequences over S is denoted by S^*.

Definition 1 *(Event, Trace, Event Log). Let A be a set of activities. A trace $\sigma \in A^*$ is a sequence of activities (also referred to as events). For $1 \le i \le |\sigma|$, $\sigma(i)$ represents the ith event of σ. $L \in \mathbf{B}(A^*)$ is an event log.*

To discover process models, we rely on event logs. An event log can be considered as a multiset of traces [2] because there can be multiple process instances (or cases) having the same trace. Each trace describes the life-cycle of a particular instance (or case).

Definition 2 *(Process Discovery). Let U_M be the universe of all process models. A process discovery approach is a function γ that maps an event log $L \in \mathbf{B}(A^*)$ to a process model $pm \in U_M$, i.e., $\gamma(L) = pm$.*

Generally speaking, a process discovery approach is able to convert an event log to a process model that is represented by *labeled Petri net*, *BPMN*, etc. Whatever representation is used, each trace in the input event log corresponds to a possible execution sequence in the discovered process model.

Definition 3 *(Event Log Sampling). An event log sampling technique is a function Δ that maps one event log $L \in \mathbf{B}(A^*)$ to another one $L' \in \mathbf{B}(A^*)$ such that $L' \subseteq L$. L is called the original log, and L' is called a sample log of L.*

According to Definition 3, a sampling technique takes as input an original log and returns one of its subsets.

4   An Approach Overview

In general, this paper aims to answer the following two research questions:

- **RQ1:** How can we find an efficient sampling technique to obtain a sample log out of a large-scale event log?
- **RQ2:** Given a sampling technique, how can we evaluate if it is effective enough from a discovery point of view?

The answer to the first question provides an efficient technique to sample a large-scale event log to a relatively small one which can be used for fast discovery. The answer to the second one is used to evaluate the effectiveness of a sampling technique. Answers to these two consecutive questions perfectly summarize the main contributions of the paper. Figure 1 shows an overview of our approach which contains the following two phases:

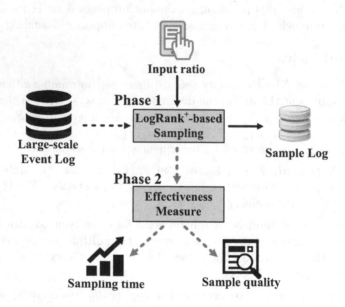

Fig. 1. An approach overview.

- **Phase 1: LogRank⁺-based Event Log Sampling Technique.** By taking an event log and a user input sampling ratio as input, we first propose the *LogRank⁺*-based approach to obtain its corresponding sample log.
- **Phase 2: Effectiveness Evaluation of Sampling Techniques.** Given a sampling technique, the effectiveness can be evaluated from the following two perspectives.

- **Quality of the sample log.** To quantify the quality of a sample log, we first discover a process model from the sample log. Note that the discovered model should guarantee 100% fitness against the sample log. Then, quality of the sample log is quantified based on the fitness of the original log and the discovered model.
- **Sampling efficiency.** The sampling efficiency is quantified by the time spent to obtain the sample log. The less time a sampling technique spent, the more efficient it is.

5 LogRank+-Based Event Log Sampling

In this section, we first introduce a novel ranking-based event log sampling approach to sample event logs, and then explain how to evaluate and compare the effectiveness of different sampling approaches.

5.1 LogRank+-Based Sampling

Event log sampling techniques aim to select a representative subset of the original log that can be analyzed with more efficiency. It is argued that a trace is more representative than others if it contains more information (or behavior) of the whole log. Therefore, the representativeness of a trace can be quantified by the similarity between the current trace and the rest of all traces. Based on this argument, we propose the *LogRank+*, a ranking-based sampling technique, to first order traces according to the similarity value, and then selecting the most representative ones to construct the sample log.

To measure similarity, we first convert a trace to a vector by selecting a typical set of features, and then compute the similarity of two vectors using the *Euclidean* distance measure [10]. Inspired by [21], we characterize traces by *profiles*, where a profile is a set of related features that describe a trace from a specific perspective. Every feature is a metric, which assigns a specific numeric value to each trace. In this way, we consider a profile with n features to be a function that maps one trace or a set of traces to a n-dimensional vector. In this paper, we use two types of profiles to create the mapping. The *activity profile* defines one feature per event/activity name, and the *directly follow profile* defines one feature for each directly follow relation.

Definition 4 (Activity Profile). *Given an event log $L \in \mathbf{B}(A^*)$. The activity profile of L is defined as: $actProfile(L) = \bigcup_{\sigma \in L} \{\sigma(i) | 1 \le i \le |\sigma|\}$.*

Definition 5 (Directly Follow Profile). *Given an event log $L \in \mathbf{B}(A^*)$. The directly follow profile of L is defined as: $dfgProfile(L) = \bigcup_{\sigma \in L} \{(\sigma(i), \sigma(i+1)) | 1 \le i \le |\sigma| - 1\}$.*

Consider event log L_T as an example. Table 1 shows the *activity profile* and *directly follow profile* of two derived logs from L_T, i.e., $\{\sigma_1\}$ is a log with only one trace, and $L_T - \{\sigma_1\}$ is a log by excluding σ_1 from L_T.

$L_T = [\sigma_1 = \langle a,c,d,e,h \rangle^{16}, \sigma_2 = \langle a,b,d,e,g \rangle^9, \sigma_3 = \langle a,b,d,e,h \rangle^8, \sigma_4 = \langle a,c,d,e,g \rangle^4]$.

Table 1. Profile examples of two logs

Log	Activity profile							Directly follow profile						
	a	b	c	d	e	g	h	(a, b)	(a, c)	(c,d)	(b,d)	(d,e)	(e,h)	(e.g)
$\{\sigma_1\}$	1	0	1	1	1	1	0	1	0	1	1	0	1	1	0
$L_T - \{\sigma_1\}$	1	1	1	1	1	1	1	1	1	1	1	1	1	1

Profiles can be represented as a n-dimensional vector where n means the number of features extracted based on the selected profiles. Therefore, event log $L_p \in \mathbf{B}(A^*)$ corresponds to vector $v_p = \langle i_{p1}, i_{p2}, ..., i_{pn} \rangle$ such that i_{pn} denotes the existence of feature n in L_p. To calculate the distance between event logs, e.g., $L_p, L_q \in \mathbf{B}(A^*)$, we use *Euclidean* distance which can be computed as follows:

$$Distance(L_p, L_q) = \sqrt{\sum_{l=1}^{n} |i_{pl} - i_{ql}|^2} \tag{1}$$

Therefore, the similarity between L_p and L_q is computed as follows:

$$Similarity(L_p, L_q) = 1 - Distance(L_p, L_q) \tag{2}$$

To sum up, the procedure of *LogRank*$^+$-based sampling by taking as input an event log and a sample ratio is described as follows:

- Step 1: for each trace $\sigma \in L$, we calculate the similarity between σ and the rest of traces in L by $Similarity(\{\sigma\}, L - \sigma)$ based on Eq. (2);
- Step 2: ranking all traces in L based on the obtained similarity values; and
- Step 3: selecting the top-N traces according to the input sample ratio.

5.2   Effectiveness Evaluation of Sampling Techniques

Given a large-scale event log, a sampling technique obtains a sample log with smaller size and less processing time. However, it is not clear how effective is a given sampling technique compared to others. This section proposes a framework to evaluate the effectiveness of a sample technique by quantifying the quality of the obtained sample logs and the sampling efficiency.

The goal of sampling is to improve process discovery efficiency without sacrificing (too much of) the quality of the discovered model. A sample log is typically not complete and may lead to discovery of low quality models. Therefore, given a sample log, the question is if it is representative enough to discover a process model of high quality compared to that discovered directly from the original log. To this end, we propose to measure the *fitness* of the process model discovered from the sample log against the original log. According to *Buijs et al.* [6], *fitness* quantifies the extent to which a process model can accurately reproduce traces in the log. The rationale behind is that if a model discovered from the sample log can replay all (or majority of) traces in the original log, we argue that the sample log is of high quality for process discovery. One of the most important factors to ensure the applicability of this quality measure is that we should guarantee the model discovered from the sample log can fully represent the behavior in the sample log, i.e., 100% fitness. Obviously, it does not make much sense to replay the original log against a model discovered from a sample log if this model cannot cover all possible behavior in the sample log. Therefore, we should select a process discovery approach that can guarantee 100% fitness. As *Inductive Miner* [12] can guarantee 100% fitness of the discovered model against the input log, it is selected in this paper.

In addition, the effectiveness and usability of a sampling technique heavily rely on its efficiency. Considering the fact that practitioners would be reluctant to select a sampling technique that needs hours to finish even if the obtained sample log is of high quality. In this paper, the efficiency of a sampling technique is quantified by the time spent to obtain the sample log. In general, the less time a sampling technique spent, the more efficient it is.

6 Tool Support

The open-source (Pro)cess (M)ining framework *ProM 6* provides a completely plugable environment for process mining and related topics. It can be extended by adding plug-ins, and currently, more than 1600 plug-ins are included. The framework can be downloaded freely.[1]

The proposed *LogRank+*-based sampling technique has been implemented as a plug-in (called *LogRank+-based Event Log Sampling*)[2] in the framework. A snapshot of the tool is shown in Fig. 2. It takes an event log and a sample ratio as input, and returns a sample log as output. Note that all sample logs in the following experiments are produced by this plug-in.

---

[1] http://www.processmining.org/.
[2] https://svn.win.tue.nl/repos/prom/Packages/SoftwareProcessMining/.

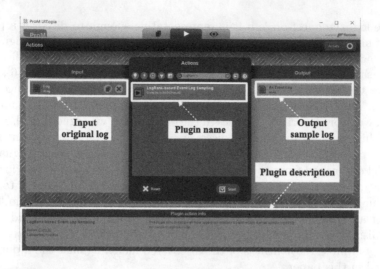

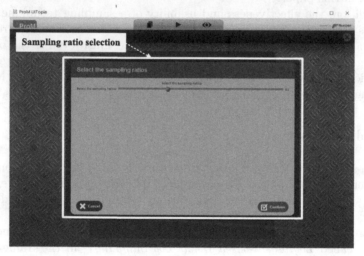

Fig. 2. A snapshot of the plugin in ProM 6.

7   Experimental Evaluation

In this section, we perform the experimental evaluation of the proposed $LogRank^+$-based sampling approach using four event logs (one synthetic log and three real-life ones). Table 2 reports the main descriptive statistics of these logs.

- **Synthetic data set:** The Synthetic data set is generated by a paper review process model, and each trace describes the procedure to review a paper.
- **Sepsis data set:** The Sepsis data set contains events of sepsis cases from a hospital. Each trace represents the pathway of an individual Sepsis patient.

- **BPI2011 data set:** The BPI2011 data set represents the gynecology department of a Dutch academic hospital, and each trace represents the medical activities that are performed by a patient.
- **BPI2012 data set:** The BPI2012 data set originates from a personal loan application process in a Dutch financial institution. The traces represent different customers applying for personal loans.

Table 2. An overview of the experimental event logs

Event log	#Traces	#Events	#Activities
Synthetic log [1]	100	2297	20
Sepsis [20]	1050	15214	16
BPI2011 [5]	1143	150291	624
BPI2012 [9]	13087	262200	36

In the following, we present the experimental results in light of the two research questions defined in Sect. 4.

RQ1: How can we find an efficient sampling technique to obtain a sample log out of a large-scale event log?

To answer this question, we propose the *LogRank+*-based sampling technique to support efficient event log sampling. This approach first ranks the importance of all traces, and then selects the top-N traces based on the input sampling ratio as the sample log. In the following experiment, we produce a set of sample logs with different ratios (from 5% to 30% with an increment of 5%) for each experiment log using the *LogRank+-based Event Log Sampling* plug-in.

RQ2: Given a sampling technique, how can we evaluate if it is effective enough from a discovery point of view?

Towards this question, we propose to measure the effectiveness of sampling techniques by quantifying (1) the quality of the sample log against the original one; and (2) the efficiency of a given sampling approach. In addition, we also compare the proposed *LogRank+*-based sampling technique with the state-of-the-art ranking-based sampling technique, i.e., *LogRank*-based approach in [19].

To measure the quality of a sample log against the original one, we propose to compute *fitness* of the process model discovered from the sample log against the original log. The rationale behind is that if a model discovered from the sample log can replay all (or majority of) the traces in the original log, we argue that the sample log is of high quality from a discovery perspective. More specifically, we first discover a process model using the *Inductive Miner* for each sample log, and then replay its original log against this model to obtain the fitness value. Detailed fitness values of *LogRank*-based and *LogRank+*-based sampling approaches using the four experimental logs are shown in Fig. 3.

The sampling efficiency is quantified by calculating the time a sampling technique required to obtain the sample log. Generally, the less time a sampling technique spent, the more efficient it is. By taking the four experimental logs with

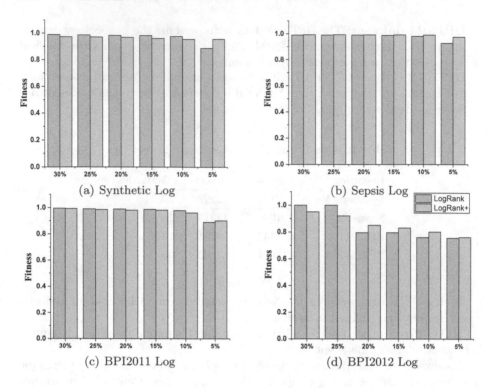

Fig. 3. Fitness comparison results.

different sampling ratios as input, the execution time of the plug-ins that implement the *LogRank*-based and *LogRank*⁺-based sampling techniques are recorded and demonstrated in Fig. 4 to compare the sampling efficiency. Note that for each experimental log with a given sampling ratio, we run the plug-in for five times, and the median values are highlighted in the box diagrams.

In general, the quality of sample logs decreases as the size of the log decreases for the two sampling techniques according to Fig. 3. It is worth noting that the quality decrease is relatively slow and can be maintained above a proper value. Considering the Synthetic, Sepsis, and BPI2011 logs as examples, the fitness values decrease only from 1 to 0.9 for sample logs with ratio range from 30% to 10%, i.e., the mainstream behavior in the original log are kept in sample logs. In addition, the quality of sample logs for both *LogRank*-based and *LogRank*⁺-based sampling techniques are similar. Differently, the sampling time spent by *LogRank*⁺-based sampling technique is far less than that of the *LogRank*-based sampling technique as shown in Fig. 4. Considering the Sepsis log as an example, compared with the *LogRank*-based sampling technique, the sampling time spent by the *LogRank*⁺-based sampling technique decreases almost 8 times from 27500 milliseconds to 3000 milliseconds while the quality of the sample logs are similar. For the BPI2012 log, compared with the *LogRank*-based technique the sampling time of the *LogRank*⁺-based sampling technique with 30% ratio is reduced from

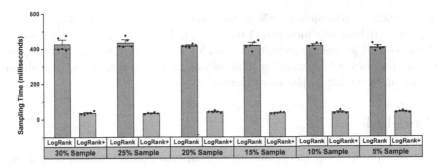

(a) Synthetic Log

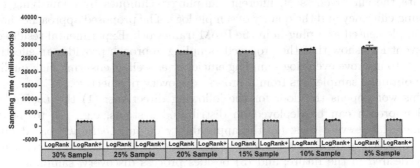

(b) Sepsis Log

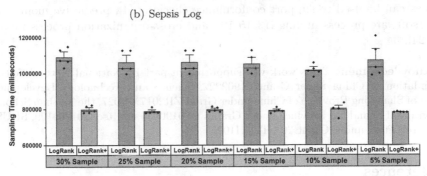

(c) BPI2011 Log

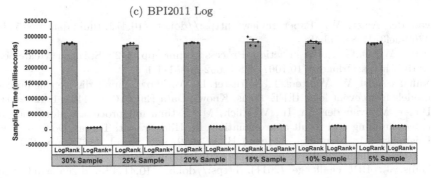

(d) BPI2012 Log

Fig. 4. Sampling efficiency comparison.

about 2750000 milliseconds to 85000 milliseconds (saving about 45 min) while their quality are basically identical (1.00 vs 0.97). Therefore, we conclude that the *LogRank*$^+$-based sampling technique provides an effective solution to improve sampling efficiency while ensuring high quality of the sample logs. In addition, the larger an event log is, the more time it saves.

8   Conclusion

To improve sampling efficiency, this paper proposes a novel ranking-based sampling technique, called *LogRank*$^+$. In addition, a framework is introduced to evaluate the effectiveness of different sampling techniques by quantifying the sampling efficiency and the quality of sample logs. The proposed approaches have been implemented as a plug-in in the ProM framework. Experimental results on four event logs show that the proposed sampling approach provides an effective solution to improve event log sampling efficiency as well as ensuring high quality of the obtained sample logs from a process discovery perspective.

This work opens the door for the following directions: (1) the *LogRank*$^+$-based approach can be deployed on distributed systems, e.g., [7], to handle extremely large event logs; (2) the applicability of our *LogRank*$^+$-based sampling approach to other real-life event logs with dedicated domain knowledge is highly desired in the future; and (3) besides process discovery, sampling techniques can be used to support conformance checking [4], predictive monitoring [22], software process mining [13,15,16], and cross-organization process mining [17,24].

Acknowledgement. This work was supported in part by National Natural Science Foundation of China under Grant 61902222, Science and Technology Development Fund of Shandong Province of China under Grant ZR2017MF027, the Taishan Scholars Program of Shandong Province under Grants ts20190936 and tsqn201909109, SDUST Research Fund under Grant 2015TDJH102.

References

1. van der Aalst, W.: Paper review. https://doi.org/10.4121/uuid:da6aafef-5a86-4769-acf3-04e8ae5ab4fe
2. Aalst, W.: Data science in action. Process Mining, pp. 3–23. Springer, Heidelberg (2016). https://doi.org/10.1007/978-3-662-49851-4_1
3. Van der Aalst, W., Weijters, T., Maruster, L.: Workflow mining: discovering process models from event logs. IEEE Trans. Knowl. Data Eng. **16**(9), 1128–1142 (2004)
4. Bauer, M., van der Aa, H., Weidlich, M.: Estimating process conformance by trace sampling and result approximation. In: Hildebrandt, T., van Dongen, B.F., Röglinger, M., Mendling, J. (eds.) BPM 2019. LNCS, vol. 11675, pp. 179–197. Springer, Cham (2019). https://doi.org/10.1007/978-3-030-26619-6_13
5. Buijs, J.: BPI challenge (2011). https://doi.org/10.4121/uuid:26aba40d-8b2d-435b-b5af-6d4bfbd7a270

6. Buijs, J.C.A.M., van Dongen, B.F., van der Aalst, W.M.P.: On the role of fitness, precision, generalization and simplicity in process discovery. In: Meersman, R. (ed.) OTM 2012. LNCS, vol. 7565, pp. 305–322. Springer, Heidelberg (2012). https://doi.org/10.1007/978-3-642-33606-5_19

7. Cheng, L., Li, T.: Efficient data redistribution to speedup big data analytics in large systems. In: 2016 IEEE 23rd International Conference on High Performance Computing (HiPC), pp. 91–100. IEEE (2016)

8. Dean, J., Ghemawat, S.: Mapreduce: simplified data processing on large clusters. Commun. ACM **51**(1), 107–113 (2008)

9. van Dongen, B.: Bpi (2012). https://doi.org/10.4121/uuid:3926db30-f712-4394-aebc-75976070e91f

10. Duda, R.O., Hart, P.E., Stork, D.G.: Pattern Classification. Wiley, Hoboken (2012)

11. Evermann, J.: Scalable process discovery using map-reduce. IEEE Trans. Serv. Comput. **9**(3), 469–481 (2016)

12. Leemans, S.J.J., Fahland, D., van der Aalst, W.M.P.: Discovering block-structured process models from event logs - a constructive approach. In: Colom, J.-M., Desel, J. (eds.) PETRI NETS 2013. LNCS, vol. 7927, pp. 311–329. Springer, Heidelberg (2013). https://doi.org/10.1007/978-3-642-38697-8_17

13. Liu, C.: Automatic discovery of behavioral models from software execution data. IEEE Trans. Autom. Sci. Eng. **99**, 1–12 (2018)

14. Liu, C.: Hierarchical business process discovery: identifying sub-processes using lifecycle information. In: International Conference on Web Services, pp. 1–5. IEEE (2020)

15. Liu, C., van Dongen, B.F., Assy, N., van der Aalst, W.M.P.: Component interface identification and behavioral model discovery from software execution data. In: International Conference on Program Comprehension, pp. 97–107. ACM (2018)

16. Liu, C., van Dongen, B., Assy, N., van der Aalst, W.M.: Component behavior discovery from software execution data. In: 2016 IEEE Symposium Series on Computational Intelligence (SSCI), pp. 1–8. IEEE (2016)

17. Liu, C., Duan, H., Qingtian, Z., Zhou, M., Lu, F., Cheng, J.: Towards comprehensive support for privacy preservation cross-organization business process mining. IEEE Trans. Serv. Comput. **12**(4), 639–653 (2019)

18. Liu, C., Pei, Y., Cheng, L., Zeng, Q., Duan, H.: Sampling business process event logs using graph-based ranking model. Concurrency and Computation: Practice and Experience XX, pp. 1–15 (2020)

19. Liu, C., Pei, Y., Zeng, Q., Duan, H.: LogRank: an approach to sample business process event log for efficient discovery. In: Liu, W., Giunchiglia, F., Yang, B. (eds.) KSEM 2018. LNCS (LNAI), vol. 11061, pp. 415–425. Springer, Cham (2018). https://doi.org/10.1007/978-3-319-99365-2_36

20. Mannhardt, F.: Sepsis. https://doi.org/10.4121/uuid:915d2bfb-7e84-49ad-a286-dc35f063a460

21. Song, M., Günther, C.W., van der Aalst, W.M.P.: Trace clustering in process mining. In: Ardagna, D., Mecella, M., Yang, J. (eds.) BPM 2008. LNBIP, vol. 17, pp. 109–120. Springer, Heidelberg (2009). https://doi.org/10.1007/978-3-642-00328-8_11

22. Verenich, I., Dumas, M., Rosa, M.L., Maggi, F.M., Teinemaa, I.: Survey and cross-benchmark comparison of remaining time prediction methods in business process monitoring. ACM Trans. Intelli. Syst. Technol. (TIST) **10**(4), 1–34 (2019)

23. Weijters, A., Ribeiro, J.: Flexible heuristics miner (FHM). In: 2011 IEEE Symposium on Computational Intelligence and Data Mining (CIDM), pp. 310–317. IEEE (2011)

24. Zeng, Q., Duan, H., Liu, C.: Top-down process mining from multi-source running logs based on refinement of Petri nets. IEEE Access **8**, 61355–61369 (2020)
25. Zeng, Q., Sun, S.X., Duan, H., Liu, C., Wang, H.: Cross-organizational collaborative workflow mining from a multi-source log. Decis. Support Syst. **54**(3), 1280–1301 (2013)

ProvONE+: A Provenance Model for Scientific Workflows

Anila Sahar Butt$^{(\boxtimes)}$ and Peter Fitch

CSIRO Land and Water, Canberra, Australia
{anila.butt, peter.fitch}@csiro.au

Abstract. The provenance of workflows is essential, both for the data they derive and for their specification, to allow for the reproducibility, sharing and reuse of information in the scientific community. Although the formal modelling of scientific workflow provenance was of interest and studied, in many fields like semantic web, yet no provenance model has existed, we are aware of, to model control-flow driven scientific workflows. The provenance models proposed by the semantic web community for data-driven scientific workflows may capture the provenance of control-flow driven workflows execution traces (i.e., retrospective provenance) but underspecify the workflow structure (i.e., workflow provenance). An underspecified or incomplete structure of a workflow results in the misinterpretation of a scientific experiment and precludes conformance checking of the workflow, thereby restricting the gains of provenance. To overcome the limitation, we present a formal, lightweight and general-purpose specification model for the control-flows involved scientific workflows. The proposed model can be combined with the existing provenance models and easy to extend to specify the common control-flow patterns. In this article, we inspire the need for control-flow driven scientific workflow provenance model, provide an overview of its key classes and properties, and briefly discuss its integration with the ProvONE provenance model as well as its compatibility to PROV-DM. We will also focus on the sample modelling using the proposed model and present a comprehensive implementation scenario from the agricultural domain for validating the model.

Keywords: Workflow provenance · Provenance model · Control-flow patterns

1 Introduction

Scientific workflows are a popular mechanism for specifying and automating repetitive experiments. One essential part of their importance lies in their capacity for reuse. If a workflow is defined and shared, it becomes a valuable building block which can be combined or changed to create new experiments. Scientific Workflow Management Systems (SWfMS) have become a critical tool for many applications, allowing complex analysis on distributed resources to be written

© Springer Nature Switzerland AG 2020
Z. Huang et al. (Eds.): WISE 2020, LNCS 12343, pp. 431–444, 2020.
https://doi.org/10.1007/978-3-030-62008-0_30

and carried out [6]. Various SWfMS are proposed and built to provide the environment for specifying and enacting workflows (e.g., Taverna, Kepler, Triana, and YAWL). These systems often adopt simple computational models, particularly a data-flow model, in which the order of execution of workflow modules is determined by the data flow through the workflow [11]. This contrasts with business workflows which provide expressive languages such as the Business Process Execution Language (BPEL) to describe complex control-flows. Additionally, unlike business workflows, scientific workflows are mostly used to perform data-intensive tasks. Nevertheless, some SWfMS support control-flow constructs (e.g., control loops), and the resulting workflows express semantics driven by the control-flow [5,8]. For example Taverna [9], using SCUFL as its modelling language, supports control constraints to define execution ordering and conditional constructs which corresponds to `if/else` or `case` structure. Triana [15] provides control-flow components to support `loops` and `triggers`. Similarly, Kepler [1] allows some level of branching and synchronisation through its supported constructs.

It is crucial for SWfMS to consider their workflow provenance, which concerns the reliability and integrity of workflows and the data as they are being routed in complex workflows. This offers valuable documentation which is a key to maintaining data, assessing the accuracy and authorship of the data, and reproducing and validating the findings. Therefore, provenance in scientific workflows is of increasing importance, due to its ability to reproduce results from earlier runs, explain unexpected results and prepare results for sharing and understanding. State-of-the-art SWfMS e.g., Kepler, Taverna, and VisTrails capture workflow provenance automatically in the form of execution traces. In addition, stand-alone approaches to provenance capture and analytics exist [12]. Most of these solutions, however, often rely on proprietary formats which make it difficult to interchange information about provenance. Scientific workflows should be published in a system-neutral language to increase their reuse across the projects and domains. To achieve the goal, a standardised description or model is required to represent scientific workflows and their traces in a more general, machine-readable, and workflow-engine-independent format so that they can be annotated, queried, and understood without relying on particular language of workflow-engines.

Over the past several years, the Semantic Web community has begun to work and propose provenance models for the formalisation and explanation of scientific workflows and their traces of execution, and has shown many applications of these models within the workflow domain. In this regard, **Wf4Ever** [2] project addresses challenges related to describing scientific experiments in a generic way through *wfdesc* ontology, and analysis and management of their execution provenance through *wfprov* ontology. The Open Provenance Model for Workflows (**OPMW**) [7] provides a framework to publish computational workflows, which includes the specification of an OWL ontology, i.e., OPMW ontology, for the description of workflow traces and their templates. As a part of this framework, the authors also published an ontology for Provenance and Plans (P-Plan)[1]. **P-Plan** extends the Provenance Ontology (PROV-O) [10] to represent the plans that guided the execution of scientific processes, describing how such plans are composed and their correspondence to provenance records that describe the execution. **ProvONE** [4] is another data model for scientific workflow provenance representation. It is built to be compatible with **PROV-DM**[2] and provides constructs to model workflow specification and workflow execution provenance.

These provenance models can capture, store, and query the provenance of workflows and their traces in a common, machine-readable, and workflow-engine independent format. However, these models lack in their ability to correctly and completely specify control-flow driven scientific workflows. This limitation of provenance models does not affect their ability to correctly capture the retrospective provenance (i.e., the information about a workflow execution). However, this results in underspecified prospective provenance (i.e., the information about the structure and static context of a workflow) that potentially leads to two major issues:

1) The publication of underspecified or incomplete workflow structure results in a scientific experiment's misinterpretation and hinders the reusability and reproducibility of the experiment and its results.
2) Conformance checking is a key to identify bottlenecks and inefficiencies in scientific workflows and to learn how to improve them. An underspecified workflow precludes conformance checking thus restricting the provenance gains.

In this paper, we are extending ProvONE to capture the provenance of control-flows driven workflows. We present **ProvONE+**[3], an enhanced provenance model for scientific workflow provenance. To this, we

- identify control-flows based workflow patterns that are supported by the existing SWfMS and could be part of a scientific workflow.
- design a generic control-flow model by reusing PROV-DM constructs to make it compatible with W3C recommended provenance solution and incorporate this pattern into ProvONE by preserving its workflow provenance semantics.
- provide provenance description of example workflow patterns and showcase a complete example to establish that our proposed model can capture control-flow patterns semantics in provenance descriptions.

The rest of the paper is organised as follows. In Sect. 2, we discuss design requirements for the provenance model. In Sect. 3, we review the existing work on modelling control flows in scientific workflows. In Sect. 4, we present ProvONE+ that includes a control-flow model, its integration with ProvONE, and ProvONE+ compliant provenance description of the control-flow patterns. In Sect. 5, we discuss a comprehensive example taken from agriculture domain. In Sect. 6, we conclude outlining future directions.

---

[1] http://www.opmw.org/model/p-plan/.

[2] https://www.w3.org/TR/prov-dm/.

[3] ProvONE+ Model: https://github.com/anilabutt/ProvONEplus.

2   Control-Flow Design Requirements

The Workflow Patterns Initiative[4] aimed to recognise the common patterns that occur when modelling business processes. The first deliverable of this project was a collection of twenty patterns that defined the control-flow perspective of workflows [16]. The revised view presented twenty-three new control-flow patterns that augment the previous patterns [13]. However, these patterns were identified for business workflows. In this paper, we are scoping control-flows that are directly supported by at least one of the existing three popular SWfMS. From these patterns, it is possible to identify requirements (listed as R1,..., Rn) of control flow enabled scientific workflow provenance model.

In this regard, firstly we identify the control-flows that are supported by the SWfMS from twenty core control-flow patterns in Table 1. Secondly, we provide a brief description of the supported patterns and their UML activity diagram as shown in Fig. 1. Finally, we consider these patterns as the core requirements for a control flow enabled scientific workflows provenance model. It is important to note, we set the requirements only for control-flow patterns in scientific workflows rather the workflows themselves.

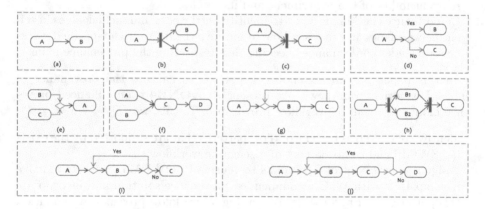

Fig. 1. Control-flow patterns UML activity diagram: (a) Sequence; (b) Parallel split; (c) Synchronisation; (d) Exclusive choice; (e) Simple merge; (f) Multi merge; (g) Arbitrary cycle; (h) Multiple instance with a prior design-time knowledge; (i) Multiple instance with a prior run-time knowledge; and (j) Milestone.

Table 1 shows that six patterns ('red') are not supported in any SWfMS and two patterns ('green') are related to the workflow execution rather specification. Here, we are extending the ProvONE workflow specification model and plan to extend its workflow execution model in future work. The remaining eleven patterns set the core requirements for our control-flow driven scientific workflow provenance model.

---

[4] http://www.workflowpatterns.com/patterns/control/index.php.

Table 1. Control patterns support in SWfMS [14]

Workflow Pattern	Kepler	Taverna	Triana
Sequence	Y	Y	Y
Parallel Split	Y	Y	Y
Synchronisation	Y	Y*	Y*
Exclusive Choice	Y	Y*	Y*
Simple Merge	Y	Y*	N
Multi choice	Y*	Y*	Y*
Synchronising merge	Y*	N	N
Multi merge	Y	N	N
Discriminator	N	N	N
Arbitrary cycles	Y	Y	Y
Implicit termination	Y	Y	Y
Multi instances without synchronisation	Y*	Y*	N
Multi instances with a prior design time knowledge	Y	Y	Y
Multi instances with a prior run time knowledge	Y	Y	N
Multi instances without a prior run time knowledge	N	N	N
Deferred choice	N	N	N
Interleaved parallel routing	Y*	N	N
Milestone	Y	Y	N
Cancel task	Y	Y	N
Cancel case	Y	N	Y

Y–The pattern is directly supported; **N**–The pattern is not supported at all; **Y***–The pattern supported with some limitations. **Red**–The pattern is not directly supported in any of SWfMS; **Green**–The pattern concern workflow execution rather structure.

- **R1: Sequence** - a task is enabled after the completion of a preceding task in the same workflow. (Fig. 1(a))
- **R2: Parallel split** - two or more parallel tasks each of which are enabled and can execute concurrently in a workflow. (Fig. 1(b))
- **R3: Synchronisation** - a subsequent task after two or more parallel tasks is enabled only when both the preceding parallel tasks have been completed. (Fig. 1(c))

- **R4: Exclusive choice** - two or more parallel tasks in a workflow such that only one of them is enabled based on a mechanism (condition) that can select a task. (Fig. 1(d))
- **R5: Simple merge** - a subsequent task after two or more parallel preceding tasks is enabled after the completion of one of the preceding tasks. (Fig. 1(e))
- **R6: Multi-merge** - a subsequent task after two or more parallel preceding tasks is enabled after the completion of each of the preceding tasks. (Fig. 1(f))
- **R7: Arbitrary cycles** - the ability to represent cycles (loops) in a workflow such that the cycle has more than one entry or exit points. (Fig. 1(g))
- **R8: Implicit termination** - a given workflow should terminate when there are no remaining tasks in the workflow. (Implicit in each workflow; therefore, not shown in Fig. 1)
- **R9: Multi instances with a prior design-time knowledge** - within a workflow, a task can be enabled multiple times. The required count is known at the design time. (Fig. 1(h))
- **R10: Multi instances with a prior run-time knowledge** - within a workflow, a task can be enabled multiple times. The required count is a run-time factor. (Fig. 1(i))
- **R11: Milestone** - a task within a workflow is only enabled when the workflow is in a specific state. (Fig. 1(j))

Considering these requirements, we design a generic and lightweight control-flow ontology pattern that most closely addresses the notions of control-flow in workflows. It is worth noticing, we are aiming for a model capable of specifying most of the listed patterns in Table 1, and easy to extend for other patterns. However, in this paper, we will discuss and provide provenance description for the eleven requirements only.

3 Related Work

First, we evaluate the existing scientific workflow provenance models and their ability to model control-flow driven workflows. Secondly, we look for some existing control-flow modelling solutions in the literature.

3.1 Scientific Workflow Provenance Models

Several recent community efforts have culminated with the development of generic models to represent the provenance of scientific workflows and to promote their reusability and reproducibility. We have evaluated ProvONE, OPMW, and Wf4Ever as the most expressive of these models [12] for their ability to support control-flows in workflows.

OPMW is a conceptual model for the representation of prospective and retrospective provenance captured from the execution of scientific workflows. It is a specialisation of PROV and OPM provenance model. *Wf4Ever* has extended PROV to present *wfdesc* and *wfprov* ontologies for the description of prospective and retrospective provenance respectively. Both *OPMW* and *Wf4Ever* provenance models are highly influenced by the data-flow driven scientific workflows; therefore, control-flow constructs are missing in these models.

ProvONE extends PROV model with an explicit representation of prospective provenance to capture the most relevant information on scientific workflows. It is designed to accommodate extensions for specific SWfMS. It includes both prospective and retrospective provenance and allows easy integration of terms from external vocabularies, including Dublin Core or SWfMS. Particularly, ProvONE model has a 'Controller' class to specify the execution orders of given tasks (i.e., Program in ProvONE) in a workflow, which allows for differing models of computation (see Sect. 4.1). For instance, in a sequential dataflow model, a given task may only start once the execution of a preceding task terminates. Similarly, parallel Split and synchronisation can be modelled using the Controller class according to the cardinality constraints defined with 'Program' in ProvONE. This implies, ProvONE fulfils R1, R2, and R3 requirements discussed in Sect. 2. Therefore, we selected *ProvONE* to extend for modelling control-flows in scientific workflows.

3.2 Control-Flow Models

The Business Process Modelling Ontology (BPMO) [3] models concepts and attributes for business process modelling, and offers control-flow constructs as integral component of the ontology. The core purpose of the authors is to resolve the heterogeneity of business process modelling language (e.g., BPEL and BPMN). BPMO and other similar ontologies tightly couple the control-flow constructs with the business process modelling concepts, and hence cannot be reused in an existing provenance model.

The control flow pattern[5] is an OWL[6] representation of control-flow patterns [16]. In control flow pattern ontology, the tasks of a workflow are categorised based on their role in the workflow. For instance, 'ActionTask'– typically data manipulation task, and 'ControlTask'– typically branching, concurrency, looping or synchronisation tasks. Additionally, four object properties namely 'follows', 'directlyFollows', 'precedes', and 'directlyPrecedes' are defined to model the execution order of the tasks. However, the control-flow pattern semantics is encoded by differentiating the control tasks based on the preceding or subsequent tasks and the responsibility to specify the actual behaviour of the task is left to the user. For example, to specify a loop control, a LoopTask is defined as a subclass of ControlTask which has an 'ActionTask' as its successor. But iteration cardinality and starting and terminating conditions could be modelled differently by a user based on the domain. In absence of an effective modelling solution for the control-flow, we develop a control-flow model for ProvONE instead of reusing an existing solution.

---

[5] http://www.ontologydesignpatterns.org/cp/owl/controlflow.owl.

[6] https://www.w3.org/OWL/.

4 ProvONE+: ProvONE Extension to Specify Control Flows

We develop an extension of ProvONE based on the control-flow requirements to capture the provenance of control-flow driven workflows. Our control-flow provenance model is consistent with the needs of the community, in particular the PROV community that cares about the *Entity*, *Activity*, and *Actor*.

4.1 Control Flow Model

Figure 2 highlights the most important classes and relationships of the control-flow model, whereas Fig. 3 presents an integration of control-flow model with ProvONE.

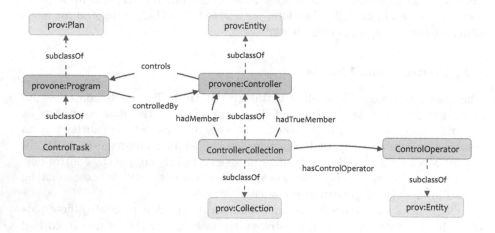

Fig. 2. The control-flow model, showing relationships to PROV-DM

The control-flow model:

- Extends `PROV` and `ProvONE` data models;
- Defines three new classes i.e., `ControlTask`, `ControllerCollection`, and `ControlOperator`. The latter specifies how the controller collection will behave;
- Defines two new object properties i.e., `hasTrueMember`, `hasControlOperator`, and the related inverses;
- Defines six instances of `ControlOperator` i.e., `AND-JOIN`, `AND-SPLIT`, `OR-JOIN`, `OR-SPLIT`, `XOR-JOIN`, and `XOR-SPLIT`. These six operators are sufficient to specify the requirements mentioned in Sect. 2 and can also specify some patterns marked red in Table 1.

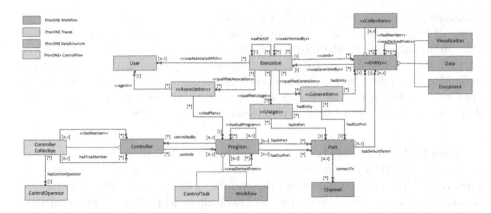

Fig. 3. ProvONE+: ProvONE with control flow model to capture provenance of control flow driven scientific workflows

Controller: The `Controller` class specifies the execution of a given task is controlled by another task. It represents a sequence of connected programs (i.e., tasks). A program is connected to two or more parallel programs through controllers in a workflow; however, the `Controller` class does not specify if all, some, or exactly one of the connected programs should execute before or after the program in the workflow.

ControllerCollection: A `ControllerCollection` is a collection of controllers, and itself a controller, that provides a structure to its member controllers. This concept is aligned with *prov:Collection*, defined as a *prov:Entity*, that provides a structure to some member constituents, which are themselves entities. A `ControllerCollection` specifies how its member Controllers are enabled. If a program is controlledBy (or controls) more than one programs through the controllers, then a `ControllerCollection`, along with its ControlOperator, specifies if all, some, or exactly one of the connected program are enabled after (or before) the program in the workflow. For each control constructs (e.g., *if/else* or *case* statements, *for* or *while* loops, and *parallel* or *synchronous* execution models) in a workflow, a `ControllerCollection` instance is used to specify the control behaviour. The object property `prov:hadMember` from PROV-DM is reused to model the members of a `ControllerCollection`. In addition, a new object property `provone+:hasTrueMember` is defined as a sub property of `prov:hadMember` to specify the program enabled if the control condition is satisfied. For instance, in-case an *if/else* decision is 'True', the controller that controls a subsequent task for 'True' case is pointed as a 'hasTrueMember' of its `ControllerCollection`.

ControlOperator: A `ControlOperator` class is created to define the behaviour of the controllers within a ControllerCollection. Each ControllerCollection has a `ControlOperator`. In ProvONE+, we model the behaviours of controllers in terms of logical operators. In current version, we have defined six operators i.e., `AND-JOIN`, `AND-SPLIT`, `XOR-JOIN`, `XOR-SPLIT`, `OR-JOIN`, and `OR-SPLIT`. These six operators are sufficient to specify many of the control patterns mentioned in Table 1 (including 'red's' e.g., **multi-choice**), and where they are not, a new operator can be created as an individual of `ControlOperator` class and can be defined as operator of a ControllerCollection of that pattern.

ControlTask: A `ControlTask` is a `Program` that specifies the decision controls (i.e., condition) in a workflow. For example, the (second) diamond in Figs. 1(d), 1(i), and 1(j) represents the condition based on which a subsequent program is enabled. We define a `ControlTask` class, which is used for specifying a conditional control as an additional program to a preceding program in the workflow.

4.2 ProvONE+ Description of Control-Flow Patterns

This section shows how to specify the control-flow patterns of Fig. 1 through the control flow model proposed in Sect. 4.1. Figure 4 provides a pictorial representation of ProvONE+ description of these patterns.

A **Sequence** pattern is specified through a single Controller between two programs, which implies the ControllerCollection is unnecessary for this patter as shown in Fig. 4(a). The **Parallel split** and **Synchronisation** patterns can be specified through a ControllerCollection and `AND-SPLIT` or `AND-JOIN` operator respectively as shown in Figs. 4(b), 4(c), 4(f), and 4(h). Each **Exclusive choice** in the workflow patterns (Figs. 1(d), 1(i), and 1(j)) is specified through a ControlTask and a ControllerCollection having `XOR-Split` ControlOperator as shown in Figs. 4(d), 4(i), and 4(j). This implies based on the output of ControlTask, only one of the Controller from a ControllerCollection is enabled. A **Merge** after exclusive choice in the workflow is specified using a ControllerCollection and an `XOR-JOIN` as shown in Figs. 4(e), 4(g), 4(i), and 4(j).

It is worth noting, two ControlOperators named `OR-JOIN` and `OR-SPLIT` are not used for specifying the given example patterns. The reason is that the proposed model can specify patterns highlighted red in Table 1 using these two control operators. For example, multi-choice, synchronisation merge and multi-instanced without synchronisation merge provenance can be specified using `OR-JOIN` and `OR-SPLIT`. A complete provenance description of these patterns is out of the scope of this paper.

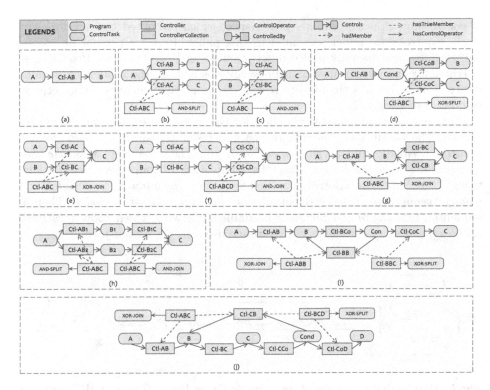

Fig. 4. ProvONE+ description of control flow patterns: (a) Sequence; (b) Parallel split; (c) Synchronisation; (d) Exclusive choice; (e) Simple merge; (f) Multi merge; (g) Arbitrary cycle; (h) Multiple instance with a prior design-time knowledge; (i) Multiple instance with a prior run-time knowledge; and (j) Milestone.

5   ProvONE+ for Agricultural Workflow

We provide a usage scenario of the proposed model from the agriculture domain. Carbon Project[7] at the CSIRO Australia[8] is developing a digital system that will enable farmers to participate profitably in greenhouse gas mitigation and maximise the benefits to the land from carbon markets. The team focuses on capturing the provenance of its workflows and their execution traces to identify bottlenecks and inefficiencies in the workflows, learn to improve them, and trust the data produced by these workflows. One of the requirements is to capture the workflow provenance in a machine-readable and workflow engine independent format to perform automatic experiments and analysis on the provenance data. To the best of our knowledge, there is no single provenance model, which facilitates the modelling of Carbon project control-flow driven workflows. We investigated the core provenance models and how they can be reused for the modelling

---

[7] https://research.csiro.au/digiscape/digiscapes-projects/digital-services-for-carbon-farming-markets/.

[8] https://www.csiro.au/.

of control-flow driven workflows provenance, which prompted the development of the control-flow model presented in this paper. Here we present an example workflow and its provenance description using ProvONE and ProvONE+. Further, we discuss the provenance descriptions of both the models to demonstrate ProvONE+ has the ablility to specify scientific experiments correctly as compared to ProvONE.

Figure 5 shows an example workflow for the green house gasses emission prediction. The workflow begins with four parallel tasks: *get-params*, *get-simulation-duration*, *drill-silo*, and *output-template-selector*. The subsequent task *ghg.soil-water-model* is enabled only after the completion of four preceding parallel tasks. Next, `if/else` control-flow enables either *ghg.annual-plant-model* and *ghg.agghgcomposite-model* sequentially or *ghg.agghgcomposite-model*. Then a sequential task *ghg.output-formater* is enabled, following which `loop` iteratively enables *ghg.display-output* for each variable listed in *ghg.output-formatter*.

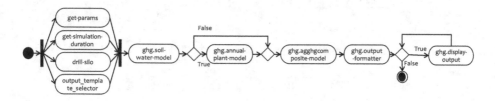

Fig. 5. UML activity diagram of an example Green House Gasses (GHG) emission prediction workflow

Figure 6(a) shows ProvONE based provenance description of the example workflow. According to the description, *ghg.annual-plant-model* is enabled all the time before *ghg.agghgcomposite-model*. Moreover, *ghg.display-output* is enabled sequentially once after *ghg.output-formatter*. This description clearly misinterprets the semantics of both the control-flows.

Figure 6(b) shows ProvONE+ based provenance description of the same workflow. The description specifies the conditional statements as additional `ControlTasks`. `if/else` control semantics is specified through `params.vegType.valuecheck` control task. Moreover, controllers that control *ghg.annual-plant-model* and *ghg.agghgcomposite-model* are collected in `ctl:split-collection` controller collection. The controller collection has an `XOR-SPLIT` control operator to specify that either *ghg.annual-plant-model* or *ghg.agghgcomposite-model* is enabled as a subsequent task after the control task. Similarly, the loop semantics is specified through `output-variableToView` control task and `XOR-JOIN` operator. Figure 6 shows that ProvONE+ based workflow specification provenance is complete and correct.

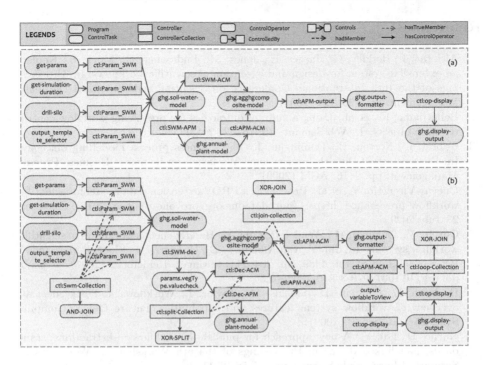

Fig. 6. Provenance descriptions of example workflow: (a) ProvONE description and (b) ProvONE+ description

6   Conclusion

In this paper, we presented a control-flow model to enhance an existing provenance solution ProvONE for capturing provenance of control-flow driven scientific workflows. The extended provenance model, named ProvONE+, overcomes the limitations of the existing provenance models. ProvONE is augmented with new classes and properties to capture the behaviour of controls in scientific workflows. Further, we provided the provenance description of the sampled control-flow patterns. For a real-world agriculture workflow, our model successfully captured the control constructs semantics in the provenance description of the workflow. The model is generic and can easily be extended for other patterns using a new ControlOperator. The proposed model is flexible and can be integrated with other provenance models (e.g., OPMW and Wf4Ever). In future research, we will consider specifying all identified control-flow patterns using or extending our model to make ProvONE+ equally effective provenance model for the scientific as well as business workflows.

References

1. Altintas, I., Berkley, C., Jaeger, E., Jones, M., Ludascher, B., Mock, S.: Kepler: an extensible system for design and execution of scientific workflows. In: Proceedings of the 16th International Conference on Scientific and Statistical Database Management, pp. 423–424 (2004)
2. Belhajjame, K., et al.: Using a suite of ontologies for preserving workflow-centric research objects. J. Web Semant. **32**, 16–42 (2015)
3. Cabral, L., Norton, B., Domingue, J.: The business process modelling ontology. In: Proceedings of the 4th International Workshop on Semantic Business Process Management, pp. 9–16. ACM (2009)
4. Cuevas-Vicenttín, V., et al.: ProvONE: a PROV extension data model for scientific workflow provenance. https://purl.dataone.org/provone-v1-dev (2016). Accessed 23 July 2020
5. Curcin, V., Ghanem, M., Wendel, P., Guo, Y.: Heterogeneous workflows in scientific workflow systems. In: Shi, Y., van Albada, G.D., Dongarra, J., Sloot, P.M.A. (eds.) ICCS 2007. LNCS, vol. 4489, pp. 204–211. Springer, Heidelberg (2007). https://doi.org/10.1007/978-3-540-72588-6_36
6. Deelman, E., Gannon, D., Shields, M., Taylor, I.: Workflows and e-science: an overview of workflow system features and capabilities. Future Gener. Comput. Syst. **25**(5), 528–540 (2009)
7. Garijo, D., Gil, Y.: A new approach for publishing workflows: abstractions, standards, and linked data. In: Proceedings of the 6th Workshop on Workflows in Support of Large-scale Science, pp. 47–56 (2011)
8. Herschel, M., Diestelkàmper, R., Ben Lahmar, H.: A survey on provenance: what for? What form? What from? VLDB J. Int. J. Very Large Data Bases **26**(6), 881–906 (2017)
9. Hull, D., et al.: Taverna: a tool for building and running workflows of services. Nucleic Acids Res. **34**(suppl_2), W729–W732 (2006)
10. Moreau, M.: World Wide Web consortium PROV-DM: the PROV data model W3C recommendation. https://www.w3.org/TR/prov-dm/ (2013). Accessed 23 July 2020
11. Moreau, L.: The foundations for provenance on the web. Found. Trends Web Sci. **2**, 99–241 (2010)
12. Oliveira, W., Oliveira, D.D., Braganholo, V.: Provenance analytics for workflow-based computational experiments: a survey. ACM Comput. Surv. **51**(3), 53:1–53:25 (2018). https://doi.org/10.1145/3184900
13. Russell, N., Ter Hofstede, A.H., Van Der Aalst, W.M., Mulyar, N.: Workflow control-flow patterns: a revised view. BPM Center Report BPM-06-22, BPMcenter. org, pp. 06–22 (2006)
14. Shiroor, A.R.: Scientific workflow management systems and workflow patterns. Ph.D. thesis, Purdue University (2009)
15. Taylor, I., Shields, M., Wang, I., Harrison, A.: The triana workflow environment: architecture and applications. In: Workflows for e-Science, pp. 320–339. Springer (2007). https://doi.org/10.1007/978-1-84628-757-2_20
16. Van Der Aalst, W.M.: Workflow patterns. In: Encyclopedia of Database Systems, pp. 3557–3558 (2009). https://doi.org/10.1007/978-0-387-39940-9_826

Data Mining and Applications

Fair Outlier Detection

Deepak P.[1,2(✉)] and Savitha Sam Abraham[2]

[1] Queen's University Belfast, Belfast, UK
deepaksp@acm.org
[2] Indian Institute of Technology Madras, Chennai, India
savithas@cse.iitm.ac.in

Abstract. An outlier detection method may be considered fair over specified sensitive attributes if the results of outlier detection are not skewed towards particular groups defined on such sensitive attributes. In this paper, we consider, for the first time to our best knowledge, the task of fair outlier detection. Our focus is on the task of fair outlier detection over multiple multi-valued sensitive attributes (e.g., gender, race, religion, nationality, marital status etc.), one that has broad applications across web data scenarios. We propose a fair outlier detection method, *FairLOF*, that is inspired by the popular *LOF* formulation for neighborhood-based outlier detection. We outline ways in which unfairness could be induced within *LOF* and develop three heuristic principles to enhance fairness, which form the basis of the *FairLOF* method. Being a novel task, we develop an evaluation framework for fair outlier detection, and use that to benchmark *FairLOF* on quality and fairness of results. Through an extensive empirical evaluation over real-world datasets, we illustrate that *FairLOF* is able to achieve significant improvements in fairness at sometimes marginal degradations on result quality as measured against the fairness-agnostic *LOF* method.

1 Introduction

There has been much recent interest in incorporating fairness constructs into data analytics algorithms, within the broader theme of algorithmic fairness [12]. The importance of fairness in particular, and democratic values in general, cannot be overemphasized in this age when data science algorithms are being used in very diverse scenarios to aid decision making that could affect lives significantly. The vast majority of fair machine learning work has focused on supervised learning, especially on classification (e.g., [17,31]). There has also been some recent interest in ensuring fairness within unsupervised learning tasks such as clustering [1], retrieval [32] and recommendations [24]. In this paper, we explore the task of fairness in outlier detection, an analytics task of wide applicability in myriad scenarios. To our best knowledge, this is the first work on embedding fairness within outlier detection. The only work so far on fairness and outlier

The original version of this chapter was revised: the first author's name was corrected to Deepak P. The correction to this chapter is available at https://doi.org/10.1007/978-3-030-62008-0_42

detection [13] focuses on analyzing outlier detection algorithms for fairness, a significantly different task, for which a human-in-the-loop method is proposed.

Outlier Detection and Fairness: The task of outlier detection targets to identify *deviant* observations from a dataset, and is usually modelled as an unsupervised task; [8] provides a review of outlier detection methods. The classical outlier characterization, due to Hawkins [16], considers outliers as 'observations that deviate so much from other observations as to arouse suspicion that they were generated by a different process'. Applications of outlier detection range across varied application domains such as network intrusions [18], financial fraud [25] and medical abnormalities [21]. Identification of non-mainstream behavior, the high-level task that outlier detection accomplishes, has a number of applications in new age data scenarios. Immigration checks at airports might want to carry out detailed checks on *'suspicious'* people, while AI is likely used in proactive policing to identify *'suspicious'* people for stop-and-frisk checks. In this age of pervasive digitization, *'abnormality'* in health, income or mobility patterns may invite proactive checks from healthcare, taxation and policing agencies. Identification of such *abnormal* and *suspicious* patterns are inevitably within the remit of outlier detection. The nature of the task of outlier detection task makes it very critical when viewed from the perspective of fairness. Even if information about ethnicity, gender, religion and nationality be hidden (they are often not hidden, and neither is it required to be hidden under most legal regulations) from the database prior to outlier identification, information about these attributes are likely inherently spread across other attributes. For example, geo-location, income and choice of professions may be correlated with ethnic, gender, religious and other identities. The identification of non-mainstream either falls out from, or entails, an analogous and implicit modelling of mainstream characteristics in the dataset. The mainstream behavior, by its very design, risks being correlated with majoritarian identities, leading to the possibility of minority groups being picked out as outliers significantly more often. Interestingly, there have been patterns of racial prejudice in such settings[1].

Outlier Detection and the Web: Web has emerged, over the past decades, as a rich source of unlabelled digital data. Thus, the web likely presents the largest set of scenarios involving outlier detection. Each user on the web leaves different cross-sections of digital footprints in different services she uses, together encompassing virtually every realm of activity; this goes well beyond the public sector applications referenced above. In a number of scenarios, identified as an outlier could lead to undesirable outcomes for individuals. For example, mobility outliers may receive a higher car insurance quote, and social media outliers may be subjected to higher scrutiny (e.g., Facebook moderation). It is important to ensure that such undesirable outcomes be distributed fairly across groups defined on protected attributes (e.g., gender, race, nationality, religion etc.) for ethical reasons and to avoid bad press[2].

---

[1] https://www.nyclu.org/en/stop-and-frisk-data.

[2] https://www.cnet.com/features/is-facebook-censoring-conservatives-or-is-modera
ting-just-too-hard/.

Our Contributions: We now outline our contributions in this paper. First, we characterize the task of fair outlier detection under the normative principle of disparate impact avoidance [4] that has recently been used in other unsupervised learning tasks [1,11]. Second, we develop a fair outlier detection method, *FairLOF*, based on the framework of *LOF* [7], arguably the most popular outlier detection method. Our method is capable of handling multiple multi-valued protected attributes, making it feasible to use in sophisticated real-world scenarios where fairness is required over a number of facets. Third, we outline an evaluation framework for fair outlier detection methods, outlining quality and fairness metrics, and trade-offs among them. Lastly, through an extensive empirical evaluation over real-world datasets, we establish the effectiveness of *FairLOF* in achieving high levels of fairness at small degradations to outlier detection quality.

2 Related Work

Given the absence of prior work on fair outlier detection methods, we cover related work across outlier detection and fairness in unsupervised learning.

Outlier Detection Methods: Since obtaining labelled data containing outliers is often hard, outlier detection is typically modelled as an unsupervised learning task where an unlabelled dataset is analyzed to identify outliers within it. That said, supervised and semi-supervised approaches do exist [8]. We address the unsupervised setting. The large majority of work in unsupervised outlier detection may be classified into one of two families. The first family, that of *global methods*, build a dataset-level model, and regard objects that do not conform well to the model, as outliers. The model could be a clustering [30], Dirichlet mixture [15] or others [14]. Recent research has also explored the usage of auto-encoders as a global model, the reconstruction error of individual data objects serving as an indication of their outlierness; *RandNet* [10] generalizes this notion to determine outliers using an ensemble of auto-encoders. The second family, arguably the more popular one, is that of *local methods*, where each data object's outlierness is determined using just its neighborhood within a relevant similarity space, which may form a small subset of the whole dataset. The basic idea is that the outliers will have a local neighborhood that differs sharply in terms of characteristics from the extended neighborhood just beyond. *LOF* [7] operationalizes this notion by quantifying the contrast between an object's local density (called local reachability density, as we will see) and that of other objects in its neighborhood. Since the *LOF* proposal, there has been much research into local outliers, leading to work such as *SLOM* [9], *LoOP* [20] and *LDOF* [33]. Schubert et al. [28] provide an excellent review of local outlier detection, including a generalized three phase meta-algorithm that most local outlier detection methods can be seen to fit in. Despite much research over the last two decades, *LOF* remains the dominant method for outlier detection, continuously inspiring systems work on making it efficient for usage in real-world settings (e.g., [3]). Accordingly, the framework of *LOF* inspires the construction of *FairLOF*.

Fairness in Unsupervised Learning: There has been much recent work on developing fair algorithms for unsupervised learning tasks such as clustering, representation learning and retrieval. Two streams of fairness are broadly used; *group fairness* that targets to ensure that the outputs are fairly distributed across groups defined on sensitive attributes, and *individual fairness* which strives to limit possibilities of similar objects receiving dissimilar outcomes. Individual fairness is typically agnostic to the notion of sensitive attributes. Our focus, in this paper, is on group fairness in outlier detection. For group fairness in clustering, techniques differ in where they embed the fairness constructs; it could be at the pre-processing step [11], within the optimization framework [1] or as a post-processing step to re-configure the outputs [6]. *FairPCA* [22], a fair representation learner, targets to ensure that objects are indistinguishable wrt their sensitive attribute values in the learnt space. Fair retrieval methods often implement group fairness as parity across sensitive groups in the top-k outputs [2]. The techniques above also differ in another critical dimension; the number of sensitive attributes they can accommodate. Some can only accommodate one binary sensitive attribute, whereas others target to cater to fairness over multiple multi-valued sensitive attributes; a categorization of clustering methods along these lines appears in [1].

In contrast to such work above, there has been no exploration into fair outlier detection. The only related effort in this space so far, to our best knowledge, is that of developing a human-in-the-loop decision procedure to determine whether the outputs of an outlier detection is fair [13]. This focuses on deriving *explanations based on sensitive attributes* to distinguish the outputs of an outlier detection method from the *'normal'* group. If no satisfactory explanation can be achieved, the black-box outlier detection method can be considered fair. The human is expected to have domain knowledge of the task and data scenario to determine parameters to identify what is unfair, and interpret explanations to judge whether it is indeed a case of unfairness.

3 Problem Definition

Task Setting: Consider a dataset $\mathcal{X} = \{\ldots, X, \ldots\}$ and an object pairwise distance function $d : \mathcal{X} \times \mathcal{X} \to \mathbb{R}$ that is deemed relevant to the outlier detection scenario. Further, each data object is associated with a set of sensitive attributes $\mathcal{S} = \{\ldots, S, \ldots\}$ (e.g., gender, race, nationality, religion etc.) which are categorical and potentially multi-valued, $V(S)$ being the set of values that a sensitive attribute, S, can take. $X.S \in V(S)$ indicates the value assumed by object X for the sensitive attribute S. Thus, each multi-valued attribute S defines a partitioning of the dataset into $|V(S)|$ parts, each of which comprise objects that take the same distinct value for S.

Outlier Detection: The task of (vanilla) outlier detection is that of identifying a small subset of objects from $\mathcal{X}$, denoted as $\mathcal{O}$, that are deemed to be outliers. Within the *local outliers* definition we adhere to, it is expected that objects in $\mathcal{O}$

differ significantly in local neighborhood density when compared to other objects in their neighborhoods. In typical scenarios, it is also expected that $|\mathcal{O}| = t$, where t is a pre-specified parameter. The choice of t may be both influenced by the dataset size (e.g., t as a fixed fraction of $|\mathcal{X}|$) and/or guided by practical considerations (e.g., manual labour budgeted to examine outliers).

Fair Outlier Detection: The task of fair outlier detection, in addition to identifying outliers, considers ensuring that the distribution of sensitive attribute groups among $\mathcal{O}$ reflect that in $\mathcal{X}$ as much as possible. This notion, referred to interchangeably as representational parity or disparate impact avoidance, has been the cornerstone of all major fair clustering algorithms (e.g., [1,6,11]), and is thus a natural first choice as a normative principle for fair outlier detection. As a concrete example, if gender is a sensitive attribute in $\mathcal{S}$, we would expect the gender ratio within $\mathcal{O}$ to be very close to, if not exactly equal to, the gender ratio in $\mathcal{X}$. Note that fairness is complementary and often contradictory to ensuring that the top neighborhood-outliers find their place in $\mathcal{O}$; the latter being the only consideration in (vanilla) outlier detection. Thus, fair outlier detection methods such as *FairLOF* we develop, much like fair clustering methods, would be evaluated on two sets of metrics: (a) *'Quality'* metrics that measure how well objects with distinct local neighborhoods are placed in $\mathcal{O}$, and (b) *Fairness* metrics that measure how well they ensure that the dataset-distribution of sensitive attribute values are preserved within $\mathcal{O}$. We will outline a detailed evaluation framework in a subsequent section. Good fair outlier detection methods would be expected to achieve good fairness while suffering only small degradations in quality when compared against their vanilla outlier detection counterparts.

Motivation for Representational Parity: It may be argued that the distribution of sensitive attribute groups could be legitimately different from that in the dataset. For example, one might argue that outlying social media profiles that correlate with crime may be legitimately skewed towards certain ethnicities since propensity for crimes could be higher for certain ethnicities than others. The notion of representational parity disregards such assumptions of skewed apriori distributions, and seeks to ensure that the inconvenience of being classed as an outlier be shared proportionally across sensitive attribute groups. This argument is compelling within scenarios of using outlier detection in databases encompassing information about humans. In particular, this has its roots in the distributive justice theory of *luck egalitarianism* [10] that distributive shares be not influenced by arbitrary factors, especially those of *'brute luck'* that manifest as membership in sensitive attribute groups (since individuals do not choose their gender, ethnicity etc.). The normative principle has been placed within the umbrella of the *'justice as fairness'* work due to John Rawls [27] that underlies most of modern political philosophy. Further, since outlier detection systems are often used to inform human decisions, it is important to ensure that outlier detection algorithms do not propagate and/or reinforce stereotypes present in society by way of placing higher burden on certain sensitive groups than others.

4   Background: Local Outlier Factor (LOF)

Our method builds upon the pioneering LOF framework [7] for (vanilla) outlier detection. LOF comprises three phases, each computing a value for each object in $\mathcal{X}$, progressively leading to LOF: (i) k-distance, (ii) local reachability density (LRD), and (iii) local outlier factor (LOF).

k-Distance: Let $N_k(X)$ be the set of k nearest neighbors[3] to X (within $\mathcal{X}$), when assessed using the distance function $d(.,.)$. The k-distance for each $X \in \mathcal{X}$ is then the distance to the kth nearest object.

$$k\text{-}distance(X) = max\{d(X, X') | X' \in N_k(X)\} \tag{1}$$

Local Reachability Density: The local reachability density of X is defined as the inverse of the average distance of X to it's k nearest neighbors:

$$lrd(X) = 1 / \left(\frac{\sum_{X' \in N_k(X)} rd(X, X')}{|N_k(X)|} \right) \tag{2}$$

where $rd(X, X')$ is an asymmetric distance measure that works out to the true distance, except when the true distance is smaller than $k\text{-}distance(X')$:

$$rd(X, X') = max\{k\text{-}distance(X'), d(X, X')\} \tag{3}$$

This lower bounding by $k\text{-}distance(X')$ - note also that $k\text{-}distance(X')$ depends on $N_k(X')$ and not $N_k(X)$ - makes the $lrd(.)$ measure more stable. $lrd(X)$ quantifies the density of the local neighborhood around X.

Local Outlier Factor: The local outlier factor is the ratio of the average lrds of X's neighbors to X's own lrd.

$$lof(X) = \left(\frac{\sum_{X' \in N_k(X)} lrd(X')}{|N_k(X)|} \right) / lrd(X) \tag{4}$$

An $lof(X) = 1$ indicates that the local density around X is comparable to that of it's neighbors, whereas a $lrd(X) >> 1$ indicates that it's neighbors are in much denser regions than itself. Once $lof(.)$ is computed for each $X \in \mathcal{X}$, the top-t data objects with highest $lof(.)$ scores would be returned as outliers.

5   FairLOF: Our Method

5.1   Motivation

In many cases, the similarity space implicitly defined by the distance function $d(.,.)$ bears influences from the sensitive attributes and grouping of the dataset

---

[3] $|N_k(X)|$ could be greater than k in case there is a tie for the kth place.

defined over such attributes. The influence, whether casual, inadvertent or conscious, could cause the sensitive attribute profiles of outliers to be significantly different from the dataset profiles. These could occur in two contrasting ways.

Under-reporting of Large/Majority Sensitive Groups: Consider the case where $d(.,.)$ is aligned with groups defined by S. Thus, across the dataset, pairs of objects that share the same value for S are likely to be judged to be more proximal than those that bear different values for S. Consider a dataset comprising 75% males and 25% females. Such skew could occur in real-world cases such as datasets sourced from populations in a STEM college or certain professions (e.g., police[4]). Let us consider the base/null assumption that *real outliers* are also distributed as 75% males and 25% females. Now, consider a male outlier (M) and a female outlier (F), both of which are equally eligible outliers according to human judgement. First, consider M; M is likely to have a *quite cohesive* and *predominantly male* kNN neighborhood due to both: (i) males being more likely in the dataset due to the apriori distribution, and (ii) $d(.,.)$ likely to judge males as more similar to each other (our starting assumption). Note that the first factor works in favour of a male-dominated neighborhood for F too; thus, F's neighborhood would be less gender homogeneous, and thus less cohesive when measured using our S aligned $d(.)$. This would yield $lrd(M) > lrd(F)$, and thus $lof(M) < lof(F)$ (Ref. Eq. 4) despite them being both equally eligible outliers. In short, *when $d(.,.)$ is aligned with groups defined over S, the smaller groups would tend to be over-represented among the outliers.*

Over-reporting of Large Sensitive Groups: Consider a domain-tuned distance function designed for a health records agency who would like to ensure that records be not judged similar just due to similarity on gender; such fine-tuning, as is often done with the intent of ensuring fairness, might often be designed with just the *'main groups'* in mind. In the case of gender, this would ensure a good spread of male and female records within the space; however, this could result in minority groups (e.g., LGBTQ) being relegated to a corner of the similarity space. This would result in a tight clustering of records belonging to the minority group, resulting in the LOF framework being unable to pick them out as outliers. Thus, *a majority conscious design of $d(.,.)$ would result in over-representation of minority groups among outliers.*

5.2   FairLOF: The Method

The construction of *FairLOF* attempts to *correct* for such kNN neighborhood distance disparities across object groups defined over sensitive attributes. *FairLOF* distance correction is based on three heuristic principles; (i) **neighborhood diversity** (object-level correction), (ii) **apriori distribution** (value-level), and

---

[4] https://www.statista.com/statistics/382525/share-of-police-officers-in-england-and-wales-gender-rank/.

(iii) **attribute asymmetry** (attribute-level). We outline these for the first scenario in Sect. 5.1, where $d(.,.)$ is aligned with the sensitive attribute, S, resulting in *minority over-representation among outliers*; these will be later extended to the analogous scenario, as well as for multiple attributes in S.

Neighborhood Diversity: Consider the case of objects that are embedded in neighborhoods comprising objects that take different values of S than itself; we call this as a S-diverse neighborhood. These would be disadvantaged with a higher k-*distance*, given our assumption that $d(.,.)$ is aligned with S. Thus, the k-*distance* of objects with highly diverse neighborhoods would need to be corrected *downward*. This is an object-specific correction, with the extent of the correction determined based on S-diversity in the object's neighborhood.

Apriori Distribution: Consider objects that belong to an S group that are very much in minority; e.g., LGBTQ groups for $S = gender$. Since these objects would have an extremely diverse neighborhood due to their low apriori distribution in the dataset (there aren't enough objects with the same S value in the dataset), the neighborhood diversity principle would correct them deeply downward. To alleviate this, the neighborhood diversity correction would need to be discounted based on the sparsity of the object's value of S in the dataset.

Attribute Asymmetry: The extent of k-*distance* correction required also intuitively depends on the extent to which $d(.,.)$ is aligned with the given S. This could be directly estimated based on the extent of minority over-representation among outliers when vanilla LOF is applied. Accordingly, the attribute asymmetry principle requires that the correction based on the above be amplified or attenuated based on the extent of correction warranted for S.

The above principles lead us to the following form for k-*distance*:

$$\left(max\{d(X,X')|X' \in N_k(X)\} \right) \left(1 - \lambda \times W_S^{\mathcal{X}} \times D_{X.S}^{\mathcal{X}} \times Div(N_k(X), X.S) \right) \quad (5)$$

where $Div(N_k(X), X.S)$, $D_{X.S}^{\mathcal{X}}$ and $W_S^{\mathcal{X}}$ relate to the three principles outlined above (respectively), $\lambda \in [0,1]$ being a weighting factor. These terms are constructed as below:

$$Div(N_k(X), X.S) = \frac{|\{X'|X' \in N_k(X) \wedge X'.S \neq X.S\}|}{|N_k(X)|} \quad (6)$$

$$D_{X.S}^{\mathcal{X}} = \frac{|\{X'|X' \in \mathcal{X} \wedge X'.S = X.S\}|}{|\mathcal{X}|} \quad (7)$$

$$W_S^{\mathcal{X}} = c + |D_{v^*}^{\mathcal{X}} - D_{v^*}^{\mathcal{R}_{LOF}}| \text{ where } v^* = \underset{v \in V(S)}{\arg\max} \ D_v^{\mathcal{X}} \quad (8)$$

Equation 6 measures diversity as the fraction of objects among $N_k(X)$ that differ from X on it's S attribute value. Equation 7 measures the apriori representation as the fraction of objects in $\mathcal{X}$ that share the same S attribute value as that of X. For Eq. 8, $D_v^{\mathcal{R}_{LOF}}$ refers to the fraction of $S = v$ objects found among the top-t results of vanilla LOF over $\mathcal{X}$. $W_S^{\mathcal{X}}$ is computed as a constant factor (i.e., c) added to the asymmetry extent measured as the extent to which the largest S-defined group in the dataset is underrepresented in the vanilla LOF results. While we have used a single S attribute so far, observe that this is easily extensible to multiple attributes in $\mathcal{S}$, yielding the following refined form for k-distance:

$$\left(max\{d(X,X')|X' \in N_k(X)\} \right) \left(1 - \lambda \sum_{S \in \mathcal{S}} (W_S^{\mathcal{X}} \times D_{X.S}^{\mathcal{X}} \times Div(N_k(X), X.S)) \right)$$
(9)

While we have been assuming the case of $d(.,.)$ aligned with S and minority over-representation among outliers, the opposite may be true for certain attributes in $\mathcal{S}$; recollect the second case discussed in Sect. 5.1. In such cases, the k-distance would need to be corrected upward, as against downward. We incorporate that to yield the final k-distance formulation for FairLOF.

$$k\text{-}distance_{FairLOF}(X) = \left(max\{d(X,X')|X' \in N_k(X)\} \right)$$
$$\times \left(1 - \lambda \sum_{S \in \mathcal{S}} (\mathbb{D}(\mathcal{X},S) \times W_S^{\mathcal{X}} \times D_{X.S}^{\mathcal{X}} \times Div(N_k(X), X.S)) \right) (10)$$

where $\mathbb{D}(\mathcal{X},S) \in \{-1,+1\}$ denotes the direction of correction as below:

$$\mathbb{D}(\mathcal{X},S) = \begin{cases} +1 & \text{if } D_{v^*}^{\mathcal{X}} > D_{v^*}^{\mathcal{R}_{LOF}} \text{ where } v^* = \arg\max_{v \in V(S)} D_v^{\mathcal{X}} \\ -1 & otherwise \end{cases} \quad (11)$$

This modification in k-distance warrants an analogous correction of $rd(.,.)$ to ensure level ground among the two terms determining $rd(.,.)$.

$$rd_{FairLOF}(X,X') = max\Big\{ k\text{-}distance_{FairLOF}(X'),$$
$$d(X,X') \times (1 - \lambda \sum_{S \in \mathcal{S}} (\mathbb{D}(\mathcal{X},S) \times W_S^{\mathcal{X}} \times D_{X.S}^{\mathcal{X}} \times \mathbb{I}(X.S \neq X'.S)\}) \Big\} (12)$$

The second term in $rd(.,.)$ is corrected in the same manner as for k-distance, except that the diversity term is replaced by a simple check for inequality, given that there is only one object that X is compared with.

These distance corrections complete the description of FairLOF, which is the LOF framework from Sect. 4 with k-distance$(.,.)$ and $rd(.,.)$ replaced by their corrected versions from Eq. 10 and Eq. 12 respectively. While we omit the whole sequence of *FairLOF* steps to avoid repetition with Sect. 4, we will use $flof(.)$ to denote the final outlier score from *FairLOF*, analogous to $lof(.)$ for LOF. The *FairLOF* hyperparameter, λ, determines the *strength* of the fairness correction applied, and could be a very useful tool to navigate the space of options *FairLOF* provides, as we will outline in the next section.

Note on Complexity: Equation 7 and Eq. 8 can be pre-computed at a an exceedingly small cost of $\mathcal{O}(|\mathcal{S}| \times m)$ where m is $max_{S\in\mathcal{S}}|V(S)|$. Equation 6 needs to be computed at a per-object level, thus multiplying the LOF complexity by $|\mathcal{S}| \times m$. With typical values of $|\mathcal{S}| \times m$ being in the 1000 s at max (e.g., 3–5 sensitive attributes with 10–20 distinct values each), and outlier detection being typically considered as an offline task not requiring real-time responses, the overheads of the *k-distance* adjustment may be considered as very light.

6   Evaluation Framework for Fair Outlier Detection

Enforcing parity along S-groups among outliers, as discussed, often contradicts with identifying high-LOF outliers. This trade-off entails two sets of evaluation measures, inspired by similar settings in fair clustering [1].

Quality Evaluation: While the most desirable quality test for any outlier detection framework would be accuracy measured against human generated outlier/non-outlier labels, public datasets with such labels are not available, and far from feasible to generate. Thus, we measure how well *FairLOF* results align with the fairness-agnostic *LOF*, to assess quality of *FairLOF* results.

$$Jacc(\mathcal{R}_{LOF}, \mathcal{R}_{FairLOF}) = \frac{|\mathcal{R}_{LOF} \cap \mathcal{R}_{FairLOF}|}{|\mathcal{R}_{LOF} \cup \mathcal{R}_{FairLOF}|} \qquad (13)$$

$$Pres(\mathcal{R}_{LOF}, \mathcal{R}_{FairLOF}) = \frac{\sum_{X\in\mathcal{R}_{FairLOF}} lof(X)}{\sum_{X\in\mathcal{R}_{LOF}} lof(X)} \qquad (14)$$

where $\mathcal{R}_{LOF}$ and $\mathcal{R}_{FairLOF}$ are top-t outliers (for any chosen k) from *LOF* and *FairLOF* respectively. $Jacc(.,.)$ computes the jaccard similarity between the result sets. Even in cases where $\mathcal{R}_{FairLOF}$ diverges from $\mathcal{R}_{LOF}$, we would like to ensure that it does not choose objects with very low $lof(.)$ values within $\mathcal{R}_{FairLOF}$; $Pres(.,.)$ computes the extent to which high $lof(.)$ scores are preserved within $\mathcal{R}_{FairLOF}$, expressed as a fraction of the total $lof(.)$ across $\mathcal{R}_{LOF}$. For both these, higher values indicate better quality of *FairLOF* results.

Fairness Evaluation: For any particular sensitive attribute $S \in \mathcal{S}$, we would like the distribution of objects across its values among outliers (i.e., $\mathcal{R}_{FairLOF}$) be similar to that in the dataset, $\mathcal{X}$. In other words, we would like the

$\mathcal{D}_S^{\mathcal{R}_{FairLOF}} = [\ldots, D_v^{\mathcal{R}_{FairLOF}}, \ldots]$ vector ($V \in V(S)$, and Ref. Eq. 7 for computation) to be as similar as possible to the distribution vector over the dataset for S, i.e., $\mathcal{D}_S^{\mathcal{X}} = [\ldots, D_v^{\mathcal{X}}, \ldots]$, as possible. We would like this to hold across all attributes in $\mathcal{S}$. Note that this fairness notion is very similar to that in fair clustering, the only difference being that we evaluate the outlier set once as against each cluster separately. Thus, we adapt the fairness metrics from [1, 29] as below:

$$ED(\mathcal{R}_{FairLOF}) = \sum_{S \in \mathcal{S}} Euclidean\_Distance(\mathcal{D}_S^{\mathcal{R}_{FairLOF}}, \mathcal{D}_S^{\mathcal{X}}) \qquad (15)$$

$$Wass(\mathcal{R}_{FairLOF}) = \sum_{S \in \mathcal{S}} Wasserstein\_Distance(\mathcal{D}_S^{\mathcal{R}_{FairLOF}}, \mathcal{D}_S^{\mathcal{X}}) \qquad (16)$$

where $ED(.)$ and $Wass(.)$ denote aggregated Euclidean and Wasserstein distances across the respective distribution vectors. Since these measure deviations from dataset-level profiles, lower values are desirable in the interest of fairness.

Quality-Fairness Trade-Off: Note that all the above metrics can be computed without any external labellings. Thus, this provides an opportunity for the user to choose different trade-offs between *quality* and *fairness* by varying the *FairLOF correction strength* hyper-parameter λ. We suggest that a practical way of using *FairLOF* would be for a user to try with progressively higher values of λ from $\{0.1, 0.2, \ldots\}$ (note that $\lambda = 0.0$ yields *FairLOF = LOF*, with higher values reducing $Jacc(.)$ and $Pres(.)$ progressively) using a desired value of *Jaccard* similarity as a pilot point. For example, we may want to retain a Jaccard similarity (i.e., $Jacc(.)$ value) of approximately 0.9 or 0.8 with the original *LOF* results. Thus, the user may stop when that is achieved. The quantum of fairness improvements achieved by *FairLOF* over *LOF* at such chosen points will then be indicative of *FairLOF*'s effectiveness.

Single Sensitive Attribute and a Quota-Based System: As noted upfront, *FairLOF* is targeted towards cases where there are multiple sensitive attributes to ensure fairness over; this is usually the case since there are often many sensitive attributes in real world scenarios (e.g., gender, ethnicity, caste, religion, native language, marital status, and even age in certain scenarios). For the case of a single sensitive attribute with a handful of possible values, there is a simple and effective strategy for fair outlier detection. Consider $S = gender$; we take the global list of objects sorted in descending order of $lof(.)$ scores and splice them into *male* list, *female* list etc. Based on the desired distribution of gender groups among outliers (as estimated from the dataset), *'quotas'* may be set for each gender value, and the appropriate number of top objects from each gender-specific list is then put together to form the outlier set of t objects. This is similar to the strategy used for job selection in India's affirmative action policy (*aka* reservation[5]). The extension of this quota-based strategy to multiple sensitive attributes by modelling them as one giant attribute taking values from the

---

[5] https://en.wikipedia.org/wiki/Reservation_in_India.

cross-product, is impractical due to multiple reasons. First, the cross-product may easily exceed t, leading to practical and legal issues across scenarios; for example, with just $\mathcal{S} = \{nationality, ethnicity\}$, we could have the cross product approaching 2000+ given there are 200+ nationalities and at least 10+ ethnicities, and practical values of t could be in the 100s due to manual perusal considerations. Thus, the quota system, by design, would exclude the large majority of rare combinations of sensitive attribute values from being represented among outliers however high their LOF scores may be; such a policy is unlikely to survive any legal or ethical scrutiny to allow practical uptake. Second, the quota based system offers no way to control the trade-off between fairness and quality, making it impractical to carefully choose trade-offs as outlined earlier. Third, even a simpler version of the extension of the quota-based system to multiple attributes has been recently shown to be NP-hard [5].

Table 1. Dataset information

Dataset	Domain	$|\mathcal{X}|$	Sensitive attributes used
Adult[a]	US 1994 Census	48842	Marital status, race, sex, nationality
CC[b]	Credit Card Default	30000	Sex, education, marital status
W4HE[c]	Wikipedia HE Use	913	Gender, disciplinary domain, uni name
St-Mat[d]	Student Maths Records	649	Gender, age

[a] http://archive.ics.uci.edu/ml/datasets/Adult
[b] https://archive.ics.uci.edu/ml/datasets/default+of+credit+card+clients
[c] https://archive.ics.uci.edu/ml/datasets/wiki4he
[d] https://archive.ics.uci.edu/ml/datasets/Student+Performance

7   Experimental Evaluation

Datasets and Experimental Setup: There are only a few public datasets with information of people, the scenario that is most pertinent for fairness analysis; this is likely due to person-data being regarded highly personal and anonymization could still lead to leakage of identifiable information[6]. The datasets we use along with details are included in Table 1. The datasets encompass a wide variety of scenarios, and vary much in sizes as well as the sensitive attributes used. We set t (to get top-t results) to 5% of the dataset size capped at 500, $k = 5$ and c (Ref. Eq. 8) to be $1/s$ where s is the number of sensitive attributes.

7.1   FairLOF Effectiveness Study

The effectiveness of *FairLOF* may be assessed by considering the quantum of fairness achieved at low degradations to quality. *It may be noted that higher*

---

[6] https://en.wikipedia.org/wiki/AOL_search_data_leak.

values are better on the quality measures (Jacc and Pres) and lower values are better on the fairness measures (ED and Wass). We follow the quality-fairness trade-off strategy as outlined in Sect. 6 with a λ search step-size of 0.1 and choose 0.9 and 0.8 as *guide points* for *Jacc*. The detailed results are presented in Table 2, with details of the result formatting outlined in the caption therein. Broadly, we observe the following:

- **Fairness Improvements:** *FairLOF* is seen to achieve significant improvements in fairness metrics at reasonable degradations to quality. The *ED* measure is being improved by 30% on an average at the chosen guide points, whereas *Wass* is improved by 12% and 22% on an average at the guide points of 0.9 and 0.8 respectively. These are evidently hugely significant gains indicating that *FairLOF* achieves compelling fairness improvements.
- **Trends on Pres:** Even at *Jacc* close to 0.9 and 0.8, the values of *Pres* achieved by *FairLOF* are seen to be only marginally lower than 1.0, recording degradations of less than 1.0% in the majority of the cases. This indicates that while the *LOF* results are being altered, *FairLOF* is being able to replace them with other objects with substantively similar $lof(.)$ values. This, we believe, is a highly consequential result, indicating that *FairLOF* remains very close in spirit to *LOF* on result quality while achieving the substantive fairness gains.

Table 2. FairLOF effectiveness study. The *FairLOF* results at *guide points* set to $Jacc = 0.9$ and $Jacc = 0.8$ are shown along with LOF results for each of the datasets. Since we use coarse steps for λ, the precise guide point value for *Jacc* may not be achieved; so we choose the closest *Jacc* that is achievable to the guide point. The deteriorations in Quality metrics and improvements in Fairness metrics are indicated in percentages. Fairness improvements of 20%+ are italicized, and those that are 30%+ are shown in bold, whereas Quality deteriorations < 1% are indicated in bold.

Dataset	Guide point	Quality				Fairness			
		Jacc	Det%	Pres	Det%	ED	Impr%	Wass	Impr%
Adult	LOF	1.0	N/A	1.0	N/A	0.2877	N/A	0.5328	N/A
	0.9	0.8939	10.61%	0.9977	**00.23%**	0.1906	**33.75%**	0.3468	**34.91%**
	0.8	0.7986	20.14%	0.9906	**00.94%**	0.1714	**40.42%**	0.2372	**55.48%**
CC	LOF	1.0	N/A	1.0	N/A	0.2670	N/A	0.2152	N/A
	0.9	0.9011	09.89%	0.9976	**00.24%**	0.2235	16.29%	0.2112	01.86%
	0.8	0.7921	20.79%	0.9879	01.21%	0.1568	**41.27%**	0.2012	06.51%
W4HE	LOF	1.0	N/A	1.0	N/A	0.2121	N/A	0.3305	N/A
	0.9	0.8776	12.24%	0.9987	**00.13%**	0.0966	**54.46%**	0.2989	09.56%
	0.8	0.8039	19.61%	0.9951	**00.49%**	0.1820	14.19%	0.2498	*24.42%*
St-Mat	LOF	1.0	N/A	1.0	N/A	0.4174	N/A	0.9196	N/A
	0.9	0.9047	09.53%	0.9970	**00.30%**	0.3467	16.94%	0.8962	02.54%
	0.8	0.8182	18.18%	0.9896	01.04%	0.3467	16.94%	0.8962	02.54%

In addition to the above, we note the following trends on *FairLOF* performance. First, $Wass$ is significantly harder to optimize for, as compared to ED; this is because $Wass$ prefers the gains to be equally distributed across sensitive attributes. Second, for small datasets where there is relatively less room for re-engineering outlier results for fairness, *FairLOF* gains are seen to saturate quickly. This is most evident for *St-Mat* in Table 2.

7.2   FairLOF Parameter Sensitivity Study

One of the key aspects is to see whether *FairLOF* effectiveness is smooth against changes in λ, the only parameter of significant consequence in *FairLOF*. In particular, we desire to see consistent decreases on each of *Jacc*, *Pres*, *ED* and *Wass* with increasing λ. On each of the datasets, such gradual and smooth trends were observed, with the gains tapering off sooner in the case of the smaller datasets, W4HE and St-Mat. The trends on Adult and CC were very similar; for Adult, we observed that the Pearson product-moment correlation co-efficient [26] against $\lambda \in [0, 1]$ to be -0.900 for *Jacc*, -0.973 for *Pres*, -0.997 for *ED* and -0.959 for *Wass* indicating a graceful movement along the various metrics with changing λ. We observed similar consistent trends for increasing c (Eq. 8) as well. *FairLOF* was also observed to be quite stable with changes of k and t.

8   Conclusions and Future Work

In this paper, for the first time (to our best knowledge), we considered the task of fair outlier detection. Fairness is of immense importance in this day and age when data analytics in general, and outlier detection in particular, is being used to make and influence decisions that will affect human lives to a significant extent, especially within web data scenarios that operate at scale. We consider the paradigm of local neighborhood based outlier detection, arguably the most popular paradigm in outlier detection literature. We outlined the task of fair outlier detection over a plurality of sensitive attributes, basing our argument on the normative notion of luck egalitarianism, that the costs of outlier detection be borne proportionally across groups defined on protected/sensitive attributes such as gender, race, religion and nationality. We observed that using a task-defined distance function for outlier detection could induce unfairness when the distance function is not fully orthogonal to all the sensitive attributes in the dataset. We develop an outlier detection method, called *FairLOF*, inspired by the construction of *LOF* and makes use of three principles to nudge the outlier detection towards directions of increased fairness. We outline an evaluation framework for fair outlier detection, and use that in evaluating *FairLOF* extensively over real-world datasets. Through our empirical results, we observe that *FairLOF* is able to deliver substantively improved fairness in outlier detection results, at reasonable detriment to result quality as assessed against *LOF*. This illustrates the effectiveness of *FairLOF* in achieving fairness in outlier detection.

Future Work: In this work, we have limited our attention to local neighborhood based outlier detection. Extending notions of fairness to global outlier detection would be an interesting future work. Further, we are considering extending *FairLOF* to the related task of identifying groups of anomalous points, and other considerations of relevance to fair unsupervised learning [23].

References

1. Abraham, S.S., Deepak, P., Sundaram, S.S.: Fairness in clustering with multiple sensitive attributes. In: EDBT, pp. 287–298 (2020)
2. Asudeh, A., Jagadish, H., Stoyanovich, J., Das, G.: Designing fair ranking schemes. In: SIGMOD (2019)
3. Babaei, K., Chen, Z., Maul, T.: Detecting point outliers using prune-based outlier factor (PLOF). arXiv preprint arXiv:1911.01654 (2019)
4. Barocas, S., Selbst, A.D.: Big data's disparate impact. Calif. Law. Rev. **104**, 671 (2016)
5. Bei, X., Liu, S., Poon, C.K., Wang, H.: Candidate selections with proportional fairness constraints. In: AAMAS (2020)
6. Bera, S.K., Chakrabarty, D., Flores, N., Negahbani, M.: Fair algorithms for clustering. In: NeurIPS, pp. 4955–4966 (2019)
7. Breunig, M.M., Kriegel, H.P., Ng, R.T., Sander, J.: LOF: identifying density-based local outliers. In: SIGMOD, pp. 93–104 (2000)
8. Chandola, V., Banerjee, A., Kumar, V.: Outlier detection: a survey. ACM Comput. Surv. **14**, 15 (2007)
9. Chawla, S., Sun, P.: SLOM: a new measure for local spatial outliers. Knowl. Inf. Syst. **9**(4), 412–429 (2006). https://doi.org/10.1007/s10115-005-0200-2
10. Chen, J., Sathe, S., Aggarwal, C., Turaga, D.: Outlier detection with autoencoder ensembles. In: SDM (2017)
11. Chierichetti, F., Kumar, R., Lattanzi, S., Vassilvitskii, S.: Fair clustering through fairlets. In: NIPS (2017)
12. Chouldechova, A., Roth, A.: A snapshot of the frontiers of fairness in machine learning. Commun. ACM **63**(5), 82–89 (2020)
13. Davidson, I., Ravi, S.: A framework for determining the fairness of outlier detection. In: ECAI (2020)
14. Domingues, R., Filippone, M., Michiardi, P., Zouaoui, J.: A comparative evaluation of outlier detection algorithms: experiments and analyses. Pattern Recogn. **74**, 406–421 (2018)
15. Fan, W., Bouguila, N., Ziou, D.: Unsupervised anomaly intrusion detection via localized Bayesian feature selection. In: ICDM (2011)
16. Hawkins, D.M.: Identification of Outliers. MSAP, vol. 11. Springer, Dordrecht (1980). https://doi.org/10.1007/978-94-015-3994-4
17. Huang, L., Vishnoi, N.K.: Stable and fair classification. arXiv:1902.07823 (2019)
18. Jabez, J., Muthukumar, B.: Intrusion detection system (IDS): anomaly detection using outlier detection approach. Procedia Comput. Sci. **48**, 338–346 (2015)
19. Knight, C.: Luck Egalitarianism: Equality, Responsibility, and Justice. EUP, Edinburgh (2009)
20. Kriegel, H.P., Kröger, P., Schubert, E., Zimek, A.: LoOP: local outlier probabilities. In: CIKM (2009)

21. Kumar, V., Kumar, D., Singh, R.: Outlier mining in medical databases: an application of data mining in health care management to detect abnormal values presented in medical databases. IJCSNS Int. J. Comput. Sci. Netw. Secur. **8**, 272–277 (2008)
22. Olfat, M., Aswani, A.: Convex formulations for fair principal component analysis. In: AAAI, vol. 33, pp. 663–670 (2019)
23. Deepak, P.: Whither fair clustering? In: AI for Social Good Workshop (2020)
24. Patro, G.K., et al.: Incremental fairness in two-sided market platforms: on updating recommendations fairly. In: AAAI (2020)
25. Pawar, A.D., Kalavadekar, P.N., Tambe, S.N.: A survey on outlier detection techniques for credit card fraud detection. IOSR J. Comput. Eng. **16**(2), 44–48 (2014)
26. Pearson, K.: Vii. Note on regression and inheritance in the case of two parents. Proc. R. Soc. London **58**(347–352), 240–242 (1895)
27. Rawls, J.: A Theory of Justice. Harvard University Press, Cambridge (1971)
28. Schubert, E., Zimek, A., Kriegel, H.-P.: Local outlier detection reconsidered: a generalized view on locality with applications to spatial, video, and network outlier detection. Data Min. Knowl. Disc. **28**(1), 190–237 (2012). https://doi.org/10.1007/s10618-012-0300-z
29. Wang, B., Davidson, I.: Towards fair deep clustering with multi-state protected variables. arXiv preprint arXiv:1901.10053 (2019)
30. Yu, D., Sheikholeslami, G., Zhang, A.: FindOut: finding outliers in very large datasets. Knowl. Inf. Syst. **4**(4), 387–412 (2002). https://doi.org/10.1007/s101150200013
31. Zafar, M.B., Valera, I., Rodriguez, M.G., Gummadi, K.P.: Fairness constraints: mechanisms for fair classification. arXiv preprint arXiv:1507.05259 (2015)
32. Zehlike, M., Bonchi, F., Castillo, C., Hajian, S., Megahed, M., Baeza-Yates, R.: FA*IR: a fair top-k ranking algorithm. In: CIKM, pp. 1569–1578 (2017)
33. Zhang, K., Hutter, M., Jin, H.: A new local distance-based outlier detection approach for scattered real-world data. In: PAKDD (2009)

A New Effective and Efficient Measure for Outlying Aspect Mining

Durgesh Samariya[1]([envelope]), Sunil Aryal[2], Kai Ming Ting[3], and Jiangang Ma[1]

[1] School of Engineering, Information Technology and Physical Sciences,
Federation University, Churchill, VIC, Australia
{d.samariya,j.ma}@federation.edu.au
[2] School of Information Technology, Deakin University, Geelong, VIC, Australia
sunil.aryal@deakin.edu.au
[3] National Key Laboratory for Novel Software Technology,
Nanjing University, Nanjing, China
tingkm@lamda.nju.edu.cn

Abstract. Outlying Aspect Mining (OAM) aims to find the subspaces (a.k.a. aspects) in which a given query is an outlier with respect to a given data set. Existing OAM algorithms use traditional distance/density-based outlier scores to rank subspaces. Because these distance/density-based scores depend on the dimensionality of subspaces, they cannot be compared directly between subspaces of different dimensionality. Z-score normalisation has been used to make them comparable. It requires to compute outlier scores of all instances in each subspace. This adds significant computational overhead on top of already expensive density estimation—making OAM algorithms infeasible to run in large and/or high-dimensional datasets. We also discover that Z-score normalisation is inappropriate for OAM in some cases. In this paper, we introduce a new score called *Simple Isolation score using Nearest Neighbor Ensemble* (SiNNE), which is independent of the dimensionality of subspaces. This enables the scores in subspaces with different dimensionalities to be compared directly without any additional normalisation. Our experimental results revealed that SiNNE produces better or at least the same results as existing scores; and it significantly improves the runtime of an existing OAM algorithm based on beam search.

Keywords: Outlying aspect mining · Dimensionality-unbiased score · Outlier explanation · Nearest neighbor ensemble

1 Introduction

Real-world datasets often have some anomalous data, a.k.a. outliers, which do not conform with the rest of the data. *Outlier Detection* (OD) is an important task in data mining that deals with detecting outliers in datasets automatically. A wide range of OD algorithms has been proposed to detect outliers in a data set [1,3,9]. While those algorithms are good at detecting outliers, they cannot

© Springer Nature Switzerland AG 2020
Z. Huang et al. (Eds.): WISE 2020, LNCS 12343, pp. 463–474, 2020.
https://doi.org/10.1007/978-3-030-62008-0_32

explain why a data instance is considered as an outlier, i.e., they cannot tell in which feature subset(s) the data instance is significantly different from the rest of the data.

Recently, researchers have started working on the problem of *Outlying Aspect Mining* (OAM) [5, 10, 12], where the task is to discover feature subset(s) for a query where it significantly deviates from the rest of the data. Those feature subset(s) are called outlying aspects of the given query. It is worth to note that OAM and OD are different—the main aim in the former is to find aspects for an instance where it exhibits the most outlying characteristics while the latter focuses on detecting all instances exhibiting outlying characteristics in the given original input space.

Identifying outlying aspects for a query data object is useful in many real-world applications. For example, (i) an insurance analyst may be interested in which particular aspect(s) an insurance claim looks suspicious; and (ii) when evaluating job applications, a selection panel wants to know in which aspect(s) an applicant is extraordinary compared to other applicants.

In the literature, the task of OAM is also referred to as *outlying subspace detection* [12] and *outlying aspect mining* [5, 10]. OAM algorithms require a score to rank subspaces based on the outlying degrees of the given query in all subspaces. Existing OAM algorithms [5, 10, 12] use traditional distance/density-based outlier scores as the ranking measure. Because distance/density-based outlier scores depend on the dimensionality of subspaces, they can not be compared directly to rank subspaces. Vinh et al. (2016) [10] used Z-score normalisation to make them comparable. However, it requires to compute outlier scores of all instances in each subspace adding significant computational overhead on already expensive density estimation. It makes OAM algorithms infeasible to run in large and/or high-dimensional datasets. In addition, we discover an issue with Z-score normalisation that makes it inappropriate for OAM in some cases.

Our main contributions in the paper are as follows.

1. We identify an issue of using Z-score normalisation of density-based outlier scores to rank subspaces and show that it has a bias towards a subspace having high variance.
2. We propose a new simple measure called *Simple Isolation score using Nearest Neighbor Ensemble* (SiNNE). It is independent of the dimensionality of subspaces and hence it can be used directly to rank subspaces. It does not require any additional normalisation.
3. We conduct extensive experiments to validate the effectiveness and efficiency of SiNNE in OAM. Our empirical results show that SiNNE can detect more interesting outlying aspects than three existing scoring measures, particularly in real-world datasets. In addition to that, it allows the OAM algorithm to run orders of magnitude faster than existing state-of-the-art scoring measures.

The rest of the paper is organized as follows. Section 2 provides a review of the previous work related to this paper. The limitation of Z-score normalisation in OAM is discussed in Sect. 3. The proposed new outlier score SiNNE is presented in Sect. 4. Empirical evaluation results are provided in Sect. 5 followed by conclusions and future work in Sect. 6.

2   Related Work

Let $\mathbb{D} = \{\mathbf{x}_1, \mathbf{x}_2, \cdots, \mathbf{x}_N\}$ be a data set of N instances, where each instance is represented as a vector, $\mathbf{x}_i \in \mathcal{R}^M$, M is the number of dimensions/features of data and $\mathcal{R}$ is a real domain. $\mathcal{F} = \{F_1, F_2, \cdots, F_M\}$ is the set of M featured. The problem of outlying aspect mining is to identify $S \subset \mathcal{F}$ for a given $\mathbf{q} \in \mathbb{D}$ in which it is most significantly different from the rest of the data.

To the best of our knowledge, Zhang et al. (2004) [12] is the earliest work that defines the problem of OAM. They introduced a framework to detect an outlying subspace called **H**igh-dimensional **O**utlying **S**ubspace **Miner** (HOS-Miner). They used a distance-based measure called 'Outlying Degree' (*OutD* in short) to rank subspaces. The *OutD* of a query $\mathbf{q}$ in subspace S is computed as sum of distance between $\mathbf{q}$ and its k - nearest neighbors (k-NN).

Instead of using the k-NN distances, **O**utlying **A**spect **Miner** (OAMiner) [5] uses density based on Kernel Density Estimator (KDE) to measure the outlierness of $\mathbf{q}$ in each subspace. In OAMiner [5], authors have reported that density is a biased measure because density decreases dramatically as the number of dimensions increases. Densities of a query point in subspaces with different dimensionality cannot be compared directly. Therefore, to eliminate the effect of dimensionality, they proposed to use density rank as an outlying measure. They used the same OAMiner algorithm by replacing the kernel density value by its rank. OAMiner searches for all possible combinations of subspaces systematically by traversing in the depth-first manner.

Recently, Vinh et al. (2016) [10] discussed the issue of using density rank as an outlier score in OAM and provided some examples where it can be counterproductive. Rather than using density rank, they proposed to use Z-score normalized density to make scores in subspaces with varying dimensionality comparable. They proposed a beam search strategy to search for subspaces. It uses the breadth-first method to search subspaces of up to a fixed number of dimensions called a beam width.

In recent work, Wells and Ting (2019) [11] proposed sGrid density estimator, which is a smoothed variant of the traditional grid-based estimator (a.k.a histogram). Authors replaced the kernel density estimator by sGrid in the Beam search OAM [10]. They also used Z-score normalisation to make the density values of a query point in subspaces with varying dimensionality comparable. Because sGrid density can be computed faster than KDE, it allows Beam search OAM to run orders of magnitude faster.

Both density rank and Z-score normalisation require to compute outlier scores of all N instances in the given data set in each subspace to compute the score of the given query. This adds significant computational overhead making the existing OAM algorithms infeasible to run in large and/or high-dimensional datasets. [10] discussed the issue of using density rank and proposed to use

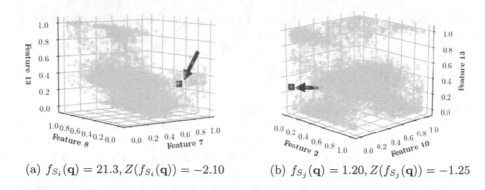

(a) $f_{S_i}(\mathbf{q}) = 21.3, Z(f_{S_i}(\mathbf{q})) = -2.10$ (b) $f_{S_j}(\mathbf{q}) = 1.20, Z(f_{S_j}(\mathbf{q})) = -1.25$

Fig. 1. Data distribution in two three-dimensional subspaces of the Pendigits data set.

Z-score normalized density. In the next section, we discuss an issue of using Z-score normalized density for OAM that makes it counter-productive in some data condition.

3 Issue of Using Z-Score Normalised Density

Because Z-score normalisation uses mean and variance of density values of all data instances in a subspace (μ_{fs_i} and σ_{fs_i}), it can be biased towards a subspace having high variation of density values (i.e., high σ_{fs_i}).

Let's take a simple example to demonstrate this. Assume that S_i and S_j ($i \neq j$), be two different subspaces of the same dimensionality (i.e., $|S_i| = |S_j|$). Intuitively, because they have the same dimensionality, they can be ranked based on the raw density (unnormalised) values of a query $\mathbf{q}$. Assuming $\mu_{fs_i} = \mu_{fs_j}$, we can have $Z(f_{S_i}(\mathbf{q})) < Z(f_{S_j}(\mathbf{q}))$ even though $f_{S_i}(\mathbf{q}) = f_{S_j}(\mathbf{q})$ if $\sigma_{fs_i} > \sigma_{fs_j}$ (i.e., S_i is ranked higher than S_j based on density Z-score normalisation just because of higher σ_{fs_i}).

To show this effect in a real-world data set, let's take an example of the Pendigits data set ($N{=}9868$ and $M{=}16$). Figure 1 shows the distribution of data in two three-dimensional subspaces $S_i = \{7, 8, 13\}$ and $S_j = \{2, 10, 13\}$. Visually, the query $\mathbf{q}$ represented by the red square appears to be more outlier in S_j than in S_i. This is consistent with its raw density values in the two subspaces, $f_{S_j}(\mathbf{q}) = 1.20 < f_{S_i}(\mathbf{q}) = 21.30$. However, the ranking is reversed after the Z-score normalisation, $(Z(f_{S_j}(\mathbf{q})) = -1.25 > Z(f_{S_i}(\mathbf{q})) = -2.10)$. This is due to the higher $\sigma_{fs_i} = 57.3 > \sigma_{fs_j} = 34.2$.

From the above example, we can say that Z-score normalisation has a bias towards a subspace having high variance. To overcome this weakness of Z-score normalisation, we proposed a new scoring measure in the next section which has no such bias in its raw form and does not require any normalisation.

4   The Proposed New Efficient Score

There are two limitations of density-based scores in OAM: (i) they are dimensionality biased and they require normalisation; and (ii) they are expensive to compute in each subspace. Being motivated by these limitations of density-based scores in OAM, we introduce a new measure which is dimensionality unbias in its raw form and can be computed efficiently.

Being motivated by the isolation using Nearest Neighbor Ensembles (iNNE) method for anomaly detection [2], we propose to use an ensemble of models where each model $\mathcal{H}_i$ ($i = 1, 2, \cdots, t$) is constructed from a small random subsample of data, $\mathcal{D}_i \subset D$, $|\mathcal{D}_i| = \psi$. In each subspace S, t models are constructed. To simply notations, we dropped S in the discussion. Each model $\mathcal{H}_i$ defines normal region as the area covered by a set hyperspheres centered at each $\mathbf{x} \in \mathcal{D}_i$, where the radius of the ball is the euclidean distance of $\mathbf{x}$ to its nearest neighbour in $\mathcal{D}_i$. The rest of the space outside of the hyperspheres is treated as the anomaly region.

In $\mathcal{H}_i$, a query $\mathbf{q}$ is considered as a normal instance if it falls in the normal region (at least in one hypersphere), otherwise it is considered as an anomaly:

$$s(\mathbf{q}|\mathcal{H}_i) = \begin{cases} 1, & \text{if } \mathbf{q} \text{ falls outside of all hyperspheres}, \\ 0, & \text{otherwise} \end{cases}$$

The final outlying score of $\mathbf{q}$ in subspace S based is defined as the average of models in the ensemble where it falls in anomaly regions (i.e., outside of all hyperspheres).

$$\text{SiNNE}(\mathbf{q}) = \frac{1}{t} \sum_{i=1}^{t} s(\mathbf{q}|\mathcal{H}_i)$$

It is interesting to note that iNNE uses a different definition of $s(\mathbf{q}|\mathcal{H}_i)$ using the radii of hyperspheres centered at the nearest neighbor of $\mathbf{q}$ and their nearest neighbor in $\mathcal{D}_i$. Our definition is a lot simpler and more intuitive as anomalies are expected to fall in anomaly regions in many models than normal instances. It is a simpler version of iNNE. Hence we call the proposed measure SiNNE, where 'S' stands for "Simple".

Because the area covered by each hypersphere decreases as the dimensionality of the space increases and so as the actual data space covered by normal instances. Therefore, SiNNE is independent of the dimensionality of space in its raw form without any normalisation making it ideal for OAM. It adapts to the local data density in the space because the sizes of the hyspheres depend on the local density. It can be computed a lot faster than the k-NN distance or density. Also, it does not require to compute outlier scores of all N instances in each subspace (which is required with existing score for Z-score normalisation) which gives it a significant advantage in terms of runtime.

The procedures to build an ensemble of models and using them to compute anomaly score of a test data in subspace S are provided in Algorithms 1 and 2.

An example of constructing $\mathcal{H}_i$ from in a two-dimensional space from a data set $\mathbb{D}$ ($N = 20$) and $\psi = 8$ is shown in Fig. 2.

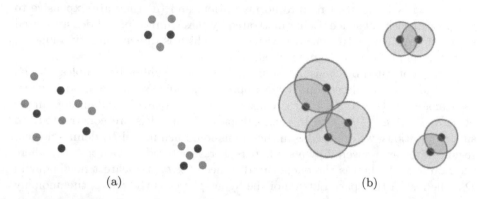

(a)   (b)

Fig. 2. (a) An example data set $\mathbb{D}$ (samples on dark black are selected to be in $\mathcal{D}_i$ to construct $\mathcal{H}_i$); and (b) Normal region defined by the area covered by hyperspheres in $\mathcal{H}_i$.

5   Experiments

In this section, we present results of our empirical evaluation of the proposed measure of SiNNE against the state-of-the-art OAM measures of Kernel density rank (R_{KDE}), Z-score normalised Kernel density (Z_{KDE}) and Z-score normalised sGrid density (Z_{sGrid}) using synthetic and real-world datasets in terms of effectiveness and efficiency. We used the synthetic dataset[1] used by Vinh et al. (2016). We also used four real-world datasets[2] available in the ELKI outlier data repository [4]. The characteristics of the datasets in terms of data size and the dimensionality of the original input space are provided in Table 1.

All measures and experimental setup were implemented in Java using WEKA platform [6]. We made the required changes in the Java implementation of iNNE provided by the authors to implement SiNNE. We implemented R_{KDE} and Z_{KDE} based on the KDE implementation available in WEKA [6]. We used the Java implementations of sGrid made available by the authors [11]. We used the same Beam search strategy for the subspace search as done in [10,11].

We used default parameters as suggested in respective papers unless specified otherwise. For SiNNE, we set $\psi = 8$ and $t = 100$. R_{KDE} and Z_{KDE} employ Kernel Density Estimator (KDE) to estimate density. KDE uses the Gaussian kernel with default bandwidth as suggested by [7]. The block size parameter (w) for bit set operation in sGrid was set as default to 64 as suggested by [11]. Parameters beam width (W) and maximum dimensionality of subspace (ℓ) in

---

[1] The synthetic data set is downloaded from https://www.ipd.kit.edu/~muellere/ HiCS/.

[2] https://elki-project.github.io/datasets/outlier.

Algorithm 1: Build Hyperspheres $(\mathbb{D}, t, \psi)$

Input: $\mathbb{D}$ - given data set; t - number of sets, ψ - number of sub samples
Output: $\mathcal{M}$ - An ensemble of t sets of ψ hyperspheres
1 initialize $\mathcal{M} = \Phi$;
2 **for** $i \leftarrow 1$ *to* t **do**
3   Generate $\mathcal{D}_i$ by randomly selecting ψ data points from $\mathbb{D}$ without replacement ;
4   initialize $\mathcal{H}_i = 0$;
5   **for** $c \in \mathcal{D}_i$ **do**
6         $\eta(c|\mathcal{D}_i) \leftarrow$ The nearest neighbour of c in $\mathcal{D}_i$;
7         $H(c) \leftarrow$ Build a hypersphere centered at c with radius $||c - \eta(c|\mathcal{D}_i)||_2$;
8         $\mathcal{H}_i = \mathcal{H}_i \cup H(c)$;
9   **end**
10     $\mathcal{M} = \mathcal{M} \cup \mathcal{H}_i$;
11 **end**
12 **return** $\mathcal{M}$;

---

* Note that, we do not require to build a hypersphere during implementation because whether a data instance falls in a hypersphere or not can be checked by comparing its distance to the center and radius of the hypersphere.

Algorithm 2: SiNNE: Computing anomaly score of a query instance

---

Input: $\mathbf{q}$ - query point, $\mathcal{M}$ - $\{\mathcal{H}^i | i = 1, \ldots, t\}$
Output: SiNNE($\mathbf{q}$)
1 initialize $s = 0$;
2 **for** $i \leftarrow 1$ *to* t **do**
3     $H \leftarrow$ search$(\mathcal{H}_i, \mathbf{q})$ {return a hypersphere H that covers $\mathbf{q}$ in $\mathcal{H}_i$} ;
4   **if** H *is NULL* **then**
5         $s += 1$;
6
7 **end**
8 **return** s/t;

Beam search procedure were set to 100 and 3, respectively, as done in [10] and [11].

All experiments were conducted on a machine with AMD 16-core CPU and 64 GB main memory, running on Ubuntu 18.03. All jobs were performed up to 10 days, and incomplete jobs were killed and marked as '$\Diamond$'.

5.1 Evaluation I: Quality of Discovered Subspaces

In this subsection, we focus on the quality of the discovered subspaces. We discussed results in synthetic and real-world datasets separately.

Table 1. Data set statistics.

Data set	#Data size (N)	#Dimension (M)
Synthetic data set	1000	10
Pendigits	6870	16
Shuttle	49097	9
ALOI	50000	33
KDDCup99	60632	38

Table 2. Comparison of SiNNE, R_{KDE}, Z_{KDE}, and Z_{sGrid} in the Synthetic data set. Discovered subspaces with the exact matches with the ground truths are bold-faced. **q**-id represent query point index; the numbers in the bracket (subspace) are attribute indices.

q-id	Ground Truth	SiNNE	R_{KDE}	Z_{KDE}	Z_{sGrid}
172	{8, 9}	**{8, 9}**	{1, 8, 9}	**{8, 9}**	**{8, 9}**
245	{2, 3, 4, 5}	**{2, 3, 4, 5}**	**{2, 3, 4, 5}**	**{2, 3, 4, 5}**	{3, 4, 5}
323	{8, 9}	**{8, 9}**	{2, 8, 9}	**{8, 9}**	**{8, 9}**
577	{2, 3, 4, 5}	**{2, 3, 4, 5}**	{0, 3, 7}	{6, 7}	**{2, 3, 4, 5}**
654	{2, 3, 4, 5}	**{2, 3, 4, 5}**	{1, 2, 3, 4, 5}	**{2, 3, 4, 5}**	**{2, 3, 4, 5}**
754	{6, 7}	**{6, 7}**	**{6, 7}**	**{6, 7}**	**{6, 7}**
765	{6, 7}	**{6, 7}**	{1, 6, 7}	**{6, 7}**	**{6, 7}**
781	{6, 7}	**{6, 7}**	**{6, 7}**	**{6, 7}**	**{6, 7}**
824	{8, 9}	**{8, 9}**	{6, 8, 9}	**{8, 9}**	**{8, 9}**
975	{8, 9}	**{8, 9}**	**{8, 9}**	**{8, 9}**	**{8, 9}**

Performance on Synthetic Datasets. We used the 10-dimensional synthetic data set provided by [8], which has 19 outliers. We reported the subspace found by SiNNE, R_{KDE}, Z_{KDE}, Z_{sGrid} and ground truths for 10 queries[3] in Table 2.

In terms of exact matches, SiNNE is the best performing measure which detected ground truth as the top outlying aspect for each query. Z_{KDE} and Z_{sGrid} produced exact matches for nine queries. R_{KDE} is the worst performing measure, which produced four exact matches.

Performance on Real-World Datasets. It is worth noting that we do not have ground truth of the real-world datasets to verify the quality of discovered subspaces. Also, there is no quality assessment measure/criteria of discovered subspaces. Thus, we compare the results of contending measures visually where the dimensionality of subspaces are up to 3. We used the state-of-the-art outlier

---

[3] We reported results of 10 queries only out of 19 because of the page limit.

detector called LOF [3][4] to find the top n ($n = 5$) outliers and used them as queries.

Table 3 shows the subspaces discovered for the top ranked outlier[5] by contending scoring measures in Pendigits, Shuttle, ALOI, and KDDCup99 datasets, respectively. Note that, we plotted all one-dimensional subspaces histograms, where the number of bins was set to 10. Visually, we can confirm that SiNNE identified better or at least similar outlying subspaces compared to existing measures of OAM.

Table 3. Visualization of discovered subspaces by SiNNE, R_{KDE}, Z_{KDE} and Z_{sGrid} on real-world datasets.

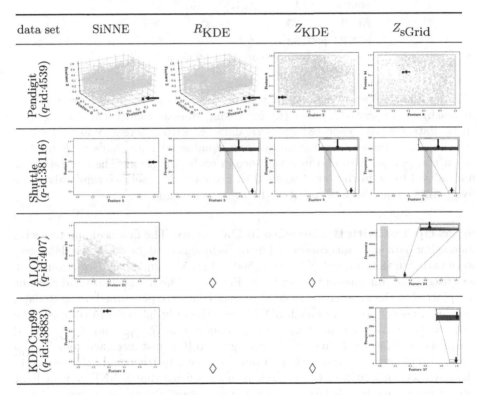

5.2 Evaluation II: Efficiency

The average runtime of randomly chosen 5 queries of the contending measures in the four real-world datasets are provided in Table 4. SiNNE and Z_{sGrid} were able to finish in all four datasets. R_{KDE} and Z_{KDE} were unable to complete within ten days in the two largest datasets - ALOI and KDDCup99. These results show

---

[4] We used the implementation of LOF available in Weka [6] and parameter $k = 50$.

[5] We present the results of the top ranked query only because of the page limit.

that SiNNE enables the existing OAM approach based on the Beam search to run orders of magnitude faster in large datasets. SiNNE was at least four orders of magnitude faster than R_{KDE} and Z_{KDE}; and an order of magnitude faster than Z_{sGrid} in the two largest datasets.

Table 4. Average runtime (in seconds) for 5 queries of SiNNE, R_{KDE}, Z_{KDE} and Z_{sGrid} on real-world datasets.

Data set	SiNNE	R_{KDE}	Z_{KDE}	Z_{sGrid}
Pendigits	1	10536	12450	9
Shuttle	1	124781	125225	34
ALOI	**25**	◊	◊	365
KDDCup99	**33**	◊	◊	524

◊ Expected to take more than 10 days.

We also conducted a scale-up test of the contending measures w.r.t. (i) increasing data sizes (N) and (ii) increasing dimensionality (M), using synthetic datasets. We generated synthetic datasets with different N and M where the data distribution is a mixture of five equal-sized Gaussian's with random mean $\mu \in [-10.0, 10.0]$ and unit variance in each dimension. The datasets were normalised to be in the same range [0,1]. For each data set, we randomly chose five points as queries and reported the average runtime.

Scale-Up Test with the Increase in Data Size. The first scale-up test with increasing data sizes was conducted using 5-dimension data set where data sizes were varied in the range of $N = 100$ k, 500 k, 1 m, 5 m, and 10 m. Note that $\ell = 3$ was used. The runtimes are presented in Fig. 3(a). The data set size and runtime are plotted in the logarithmic scale. Again all jobs were performed up to 10 days, and incomplete jobs were killed. SiNNE was the only measure to complete the task for the data set containing 10m instances. The R_{KDE} and Z_{KDE} could complete in 10 days only in datasets having up to 100k instances, whereas Z_{sGrid} could complete in the data set with 5m instances, but it couldn't complete in the data set with 10 m instances. The result confirms that SiNNE runs at least two orders of magnitude faster than existing state-of-the-art measures.

The runtime of SiNNE in the data set with 10m instances was 44 s whereas R_{KDE} and Z_{KDE} were projected to take more than 30 days, and Z_{sGrid} to take more than 15 days.

Scale-Up Test with the Increase in Dimensionality. In this scale-up test, we examined the efficiency of scoring metrics w.r.t the number of dimensions (M). A wide range of M values, {2, 5, 10, 50, 100, 200, 300, 500, 750, 1000}, were used with fixed data size $N = 100$ k. Figure 3 (b) shows the average runtimes of the contending measures. Note that the runtime is plotted using a logarithmic

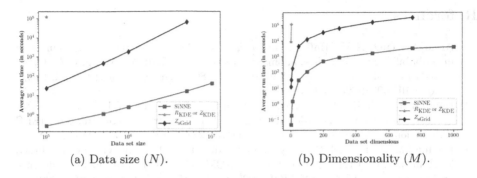

(a) Data size (N). (b) Dimensionality (M).

Fig. 3. Scale-up test.

scale. Again all jobs were performed up to 10 days, and incomplete jobs were killed. SiNNE was the only measure to complete the task for datasets with 1000 dimensions. Z_{sGrid} could only complete up to 750 dimensions, while R_{KDE} and Z_{KDE} could complete only up to 5 dimensions.

The runtimes for the 1000-dimensional data set were as follows: SiNNE: 1 h 8 min, R_{KDE}: > 100 days (projected runtime), Z_{KDE}: > 100 days (projected runtime) and Z_{sGrid} : > 15 days (projected runtime).

6 Conclusions and Future Work

In this study, we identify an issue of using Z-score normalisation of density to rank subspaces for OAM. Also, Z-score normalisation requires to compute densities of all instances in all subspaces making an OAM algorithm impossible to run in datasets with large data sizes or dimensions. We introduce an efficient and effective scoring measure for OAM called **S**imple **I**solation score using **N**earest **N**eighbor **E**nsemble (SiNNE). SiNNE uses an isolation based mechanism to compute outlierness of the query in each subspace, which is dimensionality unbias. Therefore, SiNNE does not require any normalisation to compare the scores of subspaces with different dimensions: Its raw scores can be compared directly. It runs significantly faster than existing measures because it does not require to compute scores of all instances like rank or Z-Score normalisation. By replacing the existing scoring measure with the proposed scoring measure, the existing OAM algorithm can now easily run in datasets with millions of data instances and thousands of dimensions. Our results show that SiNNE identifies more convincing outlying subspaces for queries than existing measures.

Our future work aims to investigate the theoretical properties of SiNNE and a better definition of dimensionality unbiasedness in the context of OAM.

Acknowledgments. This work is supported by Federation University Research Priority Area (RPA) scholarship, awarded to Durgesh Samariya.

References

1. Aryal, S., Ting, K.M., Haffari, G.: Revisiting attribute independence assumption in probabilistic unsupervised anomaly detection. In: Chau, M., Wang, G.A., Chen, H. (eds.) PAISI 2016. LNCS, vol. 9650, pp. 73–86. Springer, Cham (2016). https://doi.org/10.1007/978-3-319-31863-9_6
2. Bandaragoda, T.R., Ting, K.M., Albrecht, D., Liu, F.T., Wells, J.R.: Efficient anomaly detection by isolation using nearest neighbour ensemble. In: 2014 IEEE International Conference on Data Mining Workshop, pp. 698–705 (Dec 2014)
3. Breunig, M.M., Kriegel, H.P., Ng, R.T., Sander, J.: LOF: identifying density-based local outliers. In: Proceedings of the 2000 ACM SIGMOD International Conference on Management of Data, SIGMOD 2000, pp. 93–104. ACM, New York (2000)
4. Campos, G.O., et al.: On the evaluation of unsupervised outlier detection: measures, datasets, and an empirical study. Data Min. Knowl. Discovery 30(4), 891–927 (2016). https://doi.org/10.1007/s10618-015-0444-8
5. Duan, L., Tang, G., Pei, J., Bailey, J., Campbell, A., Tang, C.: Mining outlying aspects on numeric data. Data Min. Knowl. Discovery 29(5), 1116–1151 (2015). https://doi.org/10.1007/s10618-014-0398-2
6. Hall, M., Frank, E., Holmes, G., Pfahringer, B., Reutemann, P., Witten, I.H.: The WEKA data mining software: an update. SIGKDD Explor. Newsl. 11(1), 10–18 (2009)
7. Härdle, W.: Smoothing Techniques: with Implementation in S. Springer, New York (2012). https://doi.org/10.1007/978-1-4612-4432-5
8. Keller, F., Muller, E., Bohm, K.: HiCS: high contrast subspaces for density-based outlier ranking. In: Proceedings of the 2012 IEEE 28th International Conference on Data Engineering, ICDE 2012, pp. 1037–1048. IEEE Computer Society, Washington, DC (2012)
9. Liu, F.T., Ting, K.M., Zhou, Z.H.: Isolation forest. In: Proceedings of the 2008 Eighth IEEE International Conference on Data Mining, pp. 413–422 (2008)
10. Vinh, N.X., et al.: Discovering outlying aspects in large datasets. Data Min. Knowl. Discovery 30(6), 1520–1555 (2016). https://doi.org/10.1007/s10618-016-0453-2
11. Wells, J.R., Ting, K.M.: A new simple and efficient density estimator that enables fast systematic search. Pattern Recogn. Lett. 122, 92–98 (2019)
12. Zhang, J., Lou, M., Ling, T.W., Wang, H.: HOS-miner: a system for detecting outlyting subspaces of high-dimensional data. In: Proceedings of the Thirtieth International Conference on Very Large Data Bases - Volume 30, VLDB 2004, pp. 1265–1268. VLDB Endowment, Toronto (2004)

Time Series Data Cleaning Based on Dynamic Speed Constraints

Guohui Ding, Chenyang Li[(✉)], Ru Wei, Shasha Sun, Zhaoyu Liu,
and Chunlong Fan

Shenyang Aerospace University, Shenyang, China
lcy_lucky123@163.com

Abstract. Errors are ubiquitous in time series data as sensors are often
unstable. Existing approaches based on constraints can achieve good
data repair effect on abnormal values. The constraint typically refers to
the speed range of data changes. If the speed of data changes is not in
the range, it is identified as abnormal data violating the constraint and
needs repair, like if the oil consumption per hour of a sedan is negative or
greater than 15 gallons, it is probably abnormal data. However, existing
methods are only limited to specific type of data whose value change
speed is stable. They will be inefficient when handling the data stream
with sharp fluctuation because their constraints based on priori, fixed
speed range might miss most abnormal data. To make up the gap in
this scenario, an online cleaning approach based on dynamic speed constraints is proposed for time series data with fluctuating value change
speed. The dynamic constraints proposed is not determined in advance
but self-adaptive as data changes over time. A dual window mechanism is
devised to transform the global optimum of data repair problem to local
optimum problem. The classic minimum change principle and median
principle are introduced for data repair. With respect to repair invalidation of minimum change principle facing consecutive data points violating constraints, we propose to use the boundary of the corresponding
candidate repair set as repair strategy. Extensive experiments on real
datasets demonstrate that the proposed approach can achieve higher
repair accuracy than traditional approaches.

Keywords: Data cleaning · Speed constraint · Minimum change
principle

1 Introduction

Dirty values commonly exist in data streams [1,6,10,11,13], which mainly
includes three categories: incomplete data, wrong data, and repeated data. The
most important thing is that they have negative effects on the quality of many
applications, such as classification, prediction of time series data. The purpose

Supported by the National Natural Science Foundation of China under Grant 61303016.

Z. Huang et al. (Eds.): WISE 2020, LNCS 12343, pp. 475–487, 2020.
https://doi.org/10.1007/978-3-030-62008-0_33

of data cleaning is to detect and repair these dirty values in the data streams. Existing data cleaning methods [2,3,11,15] typically introduce new errors during the cleaning process [4,6]. For example, the filter cleaning algorithm [2,3] can make the erroneous data closer to the ground truth, but the original correct data is more likely to be changed during repair, which causes secondary errors in the data [15]. In order to minimize the change of the original correct data as much as possible, a series of algorithms [4,6,14,15] based on the principle of minimal change are proposed. At the same time, they introduce the speed constraint to detect data violation. SCREEN [6] is the representative one among these methods. It uses the speed constraint to determine whether the data points are correct, and selects a repair scheme based on the principle of minimal change. Although SCREEN reduces the secondary pollution of the data to a large extent, it only has a good effect on the data stream with a relatively stable changing speed of values, which is discussed in the next paragraph. Later the authors of SCREEN presented iterative minimum repairing approach [14] and statistical approach [13] for data cleaning. But the cleaning of time series data with fluctuating value change speed has not been solved.

Data with fluctuating speed are also very common in daily life. For example, the electricity consumption collected by a household electricity meter will increase slightly during daytime; however, during the period of peak demand for electricity (cooking, lighting, showering, and other activities with electricity usage together at night), the electricity consumption will increase instantly. For this kind of time series data with a large range of speed change, the existing data cleaning algorithms based on constraints have a poor cleaning effect because they employ a priori fixed speed constraint which is actually the maximum/minimum expectation of the changing speed of values in the historical data. For example, if the changing speed (maximum) of electricity consumption during the period that all electrical appliances are working together is used as the cleaning constraint for the whole data stream, then the data points deviating from the ground truth during the period of slight speed change are very difficult to be detected as dirty data as they are more likely to conform to the constraint. Consequently, the existing methods based on constraints represented by SCREEN only work well with time series data with stable speed of value changes, but do not have high repair accuracy facing data with significant speed fluctuation.

Therefore, to solve the problem discussed above, we propose an online cleaning approach based on dynamic speed constraints (CDDC). The speed constraint used in our approach constantly changes as time goes and is close to the variation trend of coming data rather than being fixed beforehand. It can be observed that the speed changes of most temporal variables in practice are typically periodic, regular and follow some rules. Thus, we exploit the Extreme Learning Machine (ELM) [5,12] to learn the rules of speed changes from the history data. Its strengths in learning speed and generalization performance will fit our online prediction task very well relative to other machine learning techniques like the popular deep learning [5,12]. We devised a dual window mechanism including detection window and prediction window. In the detection window, the idea from work [6, 7, 9] is introduced to transform the global optimal problem of the entire sequence to be repaired into local optimization problem. Then, we use the prediction window to predict the changing speed of values for each detection window. This mechanism enables our approach to conduct data cleaning online while holding the capacity of training and updating the learning model synchronously. We employ the classic minimum change principle and median principle [14, 15] to calculate the optimal repair plan. However, we observed that the minimum change principle is not valid all the time especially when consecutive data violation happened. We proposed to add a condition to judge whether it is a abnormal point of consecutive normal value, and use the boundary of the corresponding candidate repair set to repair data. Finally, extensive experiments on real datasets are conducted and the results demonstrate that the proposed approach can achieve higher repair accuracy than the traditional approaches.

The main contributions of this paper are as follows:

(1) An online cleaning approach based on dynamic speed constraints is proposed to clean time series data with fluctuating value change speed, which is a gap and can not be solved by existing methods because of their fixed speed constraints given in advance.
(2) Machine learning technology is applied to learn the rules in historical data to calculate the dynamic speed constraints of a dual window mechanism which is proposed, in order to implement cleaning task online.
(3) We conduct extensive experiments on real data sets to validate the proposed approach and the results show that it has better performance than existing methods.

2 Preliminaries

In this section, we will introduce some basic concepts and propositions from traditional methods [6, 7, 15] to formally describe the problem of time series data cleaning based on constraints.

2.1 Global Optimum

First, we assume that the current sequence is the entire sequence, and seek a solution with the minimum repair distance as the global optimal result. The sequence to be cleaned is $x = x[1], x[2], \ldots, x[n]$, where each $x[i]$ is the value of the i-th data point and the corresponding timestamp is $t[i]$. We train the historical data in order to get the predicted speed set $L = L[1], L[2], \ldots, L[n-w]$, where $L[i]$ is the predicted changing speed of values in the i-th window. Window is the smallest unit of data repair each time, whose length is w. If two points are in the same window, their time distance is lower than the length w. An error coefficient θ from the training model is introduced to map the speed from a point value to a range. A range is relatively more reasonable rather than a single point value predicted because it is impossible to get an absolutely exact prediction. How to calculate θ will be discussed later. So, we get a set of speed constraints: $s = \{(s[1]_{min}, s[1]_{max}), (s[2]_{min}, s[2]_{max}), \ldots, (s[s]_{min}, \ s[s]_{max})\}$, where $s[i]_{min} = L[i] - \theta, s[i]_{max} = L[i] + \theta$. The interval $(s[i]_{min}, s[i]_{max})$ is the changing range of speed of the i-th window. Therefore, if $0 < t[j] - t[i] \leq w$ and $i - w \leq a \leq i$, we have formula (1):

$$\frac{x[j] - x[i]}{t[j] - t[i]} \in [s[a]_{min}, s[a]_{max}] \tag{1}$$

Proposition 1. *If a data point conforms to formula (1), it is identified as correct data meeting speed constraints. If not, the point needs to be repaired.*

The repair distance is the distance between the repaired value x' and the original value x in the sequence, which is defined as:

$$\Delta(x, x') = \sum_{x[i] \in x} |x[i] - x[i]'| \tag{2}$$

Proposition 2. *If data points need to be repaired, the repair value not only needs to satisfy formula (1), but also minimizes $\Delta(x, x')$ in formula (2).*

For the evaluation criteria of repair results, root mean square value is adopted to represent the degree of deviation between the repair value and the real value (also known as loss value), which is formally described as:

$$RMS = \sqrt[2]{\frac{\sum_{i \in (1,n)} (x[i]' - r[i])^2}{n}} \tag{3}$$

where $r[i]$ is the real value at time point i.

3   Online Cleaning Based on Dual Window Mechanism

The prediction window will use the prediction set as the sliding unit. Before the data prediction starts, we need to put some historical data close to the prediction window into the training set. This part will be executed before the data cleaning process, which means that suppose the first part of the data is correct when making the first prediction. When the first prediction window ends, the data cleaning algorithm begins. The prediction window is the smallest unit for predicting dynamic constraints. It uses the local optimal results in each prediction window to splice together the global optimal results [7]. During the sliding process, the test set will be used as the training set for the next prediction window, and the original training set will discard the data of the test set size at the furthest time point, in order to ensure that the size of the training set is unchanged.

For the problem of data violation detection, the detection window [6,9] can divide the cleaning sequence and is the minimum unit for abnormal point detection. In addition, the principle of minimum change and dynamic rule constraints are used to check data points in the detection window to determine whether a point violates the constraints, and as the data points consecutively arrive, the detection window slides backwards and then keep executing the detection algorithm in order to achieve cleaning online.

There is a correspondence between the two windows; that is, each prediction window contains n speeds, and each speed must uniquely correspond to one detection window. If we use ELM to train historical data, we will obtain the prediction set $L = L[1], L[2], ..., L[n-w]$, where $L[i]$ is the predicted value of the speed of the i-th detection window. Moreover, in the execution process of the detection algorithm, the dynamic constraints correspond to the speed change domain which is calculated on the basis of $L[i]$ in the prediction window.

4   Dynamic Rule Constraints

4.1   Changing Speed of Values

The speed can directly reflect changes between data points in time series data. According to the value of the current point and its corresponding speed, we can infer and constrain the value of next data point, which means speed change can represent the range of changes of data points in the data stream. We use the range as constraints to clean the data.

In practice, there are also many time series data with uneven changes. In addition to the aforementioned data flow of electricity meters, there are also instantaneous flight speeds of aircraft, etc. These data streams are difficult to be constrained by the global constant speed, because a constraint that contain all speed changes always contains some extreme cases and it is not universal for ordinary situations. For example, when the aircraft takes off, it needs a large instantaneous speed to obtain a pressure difference, which supports it to leave the ground. During flight, there is no need for such a large speed, so a constant constraint value will cause some erroneous data points to be undetectable.

4.2   Speed Change Domain

ELM is used as the learning model in our approach. We use ELM to predict the speed of the data points in each window of the data stream. However, there is no learning model which can predict the changing speed with accuracy of 100%. Thus, we introduce an error coefficient θ to extend the speed from a single point value to a range which will make the speed constraints more reasonable in practical applications. The speed constraints with range is referred to as speed change domain.

The error coefficient θ is calculated based on the set of loss values $L_{validation} = \left\{ L_i | L_i = |y_i - y_i'|, i = 1, 2, \ldots, n \right\}$ for each predicted data point in the validation set. The loss set of the validation set can basically reflect the prediction effect, so is used as a representative to calculate the error coefficient θ. Following the principle of minimum change, we finally determine to use the maximum value in the loss set of the validation set as the error coefficient of the current prediction set.

Assuming a prediction result set $L_{test} = \{L[i]|L[1], L[2], \ldots, L[n]\}$, the result set corresponding to the loss value set of the validation set is $L_{validation} = \{L_i|L_1, L_2, \ldots, L_n\}$. Then the corresponding error coefficient is the maximum value of the loss value set, that is $\theta = max\,(L_{validation})$. θ is only valid within this prediction set. The speed change domain corresponding to the speed is $s_i\,(s_{i_{min}}, s_{i_{max}})$, that is:

$$\begin{cases} s_{i_{min}} = L[i] - \theta \\ s_{i_{max}} = L[i] + \theta \end{cases} \tag{4}$$

5 Data Repair

We employ the minimum change principle and median principle [6] to repair data violation. However, we observe that they only be effective for intermittent abnormal points. If the data points consecutively violate the constraints, they might lose efficacy. A repair strategy for consecutive data violation is proposed in this section.

5.1 Consecutive Abnormal Points

If x_k is an abnormal point, it will be repaired by the traditional intermittent abnormal point repair method. That is, the follow-up data of the abnormal point x_k is used to calculate the set of its optimal repair scheme according to the dynamic speed constraint, and the median x_k^{mid} of the set $\left(X_k^{min} \cup X_k^{max} \cup \{x_k\}\right)$ is taken as the optimal repair scheme based on the minimum change principle, where

$$\begin{aligned} X_k^{min} &= \{x_i + s[i]_{min}(t_k - t_i)|t_k < t_i < t_k + w, 1 \le i \le n\} \\ X_k^{max} &= \{x_i + s[i]_{max}(t_k - t_i)|t_k < t_i < t_k + w, 1 \le i \le n\} \end{aligned} \tag{5}$$

The optimal repair scheme should be in the candidate repair set $\left[x_k^{min}, x_k^{max}\right]$ determined by the data before the abnormal point x_k, where

$$\begin{aligned} x_k^{min} &= x_{k-1}' + s[i]_{min}(t_k - t_{k-1}) \\ x_k^{max} &= x_{k-1}' + s[i]_{max}(t_k - t_{k-1}) \end{aligned} \tag{6}$$

Generally, the optimal repair scheme x_k^{mid} is the repair value of x_k; if not, the repair value can be calculated directly according to the following formula:

$$x_k' = \begin{cases} x_k^{max} & x_k^{max} < x_k^{mid} \\ x_k^{min} & x_k^{min} > x_k^{mid} \\ x_k^{mid} & otherwise \end{cases} \tag{7}$$

We will record the anomaly as $x[i]$, if none from $x[i]$ to $x[i+n]$ $(n > 1)$ satisfies formula (1), then $x[i]$ to $x[i+n]$ $(n > 1)$ are referred as consecutive abnormal points. If $x[i] > x[i-1]'$, then $x[i]$ to $x[i+n]$ $(n > 1)$ are referred as abnormal points with consecutive normal values.

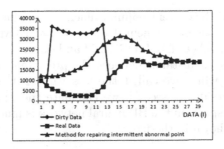

Fig. 1. Comparison of repair effects

5.2 Method for Repairing Abnormal Points of Consecutive Normal Values

For the abnormal points with consecutive normal values from Fig. 1, if the method of repairing intermittent abnormal points is used for repairing, the algorithm always takes the upper boundary value as the optimal repair solution, causing the deviation between the repaired value and the real value to increase as the number of consecutive abnormal point increases. Therefore, we believe that if the lower boundary value is used as the repair value, it should make the repair value closer to the real value. Similarly, during the repair process, a condition is added into the algorithm, i.e., when x_k is repaired, if the data point x_{k+1} is still an abnormal point, and $x_{k+1} > x_k'$, then $x_{k+1}' = x_k^{min}$, where $x_k^{min} \in \left[x_k^{min}, x_k^{max} \right]$.

5.3 The Effectiveness of the Algorithm for Repairing Consecutive Anomaly

For abnormal points of consecutive normal values, we calculate the deviation of the traditional intermittent abnormal point repair method according to Fig. 1, $x_n' = x_2' + (n-2) S_{max} = x_2' + (n-2) L[i] + (n-2) \theta$; while for the consecutive abnormal point repair method, $x_n' = x_2' + (n-2) L[i] - (n-4) \theta$. It can be seen that the repair value of the consecutive anomaly repair method is always closer to the ground truth than the traditional method by 2θ.

6 Experiment

In the experiment, we used a dataset of taxi usage in New York, which is a public dataset. We compare our approach CDDC with the traditional, representative method SCREEN [6] which is currently a better data cleaning algorithm of time series data.

First, we compared the current popular machine learning methods, including various improved versions of BP neural network and ELM [12], namely FOS-ELM [5], NFOS-ELM, NAOS-ELM, OR-ELM and itself. We would analyze from two aspects: first, it is the distance between the predicted value and the ground truth in the test set, which we call it the loss value RMS, also expressed as the sum of squared residuals; second, we calculate the time cost of the algorithm. Since the time spent on the BP neural network is much higher than other algorithms, it is not shown in the figure.

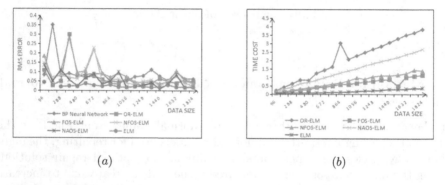

(a)   (b)

Fig. 2. Comparison of performance of prediction algorithms

As can be seen in Fig. 2(a), the prediction effects of BP Neural Network, FOS-ELM, OR-ELM and NFOS-ELM are relatively unstable. For NAOS-ELM, although the jump is small, it is difficult to find the optimal state. The loss value of the ELM is very stable and has been at the lowest position among the six algorithms; that is, the loss value is the lowest.

From Fig. 2(b), we can find that the OR-ELM algorithm takes the highest time cost, followed by NAOS-ELM. The time cost of the NFOS-ELM is slightly higher than that of FOS-ELM, but the difference is not large. The time cost of ELM is the lowest, and as the training set increases, the trend of time cost grows slowly. Therefore, the ELM algorithm is more in line with the requirements of the dynamic constraint model.

The proposed CDDC is compared with SCREEN from the time cost and detection accuracy. The detection accuracy is the ratio between the abnormal points detected in the test sequence and the actual abnormal points. During the experiment, the proportion of abnormal points in the test sequence is fixed, and the performance of the two algorithms is analyzed by changing the size of the data set to be tested.

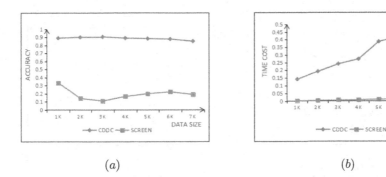

(a)   (b)

Fig. 3. Comparative analysis of CDDC and SCREEN

The data distribution trend in Fig. 3(a) clearly shows that the accuracy of CDDC is much higher than the SCREEN algorithm, which is 3 to 5 times that of SCREEN. The reason is that the existing methods represented by SCREEN only fit for time series data with stable changing speed. In Fig. 3(b), it can be seen that the time cost of the CDDC is much higher than SCREEN, but it is the global time. The local time can be adjusted through the prediction window and the overall delay can be controlled within 0.1 s. Therefore, the performance of CDDC is better than the SCREEN algorithm, and is relatively stable.

The effect of the repair sequence is still analyzed from two perspectives, namely the loss value and the repair distance. The loss value RMS is the sum of squared residuals between the repaired value and the real value, and the repair distance is the sum of the changes between the test sequence and the repaired sequence. The experiment would be launched from two aspects: (1) the proportion of abnormal points in the test sequence is fixed, and the performance of the two algorithms is analyzed by changing the size of the data set to be tested; (2) the size of the test sequence is fixed, and the proportion of abnormal points increases. The loss values and repair distances of the two algorithms are compared and analyzed with the increase of the proportion of abnormal points.

The trend in Fig. 4(a) and Fig. 4(d) clearly shows that the loss value of SCREEN is close to the loss value of the original sequence, indicating that the cleaning effect is not good while the loss value of the data repair algorithm for intermittent abnormal points in CDDC is lower than SCREEN. In Fig. 4(b) and Fig. 4(c), the repaired distance of the intermittent abnormal point repair algorithm is larger than that of SCREEN, which is due to the high accuracy of CDDC. In Fig. 4(c), SCREEN has almost no repair distance, which means that most data violation is not detected by SCREEN. Therefore, the data cleaning based on dynamic constraints is far superior to SCREEN in terms of repair performance.

It can be seen in Fig. 5(a) and Fig. 5(d) that the loss value of the data repair method for intermittent abnormal points is much smaller than the loss value of the sequence to be cleaned, but it is still greater than the loss value of the consecutive abnormal data repair method. This shows that the performance of

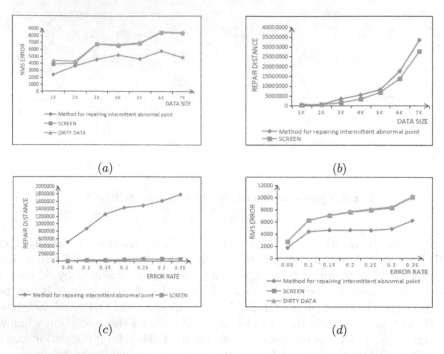

Fig. 4. Comparison of data repair performance for intermittent abnormal points

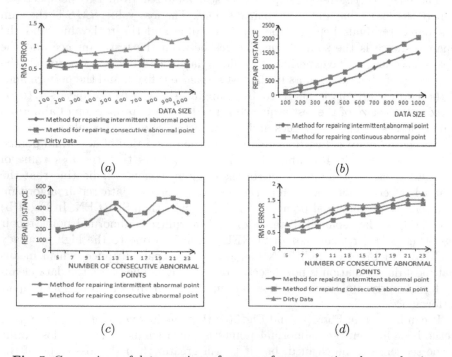

Fig. 5. Comparison of data repair performance for consecutive abnormal points

the proposed algorithm for consecutive abnormal points is better than the traditional algorithm. In Fig. 5(b) and Fig. 5(c), the repair distance of consecutive abnormal point repair method in CDDC indicates that the consecutive repair value is closer to the real value.

7 Related Work

Filter Cleaning Method. Kalman filtering [1, 2] is a data processing technique used for data denoising and cleaning. In addition, nonlinear filtering [1, 3] is also a technique for removing noise from signal data. This type of method always uses linear approximation to clean the data.

Smoothing [1, 6, 11, 13, 14]. SWAB [7] is an online cleaning method that seeks approximations in the way of linear approximation. Forecasting by the above method will always change most data of the original sequence including the correct value.

Sequential data cleaning [13] is a method based on maximum likelihood by observing the speed change of regular sequential data and modeling according to the probability distribution. However, this method only maximizes the possibility of data repair on the basis of the previous data. Time series data cleaning [7, 10, 11, 14] is a cleaning method of continuously iterative minimum repair in error prediction to ensure accuracy. SCREEN [1, 6, 15] is a cleaning method that establishes a fixed maximum/minimum speed constraint according to the changing law of data. This method can not show good performance in the face of data with large differences in speed. Therefore, this paper adopts dynamic rule constraints that are closer to the speed change of the data to solve this problem.

Clustering [8] now combines Piece-wise Aggregate Approximation and the density-based spatial clustering of applications with noise to classify medical data and uses dynamic maintenance for real-time updates. Cumulative frequency query [16] proposes a framework composed of data aggregation and dynamic maintenance to achieve the purpose of monitoring important medical data.

Load shedding is a technology that achieves considerable cost and efficiency in processing streaming data. Among them, a buffer-based QoS adaptive framework [17] composed of scheduler, adapter, and cleaner provides a load shedding technology to maintain the quality of streaming data. Sliding window joins [9] use the dual window mechanism to achieve the purpose of producing the maximum subset join by combining the front-shedding and rear-shedding.

8 Conclusions

The key point of our method is to use dynamic rule constraints instead of a constant maximum/minimum to detect abnormal points in time series data with fluctuating speed. Our approach makes up for the gap that the existing methods only fit for data with stable value change speed rather than data with obvious fluctuation. Predict the speed value through the ELM learning model, and finally a dynamic rule constraint is mapped by the error coefficient. In addition, a consecutive abnormal points repair scheme is proposed to solve the inefficiency of the existing methods facing consecutive data violation. Experiments conducted on real data sets verify the proposed approach has better performance than existing methods with respect to the time series data with unstable fluctuation.

References

1. Broeck, J.V.D., Fadnes, L.T.: Data Cleaning (2005). https://doi.org/10.1371/journal.pmed.0020267
2. Welch, G., Bishop, G.: An Introduction to the Kalman Filter (1995). https://doi.org/10.1007/978-0-387-31439-6_716
3. Challa, S., Barshalom, Y.: Nonlinear filter design using Fokker-Planck-Kolmogorov probability density evolutions. IEEE Trans. Aerosp. Electron. Syst. **36**(1), 309–315 (2000). https://doi.org/10.1109/7.826335
4. Brillinger, D.R.: Time series: data analysis and theory. IEEE Signal Process. Mag. **19**(2), 94–94 (1975). https://doi.org/10.1137/1.9780898719246
5. Zhao, J., Wang, Z., Dong, S.P.: Online sequential extreme learning machine with forgetting mechanism. Neurocomputing **87**(none), 79–89 (2012). https://doi.org/10.1016/j.neucom.2012.02.003
6. Song, S., Zhang, A., Wang, J., Yu, P.S.: SCREEN: stream data cleaning under speed constraint. In: SIGMOD 2015, vol. 06, no. 3, pp. 827–841 (2015). https://doi.org/10.1145/2723372.2723730
7. Keogh, E.J., Chu, S., Hart, D.M., Pazzani, M.J.: An online algorithm for segmenting time series. In: ICDM, pp. 289–296 (2001). https://doi.org/10.1109/ICDM.2001.989531
8. AI-Shammari, A., Zhou, R., Naseriparsa, M., Liu, C.: An effective density-based clustering and dynamic maintenance framework for evolving medical data streams. Int. J. Med. Inform. **126**, 176–186 (2019). https://doi.org/10.1016/j.ijmedinf.2019.03.016
9. Han, D., Xiao, C., Zhou, R., Wang, G., Huo, H., Hui, X.: Load shedding for window joins over streams. In: Yu, J.X., Kitsuregawa, M., Leong, H.V. (eds.) WAIM 2006. LNCS, vol. 4016, pp. 472–483. Springer, Heidelberg (2006). https://doi.org/10.1007/11775300_40
10. Volkovs, M., Fei, C., Szlichta, J., et al.: Continuous data cleaning. In: IEEE International Conference on Data Engineering (2014). https://doi.org/10.1109/ICDE.2014.6816655
11. Hong, L., Tk, A.K., Thomas, J.P., et al.: Cleaning Framework for BigData: an interactive approach for data cleaning. In: IEEE Second International Conference on Big Data Computing Service & Applications (2016). https://doi.org/10.1109/BigDataService.2016.41
12. Huang, G.B., Zhu, Q.Y., Siew, C.K.: Extreme learning machine: a new learning scheme of feedforward neural networks. In: IEEE International Joint Conference on Neural Networks (2005). https://doi.org/10.1109/IJCNN.2004.1380068
13. Zhang, A., Song, S., Wang, J.: Sequential data cleaning: a statistical approach. In: ACM SIGMOD International Conference on Management of Data (2016). https://doi.org/10.1145/2882903.2915233
14. Zhang, A., Song, S., Wang, J., Yu, P.S.: Time series data cleaning: from anomaly detection to anomaly repairing. Proc. VLDB Endow. **10**(10), 1046–1057 (2017). https://doi.org/10.14778/3115404.3115410

15. Bohannon, P.. Flaster, M., Fan, W., Rastogi. R.: A cost-based model and effective heuristic for repairing constraints by value modification. In: SIGMOD Conference, pp. 143–154 (2005). https://doi.org/10.1145/1066157.1066175
16. Al-Shammari, A., Zhou, R., Liu, C., Naseriparsa, M., Vo, B.Q.: A framework for processing cumulative frequency queries over medical data streams. In: Hacid, H., Cellary, W., Wang, H., Paik, H.-Y., Zhou, R. (eds.) WISE 2018. LNCS, vol. 11234, pp. 121–131. Springer, Cham (2018). https://doi.org/10.1007/978-3-030-02925-8_9
17. Zhou, R., Wang, G., Han, D., Gong, P., Xiao, C., Li, H.: Buffer-Preposed QoS adaptation framework and load shedding techniques over streams. In: Aberer, K., Peng, Z., Rundensteiner, E.A., Zhang, Y., Li, X. (eds.) WISE 2006. LNCS, vol. 4255, pp. 234–246. Springer, Heidelberg (2006). https://doi.org/10.1007/11912873_25

Predicting MOOCs Dropout with a Deep Model

Fan Wu, Juntao Zhang, Yuling Shi, Xiandi Yang$^{(\boxtimes)}$, Wei Song, and Zhiyong Peng

School of Computer Science, Wuhan University, Wuhan, Hubei, China
{fan2013,juntaozhang,sylyjs,xiandiy,songwei,peng}@whu.edu.cn

Abstract. With the deep integration of information technology and education, Massive Open Online Courses (MOOCs) become popular and receive high attention. Although MOOCs are popular among people, it faces a great challenge—the high dropout rate, which affects its development. Predicting the dropout rate in advance can take relevant measures to avoid as many dropouts as possible. Traditional machine learning classification prediction and single sequence label prediction methods are difficult to accurately predict complex user behaviors. To solve the problem, in this paper, we perform a deep analysis of user learning behavior to find that user activity shows a periodic distribution based on the time of course release. In addition, user gender and course category also affect users' behaviors. To this end, we propose a deep model based on recurrent network which combines the influence factors of cyclical historical behavior on the basis of a single sequence of events. Meanwhile, we combine behavior periodicity with attention mechanism to select effective historical behavior impact factors. Then we embed the attributes of user and course to predict the dropout rate. Finally, experiments on different data sets show that our approach performs better than the state-of-the art methods.

Keywords: Dropout in MOOCs · Period · Attention

1 Introduction

With the deep integration of information technology and education, large scale online education is developing rapidly under the support of artificial intelligence and big data technology. The concept of Massive Open Online Courses (MOOCs) [1, 2] first appeared in 2008, and the learning revolution represented by it is strongly impacting the ecology of traditional education. In 2012, three educational platforms, Coursera, Udacity and edX emerged, causing a MOOC wave around the world and severely impacting traditional education model. As a result, the MOOCs wave broke out in China in 2013, and top domestic universities cooperated with edX and Coursera to create a domestic online education platform-XuetangX. MOOCs, led by XuetangX, is also rapidly developing [3]. In recent years, MOOC learning has become more and more popular. Due to its strong advantages, it has broken the time and space limitations of traditional education mechanisms and an electronic device connected to the Internet can complete the course. According to Class Central's annual report[1], by the end of 2019, more than 900 colleges

---

[1] https://www.classcentral.com/report/moocs-stats-and-trends-2019/.

© Springer Nature Switzerland AG 2020
Z. Huang et al. (Eds.): WISE 2020, LNCS 12343, pp. 488–502, 2020.
https://doi.org/10.1007/978-3-030-62008-0_34

and universities had opened 135,000 MOOC courses, excluding China. The trend of courses offered on the MOOC platform from 2012 to 2019 is shown in Fig. 1, and this number is still growing rapidly. In recent months, during special virus outbreaks, online education has provided great convenience to the majority of students. The epidemic has brought MOOC to a new climax, MOOC quickly occupied the education market with unstoppable momentum again and led the education revolution.

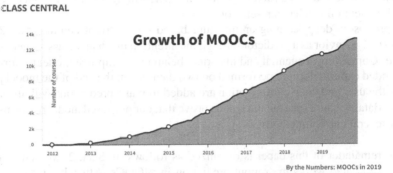

Fig. 1. The trend of the number of courses on MOOC platform from 2012 to 2019

However, with the rapid development of online education, some shortcomings have gradually emerged. The main problem is the occurrence of dropouts. Very few people can actually complete a course to obtain a certificate [4, 5], compared with the compulsory learning mechanism in traditional education, it is the openness of online education and the lack of supervision mechanism which leads to the loss of users. The reasons for users dropping out may be inappropriate learning resources, mismatched learning abilities, incorrect learning methods, or lack of communication between users, resulting in insufficient learning motivation and driving force, etc. [6]. In fact, the domestic average online school dropout rate has now reached 95.5% [3]. Facing the severe challenge, a large number of researchers have studied the learning behavior patterns and preferences of learners from multiple different perspectives and the relationship with the final learning effect [7–9]. User loss is a major challenge for MOOCs, we need to be able to predict the possibility of user dropouts in advance, then analyze the causes and take corresponding measures.

Through the deep analysis of the actual datasets, we find that most courses are published with a fixed time interval and users have a high degree of activity before or after the new course release time, and user learning behavior may be periodic. So this paper proposes a periodic attention mechanism to predict dropout rates. The dropout rate prediction is actually a sequence labeling problem [10] or a time series prediction problem. Most of the existing sequence events are predicted by using Recurrent Neural Network (RNN) or Long Short Term Memory networks (LSTM) as the model. LSTM can also be used for text context sentiment analysis [11]. The method proposed in this paper is based on the prediction of the sequence of events combined with the attention mechanism of the association period, taking the impact of historical behavior into consideration, and combining the two aspects to predict the probability can ensure accuracy.

The main contributions of this paper can be summarized as follows:

- Perform in-depth analysis on user behavior data to find demographic and behavior characteristics that have a greater impact on user behavior. At the same time, we propose a period detection algorithm to find the best user behavior period from the distribution period and structure period, and performing locating the specific target for the attention mechanism selector.
- We propose a deep learning architecture based on recurrent neural network. Take historical behavior as a predictor through the attention mechanism associated with the cycle. Combining sequential and historical behavior to improve model performance.
- Extended experiments are performed on two datasets, at the end of the model prediction, the user and course information are added to make predictions with the support of the dataset. The experimental results prove that our proposed model performs better than several current methods.

The remainder of this paper is organized as follows. In Sect. 2, we systematically review the related works in dropout prediction in MOOCs. After that we take a deep analysis about users' learning activities. Further, we introduce our predicting model in details. Section 5 we apply our model on real datasets and give the descriptions about the experiments. Finally, we conclude our paper.

2   Related Work

In this part, we make a brief summary of the research on the dropout rate prediction in the MOOCs field in the past ten years.

Many researchers study the relationship between learner learning behavior and learning effectiveness from different perspective, they use different mathematical models to predict learners' short-term learning behavior and long-term learning effectiveness. In Anderson et al. [12], learners were divided into five categories based on their learning behavior preferences, and learning effects were analyzed based on different learning models. In Kloft et al. [13], a simple linear SVM is used to predict the dropout rate. Taylor et al. [14] applies logistic regression to learn behavior characteristics and predicts student dropouts based on the students' last learning activities in the course. Ramesh et al. [15] used the discussions in the MOOCs forum and the completion of learners' homework to construct a predictive model to study learner dropout behaviors. Balakrishnan et al. [16] proposes a dropout prediction model based on Hidden Markov Model combined with support vector machines. Unlike other studies, Chanchary et al. [17] uses K-means for quantitative analysis and automatically discover inactive students by clustering students in a MOOCs environment. W Xing et al. [18] takes a combination of Bayesian Network and Decision Tree to make predictions. In addition to traditional machine learning, deep learning is also used to predict dropout rates. Fei et al. [19] believes that the prediction of dropout rates is a time series prediction problem, and proposes a temporal model which can complete predictions separately under the different definition of dropouts, they predict by using traditional RNN model with LSTM cell. Wang et al. [20] completes the prediction through a deep neural network, which is a combination of a Convolutional Neural Network and a Recurrent Neural Network. This model can automatically extract features from the original data. Scott et al. [21] adopt Natural Language Processing and other methods to analyze learners' questions and answers on the forum to predict learner completion. By combining learners' statistical information, forum behavior data and learning behaviors, a hidden dynamic factor model is proposed to predict the learning effects of learners by Qiu et al. [22].

At present, for the problem of user dropout rates on MOOCs platform, some traditional machine learning methods are used. Although the operation is simple and widely used, the internal associations of user behavior are not considered. Others use deep learning methods based on recurrent neural networks. Although they have considered the problem as a time series problem but the prediction effect will be limited if the time span is too long. Our proposed method not only introduces the influence of current sequence events, but also combines the influence of historical behavior associated with the potential period of user behavior which can improve the accuracy of prediction to some extent.

3 Datasets and Analysis

3.1 Datasets

The datasets we analyze and use in the laboratory are derived from XuetangX[2] and KDDCUP2015[3].

XuetangX is a Chinese MOOC platform developed by Tsinghua University. It was officially launched on October 10, 2013 and provides online courses to the world. As of now, there are 1800+ courses with a wide range of subject categories. This dataset contains 1,213 courses and 378,273 users. Some courses have a fixed scheduling cycle, and some courses do not have. The second dataset is from the KDDCUP competition in 2015. The KDDCUP is an annual data mining and knowledge discovery competition organized by the ACM knowledge discovery and data mining special interest group. This dataset provides a record of user behaviors within half a year of 39 online courses.

The specific categories of user behavior in the two datasets are: watching videos, doing homework, forum discussions, browsing course pages (navigate), accessing objects (access), and so on. Table 1 is the relevant statistics for these two datasets.

Table 1. Statistics of the datasets

Dataset	Courses	Users	Records
KDDCUP	1213	378273	115078786
XuetangX	39	112448	21552534

---

[2] http://moocdata.cn/data/user-activity.
[3] http://kddcup2015.com/.

3.2 Analysis

Although each data set contains multiple courses and log records, we actually use some courses and log records for data analysis and experiments.

Figure 2 statistics the user behavior activity in the course. It is calculated from the three types of users: all users in the course, users who did not drop out, and users who dropped out. We can see that when new content is released in a course, it is obvious that user activity is greatly improved. The user's activity changes periodically based on the course release, and the probability of dropping out of a user group with more regular course learning is far less than that of irregular user group.

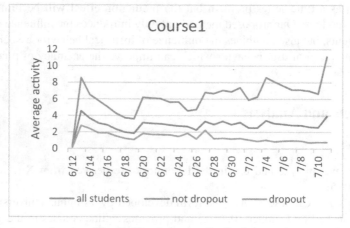

Fig. 2. User behavior distribution

Fig. 3. Comparison of user activity between Course 1 and Course 2

From Fig. 3, the record for Course 1 is from June 12 to July 11, and Course 2 is from January 17 to February 15. We know that compared with Course 1, the release of Course 2 is before and after the winter vacation, and the user activity in Course 2 is significantly lower than that of other courses. It can be seen that the number of user visits during the holidays is sharply lower than usual. During holidays, users rarely participate in learning, so if the course includes holidays, the course publisher need to adjust the course release time reasonably.

It can be seen in Fig. 4 that the dropout rate can be very different in different courses. The phenomenon of withdrawal from courses that requires a certain academic foundation is more obvious. It may be due to the mismatch of abilities and course difficulty or lack of interest. At the same time, due to the different genders in the same type of courses, there is a certain discrepancy in dropout ratios between the male and female. It can be seen in the figure that female users prefer humanities and humanities, while male users prefer social science.

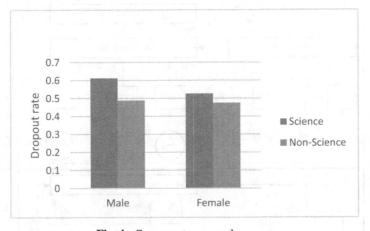

Fig. 4. Course category and user sex

4 Methodology

4.1 Formulation

Definition 1 (Behavioral Sequence). A sequence $X_u = (x_1, x_2, \ldots, x_t, \ldots, x_n)$ is defined as a series of activities that a user u has taken from the first day to the last day.

Definition 2 (Behavior). For each user u we define a m-dimensional vector of an activities sequence $x_t = (x_{t1}, x_{t2}, \ldots, x_{ti}, \ldots, x_{tm})$ which represents the user behavior series of the tth day, with $x_{ti} \in [0, 1]$. If is 0, which means the corresponding activity is not taken by the user in the tth day. On the contrary, is taken by the user.

Definition 3 (Other attributes). Continuously process discrete features such as gender, course information such as course categories, and other information in the dataset except user behavior $Z = (z_1, z_2, \ldots, z_l)$.

Our goal is to predict whether the user will drop out in the next period based on the existing behavior. If there is effective behavior, it will be recorded as not dropped out, which is represented by 0, otherwise, 1 represents dropped out.

4.2 Deep Model

Figure 5 shows the model proposed in this paper. The framework mainly includes the following parts: input module, encoding module, period detection and attention mechanism selection module, and prediction module.

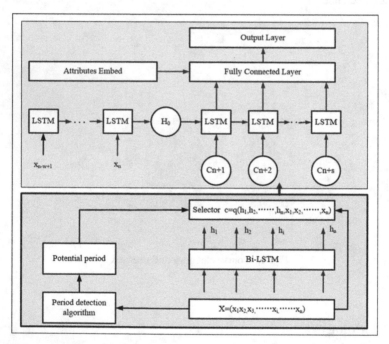

Fig. 5. Model structure

Input Module: The input module preprocesses the given user behavior data, and then selects m behavior categories that have a large impact on the dropout rate based on the hypothesis test method. Finally, the user behavior is converted into the one-hot vector as the feature vector, combining the feature vectors of each day we get the matrix $X_u = (x_1, x_2, \ldots, x_t, \ldots, x_n)$.

Encoding Module: As shown in Fig. 6, we encode each vector in the matrix in turn. The behavior is coded by using the Bi-LSTM method. There are two purposes of encoding: (1)

the Bi-LSTM method can retain the behavior characteristics before and after the cycle, and reduce the errors caused by the learner's behavior fluctuations. When introducing the influence of historical behavior, the relevant factors are selected through the cycle. Since the detected learner behavior cycle is within a certain confidence interval which means that the behavior before and after the cycle may have a certain deviation. (2) The context information captured by encoding provides a weight reference for the attention mechanism selector. When the attention mechanism selector selects historical behaviors, the weight corresponding to each behavior is obtained by calculating the similarity between the current hidden state and the result obtained by encoding. In this way, it is helpful for the attention mechanism to select more relevant behavior vectors in later prediction.

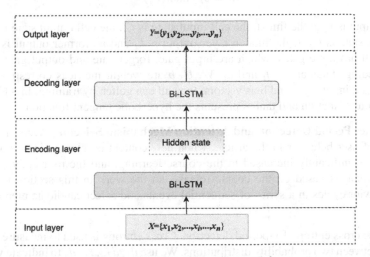

Fig. 6. Structure of encoding module

The input raw data $X = (x_1, x_2, \ldots, x_t, \ldots, x_n)$ is used as the input of the Bi-LSTM layer, and the hidden state $H' = (h_1, h_2, \ldots, h_t, \ldots, h_n)$, then passed into the decoding layer, the decoding layer is also a two-way LSTM, the hidden state is restored to the same or similar result of the original input, and the loss function is the mean square error:

$$\min \sum_{i=1}^{i=n} \left(||x_i - y_i|| \right)^2 \tag{1}$$

Bi-LSTM is composed of forward LSTM and backward LSTM, and the basic model LSTM is an improvement on the traditional RNN, it is a special RNN network in order to solve the problem of long dependence. Each unit of LSTM contains a unit state and three controlled gates to update the unit state. The specific calculation formulas are shown below:

$$i_t = \sigma(W_i h_{t-1} + U_i x_t + b_i) \tag{2}$$

$$f_t = \sigma\big(W_f h_{t-1} + U_f x_t + b_f\big) \tag{3}$$

$$o_t = \sigma(W_o h_{t-1} + U_o x_t + b_o) \tag{4}$$

$$\widetilde{C}_t = \tanh(W_a h_{t-1} + U_a x_t + b_a) \tag{5}$$

Afterwards, the cell output state can be calculated by:

$$C_t = C_{t-1} \odot f_t + i_t \odot \widetilde{C}_t \tag{6}$$

$$h_t = o_t \cdot \tanh(C_t) \tag{7}$$

The input is x_t at the time t, the cell input state is $\widetilde{C}_t$, the cell output state is C_t and its former state is C_{t-1}, the hidden layer output is h_t and its former output is h_{t-1}, a LSTM cell has three gates, which are input gate, forget gate and output gate and the corresponding states are i_t, f_t and o_t. W, U, b are weight matrices corresponding to hidden layer, input layer and bias vectors, they all can gotten by training. In addition, σ is a activation function and tanh represents the hyperbolic tangent function.

Time Series Period Detection and Attention Mechanism Selector. According to the analysis of user behavior in the article, when new content is released in a course, the activity is significantly increased in the course learning, and the user activity shows periodic changes based on the course release, so the work in this section is to find user behavior cycles in a series of sequential events and select candidate elements for attention.

We use cross entropy for period detection. Cross entropy is used to measure the difference between two probability distributions. We use $d_1, d_2, \ldots, d_n$ to indicate whether a user has a valid record of visiting the course every day. If so, d is recorded as 1 and vice versa as 0. Therefore, for each user, a binary sequence string $S = [d_1, d_2, d_3, \ldots, d_n]$ of length n is obtained, and the purpose is to analyze the sequence S to find its potential period a. Period detection is to find a suitable division from a series of 0,1, so that the elements in S are divided into k segments according to the equal length, so $S' = \{P_1, P_2, \ldots, P_k\}$, $P_i = \big[d_{a \cdot (i-1)+1}, d_{a \cdot (i-1)+2}, \ldots, d_{(a \cdot i)}\big]$. We need to find a suitable value of a such that the number of occurrences of 1 in each interval after division is the same, and the relative position of 1 in each division interval is the same. Assume that the uniform distribution is $R = \big\{\frac{1}{k}, \frac{1}{k}, \ldots, \frac{1}{k}\big\}$, and the distribution obtained according to a certain period is P. Calculate the KL (Kullback-Leibler Divergence) between two distributions by the following cross entropy. Among them, $P(i)$ refers to the ratio of the number of occurrences of '1' to the total number of times in P_i.

$$D(P\|R) = \sum_{i \in S_I} P(i) \log \frac{P(i)}{R(i)} \tag{8}$$

We calculate the similarity between actual period division and uniform distribution based on cross entropy. Through the greedy algorithm we traverse from 2 to $\left\lceil \frac{|S|}{2} \right\rceil$ in turn, based on the KL divergence distance we find the K elements with the smallest distance to form the candidate period set $KD = \{a_1, a_2, .., a_k\}$, after satisfying the distribution periodicity, the structural periodicity still needs to be satisfied, that is, in each sub-division obtained according to the periodic division of distribution, the relative position of 1 should be consistent, and we use the intra-class distance to measure, and each sub-sequence after division is regarded as a particle $P_1, P_2, .., P_k$, calculate the sum of the distances between the particles, the smaller the distance between the classes, the smaller the confidence level meets the structural periodicity. The formula for calculating the distance within a class is as follows:

$$l^2 = \frac{1}{\left\lceil \frac{|S|}{a} \right\rceil^2} \sum_{i=1}^{\left\lceil \frac{|S|}{a} \right\rceil} \sum_{j=1}^{\left\lceil \frac{|S|}{a} \right\rceil} d^2(P_i, P_j) \tag{9}$$

$$d^2(P_i, P_j) = \sum_{k=1}^{a} (d_{a\cdot(i-1)+k} - d_{a\cdot(j-1)+k})^2 \tag{10}$$

Finally, the candidate period with the smallest distance within the class is selected as the final period. The specific period detection method is shown in Algorithm 1.

Algorithm 1. The algorithm of period detection

Input:
   Behavior sequence: S
Output:
   period a
1: $KD \leftarrow \emptyset$
2: **for** $a = 2$ to $\left\lceil \frac{|S|}{2} \right\rceil$ **do**
3:   Segment the S into $\left\lceil \frac{|S|}{a} \right\rceil$ subsequences, $\left\lceil \frac{|S|}{a} \right\rceil = k$ and $S' = \{P_1, P_2, .., P_k\}$
4:   $P_i = [d_{a\cdot(i-1)+1}, d_{a\cdot(i-1)+2}, ..., d_{(a\cdot i)}]$
5:   $D(P||R) = \sum_{i \in S'} P(i) log \frac{P(i)}{R(i)}$
6:   **if** $|KD| < k$:
7:      $KD \leftarrow KD \cup a$
8:   **else if** $D(P||Q) < \min KD$:
9:      $KD \leftarrow KD \cup a$
10: **end for**
11: $A \leftarrow \infty$
12: **for** a in KD:
13:   $S' - \left\{P_1, P_2, .., P_{\left\lceil \frac{|S|}{a} \right\rceil}\right\}$ and $P_i = [d_{a\cdot(i-1)+1}, d_{a\cdot(i-1)+2}, ..., d_{(a\cdot i)}]$
14:   $l^2 = \frac{1}{\left\lceil \frac{|S|}{a} \right\rceil^2} \sum_{i=1}^{\left\lceil \frac{|S|}{a} \right\rceil} \sum_{j=1}^{\left\lceil \frac{|S|}{a} \right\rceil} d^2(P_i, P_j)$, $d^2(P_i, P_j)$ is the square of the Euclidean distance be-
     tween P_i and P_j
15:   **if** $l^2 < (\min A)^2$:
16:      a replace $(\min A)$
17: **end for**
18: **return** a

We obtain the potential period a of the user through the Algorithm 1. In encoding phase we get $H = \{h_1, h_2, \ldots, h_i, \ldots, h_n\}$, h_n represents the intermediate state corresponding to x_n, the prediction time is from t_{n+1} to t_{n+s} and assume the currently predicted moment is t_x, $k = t_x \bmod a$ and we get the set $TR_{in} = \{k + i * a\}$ of historical time periods aligned at time t_x, $i \in [0, \lfloor \frac{s-k+1}{a} \rfloor]$. The hidden layer output corresponding to each time in TR_{in} constitutes a set $H_{select} = \{h_k, h_{k+1*a}, \ldots, h_{k+\lfloor \frac{s-k+1}{a} \rfloor *a}\}$. The purpose of this selector is achieved.

In order to introduce the influence of historical behavior, we put the original behavior data with a certain weight as part of the input at the predicted moment. At the same time, avoiding the behavior deviation of the learner before and after the behavior cycle, the input also contains the encoded value corresponding to the original behavior data.

Hidden Layer State Initialization: Since the period detection takes the effects of historical behavior into account, and it is necessary to introduce the effects of sequence event. The influence of the time series requires a suitable time window size w. The selection of the initial time period of the cyclic neural network chain of the prediction module is w days before the start time of the prediction, and the initial hidden layer state is obtained from t_{n-w+1} to t_n to get the initialized hidden layer state. we set the window size w to detected period a. If the selected behavior matrix in the current time period is sparse, it is replaced by the mean value of the behavior matrix of other users in the corresponding time period. Therefore, the input of the prediction module and the state of the hidden layer initialized introduce the influence of historical period behavior and the influence of sequence events respectively.

Prediction. According to different prediction time, the selector selectively collects information from the encoding module and performs prediction. In order to introduce the influence of historical behavior, we put the original behavior data with a certain weight as part of the input at the predicted moment. At the same time, avoiding the behavior deviation of the learner before and after the behavior cycle, the input also contains the encoded value corresponding to the original behavior data. The specific calculation formula is as follows:

$$c_t = \sum w_i(\beta h_i + \gamma x_i) \tag{11}$$

$$w_i = softmax(f(h_i, h_{curr})) \tag{12}$$

where w_i is the weight for h_i, h_i is the output of the coding layer, $h_i \in H_{select}$ and h_{curr} denotes current status from the recurrent layer, f is a function which can calculate the similarity between h_i and h_{curr}. c_t takes the information collected by the input layer and encoding layer as the input of the prediction module.

For datasets with relevant user information and course information data, the prediction is completed by embedding user information and course information in binary representation through a fully connected layer, and increases the original vector by some dimensions.

5   Experiment

For the experiments in this article, we used KDDCUP's 2015 competition data, which included user behavior characteristics such as watching videos, submitting assignments, forum discussions, accessing course Wiki, browsing other course objects other than video assignments, closing web pages, etc. Among them, we predict from the known 30-day behavior logs whether users will have valid behavior records for the next 10 days. The XuetangX dataset contains specific course information, including course categories, course start and end dates, user personal information, gender, age, education level, etc. and user behavior logs including the behavior initiator, occurrence time, related objects, etc. Choose a 42-day behavioral record with a forecast period of 7 days. Ten-fold cross-validation is used during the training of the algorithm.

5.1   Performance Metrics

In order to evaluate the performance of our proposed model, it is measured by four indicators, namely Precision, Recall, and F1-score, and Area Under Receiver Operating Characteristic Curve (AUC) score. We show two representative indicators, F1score and AUC value.

Precision P:

$$P = \frac{TP}{TP + FP} \tag{13}$$

Recall R:

$$R = \frac{TP}{TP + FN} \tag{14}$$

F1-score:

$$F1 = \frac{2 * P * R}{P + R} \tag{15}$$

TP: The positive class that is correctly predicted
FP: The negative class that is predicted as positive
FN: The positive class that is predicted as negative
AUC: It is the area corresponding to the ROC curve. The larger the area, the stronger the generalization ability of the model.

5.2   Performance of Methods

We compare the proposed new model with several existing classification methods:

1) SVM: The support vector machine is a binary classification algorithm for supervised learning
2) LR: Logistic regression model is a classification algorithm that can handle binary classification and multivariate classification

3) RF: Random Forest model is ensemble learning algorithms based on decision tree
4) AdaBoost: AdaBoost is an iterative algorithm, an important ensemble learning technology.
5) LSTM: Long Short-Term Memory is a special RNN network, designed to solve the long dependency problem.

Table 2. The performance of the whole methods on KDDCUP

Methods	F1-score(%)	AUC(%)
SVM	91.07	87.81
LR	91.42	88.12
RF	92.10	88.63
AdaBoost	92.17	88.68
LSTM	92.19	88.72
Our Method	**92.78**	**89.84**

Table 3. The performance of the whole methods on XuetangX

Methods	F1-score(%)	AUC(%)
SVM	87.73	81.34
LR	87.52	81.19
RF	88.11	82.65
AdaBoost	88.77	84.06
LSTM	88.89	84.12
Our Method	**89.68**	**84.94**

Table 2 and Table 3 show the experimental results of our model and baseline methods on the KDDCUP and the XuetangX. From this, we can clearly see that all models perform better on KDDCUP than XuetangX. The former has better data quality in data processing and less noise. The same method can differ by three to five percentage points on two different datasets. At the same time, the performance of our proposed model is better than several baseline methods in F1-score and AUC values, which proves the effectiveness of our model. In baseline methods, the integrated learning algorithm, as an enhancement algorithm, is better than a single base learner. AdaBoost has achieved good results on both datasets. Compared with traditional machine learning classification algorithms, the deep learning algorithm LSTM has some advantages but it is not very obvious, may be our data is not very complicated and the time span is long, or the simple LSTM cannot fully learn the regularity of user behavior changes. The learning effect of LSTM is not very ideal. However, using it as the basic unit of our proposed model and redesigning the entire framework, the overall effect is obviously better than other methods. It can be seen that in different scenarios, although the model cannot be universally used, it may be improved according to the actual situation. In this paper, we focus on the characteristics of user learning, not only considering that user behavior is a sequence event problem, but also that user behavior will have a learning period based on course release or their own learning plan, Therefore, we have added the corresponding influencing factors of historical behavior, and various considerations make the method more effective.

6 Conclusion

In this paper, we studied the problem of predicting the dropout rate in MOOCs. Firstly, we do a deep analysis of user learning behavior to find which are the important factors the affect the dropout rate. And then we propose a novel deep model based on recurrent network. In the novel model, we combine the effects of sequential behavior over the current period with the effects of past historical behavior to predict the dropout and we also embed the attributes of user and course. Finally, we demonstrate the effectiveness of our methods by taking the experiments on two datasets, our proposed method performs better than the state-of-the art methods. In future work, we will further study the choice of sequence length in the influence of sequence behavior in the current period.

Acknowledgements. This work is supported by the key projects of the national natural science foundation of China (No. U1811263), the major technical innovation project of Hubei Province (No. 2019AAA072), the National Natural Science Foundation of China (No. 61572378), the Teaching Research Project of Wuhan University (No. 2018JG052), the Natural Science Foundation of Hubei Province (No. 2017CFB420). We also thank anonymous reviewers for their helpful reports.

References

1. Ipay, B., Ipay, C.B.: Opportunities and challenges for open educational resources and massive open online courses: the case of Nigeria. Commonwealth of Learning. Educo-Health Project. Ilorin (2013)
2. Mackness, J., Mak, S.F.J., Williams, R.: The ideals and reality of participating in a MOOC. In: Networked Learning Conference (2010)
3. Feng, W.Z., Tang, J., Liu, T.X.: Understanding dropouts in MOOCs. In: Proceedings of the 33rd AAAI Conference on Artificial Intelligence, pp. 517–524 (2019)
4. Kate, J.: Initial trends in enrolment and completion of massive open online courses. Int. Rev. Res. Open Distance Learn. **15**(1), 133–160 (2014)
5. He, J., Bailey, J., Rubinstein, B.I.P., Zhang, R.: Identifying at-risk students in massive open online courses. In: Proceedings of the 29th AAAI Conference on Artificial Intelligence, pp. 1749–1755 (2015)
6. Dalipi, F., Imran, A.S., Kastrati Z.: MOOC dropout prediction using machine learning techniques: review and research challenges. In: Global Engineering Education Conference (EDUCON), 2018 IEEE, pp. 1007–1014 (2018)

7. Shi, Y.L., Peng, Z.Y., Wang, H.N.: Modeling student learning styles in MOOCs. In: Proceedings of the 26th International Conference on Information and Knowledge Management (CIKM), pp. 979–988 (2017)

8. Natek, S., Zwilling, M.: Student data mining solution–knowledge management system related to higher education institutions. Expert Syst. Appl. **41**(14), 6400–6407 (2014)

9. Coleman, C.A., Seaton, D.T., Chuang, I.: Probabilistic use cases: discovering behavioral pattern for predicting certification. In: Proceedings of the Second ACM Conference on Learning @ Scale, pp. 141–148 (2015)

10. Graves, A.: Supervised Sequence Labelling with Recurrent Neural Networks. Springer, Heidelberg (2012). https://doi.org/10.1007/978-3-642-24797-2

11. Ito, T., Tsubouchi, K., Sakaji, H., Yamashita, T., Izumi, K.: Contextual sentiment neural network for document sentiment analysis. Data Sci. Eng. **5**(2), 180–192 (2020). https://doi.org/10.1007/s41019-020-00122-4

12. Anderson, A., Huttenlocher, D., Kleinberg, J.: Engaging with massive online courses. In: Proceedings of the 23rd International World Wide Web Conference, pp. 687–698 (2014)

13. Kloft, M., Stiehler, F., Zheng, Z., Pinkwart, N.: Predicting MOOC dropout over weeks using machine learning methods. In: Proceedings of the EMNLP 2014 Workshop on Analysis of Large Scale Social Interaction in MOOCs, pp. 60–65. Association for Computational Linguistics, Doha (2014)

14. Colin, T., Kalyan V., Una-May, O'Reilly.: Likely to stop? Predicting stopout in massive open online courses. Computer Science (2014)

15. Ramesh, A., Goldwasser, D., Huang B.: Uncovering hidden engagement patterns for predicting learner performance in MOOCs. In: Proceedings of the Second ACM Conference on Learning @ Scale, pp. 157–158 (2014)

16. Balakrishnan, D., Coetzee, D.: Predicting students retention in massive open online courses using hidden Markov models. Technical report, UC Berkeley (2013)

17. Chanchary, F.H., Haque, I., Khalid, M.S.: Web usage mining to evaluate the transfer of learning in a web-based learning environment. In: Proceedings of the First International Workshop on Knowledge Discovery and Data Mining, pp: 249–253. IEEE Computer Society (2008)

18. Stein, J., Xing, W., et al.: Temporal predication of dropouts in MOOCs: reaching the low hanging fruit through stacking generalization. Comput. Hum. Behav. (2016)

19. Fei, M., Yeung, D.Y.: Temporal models for predicting students dropout in massive open online courses. In: Proceedings of the 2011 IEEE 11th International Conference on Data Mining Workshops (ICDMW), pp. 366–372 (2011)

20. Wang, W., Yu, H., Miao, C.: Deep model for dropout prediction in MOOCs. In: Proceedings of the 2nd International Conference on Crowd Science and Engineering, pp. 26–32 (2017)

21. Crossley, S.A., McNamara, D.S.: Developing component scores from natural language processing tools to assess human ratings of essay quality. Rev. Manag. Sci. **9**(4), 1–26 (2014)

22. Qiu, J., Tang, J., Liu, T.X.: Modeling and predicting learning behavior in MOOCs. In: Proceedings of the Ninth ACM International Conference on Web Search and Data Mining, pp. 93–102 (2016)

RFRSF: Employee Turnover Prediction Based on Random Forests and Survival Analysis

Ziwei Jin[1,2], Jiaxing Shang[1,2(✉)], Qianwen Zhu[3], Chen Ling[1,2], Wu Xie[4], and Baohua Qiang[4]

[1] College of Computer Science, Chongqing University, Chongqing, China
ziweikan@foxmail.com, shangjx@cqu.edu.cn, lingchen20638@163.com
[2] Key Laboratory of Dependable Service Computing in Cyber Physical Society, Ministry of Education, Chongqing University, Chongqing, China
[3] School of Management and Engineering, Nanjing University, Nanjing, China
zhuqw@smail.nju.edu.cn
[4] Guangxi Key Laboratory of Trusted Software, Guilin University of Electronic Technology, Guilin, China
xiesixchannels@126.com, qiangbh@guet.edu.cn

Abstract. In human resource management, employee turnover problem is heavily concerned by managers since the leave of key employees can bring great loss to the company. However, most existing researches are employee-centered, which ignored the historical events of turnover behaviors or the longitudinal data of job records. In this paper, from an event-centered perspective, we design a hybrid model based on survival analysis and machine learning, and propose a turnover prediction algorithm named RFRSF, which combines survival analysis for censored data processing and ensemble learning for turnover behavior prediction. In addition, we take strategies to handle employees with multiple turnover records so as to construct survival data with censored records. We compare RFRSF with several baseline methods on a real dataset crawled from one of the biggest online professional social platforms of China. The results show that the survival analysis model can significantly benefit the employee turnover prediction performance.

Keywords: Turnover prediction · Survival analysis · Random survival forests · Professional social networks · Machine learning

1 Introduction

The employee turnover problem has been widely concerned by companies since the demission of key staff members may cause great loss to the company. Even after ordinary employees quit their job, companies will have to re-invest time and money to find new alternatives and train them. Obviously, being able to foresee whether an employee is likely to leave would benefit the company in terms of retaining talented employees and reducing losses.

© Springer Nature Switzerland AG 2020
Z. Huang et al. (Eds.): WISE 2020, LNCS 12343, pp. 503–515, 2020.
https://doi.org/10.1007/978-3-030-62008-0_35

There have been tremendous research works tackling the employee turnover problem, some of them [4,7] fall within the scope of management, sociology, psychology, etc. In recent years, with the flourish of online social networks [8], data-driven approaches began to draw researchers' attention, examples include machine learning based methods [3,13,15,21], survival analysis based methods [14,16,19], social network analysis based methods [5,6,17], etc. However, existing machine learning based methods mainly focus on feature engineering for the binary prediction task, which ignored the historical events of turnover behaviors or the longitudinal data in job records. Traditional survival analysis models usually impose strict assumptions on data distribution, and they are mainly used for factor analysis rather than turnover prediction. In this paper, we propose a hybrid model that combines survival analysis with machine learning models, based on which we further propose a turnover prediction algorithm. In our framework, we focus on turnover events rather than employees through survival analysis from the perspective of events. Specifically, we use **R**andom **S**urvival **F**orests for survival analysis and **R**andom **F**orest form turnover prediction, leading the **RFRSF** algorithm. In addition, we take strategies to handle employees with multiple turnover records so as to construct survival data with censored records. We first calculate the probability of each turnover event occurrence at a certain time point. Then, we view the probability value along with other time-invariant information as features to learn whether an employee is likely to leave at a certain time. Finally, we use these features to train machine learning models for the employee turnover prediction task.

To evaluate our proposed model, we conduct experiments on a real-world dataset crawled from one of the largest professional social platforms in China [1]. The results show that compared with other baseline models, the proposed model can achieve higher prediction accuracy.

2   Related Work

In this paper, we mainly focus on data-driven studies about employee turnover problem, and these studies can be roughly divided into three groups.

The first group are machine learning based methods, which formulate employee turnover prediction as a binary classification problem. Liu et al. [15] evaluated the importance of job skills for departure prediction. Yang et al. [21] propose a causal structure learning based feature modification method (CSFM), which helps management to retain employees who are leaving. Studies [1,3] offered experimental comparisons of various common machine learning algorithms. In general, random forests [3] and XGBoost [1] perform relatively better. De Jesus et al. [13] trained different machine learning models based on different industry characteristics.

The second group are survival analysis based methods. Studies [14,16,19] analyzed the impacting factors which contribute to employee's turnover behavior and calculated turnover probabilities with a Cox proportional hazard model.

---

[1] Due to privacy consideration, we do not disclose the name of the platform.

Zhu et al. [22] combined Cox proportional hazard model with random forest for turnover prediction.

The third group are social network based methods. Bigsby et al. [5] extracted network features to improve prediction accuracy. Oentaryo et al. [17] modeled employee mobility as a directed network, where each edge represents a job hob behavior. Based on the network they performed connectivity analysis at job and organization levels to derive insights on talent flow. Cai et al. [6] modeled the connections between employees and companies as a bipartite graph and proposed a graph embedding approach to predict employee turnover behaviors.

Besides the above methods, there are also other studies, such as the semi-Markov based algorithms [9,10].

3   Survival Analysis

Survival analysis is a branch of statistics for analyzing the expected duration of time until one or more events happen, such as death in biological organisms and failure in mechanical systems. It is straightforward to see that survival analysis can be used to analyze the employee turnover problem.

3.1   Basic Concepts

To begin, we first introduce some basic concepts about survival analysis, please refer to [2] for more details.

Event: Based on the specific problem, an event can refer to the death, failure, turnover, crime recommitment or other events of interest.

Survival Time: If an event happens, then survival time is the duration from the beginning of observation to the event occurrence. Otherwise it is the duration to the end of observation or the time when the object exits the experiment.

Censor: A record is censored if the event of interest is not observed due to the limited observation time or other reasons. Censoring can be divided into several subtypes [20], and in this paper, we only consider the right-censoring where the event is not observed due to the limited observation time.

Survival Function: It is the probability that an individual lives longer than a specific time t, which is defined as:

$$S(t) = P\{T \geq t\} \tag{1}$$

where T is event time. Survival function monotonically decreases with t, and given that all of the subjects are alive at the beginning, we have $S(0) = 1$.

Hazard Function: Consider the conditional probability of an event happening at time t, given that the event does not happen before t, i.e., $P\{X \in (t, t + dt)|X > t\}$, then the hazard function is defined as:

$$H(t) = \lim_{dt \to 0} \frac{P(t \leq T < t + dt)}{dt \cdot S(t)} \tag{2}$$

Cumulative Hazard Function: the cumulative hazard function is defined as:

$$\Lambda(t) = \int_0^t H(u)du \qquad (3)$$

3.2   Random Survival Forests Model

The RSF model, proposed by Ishwaran et al. [12], is an extension of traditional random forests (RF) model for survival analysis. Like traditional RF, RSF uses bootstrap to draw samples and grows multiple binary recursive survival trees. Compared to existing survival analysis models, RSF does not impose any assumption on data distribution.

In the RSF model, the trees are built according to the cumulative hazard function (CHF). It first calculates the CHF of the terminal nodes and then gets the CHF of each tree in the forest. Finally, to obtain an ensemble CHF, we average over all survival trees in the forest.

Harrell's Concordance Index (C-index): The RSF model can be evaluated by Harrell's Concordance Index [11]. The C-index measures whether an event occurred earlier is associated with a higher risk rate or lower survival rate. For each pair of samples $[i, j]$, to calculate C-index, we first define p_{ij} as:

$$p_{ij} = \begin{cases} 1 & i \text{ and } j \text{ are comparable and the prediction is concordant} \\ 0.5 & i \text{ and } j \text{ are comparable and the prediction is inconsonant} \\ 0 & i \text{ and } j \text{ are not comparable} \end{cases} \qquad (4)$$

Based on the above formula, the C-index is as follows:

$$C\text{-index} = \frac{\sum p_{ij}}{\sum l_{ij}} \qquad (5)$$

where l_{ij} is an indicator of whether the sample pair $[i, j]$ is comparable and 1 means comparable while 0 means not comparable.

4   The Proposed Method

4.1   Problem Formulation

In this paper, we analyze the problem of turnover prediction from an event-centered perspective. A traditional way to solve this problem is to predict an employee's turnover behavior completely based on her previous turnover behaviors, which is an employee-centered manner. However, this kind of model assumption can be "biased", because an employee's turnover behavior only relates to her turnover history. Moreover, temporal factors also play an important role because an employee's decision changes over time. Therefore, a reasonable way to investigate this problem is to fully consider the three factors: time, current status and employee's past turnover behaviors.

Based on the above analysis, we incorporate the specific prediction time into the survival function, and formulate our problem as follows:

Definition 1. *(Turnover Prediction Problem): Given an employee u's past turnover events, the current job information, the social platform information, and a specific time t, the turnover prediction problem aims at predicting whether u will quit the current job at time t.*

4.2 Our Method

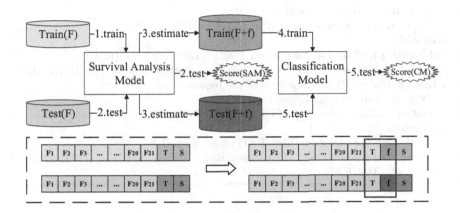

Fig. 1. The general framework of ExtTra

Framework. The framework of our method is shown in Fig. 1, which is divided into two parts: the survival model and the classification model. The survival analysis model is based on RSF, which is trained and tested to meet the preset requirements and generates survival rate (f) as the output before entering the classification model. In this part, we use 21 features and 2 tags (turnover status (S) and survival time (T)) to trained the model. Then, the classification model is trained and tested through the new testing data. In the second part, we use 23 features and 1 tag (turnover status (S)) to trained the model. It should be noted that these 23 features include the 21 features of the survival model, and the two new features of survival rate (f) generated in the survival analysis model and survival time (T) which used as a tag in the survival analysis model.

RFRSF Algorithm. According to the above framework, we propose our RFRSF employee turnover prediction algorithm, as shown in Algorithm 1. In Algorithm 1, OOBScore(RSF) and OOBScore(RF) are the out-of-bag estimation scores of RSF and RF respectively. These scores are used to find the best RSF models and RF models.

Algorithm 1. The RFRSF algorithm

Require:
   Train (F), Test (F)
Ensure:
   RSF, Score(RSF), OOBScore(RSF), RF, Score(RF), OOBScore(RF), Train(F+f),
   Test(F+f)
 1: Initialize: Score(RSF)=0.5, OOBScore(RSF)=0.5, Score(RF)=[0,0,0,0], OOB-
    Score(RF)=0
 2: **while** OOBScore(RSF) does not meet the requirements **do**
 3:   Train RSF with the algorithm Train(F)
 4:   Out-of-package estimation of RSF to get OOBScore(RSF)
 5: **end while**
 6: Substituting RSF into the algorithm Test(F), calculate Score(RSF)
 7: Substituting RSF into Train(F) and Test(F), calculate Train(F+f), Test(F+f)
 8: **while** OOBScore(RF) does not meet the requirements **do**
 9:   Train RF with the algorithm Train(F+f)
10:   Out-of-package estimation of RF to get OOBScore(RF)
11: **end while**
12: Test RF with the algorithm Test(F+f) to get Score(RF)
13: **return** RSF, Score(RSF), OOBScore(RSF), RF, Score(RF), OOBScore(RF),
    Train(F+f), Test(F+f)

---

5 Evaluation

5.1 Dataset

The dataset used in this paper are crawled from one of China's largest online professional social platform, which contains employees' personal information, educational background, work experience, and platform activities. To analyze the data with survival model, we label each sample with its survival time and turnover status. Specifically, we inspect each job record to see if it has an end time (e.g., 2016-02), if so, the survival time is the duration from start time to end time and the sample is labeled 1 (indicating the occurrence of a turnover event), otherwise the survival time is the duration from start time to the time point when the user updates this job record, the sample is labeled 0 and the record is marked as right-censored. We remove records with no start time information. After data cleaning and preprocessing, we get 287,229 samples with 119,728 positive samples and 167,501 negative ones. We randomly divide the data set into training and testing sets according to a ratio of 7:3.

5.2 Feature Extraction

From the dataset, we extract 22 features for turnover prediction, which can be divided into six categories. Table 1 summarizes the features and their detailed descriptions.

    Unlike the standard RSF survival model, in our dataset, an employee may have multiple job-hopping records. From the perspective of survival analysis, it means that the event occurs multiple times during the observation period for a single object, which is not supported by standard survival analysis models. To tackle this problem, we take the following two strategies:

i) First, we split an object with multiple events into several separate objects where each object is associated with exactly one event. For example, as shown in Fig. 2, during the observation period, employee A has two turnover events, we split the two events of A into two turnover events and marked as A1 and A2.

ii) Second, we align different events by converting absolute time into relative time. For example, as shown in Fig. 2, we align different events to make sure they have the same start time, in this way we can focus on the relative length of time before the event occurrence.

Table 1. Summary of selected features

Category	Feature	Description
Demographic	gender	Gender of employee
Job	industry_type	Employee industry type, such as education, IT, etc.
	cmp_scale	Company size, i.e., the number of employees
	position_level	Employee's position level at the company
Platform	iteractions	Number of online interactions with other user
	dongtai	Number of posts written by the employee
	guandian	Number of opinions expressed by the employee
	zhuanlan	Number of articles written by the employee
	dianping	Number of comments written by the employee
	likes	Number of likes received by the employee
	views	Number of views by other users
	recent_feeds	Number of recent feeds received by the employee
	influence	Online influence of the employee
	inf_defeat	Ratio of users defeated by the employee in terms of online influence ranking
	info_ratio	Online information integrity of the employee
	imp_tag_num	Number of impression tags given to the employee
	pro_tag_num	Number of professional tags given to the employee
Education	degree	The highest educational degree
	sch_type	The school type (in China) where the employee get the highest degree, such as 985, 211
Job change	has_turnover_num	Number of turnover records for the employee
	has_timelength	The time length the employee has worked for (including the current job)
	timelength	The time length of the current job
Survival	survival_rate	The survival rate generated through RSF model

After the aforementioned processing, we conduct a statistical analysis on the lengths of job records (measured in months), as shown in Fig. 3, where isexit=1 indicates a turnover event. The lengths of most of the job records are located between 1 to 24 months and the peaks of turnover behavior at year marks (after every 12 months). Also, we observe that the 13–24 month (1–2 year) interval has a significantly higher relative turnover rate. Therefore, we conclude that time length at the current job can be used as an important feature for turnover prediction.

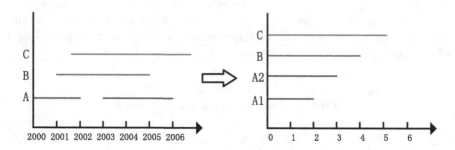

Fig. 2. The processing of objects with multiple events

5.3 Baselines

We include the following baselines for comparison:

Classic Machine Learning: The first group of baselines are classic machine learning methods without survival analysis feature (survival rate). We consider five representative machine learning models, which include Naive Bayesian (**NB**), Logistic Regression (**LR**), Decision Tree (**DT**), XGBoost (**XGB**) and Random Forest (**RF**). The model parameters are tuned to ensure a good overall performance.

Machine Learning with Cox: The second group of baselines combine classic machine learning methods with Cox proportional hazard model, which is proposed by Zhu et al. [22], we denote them as X+Cox, where X is the corresponding machine learning model.

Machine Learning with RSF: The third group of baselines combine classic machine learning methods with Random Survival Forests model, i.e., with the survival rate feature generated by RSF. We denote them as X+RSF, where X is the corresponding machine learning model.

5.4 Evaluation Metrics

We use *Accuracy, Precision, Recall, F1-score,* and *AUC,* which are widely used in classification tasks, as the evaluation metrics.

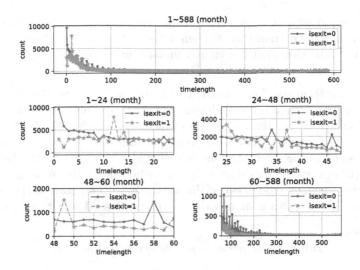

Fig. 3. The distribution of time

5.5 Results

Survival Analysis Results. We first compare the C-index scores of the RSF model with the Cox model. There are 21 covariates in the survival analysis and we perform hypothesis testing on these covariates. In the Cox model, only 7 covariates pass the testing, and it achieves the highest C-index score of 0.55 (which is unsatisfactory). We think this is mainly caused by the large amount of censored data in our dataset. On the contrary, the random survival forests model can use all the 21 covariates, and the C-index score is 0.68, which is higher than the Cox model.

To further analyze the differences between the two survival analysis models, we check their survival rate distribution, as shown in Fig. 4. It can be seen that the survival rate distribution of RSF is consistent with our common sense

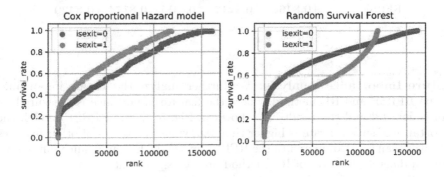

Fig. 4. The distribution of survival rate

that the survival rate of most turnover employees (isexit = 1) is significantly lower than that of the non-turnover employees. By contrast, the survival rate distribution of the Cox model obviously violates our intuition.

Classification Results. We compare the Accuracy, Precision, Recall, F1-score, and AUC index scores of different baselines, and the results are shown in Table 2, where the highest metrics are shown in bold. We see that the RFRSF model proposed in this paper achieves the best performance in terms of Accuracy, Precision, and F1-score and AUC, its Recall is also very close to the best value. Also, the RSF survival model can significantly benefit the machine learning methods than the Cox model, especially for the tree based models (DT, RF and XGBoost). We think this is mainly due to the tree-based structure of RSF.

Table 2. Classification results of different methods

Model	Accuracy	Precision	Recall	F1-score	AUC
NB	0.5181	0.4627	0.8706	0.6043	0.5653
LR	0.6400	0.6050	0.4266	0.5004	0.6114
DT	0.6246	0.5550	0.5641	0.5595	0.6165
XGBoost	0.7085	0.6845	0.5755	0.6253	0.6906
RF	0.7132	0.6945	0.5740	0.6285	0.6946
NB+Cox	0.6303	0.5616	0.5814	0.5714	0.6239
LR+Cox	0.6519	0.6210	0.4578	0.5270	0.6262
DT+Cox	0.6264	0.5585	0.5652	0.5618	0.6183
XGBoost+Cox	0.7046	0.6809	0.5700	0.6205	0.6868
RF+Cox	0.7106	0.6885	0.5787	0.6289	0.6931
NB+RSF	0.5183	0.4629	**0.8707**	0.6044	0.5655
LR+RSF	0.6497	0.6175	0.4493	0.5202	0.6228
DT+RSF	0.7915	0.7495	0.7608	0.7551	0.7873
XGBoost+RSF	0.8258	0.7941	0.7934	0.7938	0.8214
RFRSF	**0.8465**	**0.8217**	0.8315	**0.8174**	**0.8420**

Feature Importance Analysis. We further analyze the feature importance for the RFRSF and RF method, where the feature importance is measured by Gini Index [18] and the results are shown in Fig. 5. In the RFRSF model, the importance of survival rate is higher than 0.6, while the scores of other features are lower than 0.1. Compared to the RFRSF method, the importance differences between different features of RF method are less significant.

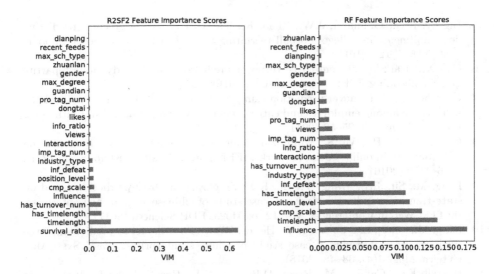

Fig. 5. Feature importance based on Gini Index

6   Conclusion

In this paper, we proposed an employee turnover prediction model by combining random survival forests with random forests. We constructed survival data based on employee's historical job records and then turned the employee turnover prediction problem into a traditional supervised binary classification problem. Experimental results on a real dataset verified the effectiveness of our model.

Acknowledgements. This work was supported in part by: National Natural Science Foundation of China (Nos. 61702059, 61966008), Fundamental Research Funds for the Central Universities (Nos. 2019CDXYJSJ0021, 2020CDCGJSJ041), Frontier and Application Foundation Research Program of Chongqing City (No. cstc2018jcyjAX0340), Guangxi Key Laboratory of Optoelectronic Information Processing (No. GD18202), Guangxi Key Laboratory of Trusted Software (No. kx201702).

References

1. Ajit, P.: Prediction of employee turnover in organizations using machine learning algorithms. Algorithms **4**(5), C5 (2016)
2. Allison, P.D.: Event History and Survival Analysis: Regression for Longitudinal Event Data, vol. 46. SAGE Publications, Thousand Oaks (2014)
3. Bao, L., Xing, Z., Xia, X., Lo, D., Li, S.: Who will leave the company?: A large-scale industry study of developer turnover by mining monthly work report. In: 2017 IEEE/ACM 14th International Conference on Mining Software Repositories (MSR), pp. 170–181. IEEE (2017)
4. Baron, J.N., Hannan, M.T., Burton, M.D.: Labor pains: change in organizational models and employee turnover in young, high-tech firms. Am. J. Sociol. **106**(4), 960–1012 (2001)

5. Bigsby, K.G., Ohlmann, J.W., Zhao, K.: The turf is always greener: predicting decommitments in college football recruiting using Twitter data. Decis. Support Syst. **116**, 1–12 (2019)
6. Cai, X., et al.: DBGE: employee turnover prediction based on dynamic bipartite graph embedding. IEEE Access **8**, 10390–10402 (2020)
7. Carraher, S.M.: Turnover prediction using attitudes towards benefits, pay, and pay satisfaction among employees and entrepreneurs in Estonia, Latvia, and Lithuania. Baltic J. Manag. **6**, 25–52 (2011)
8. Chen, H., Jin, H., Wu, S.: Minimizing inter-server communications by exploiting self-similarity in online social networks. IEEE Trans. Parallel Distrib. Syst. **27**(4), 1116–1130 (2016)
9. Fang, M., Su, J.H., Wang, T., He, R.J.: Employee turnover prediction based on state-transition and semi-Markov-a case study of Chinese state-owned enterprise. In: ITM Web of Conferences, vol. 12, p. 04023. EDP Sciences (2017)
10. Fang, M., Su, J., Liu, J., Long, Y., He, R., Wang, T.: A model to predict employee turnover rate: observing a case study of Chinese enterprises. IEEE Syst. Man Cybern. Mag. **4**(4), 38–48 (2018)
11. Harrell, F.E., Califf, R.M., Pryor, D.B., Lee, K.L., Rosati, R.A.: Evaluating the yield of medical tests. JAMA **247**(18), 2543–2546 (1982)
12. Ishwaran, H., Kogalur, U.B., Blackstone, E.H., Lauer, M.S., et al.: Random survival forests. Ann. Appl. Stat. **2**(3), 841–860 (2008)
13. de Jesus, A.C.C., Júnior, M.E.G., Brandão, W.C.: Exploiting linkedin to predict employee resignation likelihood. In: Proceedings of the 33rd Annual ACM Symposium on Applied Computing, pp. 1764–1771 (2018)
14. Li, H., Ge, Y., Zhu, H., Xiong, H., Zhao, H.: Prospecting the career development of talents: a survival analysis perspective. In: Proceedings of the 23rd ACM SIGKDD International Conference on Knowledge Discovery and Data Mining, pp. 917–925 (2017)
15. Liu, J., Long, Y., Fang, M., He, R., Wang, T., Chen, G.: Analyzing employee turnover based on job skills. In: Proceedings of the International Conference on Data Processing and Applications, pp. 16–21 (2018)
16. Mulla, Z.R., Kelkar, K., Agarwal, M., Singh, S., Sen, N.E.: Engineers' voluntary turnover: application of survival analysis. Indian J. Ind. Relat. **49**, 28–341 (2013)
17. Oentaryo, R.J., Lim, E.-P., Ashok, X.J.S., Prasetyo, P.K., Ong, K.H., Lau, Z.Q.: Talent flow analytics in online professional network. Data Sci. Eng. **3**(3), 199–220 (2018). https://doi.org/10.1007/s41019-018-0070-8
18. Shang, W., Huang, H., Zhu, H., Lin, Y., Qu, Y., Wang, Z.: A novel feature selection algorithm for text categorization. Expert Syst. Appl. **33**(1), 1–5 (2007)
19. Wang, J., Zhang, Y., Posse, C., Bhasin, A.: Is it time for a career switch? In: Proceedings of the 22nd International Conference on World Wide Web, pp. 1377–1388 (2013)
20. Wang, P., Li, Y., Reddy, C.K.: Machine learning for survival analysis: a survey. ACM Comput. Surv. (CSUR) **51**(6), 1–36 (2019)

21. Yang, Y., Zhan, D.C., Jiang, Y.: Which one will be next? An analysis of talent demission. In: The 1st International Workshop on Organizational Behavior and Talent Analytics (Held in conjunction with KDD 2018) (2018)
22. Zhu, Q., Shang, J., Cai, X., Jiang, L., Liu, F., Qiang, B.: CoxRF: employee turnover prediction based on survival analysis. In: 2019 IEEE SmartWorld, Ubiquitous Intelligence & Computing, Advanced & Trusted Computing, Scalable Computing & Communications, Cloud & Big Data Computing, Internet of People and Smart City Innovation (SmartWorld/SCALCOM/UIC/ATC/CBDCom/IOP/SCI), pp. 1123–1130. IEEE (2019)

A Deep Sequence-to-Sequence Method for Aircraft Landing Speed Prediction Based on QAR Data

Zongwei Kang[1,2], Jiaxing Shang[1,2(✉)], Yong Feng[1,2(✉)], Linjiang Zheng[1,2], Dajiang Liu[1,2], Baohua Qiang[3,4], and Ran Wei[5]

[1] College of Computer Science, Chongqing University, Chongqing, China
{shangjx,fengyong}@cqu.edu.cn
[2] Key Laboratory of Dependable Service Computing in Cyber Physical Society, Ministry of Education, Chongqing University, Chongqing, China
[3] Guangxi Key Laboratory of Optoelectronic Information Processing, Guilin University of Electronic Technology, Guilin, China
[4] Guangxi Key Laboratory of Trusted Software, Guilin University of Electronic Technology, Guilin, China
[5] Chongqing Medical Data Information Technology Co., Ltd., Chongqing, China

Abstract. Runway overrun is one of the most typical landing incidents highly concerned by airlines in the aviation industry. Previous studies have shown that high landing speed is closely related to runway overrun risks, therefore the study of landing speed prediction based on flight data has drawn attention of many scholars in recent years. However, existing methods are mainly based on traditional machine learning models and handcrafted features, which not only rely heavily on flight experts' priori knowledge, but also provide unsatisfactory prediction accuracy. To solve this problem, in this paper we propose an innovative deep encoder-decoder model for aircraft landing speed prediction based on Quick Access Recorder (QAR) data. Specifically, we first preprocess the QAR dataset through a data cleaning, Lagrange interpolation and normalization procedure. Second, based on the preprocessed QAR dataset, we use gradient boosting decision trees to select features which are most closely related to landing speed. Finally, we employ the LSTM encoder-decoder architecture where the encoder captures the pattern underlying in the past sequences while the decoder generates predictions for the future speed sequences. We evaluate our method on a dataset of 44,176 A321 flight samples. The experimental results show that the prediction accuracy of the proposed method is significantly higher than the conventional methods.

Keywords: QAR data · Landing speed prediction · Encoder-Decoder · Long short-term memory · GBDT

© Springer Nature Switzerland AG 2020
Z. Huang et al. (Eds.): WISE 2020, LNCS 12343, pp. 516–530, 2020.
https://doi.org/10.1007/978-3-030-62008-0_36

1   Introduction

It is essential for civil aviation to ensure the safety of passengers during the flight by avoiding flight accidents. The landing phase is considered to be one of the most risky stages of a flight. Statistics released by Boeing [1] have showed that landing phase (includes final approach) alone accounted for 48% of the fatal accidents occurring from 2007 to 2016, despite the fact that this phase only constitutes 4% of the whole flight time [2]. Runway overrun is one of the most typical safety accidents during the landing phase. According to the NLR Aviation Safety Analysis report [3] and the literature [4], the risk factors for runway overrun mainly include: unstable approach, visual approach, long landing, excessive landing speed, etc., among which high landing speed plays an important role. Therefore, it is of great significance to study the landing speed prediction problem.

The Quick Access Recorder (QAR), as a flight parameter recording system widely installed on aircrafts, can collect thousands of aircraft parameters during flight. These parameters dynamically record the flight process as well as the pilot operations. In recent years, with the rapid development of IoT systems and artificial intelligence technology, using QAR data to conduct flight safety research has attracted attention of many researchers, such as data mining (anomaly detection) based algorithms [5–7], risk models based methods [2,8,9], machine learning based studies [10,11], etc. However, current research on landing safety mainly uses statistical analysis or classic machine learning models. These methods suffers from the following two shortcomings: First, they rely heavily on the prior knowledge of experts to handcraft the features. Second, traditional machine learning methods are difficult to capture the temporal interdependence of time series data, resulting in insufficient method performance. A few recent studies have used deep learning models, such as the LSTM-based speed [12] and hard landing [13] prediction models proposed by Tong et al., but their methods can only generate predictions (speed or vertical acceleration) for the next moment, while the sequence or trend of the near future cannot be predicted.

In order to address the above problems, this paper propose a novel sequence-to-sequence method to predict aircraft landing speed based on sequential QAR data. Our method is able to predict the future speed sequences according to the past flight parameter sequences, so as to achieve early warning of landing events related to excessive speed. Specifically, the method proposed in this paper includes the following steps: First, we preprocess the QAR dataset through a data cleaning, Lagrange interpolation and normalization procedure. Second, we use gradient boosting decision trees to select features which are most closely related to landing speed. Finally, we employ the LSTM encoder-decoder architecture where the encoder captures the pattern underlying in the past parameter sequences while the decoder generates predictions for the future speed sequences. We conduct experiments on a large dataset of 44,176 A321 flight QAR data samples to evaluate the prediction performance of our method. Experimental results show that our method significantly outperforms the state-of-the-art baselines in terms of RMSE, MAE and MAPE metrics. In addition, the feature selection of our model does not require any prior knowledge of flight experts.

The contributions of our work are summarized as follows:

- We are the first to use a deep encoder-decoder sequence-to-sequence model to predict landing speed based on QAR data. Compared with traditional LSTM deep models which only generate predictions for the next point or predict equal length sequence, our model is able to predict the landing speed sequence a few seconds in advance, which is critical for the early warning of landing safety accidents.
- We select model features based on correlation analysis with GBDT instead of experience, which free us from the need for pilots' prior knowledge. Experimental results show that using GBDT-based feature selection can significantly benefit the model performance.
- We evaluate the proposed method on a large-scale flight dataset, the experiment results show that our method significantly outperforms the state-of-the-art methods in terms of prediction accuracy.

The rest of this paper is organized as follows. In Sect. 2, a brief overview about aircraft landing safety prediction models and methods is given. Section 3 introduce the dataset and processing technologies. In Sect. 4, our method is introduced in detail. Section 5 shows experiments and results. The paper is concluded in Sect. 6.

2   Related Work

In recent years, with the prevalence of big data and artificial intelligence, a number of studies have been conducted to investigate the landing safety prediction models and methods based on flight data. Relevant literature can be roughly divided into two groups.

The first group of research works focuses on analyzing landing safety with traditional machine learning or risk models. Wang et al. [14] proposed the ANOVA (analysis of variance) method to examine the effects of pilot flare operation on landing incidents based on QAR data. Khatwa et al. [15] found that aircraft landing risks (runway overrun and landing undershoot) may relate to multiple factors, which can be divided into environmental factors, flight control problems, and pilot operations. Eduardo et al. [4] presented a risk assessment model for runway overrun accidents based on a Bayesian network (BN) model [16], aiming at assessing the risk of runway overrun in different scenarios. Recently, Li et al. [17] studied the hard landing problem and proposed a curve clustering based approach to uncover the hard landing patterns.

The second group focuses on using deep learning models to predict or analyze landing safety accidents. The most widely used models are recurrent neural networks (RNN) model and its two variants (LSTM and GRU) [18]. In a recent study, Tong et al. [12] proposed a LSTM-based deep learning model to predict aircraft landing speed based on the QAR data. Based on similar idea, they also used the LSTM model to predict the hard landing incidents [13]. Janakiraman [19] proposed a precursor mining algorithm based on multiple instance learning and GRU deep model.

3 Data Processing

3.1 QAR Data

The dataset used in this paper is provided by a commercial airline[1] of China, covering more than forty thousand flights of Airbus A321 aircrafts that took off from or landed at two domestic airports during the time of January 2018. The decoded QAR data are stored in the form of CSV files, each of which records the multiple sequential parameters throughout a flight, i.e., from take-off to landing. Due to decoding errors and accuracy issue of sensors, some parameters and information are incorrect or missing in the original data. Therefore, it is necessary to preprocess the original data and remove those flights with incorrect information. After preprocessing, 44,176 flight samples were finally obtained, and each sample contains 39 parameters.

For each flight sample, we only consider the data of the final landing phase, and use the data of this phase to predict the landing speed sequence. Here the landing speed is the ground speed during the landing phase. In the following part, we will use the terms "landing speed" and "ground speed" interchangeably when there is no ambiguity. The reason why the entire flight data are not considered is because the final ground speed sequence mainly depends on the pilot's operations during the final landing phase. Since this phase only accounts for about 1% of the entire flight, if the entire flight data is used, the information in the landing phase can easily be overwhelmed by the redundant data of the entire flight. Based on the above consideration, we first track the radio height parameter to find when the aircraft descended to an altitude (radio height) of 50 ft during the landing phase, as shown in Fig. 1. Then we take 4 s forward and 4 s backward from this point, resulting in an 8 s interval and use parameters from this interval as input sequence of the prediction model. We further extract a 10 s interval from 4 to 14 s after the 50 ft point and use the speed parameter from this interval as the output sequence of the prediction model, as shown in Fig. 1. The reason for choosing 50 ft as a key point is because pilots usually perform flare operations around the 50 ft altitude, therefore, the information surround the 50 ft point can be critical to predict the ground speed. We generate predictions for the 4 to 14 s interval after the 50 ft point since this interval covers the touchdown time (the time from 50 ft to touchdown) of about 99.94% of the flights in our dataset.

3.2 Data Transformation and Interpolation

Most QAR parameters are numerical, while some other parameters are discrete state variables. For the state variables, we transfer them into discrete integral

---

[1] Owing to proprietary and privacy considerations, we do not disclose the airline name and data.

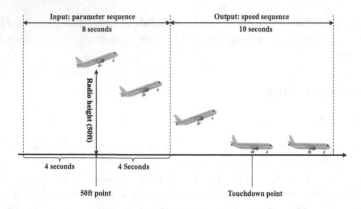

Fig. 1. Landing process.

numbers. Moreover, different QAR parameters are collected at different sampling rates, which vary from 0.5 to 8 Hz. In our experiments, we normalize the sampling rate for all parameters to 4 Hz (four frames per second) through down-sampling and up-sampling operations. Specifically, for the parameters with 8 Hz sampling rate, we perform a down-sampling operation by averaging over every two consecutive frames. For the parameters whose sampling rates are lower than 4 Hz, we further consider two up-sampling strategies. The first strategy is to replace low sampling rate parameters by calculating from the corresponding high sampling rate parameters. The second strategy is to perform an interpolation operation to add data between consecutive frames. In this paper, we use Lagrange interpolation [20] method which is widely used in many real world applications and due to the space limit, its detail is not provided in this paper.

4   Method

4.1   Problem Definition

Our purpose in this paper is to develop a proactive approach to predict landing speed (ground speed) which can provide early warning of landing incidents. We define the landing speed prediction problem as follows:

Definition 1. *(Landing Speed Prediction Problem): Given the parameter sequences extracted from the 8 s interval surrounds the 50 ft point during the landing phase, the landing speed prediction problem aims to predict the ground speed sequence during the following 10 s, as shown in Fig. 1.*

To solve the above problem, we take a two-step approach, i.e., first, we preprocess the QAR dataset through a data cleaning, Lagrange interpolation and normalization procedure, based on the preprocessed QAR dataset, we use gradient boosting decision trees to select the features which are most closely related to landing ground speed. Then with the LSTM encoder-decoder model we predict

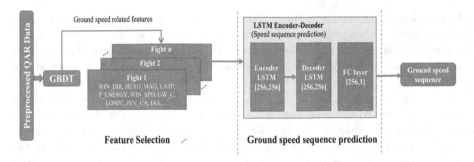

Fig. 2. Framework of the landing speed sequence prediction method.

the speed sequence. We summarize our framework in Fig. 2, and the following subsections will give detailed description.

4.2   Feature Selection

To predict the speed and height sequences, we first perform a feature selection procedure. As we have mentioned before, there are 39 flight parameters after the data process procedure, and not all parameters are beneficial to the prediction performance. In this paper, we use the gradient boost decision tree (GBDT) to find the features (parameters) which are most closely related to speed sequence. GBDT is an iterative decision tree algorithm which consists of multiple decision trees, and the conclusions of all trees are added up to make up the final model. Like other tree based algorithms, GBDT is widely used for classification, regression and feature selection. The idea of GBDT gives it a natural advantage to discover a variety of distinguishing features and feature combinations [21]. In this paper, we separately use the GBDT algorithm to calculate feature importance and rank features based on their importance scores for the speed parameter. Then the features with higher scores are selected and used as input features for the subsequent LSTM sequence-to-sequence model. The GBDT algorithm [22] for feature selection consists of the following steps:

Step one: initialization, estimate the constant value that minimizes the loss function.

Step two: (A) Calculate the value of the negative gradient of the loss function in the current model and use it as an estimation of the residual; (B) Estimate the regression leaf node area to fit the approximation of the residuals; (C) Estimating the value of the leaf node area using a linear search to minimize the loss function; (D) Update the regression tree.

Step three: Get the final model $f(x)$ of the output.

4.3   LSTM Encoder-Decoder Architecture

The LSTM encoder-decoder architecture was originally introduced for machine translation tasks [23,24]. It has the ability to read and generate sequences of arbitrary length, as shown in Fig. 5. The architecture includes two models: one to read the input sequence and encode it into a fixed-length vector, and the other to decode the fixed-length vector and output the prediction sequence. The encoder processes input sequences $x_1, x_2, ...x_T$ of length T and generates a summary of past input sequences from the LSTM cell state vector c_t. After performing T recursive updates through the LSTM cell, the encoder summarizes the entire input sequence into the final unit state vector c_T. The decoder then accepts c_T as the initial cell state of the decoder for subsequent sequence generation. As shown in Fig. 3, the decoder uses the output from the previous moment as the input for the next moment to update the current cell state, which means it is suitable for correlated sequence generation.

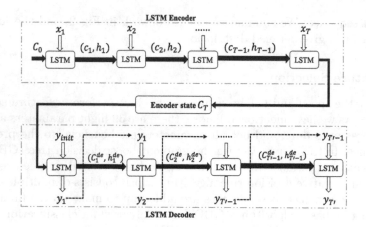

Fig. 3. LSTM encoder-decoder architecture.

In LSTM encoder-decoder architecture, the current hidden layer state h_t is determined by the hidden layer state h_{t-1} of the previous moment and the input x_t of the current moment, which is:

$$h_t = f(h_{t-1}, x_t) \tag{1}$$

For the encoder unit, the flow of information is as follows. First, the value of input gate i_t and the candidate state value $\hat{C}_t$ of LSTM cell at time t is calculated.

$$i_t = \delta(W_i(X_t, h_{t-1}) + b_i) \tag{2}$$

$$\hat{C}_t = tanh(W_c(X_t, h_{t-1}) + b_c) \tag{3}$$

Second, the activation value f_t of the forget gate at time t is calculated, based on which the updated cell state C_t at time t can be calculated as follows,

$$f_t = \delta(W_f(X_t, h_{t-1}) + b_f) \tag{4}$$

$$C_t = i_t \hat{C}_t + f_t C_{t-1} \tag{5}$$

After calculating the updated value of the cell state, the value of the output gate can be obtained.

$$O_t = \delta(W_o(X_t, h_{t-1}) + b_o) \tag{6}$$

$$h_t = O_t \tanh(C_t) \tag{7}$$

Where δ represents the sigmoid activation function, W_i, W_i, W_f, W_o are weight matrices, b_i, b_i, b_f, b_o represent bias vectors. Through the above calculations, LSTM encoder can effectively keep a long-term memory from the input sequence. After obtaining the hidden state h_t at each moment, the information is summarized to generate the final cell state vector c_T in the encoding process.

In the decoding process, the model predict the next output y_t based on c_T and the previous output sequence $y_1, y_2, ..., y_{t-1}$. Actually, the joint probability of generating the sequence $y = y_1, y_2, ...y_t$ can be decomposed into sequential conditional probabilities.

$$p(y_1, ..., y_{T'}|x_1, x_2, ..., x_T) = \prod_{t=1}^{T'} p(y_t|c_T, y_1, ..., y_{t-1}) \tag{8}$$

The decoder predict the next output y_t based on c_{t-1}^{de} and y_{t-1} of the output sequence $y_1, y_2, ..., y_{t-1}$.

$$p(y_1, ..., y_{T'}|x_1, x_2, ..., x_T) = \prod_{t=1}^{T'} p(y_t|c_{t-1}^{de}, y_{t-1}) \tag{9}$$

In this paper, we design the LSTM-based encoder-decoder architecture as follows: The encoder consists of one input layer followed by two LSTM layers stacked. And the hidden layer of LSTM is 256 dimensions, which is same as decoder. The sequence length of the encoder is set to 32.

$$h_t = LSTM_{Encoder}(x_t, h_{t-1}) \tag{10}$$

The decoder whose sequence length is set to 40 consists of the two LSTM layers stacked. The decoder LSTMs use the cell state vectors passed from the encoder as their initial cell states.

$$h_t^{de} = LSTM_{Decoder}(y_{t-1}, h_{t-1}^{de}) \tag{11}$$

We use LSTM-based encoder-decoder model to generate predictions for the ground speed sequence. The overall network structure for the model is shown in Fig. 2 and the decoder network is shown in Fig. 4.

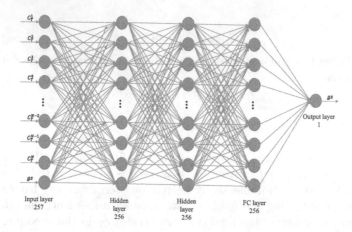

Fig. 4. Structure of decoder.

5 Experimental Evaluation

5.1 Dataset

As we have mentioned previously, the dataset used in this paper consists of 44,176 flights of Airbus A321 that took off from or landed at two domestic airports during the time of January 2018. Each flight has 39 flight parameters. We randomly select 4,417 flight samples, i.e., 10% as the testing set, and the rest 39,759 flight samples as the training set.

5.2 Baselines

We compare the proposed method with the following baselines:

- **LSTM**: this algorithm is proposed by Tong et al. [12], and based on traditional LSTM deep model, which is used for landing speed prediction. We incorporate this baseline for comparison in the landing speed sequence prediction.
- **Traditional Machine Learning Methods**: We consider classic machine learning methods for comparison, including Decision Tree (DT), Linear Regression (LR), Gradient Boosting Decision Tree (GBDT), Random Forest (RF), Neural Networks (NN) and SVM. In our experiments, we first considered 21 parameters in 32 time steps (8 s and 4 frames per second), resulting a total number of 672 features to train the models. However, the model performances were very poor, so we finally selected 21 features at the 50 ft point as input to machine learning models. The model parameters are tuned manually to make sure that each method gives its best performance.
- **Experience LSTM Encoder-Decoder**: This baseline is the same as the method proposed in this paper, except that the features are selected based on experience instead of GBDT. Specifically, we consider as much flight parameters as possible, resulting in 32 features.

5.3 Evaluation Metrics

We evaluate the performance of different prediction models in terms of RMSE, MAE and MAPE, which are the most commonly used metrics to measure the prediction accuracy for regression problems. RMSE and MAE are used to evaluate the absolute error, while MAPE is used to evaluate to relative error. The three metrics are defined as follows:

$$RMSE = \sqrt{\frac{1}{m} \sum_{i=1}^{m} (y_i - \hat{y}_i)^2} \tag{12}$$

$$MAE = \frac{1}{m} \sum_{i=1}^{m} |y_i - \hat{y}_i| \tag{13}$$

$$MAPE = \frac{100}{m} \sum_{i=1}^{m} \left| \frac{y_i - \hat{y}_i}{y_i} \right| \tag{14}$$

where y_i and $\hat{y}_i$ are actual and predicted values, respectively, and m is the number of the testing samples.

5.4 Result of Feature Selection

We apply the GBDT algorithm on the 39 candidate flight parameters to select features that are most closely related to ground speed, and these features are ranked according to feature importance. The GBDT algorithm calculates the weight of 39 parameters on each sample, and finally adds the weights of 44,176 samples to rank the importance for them. In our experiments, the regression accuracy of GDBT on ground speed is about 82.8%.

Result of Feature Selection for Landing Speed. We first pick the features that are important for speed. Figure 5 shows the rank of the most important features for landing speed (the insignificant features are removed), from which we see that wind direction *(WIN_DIR)* and magnetic heading of the aircraft *(HEAD_MAG)* play significantly more important roles than the other parameters. Surprising, the gravitational potential energy *(P_ENERGY)* also play an important role for landing speed, this parameter is not included in the original flight parameters and was rarely considered in previous works. We calculate it through the formula:

$$P\_ENERGY = GW\_C \times g \times RALT \tag{15}$$

where GW_C is the gross weight of the aircraft, $RALT$ is the radio height, g is the acceleration of gravity near the ground. Similarly, we also consider the kinetic energy of aircraft, which is calculated as follows:

$$K\_ENERGY = \frac{1}{2} \times GW\_C \times GS \times GS \tag{16}$$

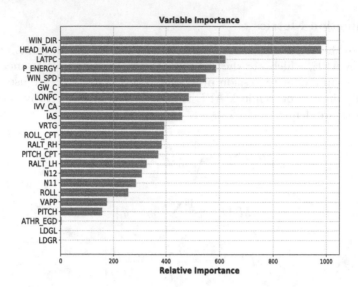

Fig. 5. Feature importance for landing speed.

where *GS* is the ground speed of the aircraft. However, the kinetic energy parameter is not included in the feature selection for ground speed since this parameter implicitly contains the ground speed parameter.

Based on the results of GBDT, 21 parameters were finally selected from the 39 parameters to predict the ground speed sequence. The top 21 parameters are summarized and shown in Table 1.

Result of Landing Speed Prediction. We train the ground speed sequence prediction model with the 44,176 processed training flight samples and evaluate the prediction accuracy on the 4,417 testing samples. The model parameters are learned with the Adam optimization algorithm. As we have mentioned before, the input parameter sequence of the encoder are extracted from a 8 s interval, and all parameters are sampled at 4 Hz sampling rate, so the length of parameter sequence input to the encoder is set to 32. Similarly, the length of speed sequence output by the decoder is set to 40 (10 s by 4 Hz).

The batch size of model training is set to 256 and the number of iterations is set to 6000. Based on the parameter setting we pick the model that gives the best performance on the test set. Figure 6 shows the predicted ground speed versus the actual values, where x-axis represents the real ground speed values while y-axis represents the predicted values. Since the test set include 4,417 samples and the sequence length of each sample is 40, there are totally 176,680 points shown on the figure. From the figure we see that the data points are closely gathered around the $y = x$ ideal line, indicating that the predicted ground speed is very close to the real value, especially when the real ground speed is larger than 110 knots.

Table 1. Top 21 selected features in gbdt

Parameter	Explanation	Parameter	Explanation
VAPP	Reference landing speed	N11	No. 1 engine speed ratio
HEAD_MAG	Magnetic heading	N12	No. 2 engine speed ratio
WIN_DIR	Wind direction	WIN_SPD	Wind speed
LONPC	Longitude	LATPC	Latitude
ROLL	Roll	PITCH_CPT	Captain pitch control
ROLL_CPT	Captain roll control	K_ENERGY	Aircraft kinetic energy
P_ENERGY	Aircraft potential energy	GW_C	Aircraft gross weight
PITCH	Aircraft pitch	VRTG	Vertical load
IAS	Lifting wheel speed	RALT_LH	Radio height (left)
RALT_RH	Radio height (right)	GS	Ground speed
IVV_CA	Decline rate		

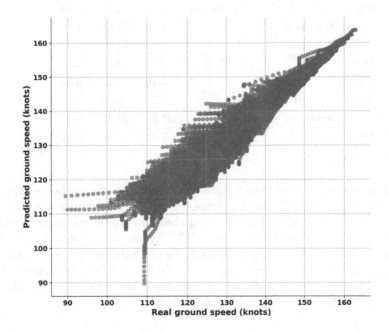

Fig. 6. Actual and predicted ground speed on test set.

Table 2 shows the comparison of prediction performance for ground speed in terms of RMSE, MAE and MAPE for different methods, where the best results are shown in bold. From the results we see that our method (GBDT LSTM Encoder-Decoder) exhibits significantly higher prediction accuracy than all the baselines, especially over those traditional machine learning methods. The five layer shallow neural networks gives the worst performance, indicating the necessity of using deep models. Among the traditional machine learning methods, the tree-based methods (DT and GBDT) perform better than the non-tree-based methods, we think this is mainly due to the generalization ability and non-linearity nature of the tree-based models. The most competitive baseline is the traditional LSTM-based method proposed by Tong et al. [12], however, its performance is still significantly inferior to ours, which shows the superiority of sequence-to-sequence model over traditional LSTM model. Moreover, by comparing GBDT LSTM Encoder-Decoder with Experience LSTM Encoder-Decoder, it shows that features selected by GBDT instead of experience can largely improve the prediction performance.

Table 2. Comparison results of rmse and mae, mape on the test set for ground speed

Model	RMSE	MAE	MAPE (%)
GBDT LSTM Encoder-Decoder	**1.71**	**1.07**	**0.83**
Experience LSTM Encoder-Decoder	2.20	1.32	1.03
LSTM Tong et al.	1.96	1.50	1.13
Linear Regression	12.57	11.01	8.43
Decision Tree	5.70	4.66	3.51
GBDT	5.79	5.22	3.95
Random Forest	9.75	8.48	6.55
Neural Networks	21.86	21.17	16.11
SVM	7.83	6.55	5.02

In summary, we can see from the above results that our method significantly outperforms the state-of-the-art baselines. There are two reasons that contribute to the success of our method. First, compared to traditional machine learning methods and the original LSTM, the LSTM encoder-decoder architecture can better learn the patterns of input sequence to generate predictions for the variable length output sequence. Second, we use GBDT for feature selection, which can help find a variety of distinguishing features and feature combinations to further improve the prediction performance.

6   Conclusion

Landing speed prediction is an important research issue for aviation safety. In this paper, we proposed a novel ground speed sequence prediction method based on the LSTM Encoder-Decoder architecture. We first propose to use gradient boosting decision trees to select the features which are most closely related to landing ground speed. Then we use LSTM Encoder-Decoder model to predict ground speed sequence. Experimental results on a real QAR dataset validate the effectiveness of the proposed method. Besides, our methods can provide earlier warnings of long landing events compared to the traditional machine learning and LSTM based methods. Our method provides a new proactive solution to reduce or avoid runway overrun accidents.

Acknowledgements. We thank Dongcheng Chen, Xuan Ding, Lilin Yu, Hao Xie, Lin Qi, Liu Liu, Zhe Huang and Ziwen Liao for their professional opinions and discussions. This work was supported in part by: National Natural Science Foundation of China (Nos. 61702059, 61966008), Fundamental Research Funds for the Central Universities (Nos. 2019CDXYJSJ0021, 2020CDCGJSJ041), Frontier and Application Foundation Research Program of Chongqing City (No. cstc2018jcyjAX0340), Key Research and Development Program of Chongqing (No. cstc2019jscx-fxydX0071), Guangxi Key Laboratory of Optoelectronic Information Processing (No. GD18202), Guangxi Key Laboratory of Trusted Software (Nos. kx201701, kx201702).

References

1. Boeing Commercial Airplanes: Statistical summary of commercial jet airplane accidents. In: Worldwide Operations, vol. 2008 (1959)
2. Lv, H., Yu, J., Zhu, T.: A novel method of overrun risk measurement and assessment using large scale QAR data. In: 2018 IEEE Fourth International Conference on Big Data Computing Service and Applications (BigDataService), pp. 213–220. IEEE (2018)
3. Van Es, G.W.H., Tritschler, K., Tauss, M.: Development of a landing overrun risk index, 21st annual European Aviation Safety Seminar (EASS) Nicosia, Cyprus, NLR Air Transport Safety Institute, Report NLR-TP-280 (2009)
4. Ayra, E.S., Ríos Insua, D., Cano, J.: Bayesian network for managing runway overruns in aviation safety. J. Aerosp. Inf. Syst. **16**(12), 546–558 (2019)
5. Valdés, R.M.A., Comendador, V.F.G., Sanz, L.P., Sanz, A.R.: Prediction of aircraft safety incidents using Bayesian inference and hierarchical structures. Saf. Sci. **104**, 216–230 (2018)
6. Sheridan, K., Puranik, T.G., Mangortey, E., Pinon-Fischer, O.J., Kirby, M., Mavris, D.N.: An application of DBSCAN clustering for flight anomaly detection during the approach phase. In: AIAA Scitech 2020 Forum, p. 1851 (2020)
7. Basora, L., Olive, X., Dubot, T.: Recent advances in anomaly detection methods applied to aviation. Aerospace **6**(11), 117 (2019)
8. Calle-Alonso, F., Pérez, C.J., Ayra, E.S.: A Bayesian-network-based approach to risk analysis in runway excursions. J. Navig. **72**(5), 1121–1139 (2019)
9. Wang, L., Ren, Y., Sun, H., Dong, C.: A landing operation performance evaluation system based on flight data. In: Harris, D. (ed.) EPCE 2017. LNCS (LNAI), vol. 10276, pp. 297–305. Springer, Cham (2017). https://doi.org/10.1007/978-3-319-58475-1_22
10. Wang, L., Wu, C., Sun, R.: An analysis of flight quick access recorder (QAR) data and its applications in preventing landing incidents. Reliab. Eng. Syst. Saf. **127**, 86–96 (2014)
11. Hu, C., Zhou, S.-H., Xie, Y., Chang, W.-B.: The study on hard landing prediction model with optimized parameter SVM method. In: 2016 35th Chinese Control Conference (CCC), pp. 4283–4287. IEEE (2016)
12. Tong, C., Yin, X., Wang, S., Zheng, Z.: A novel deep learning method for aircraft landing speed prediction based on cloud-based sensor data. Future Gener. Comput. Syst. **88**, 552–558 (2018)

13. Tong, C., et al.: An innovative deep architecture for aircraft hard landing prediction based on time-series sensor data. Appl. Soft Comput. **73**, 344–349 (2018)
14. Wang, L., Ren, Y., Wu, C.: Effects of flare operation on landing safety: a study based on ANOVA of real flight data. Saf. Sci. **102**, 14–25 (2018)
15. Khatwa, R., Helmreich, R.L.: Analysis of critical factors during approach and landing in accidents and normal flight. Flight Saf. Digest **17**(11–12), 1–2 (1999)
16. Nielsen, T.D., Jensen, F.V.: Bayesian Networks and Decision Graphs. Springer, New York (2007). https://doi.org/10.1007/978-0-387-68282-2
17. Li, X., Shang, J., Zheng, L., Liu, D., Qi, L., Liu, L.: CurveCluster: automated recognition of hard landing patterns based on QAR curve clustering. In: IEEE 16th International Conference on Ubiquitous Intelligence and Computing (UIC), pp. 602–609. IEEE (2018)
18. Park, S.H., Kim, B., Kang, C.M., Chung, C.C., Choi, J.W.: Sequence-to-sequence prediction of vehicle trajectory via LSTM encoder-decoder architecture. In: 2018 IEEE Intelligent Vehicles Symposium (IV), pp. 1672–1678. IEEE (2018)
19. Janakiraman, V.M.: Explaining aviation safety incidents using deep temporal multiple instance learning. In: Proceedings of the 24th ACM SIGKDD International Conference on Knowledge Discovery & Data Mining, pp. 406–415 (2018)
20. Berrut, J.-P., Trefethen, L.N.: Barycentric Lagrange interpolation. SIAM Rev. **46**(3), 501–517 (2004)
21. He, X., et al.: Practical lessons from predicting clicks on ads at Facebook. In: Proceedings of the Eighth International Workshop on Data Mining for Online Advertising, pp. 1–9 (2014)
22. Hastie, T., Friedman, J., Tibshirani, R.: The Elements of Statistical Learning, vol. 1, no. 10. SSS, Springer, New York (2001). https://doi.org/10.1007/978-0-387-21606-5
23. Cho, K., et al.: Learning phrase representations using RNN encoder-decoder for statistical machine translation. arXiv preprint arXiv:1406.1078 (2014)
24. Luong, M.-T., Pham, H., Manning, C.D.: Effective approaches to attention-based neural machine translation. arXiv preprint arXiv:1508.04025 (2015)

Fine-grained Multi-label Sexism Classification Using Semi-supervised Learning

Harika Abburi[1]([⊠]), Pulkit Parikh[1], Niyati Chhaya[2], and Vasudeva Varma[1]

[1] IIIT-Hyderabad, Hyderabad, India
{harika.a,pulkit.parikh}@research.iiit.ac.in,vv@iiit.ac.in
[2] Adobe Research, Bangalore, India
nchhaya@adobe.com

Abstract. Sexism, a pervasive form of oppression, causes profound suffering through various manifestations. Given the rising number of experiences of sexism reported online, categorizing these recollections automatically can aid the fight against sexism, as it can facilitate effective analyses by gender studies researchers and government officials involved in policy making. In this paper, we explore the fine-grained, multi-label classification of accounts (reports) of sexism. To the best of our knowledge, we consider substantially more categories of sexism than any related prior work through our 23-class problem formulation. Moreover, we present the first semi-supervised work for the multi-label classification of accounts describing any type(s) of sexism wherein the approach goes beyond merely fine-tuning pre-trained models using unlabeled data. We devise self-training based techniques tailor-made for the multi-label nature of the problem to utilize unlabeled samples for augmenting the labeled set. We identify high textual diversity with respect to the existing labeled set as a desirable quality for candidate unlabeled instances and develop methods for incorporating it into our approach. We also explore ways of infusing class imbalance alleviation for multi-label classification into our semi-supervised learning, independently and in conjunction with the method involving diversity. Several proposed methods outperform a variety of baselines on a recently released dataset for multi-label sexism categorization across several standard metrics.

Keywords: Sexism classification · Semi-supervised learning · Multi-label classification · Fine-grained categorization

1 Introduction

Sexism, defined as prejudice, stereotyping, or discrimination based on a person's sex, occurs in various overt and subtle forms, permeating personal as well as professional spaces. While men and boys are also harmed by sexism, women and

© Springer Nature Switzerland AG 2020
Z. Huang et al. (Eds.): WISE 2020, LNCS 12343, pp. 531–547, 2020.
https://doi.org/10.1007/978-3-030-62008-0_37

Table 1. An Instance of sexism associated with multiple categories

Account	"A colleague once saw me washing my coffee mug before leaving the office and 'joked' if I was practicing for my 'home duties'. It's sad that he doesn't see the problem with men not bearing half the load of household work"
Associated categories of sexism	Role stereotyping: False generalizations about some roles being more suitable for women; also applies to similar mistaken notions about men
	Moral policing: The promotion of discriminatory guidelines for women under the pretense of morality; also applies to statements that feed into such narratives
	Hostile work environment: Sexism suffered at the workplace; also applies when sexism perpetrated by a colleague elsewhere makes working worrisome for the victim

girls suffer the brunt of sexist mindsets and resultant wrongdoings. With increasingly many people sharing recollections of sexism experienced or witnessed by them, the automatic classification of these accounts into well-conceived categories of sexism can help fight this oppression, as it can better equip authorities formulating policies and researchers of gender studies to analyze sexism.

The detection of sexism differs from and can complement the classification of sexism. In a forum where instances of sexism are mixed with other posts unrelated to sexism, sexism detection can be used to identify the posts on which to perform sexism classification. Moreover, we observe the distinction between sexist statements (e.g., posts whereby one perpetrates sexism) and the accounts of sexism suffered or witnessed (e.g., personal recollections shared as part of the #metoo movement). We also note the prior work on detecting or classifying personal stories of sexual harassment and/or assault [6,13]. In this paper, we focus on classifying an account (report) of sexism involving any set of categories of sexism.

Most of the existing research on sexism classification [3,11,12] considers at most five categories of sexism. Further, the majority of prior approaches associate only one category of sexism with an instance of sexism. Having mutually exclusive categories of sexism is unreasonable and limiting, as substantiated by Table 1.

To the best of our knowledge, [17] is the only work that explores the multi-label categorization of accounts of sexism using machine learning and considers more than three categories of sexism. It provides the largest dataset containing accounts drawn from 'Everyday Sexism Project'[1], where experiences of sexism are shared from all over the world. The textual accounts are annotated using 23 categories of sexism formulated with the help of a social scientist. However, they perform sexism classification among 14 categories derived by merging some sets of categories. This prohibits distinguishing within category pairs

---

[1] https://everydaysexism.com.

such as {moral policing, victim blaming} and {motherhood-related discrimination, menstruation-related discrimination}. We overcome this limitation by carrying out a fine-grained (23-class) classification by building on the same labeled dataset.

Most existing approaches for the categorization of sexism are entirely supervised in nature (use no unlabeled data). The biggest labeled dataset available for sexism classification [17] comprises around 13,000 accounts. In contrast, the accounts of sexism narrated on just the 'Everyday Sexism Project' website comfortably number in several hundred thousands. Effectively tapping this large volume of data has the potential to enhance the classification performance by overcoming potential weaknesses stemming from the limited training data, especially because of the fine-grained aspect of the problem. As far as we are aware, the only existing approach that uses unlabeled data for sexism classification does so for merely fine-tuning a pre-trained model used for computing sentence representations [17]. We formulate the first set of methods for utilizing unlabeled instances in a more involved manner for sexism classification. The proposed techniques are broadly based on self-training, a semi-supervised learning paradigm, in terms of the workflow. We augment the existing labeled data through the selective addition of pseudo-multi-labeled unlabeled samples.

We develop our semi-supervised methods keeping in mind the fact that, unlike in single-label multi-class (or binary) classification, a single instance can be tagged with up to 23 categories in our multi-label multi-class classification. We identify textual diversity with reference to (a subset of) the original labeled set as a useful constituent in the evaluation of candidate unlabeled instances and construct mechanisms of building it into our approach. We also seek to incorporate the desirable quality of low class imbalance into it in ways suited for multi-label semi-supervised classification. We present multiple procedures for combining these elements into a unified semi-supervised learning approach.

Our key contributions are summarized below.

- To the best of our knowledge, this is the first work to consider as many as 23 categories for sexism classification.
- We introduce a set of semi-supervised methods to augment the labeled data for multi-label multi-class sexism classification.
- We devise mechanisms aimed at enhancing the textual diversity in the resultant expanded labeled set and alleviating the skew in the original class distribution.
- A number of proposed methods outperform numerous baselines, including the existing state-of-the-art, across various established metrics.

The rest of our paper is structured as follows. Section 2 describes related prior work. Section 3 discusses the semi-supervised methods that we propose for multi-label sexism classification. Experimental results and observations are provided in Sect. 4. We conclude with a summary in Sect. 5.

2   Related Work

In this section, we review the work on the classification of sexism after noting some distantly related work. Though our work involves classifying accounts of sexism, prior work on the classification of sexist or misogynous statements (e.g., tweets wherein one perpetrates sexism or misogyny) is also included in this review.

[16] applies topic modeling to data obtained from The Everyday Sexism Project and maps the semantic relations between topics. [9] studies user engagement with posts related to gender based violence and their language nuances. We note that sexism detection can complement sexism classification by preceding it to remove posts unrelated to sexism. The detection of sexism is performed by some hate speech classification approaches that include sexism as a category of hate [4,7,23,29]. [10] presents an approach for detecting sexism and misogyny from tweets. Given our focus on sexism classification, we do not delve into prior work related to hate speech or cyber-bullying [2,21].

[13] explores CNN, RNN, and a combination of them for categorizing personal experiences of sexual harassment into one or more of three classes. In [25], a density matrix encoder inspired by quantum mechanics is used for the classification of personal stories of sexual harassment. [14] employ deep learning methods to classify sexual violence into one of four categories. In [3], tweets identified as misogynist are classified as stereotype and objectification, discredit, sexual harassment and threats of violence, dominance, or derailing using features involving Part of Speech (POS) tags, n-grams, and text embedding. [11] performs a 4-class categorization of sexist tweets. In [12], tweets are classified as benevolent, hostile, or non-sexist using biLSTM with attention, SVM, and fastText. While its categorization of sexism pertains to how it is stated, our work concentrates on aspects such as what an instance of sexism involves, where it occurs, and who perpetrates it.

[17] explores multi-label categorization of accounts reporting any kind(s) of sexism. They provide the largest dataset for sexism classification and the state-of-the-art classifier for it. The classifier combines sentence embeddings generated using a BERT [8] model with those generated from ELMo [20] and GloVe [19] embeddings using biLSTM with attention and CNN. As far as we know, our work presents the first semi-supervised approach for the multi-label classification of accounts describing any type(s) of sexism that goes further than using unlabeled instances only for fine-tuning pre-trained models.

3   Semi-supervised Multi-label Sexism Classification

This section presents proposed methods that employ semi-supervised learning for classifying an account of sexism (also referred to as a post henceforth) such that the categories can co-occur. We begin by laying the groundwork for the description of our methods.

3.1   Basic Self-training

Self-training [1, 27] is a semi-supervised learning approach that helps augment the set of labeled instances by selectively adding unlabeled samples. For performing a task such as classification (or regression) in the presence of unlabeled and labeled data, a typical self-training method first trains a model (e.g., classifier) on the labeled instances. Next, it applies the model to the unlabeled instances and identifies a subset of them to be added to the training set, along with the predicted labels, based on criteria such as the confidence scores associated with the model's predictions. After expanding the training set by adding this pseudo-labeled subset, a new classifier is trained on the augmented set. This process is repeated until some stopping criteria such as the stabilization of model parameters and the number of iterations completed are satisfied.

3.2   Baseline Model

We use the best-performing model by [17] as our baseline classifier. Its architecture is detailed here. The neural model concatenates sentence representations obtained using a BERT [8] model tuned using unlabeled instances of sexism with those generated from ELMo [20] and GloVe [19] embeddings separately using biLSTM with an attention scheme. The combined sentence vectors are passed through biLSTM with attention to produce the post representation, which is fed to a fully connected layer with a sigmoid non-linearity, generating the output probabilities. The loss function used for training the model is a weighted mean of label-wise binary cross entropy values.

3.3   Proposed Approach

While there exists some prior work on the classification of textual records involving sexism, most methods are supervised. We observe that the accounts of sexism reported on the 'Everyday Sexism Project' website alone hugely outnumber those provided in the biggest existing labeled dataset for sexism classification [17]. The performance of a sexism classification method can be improved by leveraging this sizable chunk of unlabeled data. We explore semi-supervised techniques based on self-training to utilize unlabeled accounts of sexism. We devise multiple methods of expanding the set of labeled data using unlabeled instances befitting the multi-label nature of instances of sexism. We first formulate a basic method based on self-training tailor-made for the multi-label problem configuration. We also propose other methods built on top of it with a view to (1) improving the proportions of positive (relevant) samples across categories, (2) improving the class balance keeping in mind the mutual non-exclusivity of category labels, and (3) encouraging textual diversity in the newly labeled (pseudo-labeled) data relative to the original training set. We also develop some combinations involving these proposed methods. The augmented data generated by any of our methods can be used for training any supervised classification model. We now describe the proposed semi-supervised methods in depth.

Basic Self-training for Multi-label Classification. The most fundamental, indispensable factor in determining which unlabeled instances should be considered to be added to the original labeled set during self-training is the confidence of correctness associated with each prediction. In single-label (multi-class) classification, one can simply treat the classification probability corresponding to the one relevant class as this confidence. This procedure is inapplicable to a multi-label case wherein the baseline classifier outputs a probability of applicability for each label. We observe that our baseline multi-label classifier generates the probabilities following sigmoid (as opposed to softmax) nonlinearities and that predictions need to be made by rounding the per-label probabilities p_j. Since this implies that the probability linked with the prediction is either p_j or $1 - p_j$, we mandate that at least one of these two quantities exceed a threshold (hyper-parameter T) for any unlabeled instance to qualify for being added to the labeled set.

Improving Positive Sample Proportions across Categories (*IPSPC*). In addition to the basic confidence-based check, this method subjects unlabeled instances to another qualifying test relating to the number of predicted labels using the baseline classifier. The intuition is that the higher the number of predicted labels, the greater the number of labels for which relevant (positive) samples are contributed. Since the number of positive samples is outweighed by the negative counterpart by a substantial margin across most labels, we attempt to counter this skew by picking the unlabeled instances with a certain minimum number of predicted labels (hyper-parameter P_{min}). In order to maximize the label correctness of the chosen pseudo-labeled set, we also avoid candidate samples with an unreasonably high number of predicted labels (hyper-parameter P_{max}).

Favoring Low-Support Labels. While the previous method seeks to improve the per-category ratios of positive to negative sample counts generally, this method attempts to correct the class imbalance between categories while creating the augmented dataset. We present two methods that order unlabeled samples (adhering to the checks proposed earlier) based on notions of *support* that we design for multi-label classification. We then pick the lowest-support Top_p percent of the samples as the pseudo-labeled set in each iteration, where hyper-parameter Top_p is empirically determined. For the first method (*Support.uniform*), support for an unlabeled sample $u_k \in U$ is defined as

$$support\_uniform(u_k) = \frac{\sum_{j \in P_k^+} \sum_{i=1}^{M^*} y_{ij}}{|P_k^+|}, \qquad (1)$$

where, P_k^+ is the set of labels predicted for u_k, U denotes the unlabeled data, M^* is the number of labeled samples in the current iteration, and y_{ij} is 1 if category l_j is given for sample x_i in the labeled data and 0 otherwise.

Since this method considers the average coverage in the labeled data across all predicted labels for a sample, a sample linked with some weak labels (labels with low frequencies in the labeled data) and some extremely strong labels may get rated lower than one linked with no weak label and some moderately strong labels. In *Support.weakest*, we explicitly take into account only the weak labels for the notion of support. In each iteration. we determine weak labels based on the coverage (frequency) of each label in the current labeled data. Specifically, a label l_j is weak if $\sum_{i=1}^{M^*} y_{ij} < S_m$ and strong otherwise, where S_m is a hyper-parameter. We disregard all strong labels while calculating the support for an unlabeled sample with at least one weak predicted label. For samples with predictions involving no weak classes, we resort to the previous support computation. For the rest, we compute the support as follows.

$$support\_weakest(u_k) = \frac{\sum_{j \in P_k^+} v_j}{|\{z \mid \sum_{i=1}^{M^*} y_{iz} < S_m, z \in P_k^+\}|}, \tag{2}$$

where $v_j = \sum_{i=1}^{M^*} y_{ij}$ if $\sum_{i=1}^{M^*} y_{ij} < S_m$ and 0 otherwise.

Seeking Textual Diversity. We identify the utility of selecting pseudo-labeled data such that it complements the existing labeled data as opposed to being an expanded version of it, especially in terms of linguistic characteristics. This family of methods aims to introduce greater textual diversity with respect to (a subset of) the current labeled data in the set of samples of being added in each iteration. Candidate unlabeled samples (meeting the qualifying criteria proposed in the first two methods) are ranked as per how far they are from existing labeled samples in terms of the corresponding vector representations. We create a variant of the state-of-the-art deep learning model for the original labeled dataset given in [17] for generating the embedding for the text of a given sample. Cosine distance is used as the distance metric. The highest-rank Top_p percent of the samples are added to the labeled data in each iteration. In *Diversity.uniform*, diversity for a sample $u_k \in U$ is given by,

$$diversity\_uniform(u_k) = \frac{\sum_{i=1}^{M^*} cos\_dist(post\_rep(u_k), post\_rep(x_i))}{M^*}, \tag{3}$$

where *post_rep* refers to the vector representation for a sample. *Diversity.uniform* picks the most distinct unlabeled samples w.r.t. the current labeled set in a label-independent manner. We develop the *Diversity.label* method to incorporate per label diversity, avoiding indiscriminate comparisons against all labeled samples. For a candidate sample, for each label predicted for it, we compute the average of the distances against only the labeled samples bearing that label. Each candidate is scored using the average of the these label-wise averages. Our formulation can be expressed as,

$$diversity\_label(u_k) = \frac{\sum_{j \in P_k^+} \frac{\sum_{i=1}^{M^*} y_{ij} cos\_dist(post\_rep(u_k), post\_rep(x_i))}{\sum_{i=1}^{M^*} y_{ij}}}{|P_k^+|} \tag{4}$$

Combining Previously Proposed Methods. We develop two ways of integrating our methods favoring low-support labels and seeking greater textual diversity to explore if their individual strengths combine well.

1. Score computation: We calculate the combined score for a candidate u_k from unlabeled data U as follows.

$$score(u_k) = \frac{diversity\_uniform(u_k) \ \ (or \ \ diversity\_label(u_k))}{support\_uniform(u_k) \ \ (or \ \ support\_weakest(u_k))} \tag{5}$$

The *or* in the equation above simply indicates that we consider all four (2X2) combinations stemming from the previous two proposed families of methods. From the pseudo-labeled candidate instances which pass the screenings previously described, Top_p percent of the instances with the highest combined scores are chosen for labeled data augmentation.

2. Intersection: In each iteration, we employ a method favoring low-support labels and a method seeking greater textual diversity each to pick the pseudo-labeled samples. We then intersect the sets of pseudo-labeled samples selected by these two methods and augment the current labeled set with the resultant set of samples. In this way of combining the previously proposed methods also, we explore pairing each of the two methods favoring low-support labels with each of the two methods seeking greater textual diversity.

4 Experiments

This section presents the experimental evaluation of the proposed methods in comparison to a number of baseline methods along with analyses related to our method as well as the dataset used. Our code and all hyper-parameter values are available on GitHub[2].

4.1 Dataset

[17] introduces a dataset comprising 13,023 accounts of sexism, each labeled with at least one category of sexism. The diverse categories of sexism, derived in consultation with a social scientist, range from body shaming and menstruation-related discrimination harassment to role stereotyping and victim blaming. Most of the annotators involved had studied topics pertaining to gender and/or sexuality formally. We use this original labeled dataset (only) to train all supervised baselines.

We provide a linguistic analysis of this labeled data using Linguistic Inquiry and Word Count (LIWC), a text analysis tool [18]. We focus on the LIWC scores for the Work, Money, Religion, and Body categories. Details concerning how the LIWC scores are computed can be found in the LIWC 2015 [18]. Table 2

---

[2] https://github.com/Harikavuppala1a/semisupervised_sexism_classification.

Table 2. Linguistic analysis of the labeled dataset

ID	Category of sexism	LIWC Work	LIWC Money	LIWC Religion	LIWC Body	4-grams
1	Role stereotyping	5.466	1.059	0.110	0.374	'man of the house', 'to be a nurse', 'to cook and clean'
2	Attribute stereotyping	3.543	0.895	0.071	0.739	'drive like a girl', 'funny for a girl', 'pink is for girls'
3	Body shaming	1.765	0.188	0.085	**3.322**	'im too ugly to', 'if you lost weight', 'me a fat bitch'
4	Hyper-sexualization (excluding body shaming)	3.102	0.484	0.071	2.589	'the tits on that', 'looking me up and', 'a piece of meat'
5	Internalized sexism	3.079	0.453	0.078	1.168	'told by my mother', 'she suggested that my', 'told by my female'
6	Pay gap	**10.081**	**3.806**	0.128	0.203	'being paid more than', 'the exact same job', 'several thousand more than'
7	Hostile work environment (excluding pay gap)	8.070	0.764	0.063	1.042	'first day working at', 'the only female in', 'my boss called me'
8	Denial or trivialization of sexist misconduct	2.514	0.309	0.091	1.135	'i should be flattered', 'said it was just', 'a bit of fun'
9	Threats	1.647	0.401	0.041	0.948	'followed by a man', 'he would rape me', 'to follow me home'
10	Rape	2.019	0.208	0.102	0.823	'on top of me', 'raped at a party', 'been raped twice and'
11	Sexual assault (excluding rape)	2.023	0.318	0.043	3.251	'his crotch against my', 'i was groped by', 'i felt a hand'
12	Sexual harassment (excluding assault)	2.378	0.413	0.039	1.763	'whistled at while walking', 'looking me up and', 'asking me to come'
13	Tone policing	2.771	0.208	0.000	0.887	'me not to swear', 'that it was unladylike', 'tell me not to'
14	Moral policing (excluding tone policing)	3.566	0.501	0.164	0.988	'back to the kitchen', 'should be at home', 'should go out with'
15	victim blaming	1.767	0.126	0.076	1.916	'was asking for it', 'fault for wearing tight', 'it was her fault'
16	Slut shaming	1.977	0.575	0.042	1.479	'shouted whore at me', 'how much i cost', 'called a slag by'
17	Motherhood-related discrimination	5.095	0.576	0.189	0.604	'planning on having children', 'get married and have', 'are you a pregnant'
18	Menstruation-related discrimination	2.634	0.213	0.074	0.795	'time of the month', 'you on your period', 'must be on her'
19	Religion-based sexism	3.173	0.525	**4.675**	0.848	'my religion only males', 'that in my religion', 'piece of clothing that'
20	Physical violence (excluding sexual violence)	1.156	0.233	0.094	2.198	'spat in my face', 'pushed me into a', 'murdered by her husband'
21	Mansplaining	3.585	0.599	0.018	0.234	'cut off by a', 'to mansplain to me', 'he was embarrassing me'
22	Gaslighting	0.000	0.453	0.220	0.800	'boyfriend would guilt me', 'guilt me into agreeing', 'agreeing to various sexual'
23	Other	3.846	0.568	0.079	0.669	'to me as mrs', 'being referred to as', 'calling me young lady'

shows the scores for these LIWC categories for all the categories of sexism in the dataset. For each class of sexism, we compute the LIWC scores for all posts tagged with that class label and take the mean of all those scores to obtain the category-level scores reported. We observe the highest scores for the Work and Money LIWC categories for Pay gap. As expected, the maximum score for the Religion LIWC category is found for Religion-based sexism. For the Body LIWC category, we find the highest score for Body shaming. Table 2 also lists 4-grams from the textual accounts of sexism associated with each category of sexism.

In this paper, we devise semi-supervised methods to automatically expand this dataset. We obtain unlabeled instances of sexism from 'Everyday Sexism Project', which has already received several hundred thousand accounts of sexism from survivors and observers. We shortlist 70,000 shortest instances containing a minimum of 7 words each. Short posts are preferred in order to maximize the resemblance to the labeled data [17] on which the baseline model used for the proposed semi-supervised approach is trained. Our methods select a subset of these 70,000 accounts of sexism for augmenting the training data.

4.2   Evaluation Metrics

Evaluation metrics for multi-label classification problems differ from the standard metrics used in cases where the classes can not co-occur. We report results for a number of established metrics, namely Subset Accuracy (SA), instance-based F1 (F_{ins}), instance-based accuracy (Acc), F1 macro (F_{mac}), and F1 micro (F_{mic}) [17,28]. Subset Accuracy, which measures the fraction of the exact matches, is the strictest metric.

4.3   Baselines

Traditional Machine Learning (TML)
We report the performance using Support Vector Machine (SVM), Logistic Regression (LR), and Random Forests (RF), each applied on two feature sets, namely the average of the ELMo vectors [20] for a post's words (ELMO) and TF-IDF on word unigrams and bigrams (Word-ngrams). This gives rise to six combinations: ELMO with SVM (ELMO-SVM), ELMO with LR (ELMO-LR), ELMO with RF (ELMO-RF), word-ngrams with SVM (word-ngrams-SVM), word-ngrams with LR (word-ngrams-LR), and word-ngrams with RF (word-ngrams-RF)

Deep Learning (DL)
biLSTM and biLSTM-Attention: The word embeddings for a post are passed through a bidirectional LSTM with and without the attention scheme from [26].

Hierarchical-biLSTM-Attention: In an architecture similar to [26] with GRUs replaced with LSTMs, the word embeddings are first fed to biLSTM with attention to create a representation for each sentence. These sentence embeddings are then passed through another instance of biLSTM with attention.

BERT-biLSTM-Attention and USE-biLSTM-Attention: Sentence representations are generated using BERT via bert-as-service [24] and USE [5] each and fed to a biLSTM with attention.

CNN-biLSTM-Attention: In this baseline similar to [22], each sentence's word embeddings are passed through convolutional and max-over-time pooling layers. The resultant representations are then passed through a biLSTM with attention.

CNN-Kim: Convolutional and max-over-time pooling layers are applied to the word vectors for a post in this method similar to [15].

C-biLSTM: This is a variant of the C-LSTM architecture [30] somewhat related to a method used in [13]. After applying convolution on a post's word vectors, the feature maps are stacked along the filter dimension to generate a series of window vectors, which are subsequently fed to biLSTM.

Semi-supervised

tBERT-biLSTM-Attention: Same as BERT-biLSTM-Attention except that the pre-trained BERT model used is fine-tuned using unlabeled instances of sexism [17].

Opti-DL: The same as the baseline model described in Sect. 3.2 achieving the state-of-the-art results [17]

Random Sampling + Opti-DL: We randomly pick the same number of samples as those generated by our best method from the unlabelled data, label them using the baseline model, expand the original labeled set with them, and train Opti-DL on the expanded set.

4.4   Results

Table 3 provides the results produced by various proposed semi-supervised methods and the baselines. We set aside 15% from original labeled data for validation and testing each. The validation set was merged into the training set during the testing phase. For the semi-supervised methods augmenting the original labeled data too, the baseline classifier was trained on the union of the augmented data and validation set. For each deep learning method, for each metric, the mean of the results obtained over three runs is given. For each proposed method, data augmentation is carried out over 3 iterations. Out of the cumulative datasets at the ends of these 3 iterations, the one resulting in the highest F_{mac} on the validation set is picked as the final augmented set.

Among the combinations of classifiers and features experimented with for traditional machine learning, ELMo embeddings with logistic regression yields the best results. The best deep learning baseline is Hierarchical-biLSTM-Attention, and it outperforms its traditional ML counterpart. Overall, the semi-supervised Opti-DL [17] method emerges as the best baseline. Adding randomly sampled unlabeled instances (labeled using the baseline model described in Sect. 3.2) to Opti-DL worsens its performance to a degree.

Several proposed methods outperform all baselines across all metrics. The maximum performance improvement is observed for subset accuracy, which is the most stringent metric. Our best-performing method for data augmentation is Diversity.label ∩ support.weakest, which prioritizes samples that are the most distinct compared to the existing labeled data and have the the lowest-support weak (low-coverage) predicted label sets. Among the methods that seek greater textual diversity, Diversity.uniform produces the best results for most metrics.

Table 3. Results for the proposed semi-supervised methods as well as traditional machine learning (TML), deep learning (DL), and semi-supervised baselines

	Approach	F_ins	F_mac	F_mic	Acc	SA
TML baselines	Word-ngrams-SVM	0.453	0.227	0.413	0.331	0.107
	Word-ngrams-LR	0.544	0.188	0.492	0.454	0.287
	Word-ngrams-RF	0.538	0.246	0.482	0.444	0.272
	ELMO-SVM	0.546	0.261	0.501	0.431	0.206
	ELMO-LR	0.576	0.261	0.535	0.475	0.279
	ELMO-RF	0.374	0.100	0.330	0.307	0.185
DL baselines	biLSTM	0.627	0.451	0.577	0.472	0.147
	biLSTM-Attention	0.648	0.445	0.597	0.499	0.176
	Hierarchical-biLSTM-Attention	0.664	0.485	0.616	0.516	0.191
	BERT-biLSTM-Attention	0.591	0.397	0.546	0.431	0.089
	USE-biLSTM-Attention	0.566	0.398	0.525	0.402	0.061
	CNN-biLSTM-Attention	0.421	0.284	0.387	0.278	0.035
	CNN-Kim	0.658	0.481	0.617	0.513	0.195
	C-bilstm	0.485	0.316	0.446	0.328	0.038
Semi-supervised baselines	tBERT-biLSTM-Attention	0.632	0.435	0.580	0.472	0.127
	Opti-DL	0.714	0.546	0.665	0.572	0.242
	Random Sampling + Opti-DL	0.712	0.530	0.655	0.570	0.248
Proposed semi-supervised methods	Basic	0.721	0.539	0.669	0.582	0.257
	IPSPC	0.723	0.540	0.673	0.588	0.276
	Diversity.label	0.726	0.552	0.679	0.594	0.288
	Diversity.uniform	0.733	0.545	0.686	0.603	0.304
	Support.weakest	0.726	0.544	0.678	0.591	0.279
	Support.uniform	0.735	0.547	0.684	0.603	0.299
	S(Diversity.label, Support.uniform)	0.732	0.544	0.683	0.601	0.302
	S(Diversity.uniform, Support.weakest)	0.727	0.553	0.677	0.593	0.287
	Diversity.label ∩ Support.uniform	0.734	0.551	0.686	0.606	0.315
	Diversity.label ∩ Support.weakest	**0.738**	**0.554**	**0.691**	**0.610**	**0.316**

Among the methods favoring low-support labels, Support.uniform produces the best results for most metrics. From all four (2X2) combinations involving our diversity-based and support-based methods through the computation of combined scores and intersection each, we report the two best-performing combinations for each.

We analyze the performance of our best-performing semi-supervised method for augmenting the labeled data across different neural baseline classifiers. Figure 1 depicts the F_{ins} and F_{mac} for four supervised baselines (that use only labeled data). For each of them, we also show the performance of the variant of the best proposed semi-supervised method using it as the baseline classifier. The figure demonstrates the relative efficacy of our method across different baseline classifiers.

Figure 2 highlights the coverage of positive and negative samples for each of the 23 labels in the original labeled data and contrasts it against the improved

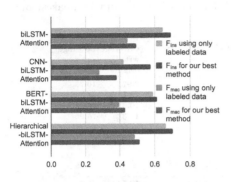
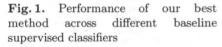

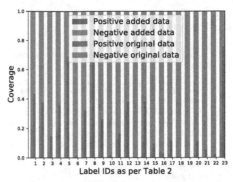

Fig. 1. Performance of our best method across different baseline supervised classifiers

Fig. 2. Coverage of positive and negative samples per label for original and added data

positive sample proportion in the data contributed (added data) by our best-performing model (*Diversity.label* ∩ *support.weakest*). The ratio of the standard deviation and mean for the (positive) label coverage for the original data is 1.074, whereas its added data counterpart is 1.014, indicating that our method also reduces the class imbalance to a degree.

Table 4 shows labeled instances of sexism from the test set for which our best method makes all correct predictions but the best baseline (Opti-DL) does not. It also provides the average cosine similarity scores w.r.t. the original labeled data and the new pseudo-labeled data produced by our best-proposed method, computed using vector representations of the posts given by the baseline model. We also report the per-label coverage, defined as the fraction of samples bearing the label, for the two sets. The higher coverage values seen for our approach could partly account for its state-of-the-art performance.

For each method, all the hyper-parameter tuning is done using the validation data. The hyper-parameters values for our proposed semi-supervised approaches are as follows. P_{min} and P_{max} are set to 4 and 7 respectively. The minimum labeled data coverage (S_m) is set to 1300 by observing the class distribution in the original labeled data. For each method, we pick the optimal confidence threshold T and Top_p based on F_{macro} on the validation set. For Diversity.label ∩ support.weakest, our best-performing method, these values are 0.75 and 0.9, respectively.

The amount of pseudo-labeled data chosen to be added by each of our methods varies according to hyper-parameter values and the iteration (stage) where the best performance on the validation set is observed. For the best proposed approach (Diversity.label ∩ support.weakest), the highest F_{macro} is seen at the third iteration, and the corresponding data generated cumulatively till that point amounts to 4632 samples, resulting in the augmented dataset consisting of 17655 labeled samples. The added data sizes for the other proposed methods at the same iteration with comparable hyper-parameter configuration range 4 K to 6 K.

Table 4. Test samples correctly classified with Diversity.label ∩ support.weakest but not with Opti-DL

Account of sexism	Categories	Similarity to original	Similarity to new	Coverage wrt original	Coverage wrt new
Every time I mention that I do not want children, I am either told that 'It's different when they're your own', or 'You just need to find the right man'	Motherhood-related discrimination, Role stereotyping, Moral policing (excluding tone policing), Other	0.209	0.231	2.97, 19.74, 10.32, 15.3	10.75, 43.5, 38.08, 75.95
A colleague of my boyfriend's asked him what he was up to at the weekend. He replied, 'My girlfriend's coming over and I'm going to cook us a nice meal.' His colleague responded with 'You're COOKING for your GIRLFRIEND??? I'm sorry, mate, but that's REALLY gay.'	Role stereotyping, Attribute stereotyping, Hostile work environment (excluding pay gap), Moral policing (excluding tone policing)	0.297	0.309	19.74, 21.15, 19.95, 10.32	43.5, 37.39, 31.93, 38.08
After a session at the gym, I was told by a colleague that girls shouldn't have muscles, because it's ugly. Apparently, girls should have lady lumps	Attribute stereo-typing, Hostile work environment (excluding pay gap), Body shaming	0.219	0.22	21.15, 19.95, 4.25	37.39, 31.93, 14.79
I just got followed and harassed by a group of men in a car (I was on my bike). Felt shitty, and I was afraid. I managed to ditch them at the traffic lights	Threats, Sexual harassment (excluding assault)	0.253	0.255	3.93, 32.25	26.12, 38.23

Our best method produces less data compared to some other proposed methods, confirming the importance of keeping the quality of the pseudo-labeled set high through effective sample selection methods and other mechanisms.

5   Conclusion

We investigated semi-supervised learning for the fine-grained classification of accounts of sexism using 23 categories of sexism. We proposed a set of methods based on self-training, designed for the multi-label formulation, for capitalizing on unlabeled instances of sexism. Most of our methods perform superior to a variety of baselines across many standard metrics, especially subset accuracy, the strictest metric. Our best-performing method for augmenting the labeled data with pseudo-labeled samples picked from unlabeled data seeks to enhance textual diversity and improve class imbalance.

References

1. Abney, S.: Semisupervised Learning for Computational Linguistics. Chapman and Hall/CRC, Boca Raton (2007)
2. Agrawal, S., Awekar, A.: Deep learning for detecting cyberbullying across multiple social media platforms. In: Pasi, G., Piwowarski, B., Azzopardi, L., Hanbury, A. (eds.) ECIR 2018. LNCS, vol. 10772, pp. 141–153. Springer, Cham (2018). https://doi.org/10.1007/978-3-319-76941-7_11
3. Anzovino, M., Fersini, E., Rosso, P.: Automatic identification and classification of misogynistic language on twitter. In: Silberztein, M., Atigui, F., Kornyshova, E., Métais, E., Meziane, F. (eds.) NLDB 2018. LNCS, vol. 10859, pp. 57–64. Springer, Cham (2018). https://doi.org/10.1007/978-3-319-91947-8_6
4. Badjatiya, P., Gupta, S., Gupta, M., Varma, V.: Deep learning for hate speech detection in tweets. In: Proceedings of the 26th International Conference on World Wide Web Companion, pp. 759–760. International World Wide Web Conferences Steering Committee (2017)
5. Cer, D., et al.: Universal sentence encoder (2018). arXiv preprint arXiv:1803.11175
6. Chowdhury, A.G., Sawhney, R., Shah, R., Mahata, D.: #YouToo? Detection of personal recollections of sexual harassment on social media. In: Proceedings of the 57th Annual Meeting of the Association for Computational Linguistics, pp. 2527–2537 (2019)
7. Davidson, T., Warmsley, D., Macy, M., Weber, I.: Automated hate speech detection and the problem of offensive language. In: Eleventh International AAAI Conference on Web and Social Media (2017)
8. Devlin, J., Chang, M.W., Lee, K., Toutanova, K.: Bert: pre-training of deep bidirectional transformers for language understanding (2018). arXiv preprint arXiv:1810.04805
9. ElSherief, M., Belding, E., Nguyen, D.: #NotOkay: understanding gender-based violence in social media. In: Eleventh International AAAI Conference on Web and Social Media (2017)
10. Frenda, S., Ghanem, B., Montes-y Gómez, M., Rosso, P.: Online hate speech against women: automatic identification of misogyny and sexism on twitter. J. Intell. Fuzzy Syst. **36**(5), 4743–4752 (2019)
11. Jafarpour, B., Matwin, S., et al.: Boosting text classification performance on sexist tweets by text augmentation and text generation using a combination of knowledge graphs. In: Proceedings of the 2nd Workshop on Abusive Language Online (ALW2). pp. 107–114 (2018)

12. Jha, A., Mamidi, R.: When does a compliment become sexist? Analysis and classification of ambivalent sexism using twitter data. In: Proceedings of the Second Workshop on NLP and Computational Social Science, pp. 7–16 (2017)
13. Karlekar, S., Bansal, M.: Safecity: understanding diverse forms of sexual harassment personal stories. In: Proceedings of the 2017 Conference on Empirical Methods in Natural Language Processing (EMNLP), pp. 2805–2811 (2018)
14. Khatua, A., Cambria, E., Khatua, A.: Sounds of silence breakers: exploring sexual violence on twitter. In: 2018 IEEE/ACM International Conference on Advances in Social Networks Analysis and Mining (ASONAM), pp. 397–400 (2018)
15. Kim, Y.: Convolutional neural networks for sentence classification. In: Proceedings of the 2017 Conference on Empirical Methods in Natural Language Processing (EMNLP), pp. 1746–1751 (2014)
16. Melville, S., Eccles, K., Yasseri, T.: Topic modelling of everyday sexism project entries. Front. Dig. Hum. **5**, 28 (2018)
17. Parikh, P., et al.: Multi-label categorization of accounts of sexism using a neural framework. In: Proceedings of the 2019 Conference on Empirical Methods in Natural Language Processing and the 9th International Joint Conference on Natural Language Processing (EMNLP-IJCNLP), pp. 1642–1652 (2019)
18. Pennebaker, J.W., Boyd, R.L., Jordan, K., Blackburn, K.: The development and psychometric properties of LIWC2015. Technical report (2015)
19. Pennington, J., Socher, R., Manning, C.: GloVe: global vectors for word representation. In: Proceedings of the 2014 Conference on Empirical Methods in Natural Language Processing (EMNLP), pp. 1532–1543 (2014)
20. Peters, M.E., et al.: Deep contextualized word representations. In: Proceedings of NAACL (2018)
21. Van Hee, C., et al.: Detection and fine-grained classification of cyberbullying events. In: Proceedings of the International Conference Recent Advances in Natural Language Processing, pp. 672–680 (2015)
22. Wang, J., Yu, L.C., Lai, K.R., Zhang, X.: Dimensional sentiment analysis using a regional CNN-LSTM model. In: Proceedings of the 54th Annual Meeting of the Association for Computational Linguistics (Volume 2: Short Papers), vol. 2, pp. 225–230 (2016)
23. Waseem, Z., Hovy, D.: Hateful symbols or hateful people? Predictive features for hate speech detection on twitter. In: Proceedings of the NAACL Student Research Workshop, pp. 88–93 (2016)
24. Xiao, H.: bert-as-service (2018). https://github.com/hanxiao/bert-as-service
25. Yan, P., Li, L., Chen, W., Zeng, D.: Quantum-inspired density matrix encoder for sexual harassment personal stories classification. In: 2019 IEEE International Conference on Intelligence and Security Informatics (ISI), pp. 218–220. IEEE (2019)
26. Yang, Z., Yang, D., Dyer, C., He, X., Smola, A., Hovy, E.: Hierarchical attention networks for document classification. In: Proceedings of the 2016 Conference of the North American Chapter of the Association for Computational Linguistics: Human Language Technologies, pp. 1480–1489 (2016)
27. Yarowsky, D.: Unsupervised word sense disambiguation rivaling supervised methods. In: 33rd Annual Meeting of the Association for Computational Linguistics, pp. 189–196 (1995)

28. Zhang, M.L., Zhou, Z.H.: A review on multi-label learning algorithms. IEEE Trans. Knowl. Data Eng. **26**(8), 1819–1837 (2014)
29. Zhang, Z., Luo, L.: Hate speech detection: a solved problem? The challenging case of long tail on twitter. In: Semantic Web, pp. 1–21 (2018)
30. Zhou, C., Sun, C., Liu, Z., Lau, F.: A C-LSTM neural network for text classification (2015). arXiv preprint arXiv:1511.08630

aDFR: An Attention-Based Deep Learning Model for Flight Ranking

Yuan Yi[1], Jian Cao[1(✉)], YuDong Tan[2], QiangQiang Nie[2], and XiaoXi Lu[2]

[1] Department of Computer Science and Engineering, Shanghai Jiao Tong University,
Shanghai, China
{candylab,cao-jian}@sjtu.edu.cn
[2] Trip.com Group Ltd. Corporation, Singapore, Singapore
{ydtan,qqnie}@trip.com

Abstract. Although there are numerous available flights for travelers on the online ticket reservation platforms after entering a query, they often make decisions only based on the information of a few flights that are ranked high on the list. Therefore, it has become an important strategy for online travel agencies to rank flights according to their potential popularity. However, to build a ranking function for flights is not an easy task because the preferences of passengers vary from flight routes and change with time. In addition, the selection of a flight is affected by the information of other flights in the same returned list. Traditional Learning to Rank (L2R) methods fail to model the distinct preferences of passengers adaptively on each query. They are also insufficient to capture the complex relationships among flights in the same list. To cope with these challenges, we design an attention-based deep neural network for flight ranking. It adopts two kinds of attention mechanisms to model the dynamic preferences of passengers and the relevance between flights respectively. Extensive experiments on real-world flight order datasets demonstrate the superiority of our method against other competitive ones.

Keywords: Flight ranking · Learning to Rank · Deep learning · Attention mechanism

1 Introduction

Today, more and more people book flights from online travel agencies (OTA). On the flight ticket booking platforms of OTAs, after providing the departure city, arrival city, and departure date, passengers can select flights from the returned flight list. It is noticed that most people look for flights only from the ones ranked in the front of the returned list. However, there are usually numerous available flights in the returned list occupying multiple pages. When people can not find an appropriate flight on the first few pages, they may seek other OTAs for help. Therefore, OTAs are motivated to optimize the ranking results of flights to

© Springer Nature Switzerland AG 2020
Z. Huang et al. (Eds.): WISE 2020, LNCS 12343, pp. 548–562, 2020.
https://doi.org/10.1007/978-3-030-62008-0_38

prevent passengers from being overwhelmed by enormous options, consequently increasing their revenues.

Intuitively, flight searching can be regarded as an information retrieval task, which is traditionally solved by a *Learning to Rank* (L2R) framework, which has been widely studied and successfully applied to ranking problems. The conventional method of L2R is to learn a global ranking function from labeled query results. Unfortunately, the passengers' preferences for flight booking on different routes are heterogeneous. For instance, on short-distance routes (flight duration is less than 6 h), around 90% of passengers will choose direct flights rather than transit flights. Contrarily, above 58% of passengers select transit flights on long-distance routes (flight duration is more than 6 h). Besides, the distribution of passenger types changes violently on different routes. For example, on the Beijing-Hongkong route, passengers are principally business travelers with lower price sensitivity. While on the Qingdao-Seoul route, vacation travelers account for a high proportion, for whom price plays a significant role in the process of choosing flights. The preference of passengers for flight booking changes with the availability of air tickets, ticket prices, and the departure time. Furthermore, travelers make decisions by comparing the flights in the returned list rather than looking into one flight without considering the advantages of the others. Although pairwise and listwise methods of L2R model the relative relevance between items by optimizing the loss function defined on item pairs and item lists respectively, they fail to model the variability of passenger preferences, as well as the feature interactions among items. In recent years, some models based on RNN [1, 2, 38] have better performance in capturing the mutual influence between a sequence of items. However, they are inadequate to model the dynamic interactions among a long list of items.

To further resolve the difficulties above, we propose an attention-based deep learning model for flight ranking named aDFR in this paper. aDFR adopts attention mechanism, which can dynamically allocate weights to different aspects of the input [3, 29, 33]. We divide the input for flight ranking into different categories, such as price-related features, query-related features, flight attribute-related features, etc. Then the importance of each category is calculated via a multi-layer perceptron attention algorithm. According to these importance scores, weighted summation is performed to get a fair evaluation. Furthermore, the output of this layer is feed to a self-attention based layer that inspired by [24, 34] to compute the mutual influence between flights in the list. Finally, each flight is scored through a fully connected layers. To optimize the proposed model, we define a mixed ranking loss that combines pointwise ranking loss and pairwise ranking loss to pursue a better performance [15]. In general, the key contributions of this paper can be summarized as follows:

- We design a deep neural network aDFR to tackle the problem of flight ranking.
- We introduce an attention layer to dynamically calculate passengers' attention to different categories of features for flight ranking on different routes.
- We model the mutual influence between flights through a self-attention based layer.

We conduct extensive experiments on three real-world datasets, which demonstrate the effectiveness of our model against other competitive models for flight list optimization.

The rest of the paper is organized as follows. In Sect. 2, we discuss the related work. In Sect. 3, we elaborate on the proposed method, which handles the flight ranking problem in Learning to Rank framework. Section 4 illustrates the datasets and experiments respectively. The conclusion is presented in Sect. 5.

2   Related Work

L2R algorithms are applied to ranking problems and most of them are based on deep neural networks recently [1,13,15,29]. In deep learning models, there are different components designed for specific functions, where the attention layer is designed to capture unique interaction in the input and demonstrates outstanding performances [3,12,14,29,33].

2.1   Learning to Rank, L2R

From statistical machine learning to deep learning methods, research on L2R has experienced a long period of development and these methods can be categorized into three major categories, i.e., pointwise, pairwise, and listwise [9,23]. The pointwise approaches treat all items as independent instances. One of the most popular pointwise methods is the logistic regression (LR) [16], which predicts the probability of an event, such as a click on the item, and then rank the items according to event probability values. Another well-known model is the gradient boosting decision tree (GBDT) [22]. The combination method of LR and GBDT [18] is widely adopted in practical application scenarios. Although the pointwise methods have achieved some progress, they ignore the order of items in the list and may distort the true distribution of the data. The pairwise approaches attempt to correctly distinguish the relative orders of two items, such as Ranking SVM [20] based on the support vector machine, which adopts hinge loss as the loss function. Additionally, the RankNet [5] proposed by Burges C et al. constructs item pairs as the input instances. Subsequently, they proposed the LambdaRank [6] algorithm, which introduces position-based weights into pairwise loss function and trains the model with a non-smooth cost function. The listwise approaches regard all items in a list as an instance. The typical algorithm includes ListNet [10], ListMLE [32], and LambdaMart [7]. Unfortunately, most of the L2R methods can not learn the preferences of passengers varying from flight routes and changed with time and fail to model the feature relevance explicitly between items.

2.2   Flight Ranking

There are two kinds of research for flight ranking problems i.e., personalized recommendation of flights and query results ranking. The first one targets at recommending k flights that are of most interest to users [8]. Unfortunately, it is challenging for personalized recommendations to achieve satisfying results in practice due to the extremely sparse flight order data. Another more flexible way is to rank searching results using a ranking function. For example, one approach proposed by Gaby Budel et al. in 2018 [4] handles the flight ranking problem with a Ranking SVM model. Our model is also for flight ranking. Comparing with their model, our model considers the heterogeneity of passengers' preferences on different flight routes.

2.3   Attention Mechanism

As a technique of performing a weighted summation of various parts of the inputs, the attention mechanism has made excellent advancements in natural language processing [3,26], computer vision [30,35,37], and recommendation [11,27,36] field. Moreover, the self-attention proposed by Google has achieved great success in machine translation scenarios [28], which can assign different weights to each part of the object itself without any external information. Afterward, the self-attention mechanism have been made new progress in the field of computer vision for obtaining long-range dependencies by calculating the similarity between any two positions [31].

Although the pairwise methods and listwise methods of L2R consider the relative orders of items, they fail to dynamically model the difference in feature distributions of different routes and neglect the complex mutual influence between items in the feature space. Taking the advantages of the previous methods into account, we adopt the attention mechanism and self-attention mechanism in aDFR.

3   aDFR: A Attention-Based Deep Learning Model for Flight Ranking

3.1   Problem Formulation

The model in this paper aims to learn an adaptive ranking function based on attention mechanism to optimize the ranking quality of the result list of flights under each query condition. Let $Q = [q_1, q_2, ...q_n]$ as a query set, $Q \in \mathbb{R}^{n \times k}$, and $V_i = [v_i^{(1)}, v_i^{(2)}, ...v_i^{(m)}]$ denotes a group of flights under the i-th query condition, where n is the number of query conditions, m represents the number of flights under each query, and $v_i^{(j)} \in \mathbb{R}^d$ indicates the j th flight of the i-th query. In order to identify passengers' flight preferences, we combine query condition and query results, so that we get a new input instance in the form of a matrix $X^{(j)} = [x_1^{(j)}, x_2^{(j)}, ...x_m^{(j)}], x_i = [q^{(j)}; v_i^{(j)}], x_i^{(j)} \in \mathbb{R}^{k+d}, X^{(j)} \in \mathbb{R}^{m \times (k+d)}$. After that, these features are divided into different , i.e., price-related features $P^{(j)} = [p_1^{(j)}, p_2^{(j)}, ...p_m^{(j)}]$, take-off time-related features $T^{(j)} = [t_1^{(j)}, t_2^{(j)}, ...t_m^{(j)}]$, flight attribute features $A^{(j)} = [a_1^{(j)}, a_2^{(j)}, ...a_m^{(j)}]$, take-off date-related features $D^{(j)} = [d_1^{(j)}, d_2^{(j)}, ...d_m^{(j)}]$ etc. Our model takes these features as the input.

3.2   aDFR Model

Figure 1 illustrates the architecture of aDFR. As shown in the figure, the model is mainly composed of four parts: 1) Feature embedding layer helps to embed discrete features; 2) Attention-aware aspect weighted module aims to dynamically combine different category features; 3) Relation attention layer models the mutual influence between flights; 4) The ranking layer is responsible for assigning a score to each flight. In this section, we will elaborate on how they are constructed in detail.

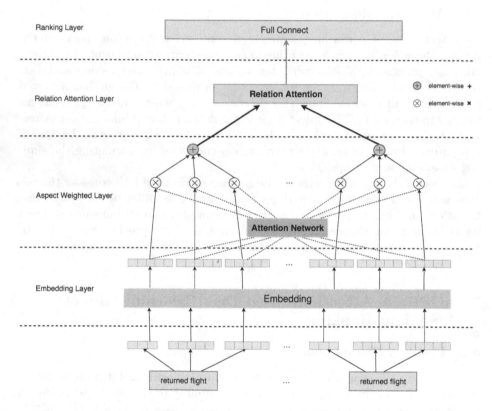

Fig. 1. The structure of aDFR

Feature Embedding Layer. Features can be typically classified into discrete features and continuous features. Since the original discrete features (ID features) have poor representation capabilities and will lead the network to learn biased representations of ID features, we utilize the embedding method in natural language processing to embed discrete features into a continuous space. Typically, we concatenate the embeddings and the normalized continuous features to get the final representation. Another dilemma is that the feature dimensions of different categories are inconsistent, and it is difficult for us to perform weighted summation directly based on its attention weight. We pad each part to align the feature dimensions to handle the dilemma.

Attention-Aware Aspect Weighted Layer. In order to adaptively estimate the passengers' preferences under each query condition, we employ a multi-layer perceptron to model the importance of the features of each category and then utilize the softmax function to calculate the normalized weights. Subsequently, we compute the weighted sum of each category according to the normalized weights as an output vector $f^{(i)}$. In a more specific way, assume $QX = [G^{(1)}, G^{(2)}, ..., G^{(m)}]$ is the feature space of a group of flights that are embedded by the feature embedding layer. $G^{(i)}$ is a set of vectors, which consists of the feature vectors of each category of the i-th flight, $G^{(i)} = [g_1^{(i)}, g_2^{(i)}, ..., g_t^{(i)}], g_j^{(i)} \in \mathbb{R}^s$, where $g_j^{(i)}$ represents the features of the j-th category of the i-th flight. As illustrated in Fig. 2, the attention mechanism in this module can be formulated as follows:

$$u_j^{(i)} = h^T Relu(W g_j^{(i)} + b)$$

$$\alpha_j^{(i)} = \frac{exp(u_j^{(i)})}{\sum_{j=1}^t exp(u_j^{(i)})} \tag{1}$$

$$f^{(i)} = \sum_{j=1}^t \alpha_j^{(i)} g_j^{(i)}$$

and $h \in \mathbb{R}^a, W \in \mathbb{R}^{a \times s}, f^{(i)} \in \mathbb{R}^s, b \in \mathbb{R}^a$. t represents the number of categories, a denotes attention size. W and b are the parameters of attention layer that are dynamically learned during the training process. The final output is $F = [f^{(1)}, f^{(2)}, ..., f^{(m)}], F \in \mathbb{R}^{m \times s}$.

In particular, the ReLU function in the formula above can be substituted by other activation functions. We have tried Tanh, Sigmoid, ReLU, and PReLU, where PReLU with dynamic coefficients set in the positive part of the ReLU function [17] presents the best performance. The target of this layer is to combine multiple sets of features fed by the embedding layer according to weights learned in this layer and input the weighted result to the next layer.

Relation Attention Layer. We introduce a self-attention based network to capture the mutual influence between flights as mentioned before. The self-attention approach we implement here is similar to the Dot product method of Non-local Neural Network [31]. The same implementation is employed in SAGAN [34]. Relation attention layer allocates mutual influence weights based on the similarity between flights. Comparing with the attention mechanism of the previous section, self-attention mechanism in this section aims to model the relevance (mutual influence) between each flight in the same list, while the former is to combine multiple parts of a flight according to their importance. Given

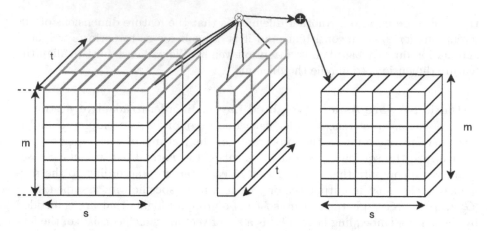

Fig. 2. Graphical representation of aspect weighted layer

an input feature matrix F, indicating a set of flights under a query condition. $F \in \mathbb{R}^{m \times h}$, where m is the number of flights under a query condition, and h represents the output dimension of the previous layer. The calculation of mutual influence weights between flights is defined as follows:

$$
\begin{aligned}
Q &= PRelu(FA^T + a) \\
K &= PRelu(FB^T + b) \\
V &= PRelu(FC^T + c) \\
S &= Q \cdot K^T \\
W &= softmax(S) \\
O &= \beta W \cdot V + F
\end{aligned}
\tag{2}
$$

A, B, C and a, b, c are the weight matrices and biases of F respectively. $A \in \mathbb{R}^{z \times h}$, $B \in \mathbb{R}^{z \times h}$, $C \in \mathbb{R}^{m \times h}$, $a \in \mathbb{R}^{m \times z}$, $b \in \mathbb{R}^{m \times z}$, $c \in \mathbb{R}^{m \times h}$, z is attention size, Q, K, V are nonlinear mappings of the input feature space all of which come from the feature space F, where Q, K are used to compute the attention matrix W. It is worth to mention that β is a parameter of the network, balancing the proportion of the weighted result WV.

Ranking Layer. In the field of advertising, click-through rate is an important indicator to evaluate the effectiveness. Whereas in the case of air ticket sales, we pay more attention to successful transactions, and the values of user click behavior are limited. Based on this observation, our model is committed to placing flights that users are more inclined to make a reservation to the front of the list. The number of orders of each flight in a day is regarded as its ground-truth. Inspired by [15], we adopt a mixed ranking-aware loss function that combines a pointwise loss and a pairwise loss. Consider that $y = [r_1, r_2, ... r_m]$ is the ground-truth of the flight list under a query and $\hat{y} = [\hat{r_1}, \hat{r_2}, ... \hat{r_m}]$ denotes the ranking score predicted by the network. The loss function of our model is formulated as follows:

$$L(y, \hat{y}) = \frac{1}{m}(\sum_{i=0}^{m}(y - \hat{y})^2 + \eta \sum_{i=0}^{m} \sum_{j=0}^{m} max(0, -(r_i - r_j)(\hat{r_i} - \hat{r_j}))) \quad (3)$$

where the first term is the mean square error (MSE), regularly used in regression prediction problems. And the second term is the Hinge loss function, which is commonly used in the maximum-margin algorithm, such as SVM. η is a hyper-parameter to adjust the two terms. Since aDFR takes the ranked lists as instances and defines the ranking loss on the lists, aDFR belongs to listwise approaches rather than pairwise approaches.

4 Experiments

In this section, we introduce the experiments performed on three real-world flight order datasets obtained from an OTA to evaluate the performance of our model.

4.1 Experimental Setting

Datasets. We select flight order data from November 26, 2019 to January 27, 2020 of 43 popular international routes. Consequently, we take the number of successful transaction orders for a flight in a day as its ground-truth. According to statistics, the click-through rate of the top 20 items in the returned lists exceeds 98% in the ticket sales scenario. Therefore, we collected the information of the first 20 flights per query to construct the dataset. The flights under a query will be fed to the network as a sample. Additionally, considering that the ticket price fluctuates up and down throughout a day, we take the minimum price and the median price of the flight tickets in one day as price features. In order to verify the effectiveness of the relation attention layer, we carry out additional experiments on two typical routes(city pairs). Precisely, the datasets used in the experiment are as follows (Table 1):

Table 1. Statistics of datasets used in our experiments.

Datasets	Routes			Queries		
	Train	Validation	Evaluation	Train	Validation	Evaluation
Beijing-Seoul	1	1	1	2471	144	103
Shanghai-Hongkong	1	1	1	2248	136	117
Whole Dataset	43	43	43	54968	3269	2803

- **Beijing-Seoul:** It contains two months of flight data from Beijing to Seoul.
- **Shanghai-Hongkong:** It contains two months of flight data from Shanghai to Hong Kong.
- **Whole Dataset:** This dataset includes all data of 43 popular routes within two months.

The three datasets include the attributes of flights each day and the number of orders. The queries that return less than 20 flights and those that have less than one successful transaction record in the whole day are removed from the dataset.

Baselines. To validate the performance of our approach, we generally apply the following baseline models.

- **LambdaMART:** This method is regarded as one of the state-of-art Learning to Rank algorithm [1], and it is the LambdaRank algorithm implemented with Gradient Boosting Decision Tree.
- **MLP:** It represents Multi-layer Perceptron that is trained with a ranking-aware loss function illustrated in Sect. 3.2.
- **Attention MLP:** In this model, a relation attention layer is added on the basis of the MLP approach.
- **Aspect MLP:** It combines the attention-aware aspect weighted layer we proposed in our work with basic MLP model.

The first model is set for the controlled experiment. The last three models which have the same hyper-parameters with aDFR are used to evaluate the effectiveness of the relation attention layer and attention-aware aspect weighted layer.

Evaluation Metrics

- **HR@K:** HR@K, named hit ratio, is a commonly accepted indicator to measure the recall rate, which is defined as below:

$$HR@K = \frac{HitAmount@K}{TotalAmount@K} \tag{4}$$

- **NDCG@K:** Normalized Discounted Cumulative Gain, which can be form as follows:

$$DCG@K = \sum_{i=1}^{k} \frac{2^{rel_i} - 1}{\log_2 (i + 1)}$$
$$NDCG@K = \frac{DCG@K}{IDCG@K} \tag{5}$$

where $IDCG@K$ denotes the value of DCG under ideal situations, in other words, if the predicted sorting result is the same as the real sorting result, then DCG is equal to IDCG.

- **OR@K:** This evaluation metric is particularly designed for the flight ranking problem, which means under N queries, the mean percentage of the successful transactions of top K flights in the predicted ranking results to the total amount of orders in the returned list of each query.

$$OR@K = \frac{1}{N} \sum_{i=1}^{N} \frac{OrderAmount@K}{TotalOrderAmountofQuery_i} \qquad (6)$$

Parameter Settings. The ranking layer consists of three fully connected layers. We use the regularization technique Dropout [25] to prevent the model from overfitting, where the dropout ratio is set to 0.5. The normalization technique Batchnormm [19] is applied to improving the stability of the network. The Adam optimization algorithm [21] is adopted during training with the batch size 20, the weight decay coefficient of the optimizer is set to 0.01 to balance the influence of model complexity on the cost function, and the initial learning rate is set to 0.0011. Lastly, the reduced learning rate on plateau approach is applied as the learning rate scheduler.

4.2 Performance Comparison

In this section, we discuss the performance of our model and conduct a detailed analysis of the experimental results which show the improvements of our model compared with the existing L2R methods.

Fig. 3. Performance comparison on order ratio (measured by OR@K)

Overall Performance. aDFR has the best performance in the evaluation criterion of the order ratio which is the most concerned evaluation metrics in practice. In terms of the ranking quality measure, NDCG, our model surpasses all other competitive models with 10.07% average improvements over the LambdaMART and 14.93% increase over the MLP on average. Figure 3 and Table 2 show that:

- In these models, MLP shows the worst performance on various indicators since simply superimposing multiple linear layers cannot capture highly nonlinear information. Meanwhile, it shows those approaches that rank the query results by statically learning a global ranking function are not applicable to different routes that are highly heterogeneous in feature distribution.
- The Attention MLP also exceeds the state-of-art model LambdaMART on the basis of order ratio and achieves similar results in NDCG, which demonstrates that we can calculate the interactions between various objects through a self-attention based layer.
- The Aspect MLP outperforms the other baseline models in most cases, especially in ranking related metrics, which infers that the attention-aware aspect weighted layer we propose can successfully capture potential information on each route. Whereas it lacks the capacity to capture the mutual influence between flights, resulting in comparatively worse performance in terms of order ratio.
- aDFR presents the best performance in terms of transaction-related evaluation metric OR@5. In Fig. 4, we found that aDFR still achieves a 2.59% increase in OR@5 compared with Aspect MLP.

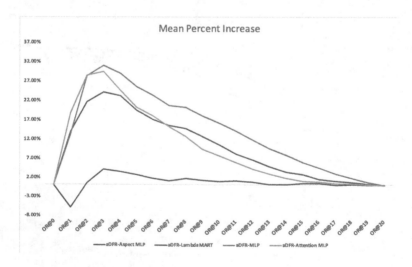

Fig. 4. Mean percent increase from baselines

Table 2. Overall performance on whole dataset.

Models	OR@5	OR@1	HR@5	NDCG@20	NDCG@10	NDCG@5
LambdaMART	0.5684	0.1594	0.5694	0.6152	0.6763	0.7571
MLP	0.5355	0.1506	0.5517	0.5839	0.6649	0.7586
Attention MLP	0.5648	0.1529	0.5893	0.6127	0.6986	0.7825
Aspect MLP	0.6614	**0.1930**	**0.6524**	**0.7385**	**0.7984**	**0.8626**
aDFR	**0.6785**	0.1815	0.5657	0.7186	0.7647	0.8244

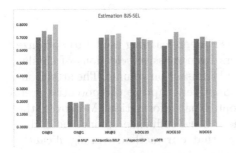

Fig. 5. Offline evaluation metric of Beijing to Seoul

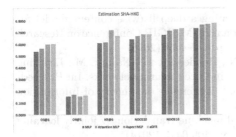

Fig. 7. Offline evaluation metric of Shanghai to Hongkong

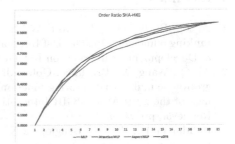

Fig. 6. Performance comparison on order ratio (Beijing to Seoul)

Fig. 8. Performance comparison on order ratio (Shanghai to Hongkong)

Impacts of Attention-Aware Aspect Weighted Layer and Relation Attention Layer. To elaborate precisely on the attention-aware aspect weighted layer and relation attention layer, we conduct ablation studies on the datasets of two separate routes and compare the results of them, from Figs. 5, 6, 7 and 8 we can find that:

– In the evaluation metric of order ratio OR@K, the models with an attention-aware aspect weighted layer exceed those without and MLP is inferior to other models that applied the attention layers. Specifically, our model aDFR outperforms other comparative ones in the dataset of Beijing-Seoul, whereas the Aspect MLP which without the relation attention layer shows the best performance in the dataset of Shanghai-Hongkong. It implies that our model aDFR possesses greater advantages in multi-route datasets rather than one specific route.

– In terms of ranking-related evaluation indicators, aDFR surpasses those without attention-aware aspect weighted layer and increases the successful transactions without grievous loss of the ranking performance. It further proves that the attention-aware aspect weighted layer can capture the attention that travelers pay on flight features.

5   Conclusions

In this paper, we introduce an attention-based deep learning model to optimize the flight list, which aims to free customers from massive selections of similar flights and also tries to enhance the successful transaction volume. The attention-aware aspect weighted layer is proved to adaptively capture the potential preferences on flight selection according to contextual information. Moreover, the self-attention based module can model the mutual influence among the flights of the returned list. The experiments show that our method can significantly improve the quality of the flight search results.

References

1. Ai, Q., Bi, K., Guo, J., Croft, W.B.: Learning a deep listwise context model for ranking refinement. In: The 41st International ACM SIGIR Conference on Research & Development in Information Retrieval, pp. 135–144 (2018)
2. Ai, Q., Wang, X., Bruch, S., Golbandi, N., Bendersky, M., Najork, M.: Learning groupwise multivariate scoring functions using deep neural networks. In: Proceedings of the 2019 ACM SIGIR International Conference on Theory of Information Retrieval, pp. 85–92 (2019)
3. Bahdanau, D., Cho, K., Bengio, Y.: Neural machine translation by jointly learning to align and translate. arXiv preprint arXiv:1409.0473 (2014)
4. Budel, G., Hoogenboom, L., Kastrop, W., Reniers, N., Frasincar, F.: Predicting user flight preferences in an airline E-shop. In: Mikkonen, T., Klamma, R., Hernández, J. (eds.) ICWE 2018. LNCS, vol. 10845, pp. 245–260. Springer, Cham (2018). https://doi.org/10.1007/978-3-319-91662-0_19
5. Burges, C., et al.: Learning to rank using gradient descent. In: Proceedings of the 22nd International Conference on Machine Learning, pp. 89–96 (2005)
6. Burges, C.J., Ragno, R., Le, Q.V.: Learning to rank with nonsmooth cost functions. In: Advances in Neural Information Processing Systems, pp. 193–200 (2007)
7. Burges, C.J.: From RankNet to LambdaRank to LambdaMART: an overview. Learning 11(23–581), 81 (2010)
8. Cao, J., Yang, F., Xu, Y., Tan, Y., Xiao, Q.: Personalized flight recommendations via paired choice modeling. In: 2017 IEEE International Conference on Big Data (Big Data), pp. 1265–1270. IEEE (2017)
9. Cao, Z., Qin, T., Liu, T.Y., Tsai, M.F., Li, H.: Learning to rank: from pairwise approach to listwise approach. In: International Conference on Machine Learning (2007)

10. Cao, Z., Qin, T., Liu, T.Y., Tsai, M.F., Li, H.: Learning to rank: from pairwise approach to listwise approach. In: Proceedings of the 24th International Conference on Machine Learning, pp. 129–136 (2007)
11. Chen, J., Zhang, H., He, X., Nie, L., Liu, W., Chua, T.S.: Attentive collaborative filtering: multimedia recommendation with item-and component-level attention. In: Proceedings of the 40th International ACM SIGIR Conference on Research and Development in Information Retrieval, pp. 335–344 (2017)
12. Chen, L., et al.: SCA-CNN: spatial and channel-wise attention in convolutional networks for image captioning. In: Proceedings of the IEEE Conference on Computer Vision and Pattern Recognition, pp. 5659–5667 (2017)
13. Cheng, H.T., et al.: Wide & deep learning for recommender systems. In: Proceedings of the 1st Workshop on Deep Learning for Recommender Systems, pp. 7–10 (2016)
14. Cheng, Z., Ding, Y., He, X., Zhu, L., Song, X., Kankanhalli, M.S.: A^3NCF: an adaptive aspect attention model for rating prediction. In: IJCAI, pp. 3748–3754 (2018)
15. Feng, F., He, X., Wang, X., Luo, C., Liu, Y., Chua, T.S.: Temporal relational ranking for stock prediction. ACM Trans. Inf. Syst. (TOIS) **37**(2), 1–30 (2019)
16. Gey, F.C.: Inferring probability of relevance using the method of logistic regression. In: Croft, B.W., van Rijsbergen, C.J. (eds.) SIGIR 1994, pp. 222–231. Springer, London (1994). https://doi.org/10.1007/978-1-4471-2099-5_23
17. He, K., Zhang, X., Ren, S., Sun, J.: Delving deep into rectifiers: surpassing human-level performance on ImageNet classification. In: Proceedings of the IEEE International Conference on Computer Vision, pp. 1026–1034 (2015)
18. He, X., et al.: Practical lessons from predicting clicks on ads at Facebook. In: Proceedings of the Eighth International Workshop on Data Mining for Online Advertising, pp. 1–9 (2014)
19. Ioffe, S., Szegedy, C.: Batch normalization: accelerating deep network training by reducing internal covariate shift. arXiv preprint arXiv:1502.03167 (2015)
20. Joachims, T.: Optimizing search engines using clickthrough data. In: Proceedings of the Eighth ACM SIGKDD International Conference on Knowledge Discovery and Data Mining, pp. 133–142 (2002)
21. Kingma, D.P., Ba, J.: Adam: a method for stochastic optimization. arXiv preprint arXiv:1412.6980 (2014)
22. Li, P., Wu, Q., Burges, C.J.: McRank: learning to rank using multiple classification and gradient boosting. In: Advances in Neural Information Processing Systems, pp. 897–904 (2008)
23. Liu, T.Y., et al.: Learning to rank for information retrieval. Found. Trends® Inf. Retrieval **3**(3), 225–331 (2009)
24. Pei, C., et al.: Personalized re-ranking for recommendation. In: Proceedings of the 13th ACM Conference on Recommender Systems, pp. 3–11 (2019)
25. Srivastava, N., Hinton, G., Krizhevsky, A., Sutskever, I., Salakhutdinov, R.: Dropout: a simple way to prevent neural networks from overfitting. J. Mach. Learn. Res. **15**(1), 1929–1958 (2014)
26. Sutskever, I., Vinyals, O., Le, Q.V.: Sequence to sequence learning with neural networks. In: Advances in Neural Information Processing Systems, pp. 3104–3112 (2014)
27. Tay, Y., Luu, A.T., Hui, S.C.: Multi-pointer co-attention networks for recommendation. In: Proceedings of the 24th ACM SIGKDD International Conference on Knowledge Discovery & Data Mining, pp. 2309–2318 (2018)

28. Vaswani, A., et al.: Attention is all you need. In: Advances in Neural Information Processing Systems, pp. 5998–6008 (2017)
29. Wang, B., Klabjan, D.: An attention-based deep net for learning to rank. arXiv preprint arXiv:1702.06106 (2017)
30. Wang, X., Girshick, R., Gupta, A., He, K.: Non-local neural networks (2017)
31. Wang, X., Girshick, R., Gupta, A., He, K.: Non-local neural networks. In: Proceedings of the IEEE Conference on Computer Vision and Pattern Recognition, pp. 7794–7803 (2018)
32. Xia, F., Liu, T.Y., Wang, J., Zhang, W., Li, H.: Listwise approach to learning to rank: theory and algorithm. In: Proceedings of the 25th International Conference on Machine Learning, pp. 1192–1199 (2008)
33. Ying, H., et al.: Sequential recommender system based on hierarchical attention network. In: IJCAI International Joint Conference on Artificial Intelligence (2018)
34. Zhang, H., Goodfellow, I., Metaxas, D., Odena, A.: Self-attention generative adversarial networks. arXiv preprint arXiv:1805.08318 (2018)
35. Zhao, H., Jia, J., Koltun, V.: Exploring self-attention for image recognition. arXiv preprint arXiv:2004.13621 (2020)
36. Zhou, C., et al.: ATRank: an attention-based user behavior modeling framework for recommendation. In: Thirty-Second AAAI Conference on Artificial Intelligence (2018)
37. Zhu, Z., Xu, M., Bai, S., Huang, T., Bai, X.: Asymmetric non-local neural networks for semantic segmentation. In: The IEEE International Conference on Computer Vision (ICCV), October 2019
38. Zhuang, T., Ou, W., Wang, Z.: Globally optimized mutual influence aware ranking in e-commerce search. arXiv preprint arXiv:1805.08524 (2018)

Dealing with Ratio Metrics in A/B Testing at the Presence of Intra-user Correlation and Segments

Keyu Nie[1(✉)], Yinfei Kong[2], Ted Tao Yuan[1], and Pauline Berry Burke[1]

[1] eBay Inc., San Jose, USA
{knie,teyuan}@ebay.com, pmburke10@gmail.com
[2] California State University Fullerton, Fullerton, USA
yikong@fullerton.edu

Abstract. We study ratio metrics in A/B testing at the presence of correlation among observations coming from the same user and provides practical guidance especially when two metrics contradict each other. We propose new estimating methods to quantitatively measure the intra-user correlation (within segments). With the accurately estimated correlation, a uniformly minimum-variance unbiased estimator of the population mean, called correlation-adjusted mean, is proposed to account for such correlation structure. It is proved theoretically and numerically better than the other two unbiased estimators, naive mean and normalized mean (averaging within users first and then across users). The correlation-adjusted mean method is unbiased and has reduced variance so it gains additional power. Several simulation studies are designed to show the estimation accuracy of the correlation structure, effectiveness in reducing variance, and capability of obtaining more power. An application to the eBay data is conducted to conclude this paper.

Keywords: A/B testing · Repeated measures · Uniformly minimum-variance unbiased estimator · Sensitivity · Variance reduction

1 Introduction

The A/B testing is an online randomized controlled experiment that compares the performance of a new design B (treatment) of web service with the current design A (control). It randomly assigns traffic/experiment units (users, browse id/guid, XID and so on) into one of the two groups, collects metrics of interests, conducts hypothesis testing to claim significant treatment effect and estimates the average treatment effect/lift (ATE) over the whole targeting group. In this paper, we would call a random experiment unit as a user. Statistically, the simple A/B test is a two-sample t-test but there are many situations that the data

---

Y. Kong—Both authors contributed equally to this research.
This work was done when Pauline Berry Burke worked at eBay Inc.

structure is no longer simply two independent groups of observations with independent and identical distribution (i.i.d.). For example, a user may visit a website multiple times so those visits can be considered as repeated measures of the same user. These repeated measures are identical but not independent, as there should be a certain level of correlation among repeated measures within the user. Such intra-user correlation has big impacts when dealing with ratio metrics, where ratio metrics involve the analysis units having smaller granular level comparing with experiment units, such as click-through rate (CTR), average selling price (ASP), search result page to view item (SRP to Vi), search result page exit rate (SRP Exit Rate) and so on.

At the same time, there can be various types of users such as cell phone users, laptop users, and desktop users. In this paper, we call a type of user as a segment. Those users in the same segment shall share some level of behavioral similarities, such as independent and identical distribution. Thus, we consider a three-level data structure, segment-user-observation, as follows. Observations within a user are identically and commonly correlated. However, observations from different users but the same segment shall be identically and independently distributed. We further assume that the distribution of all measures (i.e., observations) from the same segment, whichever users they are from, is identical. Across different segments, such distribution could be different in means. The main purpose of this article is to shed light on ratio metrics in A/B test at the presence of repeated measures of users as well as multiple segments of users.

Many researchers have considered the same kind of data structure [1,3], and delta method is suggested to estimate the variance of naive mean estimators. Here naive mean is to take the average of all repeated measures, and it is common to use the difference of naive means as an unbiased estimator of ATE. The supporters [1,3] argue that it naturally matches the definition of the metrics, like CTR as the summation of all clicks divided by the summation of all impressions, taking the expected number of repeated measures as part of the metrics. On the other hand, we notice that there is another way of computing this ratio metrics called normalized mean: compute the ratio metrics for each user/experiment-unit and then take the average of these user-level normalized metrics. Researchers [4] observed that the two ratio metrics could lead to contradicted conclusions via a two-sample t-test. However, in Kohavi et al.'s new book [6], he discussed two estimators (naive/normalized mean) and argued that both are useful definitions. Normalized mean is generally recommended in [6], as it is "more robust to outliers, such as bots having many page views or clicking often". More other recent theoretical investigation of the variance of ratio metrics can be found in [9].

Despite the various opinions among different researchers in this area, this article provides another view of this problem and proposes practical guidance especially when two metrics contradicted each other. We offer new estimating methods to quantitatively measure the intra-user correlation (within segments). With the accurately estimated correlation, we can construct a uniformly minimum-variance unbiased estimator (UMVUE) of the population mean called "correlation-adjusted mean" for the data with repeated measures. Hence, we shall have improved power in the A/B test. We formulate the problem in detail in Sect. 2, deploy the proposed method in Sect. 3, and support it with numerical analysis in Sect. 4.

2 Problem Formulation

Suppose we are conducting a randomized controlled experiment and denote each experiment unit by i. We observe repeated measurements of metrics $(X_{i1}, \ldots, X_{in_i})$ where n_i denotes the number of repetitions for unit i. This data structure is used in this paper to describe ratio metrics. In the context of A/B testing for website comparison, an experiment unit is a user while repeated measures refer to the multiple website visits from the same user. We are interested in testing if the treatment effect is significant by the means of $E(X)$ per event, like click through rate (CTR) or average sold price (ASP). Naturally there are two ratio metrics describing this [6]:

$$\bar{R}^A = \frac{\sum_{i=1}^{N} \sum_{j=1}^{n_i} X_{ij}}{\sum_{i=1}^{N} n_i}, \quad \bar{R}^B = N^{-1} \sum_{i=1}^{N} \frac{\sum_{j=1}^{n_i} X_{ij}}{n_i}.$$

Both ratios can be used to estimate the population mean but which one is better. What are the differences between the two? From many people, $\bar{R}^A$ is more natural comparing with $\bar{R}^B$ and many papers from Linkedin, Microsoft [3], Uber [10], Yandex [1,4] follow this definition. Yandex [4] reported that these two metrics can have different directional indications. In this paper, we want to be inline with [6] and argue both metrics are natural estimates of the population mean but they are different in many aspects. For notation convenience, we call $\bar{R}^A$ as naive mean and $\bar{R}^B$ as normalized mean.

3 Approach

We consider the problem of choosing between these two ratios in a randomized controlled experiment with repeated measures under special designs as described above. Under this structure, instead of assuming i.i.d. of X_{ij}, we should expect the repeated measures share a common mean with constant intra-user correlation. The X_{ij} are identical distributed with correlation ρ and follow the same distribution F for each unit/user i. Let $E(X_{ij}) = \mu$ and $Var(X_{ij}) = \sigma^2$. It is easy to show that $r_i = \sum_{j=1}^{n_i} X_{ij}/n_i$ is an unbiased estimator of μ with variances $\sigma_{r_i}^2 = \sigma^2\{(1-\rho)/n_i + \rho\}$ (please see Appendix A for proof).

Theorem 1. *With the model setting above, we define*

$$S_1 = \frac{1}{\sum_{i=1}^{N} n_i - 1} \sum_{i=1}^{N} \sum_{j=1}^{n_i} (X_{ij} - \bar{R}^A)^2, S_2 = \frac{1}{\sum_{i=1}^{N} n_i - 1} \sum_{i=1}^{N} \sum_{j=1}^{n_i} (X_{ij} - \bar{R}^B)^2,$$

$$S_3 = \frac{1}{\sum_{i=1}^{N}(n_i - 1)} \sum_{i=1}^{N} \sum_{j=1}^{n_i} \sum_{j'=1 \text{ and } j' \neq j}^{n_i} (X_{ij} - r_i)(X_{ij'} - r_i).$$

Then we have

$$E(S_1) \approx \sigma^2, \quad E(S_2) \approx \sigma^2, \quad E(S_3) = (\rho - 1)\sigma^2,$$

The proof of Theorem 1 can be found in the Appendix D. We can get the estimates of ρ and σ^2 from above. Review that in the literature of univariate repeated measures, it follows the designs as $Y_{ij} = \mu + \pi_i + \tau_j + e_{ij}$ with $\pi_i \sim N(0, \sigma_\pi^2)$ representing the individual difference component for observation i, $\tau_j \sim N(0, \sigma_\tau^2)$ representing random deviation due to repeated measures, and $e_{ij} \sim N(0, \sigma_e^2)$ representing error. With such formulation, $\mathrm{corr}(Y_{ij}, Y_{ij'}) = \sigma_\pi^2/(\sigma_\pi^2 + \sigma_\tau^2)$ is a parameter independent of i or j. This correlation is usually called intra-class correlation. Although our data design does not necessarily follow this formulation, we argue that it is reasonable to assume $\mathrm{corr}(Y_{ij}, Y_{ij'})$ is a constant for the problem of repeated measures among users.

3.1   Unified View of Two Ratio Metrics

Suppose we have a total of N units/users. Then for an arbitrary weight vector $W = (w_1, \ldots, w_N)^T$ with $\sum_{i=1}^N w_i = 1$, the following quantity is also an unbiased estimator of μ:

$$\bar{R}^W = \sum_{i=1}^N w_i r_i.$$

With such definition, we see that if $W = (1/N, \ldots, 1/N)^T$, then $\bar{R}^W = \bar{R}^B$. It is also clear that if $W = (n_1/\sum_{i=1}^N n_i, \ldots, n_N/\sum_{i=1}^N n_i)^T$, then $\bar{R}^W = \bar{R}^A$. Both $\bar{R}^A$ and $\bar{R}^B$ are special cases and unbiased estimators of μ under repeated measures with common mean data structure. The difference is that $\bar{R}^A$ (naive mean) utilizes weights proportional to repeated measure count n_i over r_i per unit, but $\bar{R}^B$ (normalized mean) equally weights r_i from each unit.

In fact, the uniformly minimum-variance unbiased estimator (UMVUE) [2] of μ (proof in Appendix B) is by setting

$$w_i \propto \frac{1}{\sigma_{r_i}^2} \propto \frac{n_i}{1 + (n_i - 1)\rho}. \tag{1}$$

Theorem 1 provides a quantitative way to measure the intra-user correlation as $\hat{\rho}$. Based on the magnitude of $\hat{\rho}$, we could compare the variance between naive mean $(\bar{R}^A)$ and normalized mean $(\bar{R}^B)$. Specifically, when $\rho = 1$, $\bar{R}^B$ is the UMVUE of μ, hence $Var(\bar{R}^B) < Var(\bar{R}^A)$. Similarly when $\rho = 0$ we have $Var(\bar{R}^A) < Var(\bar{R}^B)$. We suggest following the rule based on our observation:

if $\hat{\rho} < \dfrac{\frac{\sum_{i=1}^N \frac{1}{n_i}}{N^2} - \frac{1}{\sum_{i=1}^N n_i}}{\frac{\sum_{i=1}^N n_i^2}{(\sum_{i=1}^N n_i)^2} + \frac{\sum_{i=1}^N \frac{1}{n_i}}{N^2} - \frac{1}{\sum_{i=1}^N n_i} - \frac{1}{N}}$, then $Var(\bar{R}^A) < Var(\bar{R}^B)$; Otherwise,

$Var(\bar{R}^B) < Var(\bar{R}^A)$.

Remark 1. (1) Normalized mean($\bar{R}^B$) is a fairness metric, since it treats each user equally. In the meanwhile, naive mean($\bar{R}^A$) could easily attribute the treatment changes (in percent) to numerator and denominator. Both ratio metrics are useful in final reports. In practice, $\bar{R}^A$ is slightly preferred in search-related experiments and $\bar{R}^B$ is preferred in advertisement experiments.

(2) As mentioned in [3], delta method should be adapted to estimate the variance of the naive mean($\bar{R}^A$). However, we could directly utilize sample variance to estimate the variance of the normalized mean($\bar{R}^B$).

3.2 Optimized $\bar{R}^W$ Under UMVUE

It is shown in Eq. 1 that the optimal (UMVUE) weight $w_i = \frac{n_i}{1+(n_i-1)\rho} \approx n_i^{1-\rho}$ (in Appendix C), where $n_i^{1-\rho}$ represents the effective sample size. The optimized $\bar{R}^W$ (UMVUE of μ) would be:

$$\bar{R}^\rho = \sum_{i=1}^{N} \frac{n_i^{1-\rho}}{\sum_{i=1}^{N} n_i^{1-\rho}} r_i. \tag{2}$$

We denote it as correlation-adjusted mean. By plugging in the estimator of ρ in Theorem 1, this correlation-adjusted mean would have the smallest variance. Of course, Delta method should be applied to estimate its variance. We are primarily focusing on Improving metric sensitivity (aka. variance reduction) in our large-scale trustworthy experimentation platform. The option proposed here follow this guideline: we always prefer metric with smaller variance.

3.3 What to Do When Two Ratio Metrics Have Contradiction

Both the naive mean and normalized mean are unbiased estimators of the common mean μ. Ideally, we should expect the two matches to each other in direction. In practice, it is not always like that. Actually, at eBay, we also found similar cases mentioned in [4] that the signs of treatment lift from naive mean and normalized mean are contradicted with each other. We utilize a simplified example with Simpson's paradox to illustrate the contradictory conclusions can even be possible using these two ratio metrics. A randomized controlled experiment is conducted in which two observations are repeatedly measured multiple times. Table 1 summarizes the experiment results. We calculate the two ratio metrics:

Table 1. Simpson's paradox

	Control		Treatment	
	$\sum_{j=1}^{n_i} X_{ij}$	n_i	$\sum_{j=1}^{m_i} Y_{ij}$	m_i
Obs. 1 (Segment 1)	200	300	20	24
Obs. 2 (Segment 2)	10	30	100	200

$$\bar{R}^A_{ctr} = \frac{200 + 10}{300 + 30} = \frac{7}{11}, \qquad \bar{R}^A_{trt} = \frac{20 + 100}{24 + 200} = \frac{15}{28},$$

$$\bar{R}^B_{ctr} = \frac{1}{2}\left(\frac{200}{300} + \frac{10}{30}\right) = \frac{1}{2}, \qquad \bar{R}^B_{trt} = \frac{1}{2}\left(\frac{20}{24} + \frac{100}{200}\right) = \frac{2}{3},$$

where the sub-indices ctr and trt are introduced to denote the control and treatment group respectively. Therefore, the treatment effect can be estimated as:

$$\hat{\Delta}^A = \bar{R}^A_{trt} - \bar{R}^A_{ctr} = -0.10, \quad \hat{\Delta}^B = \bar{R}^B_{trt} - \bar{R}^B_{ctr} = 0.17.$$

Since the signs of this two estimates are different, the conclusion of comparison between treatment and control is inconsistent. The example explains the contradictory of the two ratio metrics potentially accounts for Simpson's paradox.

Usually, Simpson's paradox happens when there are heterogeneous treatment effects, or the ratio-metric R has differences in segments. The contradicted conclusion from naive mean and normalized mean would be a sign of Simpson's paradox. Further analysis in each user segment separately is usually recommended for this situation, we would discuss it in more details in Sect. 3.4. Continuously monitoring differences of sample count (N) and repeated measure count ($\sum_{i=1}^{N} n_i$) from test and control in user segments will help us alert such situation (two metrics with contradiction) in the early stage.

3.4 Generalization to Multiple User Segments

In the context of website comparison, most of the time, experiment units, i.e. users, could be from different segments. Common user segments include device type (desktop vs mobile), browser type, country, and so on. Metrics in different user segments usually are different. For instance, it is reasonable to assume desktop users have different CTR from mobile users, as they may experience different UI designs. To avoid Simpson's paradox, as well as possible contradicted signs, it is always better to analyze it in segment view, and follow the guidance in Sects. 3.1 and 3.2.

To integrate the estimators from multiple segments, we can use the weighted sum of these estimators from segments with the segment weight proportional to either the sample size of each segment (weight on users) or the repeated measure size of each segment (weight on the repeated measure). The choice of segment weights is beyond the scope of this article but we refer to [5, 7, 8] for more discussion on it. We can show that naive mean ($\bar{R}^A$) of whole users is the same as integrating multiple naive means of each segment with segment-weights on the number of repeated measure count of each segment; normalized mean ($\bar{R}^B$) integrate multiple normalized means of each segment with segment-weights on the number of users (sample count) of each segment.

4 Numerical Analysis

We illustrate the effectiveness of the proposed methods with several simulation studies and real data analysis. In the simulation studies, we present four examples to demonstrate the following points.

First, we want to evaluate the accuracy of our proposed estimator (in Theorem 1) for the intra-user correlation ρ and show that our estimator is accurate. Second, we investigate how the standard errors of $\bar{R}^A$ and $\bar{R}^B$ vary with ρ. We will show that the true standard deviation of $\bar{R}^\rho$ is smaller than that of $\bar{R}^A$ and $\bar{R}^B$. We estimate ρ from Theorem 1 and plug it in Eq. (2) to obtain the correlation-adjusted mean. We empirically calculate the true standard deviation of $\bar{R}^\rho$, $\bar{R}^A$, and $\bar{R}^B$ with bootstrapping (1000 iterations). Third, we demonstrate that more power can be obtained in the A/B test by combining the proposed estimator $\bar{R}^\rho$ obtained from different segments in treatment and control groups. We simulation an additional example that mimics the real A/B testing where the test group and control group are both comprised of users coming from different segments. Finally, in the fourth example, we consider a more general correlation structure instead of a common correlation as assumed in our method. But we will see that the correlation adjustment idea still works in the general situation. It is also implying the robustness of common correlation assumption.

In addition to the simulation studies, we apply it to an eBay real data set. We find that the intra-user correlation ρ is presented in our data set, and provide estimates of the correlations for different metrics. We also conclude that the proposed estimator $\bar{R}^\rho$ does provide a more sensitive metric to measure the success of an A/B test. To simplify the process and highlight the scope of the paper, we only evaluate the standard errors of three different estimators within one experiment group (treatment or control) using bootstrap.

As we explained above that we design four simulation examples for four different purposes. The first example only contains one segment of users and we want to show the estimation accuracy of ρ as below.

Example 1. We simulate 1000 users in the following manner. For the i-th user, we generate the number of observations n_i from Poisson(10) + 1. Denote each observation by $X_{i,j}$ with $j \in \{1, \cdots, n_i\}$. Let $X_{i,j} \sim$ Bernoulli(p_i) with $p_i \sim N(0.3, 0.05)$. Note that observations belonging to the same user may not necessarily independent of each other. We set corr($X_{ij}, X_{ij'}) = \rho$ for $j \neq j'$. There are a number of ways to introduce correlation structure for Bernoulli distribution. We choose to firstly generate multivariate normal observations $\widetilde{\mathbf{x}}_{i\cdot} = (\widetilde{X}_{i1}, \cdots, \widetilde{X}_{in_i})^T \sim N(\mathbf{0}, \Sigma(\widetilde{\rho}))$ with diagonal elements of $\Sigma(\widetilde{\rho})$ being 1 and off-diagonal elements being $\widetilde{\rho}$. We then dichotomize $\widetilde{X}_{ij}$ to $\{0, 1\}$ at $\Phi^{-1}(1 - p_i)$ where $\Phi(\cdot)$ is the cumulative density function of a standard normal distribution. In other words, we let $X_{ij} = 1$ if $\widetilde{X}_{ij} > \Phi^{-1}(1 - p_i)$ and $X_{ij} = 0$ otherwise. We consider $\widetilde{\rho} \in \{0, 0.2, 0.4, 0.6, 0.8\}$ and present the estimation results in columns 3–5 of Table 2. The column 3 presents the true correlation between dichotomized $X_{i,j}$ and $X_{ij'}$. Note that the new correlation (ρ) is different from their correlation ($\widetilde{\rho}$) before dichotomization. The experiments are repeated 1000 times.

The estimation accuracy of ρ for Example 1 is presented in Table 2. The S3/S1 and S3/S2 refer to the estimation methods of ρ by calculating the sample version of $E(S_3)/E(S_1) + 1$ and $E(S_3)/E(S_2) + 1$, respectively. The columns $\hat{\rho}$ and SD($\hat{\rho}$) denotes the mean and standard deviation of estimated values based

Table 2. Estimation Accuracy of ρ in Example 1

$\widetilde{\rho}$	ρ	Method	$\hat{\rho}$	SD($\hat{\rho}$)
0	0.0022	S3/S1	0.0117	0.0050
0.2	0.1204	S3/S1	0.1292	0.0093
0.4	0.2488	S3/S1	0.2559	0.0126
0.6	0.3947	S3/S1	0.4015	0.0150
0.8	0.5773	S3/S1	0.5813	0.0161
0	0.0022	S3/S2	0.0117	0.0050
0.2	0.1204	S3/S2	0.1292	0.0093
0.4	0.2488	S3/S2	0.2559	0.0126
0.6	0.3947	S3/S2	0.4015	0.0150
0.8	0.5773	S3/S2	0.5813	0.0161

on 1000 repetitions. We see from the Table 2 that our methods estimate ρ very accurately since $\hat{\rho}$ is fairly close to the true value ρ.

Example 2. The capability of accurately estimating ρ is shown in Example 1. We now proceed to show how the change of intra-user correlation ρ interacts with the variances of naive mean $(\bar{R}^A)$ and normalized mean $(\bar{R}^B)$. We also show the optimized estimator of μ, $\bar{R}^\rho$, will have a reduced variance compared to others. We want to show it for more values of ρ and show the robustness of our method under various settings. Therefore, we simulate a separate example here. Following the notation in Example 1, we generate 1000 users with $n_i \sim \text{Poisson}(2) + 1$, $p_i \sim N(0.3, 0.04)$, and $\rho \in \{0.1, 0.2, \cdots, 0.9\}$.

We then present the results of Example 2 in Table 3. We only choose S3/S1 for illustration purposes in Example 2. Results should be similar if the other estimator is used. In Table 3, column $\bar{y}$ gives the true value of μ; column $\hat{y}$ gives the estimated value of μ; column $\hat{\sigma}_{\bar{y}}$ gives the estimated standard deviation for each method calculated based on values of 1000 repetitions. Three methods, naive mean $(\bar{R}^A)$, normalized mean $(\bar{R}^B)$, and the proposed method, weighted mean $(\bar{R}^\rho)$ are included in the table. It is clear that the correlation-adjusted mean $(\bar{R}^\rho)$ has the smallest standard deviation across all values of ρ. In the meanwhile, the standard error of $\bar{R}^A$ is smaller than $\bar{R}^B$ if $\rho < 0.3$, and the standard error of $\bar{R}^A$ is bigger than $\bar{R}^B$ if $\rho > 0.3$.

Example 3. We further illustrate that additional power can be gained if using the proposed weight mean method. We design an example where users in the treatment and control groups come from different segments. Let us focus on the control group first. Consider three segments of subjects/users following the multi-nomial distribution $(C_1, C_2, C_3)^T \sim \text{Multi-nomial}(1/3, 1/2, 1/6)$. For segment 1, $n_i \sim \text{Poisson}(2) + 1$ and $X_{i,j} \sim \text{Bernoulli}(p_i)$ with $p_i \sim N(0.3, 0.04)$. For segment 2, $n_i \sim \text{Poisson}(5) + 1$ and $X_{i,j} \sim \text{Bernoulli}(p_i)$ with $p_i \sim N(0.5, 0.08)$. For

Table 3. Variance reduction compared to naive and normalized mean in Example 2

	$\bar{y}$ (truth)	$\hat{y}$ (estimate)	$\hat{\sigma}_{\bar{y}}$ (estimate)	ρ
Naive	0.3	0.29971	0.00947	0.1
Normalized	0.3	0.29953	0.01016	0.1
Corr. Adj	0.3	0.29972	0.00942	0.1
Naive	0.3	0.30006	0.01041	0.2
Normalized	0.3	0.29997	0.01074	0.2
Corr. Adj	0.3	0.30007	0.01026	0.2
Naive	0.3	0.29968	0.01126	0.3
Normalized	0.3	0.29947	0.01126	0.3
Corr. Adj	0.3	0.29966	0.01100	0.3
Naive	0.3	0.30015	0.01206	0.4
Normalized	0.3	0.30009	0.01177	0.4
Corr. Adj	0.3	0.30016	0.01160	0.4
Naive	0.3	0.30020	0.01280	0.5
Normalized	0.3	0.30005	0.01226	0.5
Corr. Adj	0.3	0.30015	0.01219	0.5
Naive	0.3	0.29998	0.01349	0.6
Normalized	0.3	0.29965	0.01274	0.6
Corr. Adj	0.3	0.29982	0.01270	0.6
Naive	0.3	0.30064	0.01417	0.7
Normalized	0.3	0.30060	0.01321	0.7
Corr. Adj	0.3	0.30062	0.01321	0.7
Naive	0.3	0.29952	0.01479	0.8
Normalized	0.3	0.29953	0.01364	0.8
Corr. Adj	0.3	0.29955	0.01362	0.8
Naive	0.3	0.29999	0.01540	0.9
Normalized	0.3	0.29974	0.01406	0.9
Corr. Adj	0.3	0.29977	0.01407	0.9

segment 3, $n_i \sim$ Poisson(30) $+1$ and $X_{i,j} \sim$ Bernoulli(p_i) with $p_i \sim N(0.7, 0.04)$. In any of the three segments, we set corr($X_{ij}, X_{ij'}$) = 0.3. Then we simulate the data for the treatment group in a very similar way but the only difference is p_i follows $N(0.3 + d, 0.04)$, $N(0.5 + d, 0.08)$, and $N(0.7 + d, 0.04)$, respectively. We consider $d \in \{0.01, 0.02, \cdots, 0.08\}$. The experiment is repeated 1000 times for each d. Figure 1 shows the comparison of power in this example.

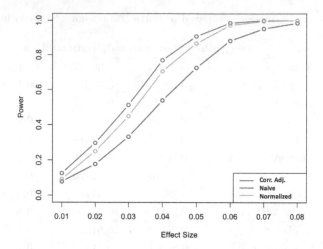

Fig. 1. Power analysis in Example 3

Example 4. We explore the effectiveness of the proposed method in handling the unequal correlation structure. To be more specific, we borrow the setting in Example 2 but the correlation between X_{ij} and $X_{ij'}$ is $\rho^{|j-j'|}$ for any $j \neq j'$. We set $\rho = 0.9$ as it is common to have a high correlation for repeated measures next to each other but low correlation when they are far from each other. Note that the true correlation structure is no longer equal between observations but in a more general autoregressive pattern.

The results for this example are presented in Table 4. The standard deviation is smaller based on our correlation adjusted method. It also suggests that assuming constant intra-user correlation is robust when the true data structure violates the assumption.

Table 4. Variance reduction compared to naive and normalized mean in Example 4

	$\bar{y}$ (truth)	$\hat{y}$ (estimate)	$\hat{\sigma}_{\bar{y}}$ (estimate)
Naive	0.3	0.28557	0.01394
Normalized	0.3	0.29205	0.01355
Corr. Adj	0.3	0.29074	0.01342

4.1   Validation in eBay Data

We extend our validation into eBay real data to verify how the significance of $\hat{\rho}$ guides the relevant variance compare between naive mean and normalized mean, as well as how much improvement (with correlation-adjusted mean) can be achieved. We randomly sampled search activities from users with each size of

500,000 users on eBay global site with primary metrics as "Ratio R_1" (alike Exit Rate) for UK and "Ratio R_2" (alike CTR) for US in some treatment groups. We further repeated 5 replicas of the previous process with equal size 500,000 and showed the average estimation results at Table 5. The reported metric estimations were added a constant to hide the real information, but it should not change our conclusion in follows. To properly evaluate the estimator standard errors, we use bootstrap (1000 iterations) with re-sampling id being the experiment unit id to be an educated guess at each replica.

Our results in Table 5 compared the standard errors of naive mean ($\bar{R}^A$), normalized mean ($\bar{R}^B$) and correlation-adjusted mean ($\bar{R}^\rho$). The method S3/S1 is used to accurately estimate the intra-user correlation ρ. It clearly revealed that the intra-user correlation coefficient ρ did present in our data set, and varied from 9.6% to 56% for "Ratio R_1" at UK site and "Ratio R_2" at US site respectively. The root cause of different intra-user correlated behaviors (ρ) between the UK and US is actually due to a search feature launched in the US site only. The dependence among repeated measures within users could not be ignored and should be carefully addressed to pick the right estimation method. For small ρ (9.6%), $\bar{R}^A$ showed 3.3% improvement on standard error comparing with $\bar{R}^B$, which in hence saved $1-(1-3.3\%)^2 = 6.6\%$ on sample sizes. On the contrast, for big ρ (56%), $\bar{R}^B$ showed 16.7% improvement on sample sizes comparing with $\bar{R}^A$. For both cases, $\bar{R}^\rho$ saved up to 15.2% and 40.3% sample sizes correspondingly.

We noticed the bias between the naive mean and normalized mean for "Ratio R_2" could be a sign of heterogeneous user effects presence. A proper segment classification algorithm is needed to further analyze this heterogeneous effect, which is beyond the scope of this paper.

Table 5. Estimation average of 5 replicas with each sample size = 500,000 users from eBay data

Method	Metric Name	Site	Estimate	Std. Error	$\hat{\rho}$
Naive ($\bar{R}^A$)	Ratio R_1	UK	0.42850048	0.00061878	0.09618527
Normalized ($\bar{R}^B$)	Ratio R_1	UK	0.43028952	0.00064033	0.09618527
Corr. Adj. ($\bar{R}^\rho$)	Ratio R_1	UK	0.42839652	0.00058949	0.09618527
Naive ($\bar{R}^A$)	Ratio R_2	US	0.49476930	0.00129048	0.56069220
Normalized ($\bar{R}^B$)	Ratio R_2	US	0.52730064	0.00107419	0.56069220
Corr. Adj. ($\bar{R}^\rho$)	Ratio R_2	US	0.51028800	0.00077074	0.56069220

5   Conclusion and Restriction

In this paper, we studied how to estimate ratio metrics in a randomized controlled experiment with repeated measures within each experiment unit. As the intra-user correlation coefficient ρ could not be ignored, we established a way to accurately measure the severity of intra-user correlation coefficients ρ. We showed the naive mean $(\bar{R}^A)$ and normalized mean $(\bar{R}^B)$ were both weighted user means with different weights. We proved there is no clear winner between naive mean and normalized mean when considering estimator variance, as it highly depends on the severity of intra-unit correlation coefficients ρ. We further proposed a correlation-adjusted mean $(\bar{R}^\rho)$, which adopted the optimal weights depending on ρ. Our simulation and real data empirical analysis validated that we can accurately estimate ρ and built more sensitive ratio-metric estimators based on ρ. Due to the variance reduction technique, we shall improve the power as well as requiring less sample size for A/B testing, which improves experiment efficiency.

The main restriction of the proposed method is on its assumption of a common correlation for different users within the same segment. It is possible that there is a variation of such correlation for the same type of users (from the same segment). Such observation motivates us to consider more flexible settings where the correlation for a user within the same segment may follow a random distribution, and the random distribution can be different across segments. However, this type of setting is the subject of our future work.

A   Variance Formula of Sample Mean with Correlated Measure

Claim: The X_j are identical distributed with correlation ρ and follow the same distribution $\boldsymbol{F}$. Let $E(X_j) = \mu$ and $Var(X_j) = \sigma^2$. Then:

a. $r = \sum_{j=1}^{n} X_j/n$ is an unbiased estimator of μ.
b. The variance of r is $\sigma_r^2 = \sigma^2\{(1-\rho)/n + \rho\}$.

Proof. Since

$$E(r) = E(\sum_{j=1}^{n} X_j/n) = \sum_{j=1}^{n} E(X_j)/n = \sum_{j=1}^{n} \mu/n = \mu,$$

then $r = \sum_{j=1}^{n} X_j/n$ is an unbiased estimator of μ.
    The variance of r is

$$\sigma_r^2 = E(r - E(r))^2 = E(r - \mu)^2 = E(\sum_{j=1}^{n} X_j/n - \mu)^2 = \frac{1}{n^2}E(\sum_{j=1}^{n}(X_j - \mu))^2$$

$$= \frac{1}{n^2}(\sum_{j=1}^{n} E(X_j - \mu)^2 + \sum_{k=1}^{n}\sum_{m=1,m\neq k}^{n} E(X_k - \mu)(X_m - \mu))$$

$$= \frac{1}{n^2}((\sum_{j=1}^{n}\sigma^2 + \sum_{k=1}^{n}\sum_{m=1,m\neq k}^{n}\sigma^2\rho) = \frac{1}{n}\sigma^2 + \frac{n-1}{n}\sigma^2\rho$$

$$= \sigma^2\{(1-\rho)/n + \rho\}.$$

B   Proof of UMVUE

Claim: Suppose r_i with $i = 1, \ldots, N$ are independent with each other and $E(r_i) = \mu$, $Var(r_i) = \sigma_{r_i}^2$. Then

a. for an arbitrary weight vector $W = (w_1, \ldots, w_N)^T$ with $\sum_{i=1}^{N} w_i = 1$, the following quantity is also an unbiased estimator of μ: $\bar{R}^W = \sum_{i=1}^{N} w_i r_i$.

b. the UMVUE of μ is by setting: $w_i \propto \frac{1}{\sigma_{r_i}^2}$.

Proof

$$E(\bar{R}^W) = E(\sum_{i=1}^{N} w_i r_i) = \sum_{i=1}^{N} w_i E(r_i) = \sum_{i=1}^{N} w_i \mu = \mu.$$

Thus, $\bar{R}^W$ is an unbiased estimator of μ.

To find the minimum value of $Var(\bar{R}^W) = \sum_{i=1}^{N} w_i^2 \sigma_{r_i}^2$ under restriction $\sum_{i=1}^{N} w_i = 1$, we use method of Lagrange multiplier: $\mathcal{L} = \sum_{i=1}^{N} w_i^2 \sigma_{r_i}^2 - \lambda(\sum_{i=1}^{N} w_i - 1)$. Since $\frac{\partial^2 \mathcal{L}}{\partial w_i^2} = 2\sigma_{r_i}^2 > 0$, so $\mathcal{L}$ is a concave function of w_i. The minimum value of $Var(\bar{R}^W)$ is equal to the minimum of $\mathcal{L}$, and it achieves at:

$$\frac{\partial \mathcal{L}}{\partial w_i} = 2 w_i \sigma_{r_i}^2 - \lambda w_i = 0 \; for \; i = 1, \ldots, N,$$

$$\frac{\partial \mathcal{L}}{\partial \lambda} = \sum_{i=1}^{N} w_i - 1 = 0.$$

By solving equations above we have $w_i = \frac{\frac{1}{\sigma_{r_i}^2}}{\sum_{i=1}^{N} \frac{1}{\sigma_{r_i}^2}} \propto \frac{1}{\sigma_{r_i}^2}$.

C   Approximation

Claim: $\frac{n}{1+(n-1)\rho} \approx n^{1-\rho}$.

Proof. From Taylor series expansion, the 1st degree polynomial approximation of function $f(n) = n^\rho$ at $n = 1$ is $n^\rho \approx 1 + (n-1)\rho$. Therefore,

$$n^{1-\rho} \approx \frac{n}{1+(n-1)\rho}.$$

D   Proof of Theorem 1

Without loss of generality, we set $\mu = 0$. It is obvious that:

$$E(r_i^2) = (\frac{1}{n_i} + \frac{n_i - 1}{n_i}\rho)\sigma^2,$$

$$E(\bar{R}^{A^2}) = (\frac{1}{\sum_{i=1}^N n_i})^2 \sum_{i=1}^N n_i^2 E(r_i^2) = (\frac{1}{\sum_{i=1}^N n_i})^2 \sum_{i=1}^N n_i\{1 + (n_i - 1)\rho\}\sigma^2$$

$$= \left\{\frac{1}{\sum_{i=1}^N n_i} + \frac{\sum_{i=1}^N n_i(n_i - 1)}{(\sum_{i=1}^N n_i)^2}\rho\right\}\sigma^2,$$

$$E(\bar{R}^{B^2}) = (\frac{1}{\sum_{i=1}^N 1})^2 \sum_{i=1}^N E(r_i^2) = (\frac{1}{N})^2 \sum_{i=1}^N n_i^{-1}\{1 + (n_i - 1)\rho\}\sigma^2.$$

We also have the following for any i, j:

$$E(r_i X_{i'j}) = \begin{cases} 0 \ \ if \ i \neq i' \\ (\frac{1}{n_i} + \frac{n_i - 1}{n_i}\rho)\sigma^2 \ \ if \ i = i', \end{cases}$$

$$E(\bar{R}^A X_{ij}) = \frac{n_i E(r_i X_{ij})}{\sum_{i=1}^N n_i} = \frac{1 + (n_i - 1)\rho}{\sum_{i=1}^N n_i}\sigma^2,$$

$$E(\bar{R}^B X_{ij}) = \frac{1}{N}E(r_i X_{ij}) = \frac{1 + (n_i - 1)\rho}{Nn_i}\sigma^2.$$

Thus,

$$E\{(X_{ij} - \bar{R}^A)^2\} = E(X_{ij}^2 - 2\bar{R}^A X_{ij} + \bar{R}^{A^2})$$

$$= \sigma^2 - 2\frac{1 + (n_i - 1)\rho}{\sum_{i=1}^N n_i}\sigma^2 + (\frac{1}{\sum_{i=1}^N n_i} + \frac{\sum_{i=1}^N n_i(n_i - 1)}{(\sum_{i=1}^N n_i)^2}\rho)\sigma^2,$$

$$E\{(X_{ij} - \bar{R}^B)^2\} = E(X_{ij}^2 - 2\bar{R}^B X_{ij} + \bar{R}^{B^2})$$

$$= \sigma^2 - 2\frac{1 + (n_i - 1)\rho}{Nn_i}\sigma^2 + (\frac{1}{N})^2 \sum_{i=1}^N n_i^{-1}(1 + (n_i - 1)\rho)\sigma^2,$$

$$E\{(X_{ij} - r_i)(X_{ij'} - r_i)\} = E(X_{ij}X_{ij'} - r_i X_{ij} - r_i X_{ij'} + r_i^2)$$

$$= \rho\sigma^2 - (\frac{1}{n_i} + \frac{n_i - 1}{n_i}\rho)\sigma^2 = \frac{1}{n_i}(\rho - 1)\sigma^2,$$

Therefore, we have

$$E(S_1) = \frac{1}{\sum_{i=1}^N n_i - 1} \sum_{i=1}^N \sum_{j=1}^{n_i} \left[\sigma^2 - 2\frac{1 + (n_i - 1)\rho}{\sum_{i=1}^N n_i}\sigma^2 + \left\{\frac{1}{\sum_{i=1}^N n_i} + \frac{\sum_{i=1}^N n_i(n_i - 1)}{(\sum_{i=1}^N n_i)^2}\rho\right\}\sigma^2\right]$$

$$= \frac{\sigma^2}{\sum_{i=1}^N n_i - 1} \sum_{i=1}^N \left[n_i - \frac{2n_i + n_i(n_i - 1)\rho}{\sum_{i=1}^N n_i} + \left\{\frac{n_i}{\sum_{i=1}^N n_i} + n_i\frac{\sum_{i=1}^N n_i(n_i - 1)}{(\sum_{i=1}^N n_i)^2}\rho\right\}\right]$$

$$= \frac{\sigma^2}{\sum_{i=1}^N n_i - 1} \left[\frac{(\sum_{i=1}^N n_i)^2 - \sum_{i=1}^N n_i}{\sum_{i=1}^N n_i} - \frac{\{\sum_{i=1}^N n_i(n_i - 1)\}\rho}{\sum_{i=1}^N n_i}\right]$$

$$= \sigma^2\left\{1 - \frac{\sum_{i=1}^N n_i(n_i - 1)}{(\sum_{i=1}^N n_i)(\sum_{i=1}^N n_i - 1)}\rho\right\}$$

$$\approx \sigma^2,$$

$$E(S_2) = \frac{1}{\sum_{i=1}^{N} n_i - 1} \sum_{i=1}^{N} \sum_{j=1}^{n_i} \left[\sigma^2 - 2\frac{1+(n_i-1)\rho}{Nn_i}\sigma^2 + (\frac{1}{N})^2 \sum_{i=1}^{N} n_i^{-1}\{1+(n_i-1)\rho\}\sigma^2 \right]$$

$$= \frac{\sigma^2}{\sum_{i=1}^{N} n_i - 1} \sum_{i=1}^{N} \left[n_i - \frac{2+(n_i-1)\rho}{N} + n_i(\frac{1}{N})^2 \sum_{i=1}^{N} n_i^{-1}\{1+(n_i-1)\rho\} \right]$$

$$= \frac{\sigma^2}{\sum_{i=1}^{N} n_i - 1} \left[\left\{ \sum_{i=1}^{N} n_i - 2 + \frac{(\sum_{i=1}^{N} n_i)(\sum_{i=1}^{N} n_i^{-1})}{N^2} \right\} + \frac{(\sum_{i=1}^{N} n_i)(N - \sum_{i=1}^{N} n_i^{-1}) - N\sum_{i=1}^{N}(n_i-1)}{N^2}\rho \right]$$

$$= \frac{\sigma^2}{\sum_{i=1}^{N} n_i - 1} \left[\left\{ \sum_{i=1}^{N} n_i - 2 + \frac{(\sum_{i=1}^{N} n_i)(\sum_{i=1}^{N} n_i^{-1})}{N^2} \right\} + \frac{N^2 - (\sum_{i=1}^{N} n_i)(\sum_{i=1}^{N} n_i^{-1})}{N^2}\rho \right]$$

$$= \sigma^2 \left\{ 1 + \frac{N^2 - (\sum_{i=1}^{N} n_i)(\sum_{i=1}^{N} n_i^{-1})}{N^2(\sum_{i=1}^{N} n_i - 1)}(\rho-1) \right\}$$

$$\approx \sigma^2,$$

$$E(S_3) = \frac{1}{\sum_{i=1}^{N}(n_i-1)} \sum_{i=1}^{N} \sum_{j=1}^{n_i} \sum_{j'=1 \ and \ j' \neq j}^{n_i} \{\frac{1}{n_i}(\rho-1)\sigma^2\} = \frac{1}{\sum_{i=1}^{N}(n_i-1)} \sum_{i=1}^{N}(n_i-1)(\rho-1)\sigma^2$$

$$= (\rho-1)\sigma^2,$$

References

1. Budylin, R., Drutsa, A., Katsev, I., Tsoy, V.: Consistent transformation of ratio metrics for efficient online controlled experiments. In: Proceedings of the Eleventh ACM International Conference on Web Search and Data Mining, pp. 55–63. ACM (2018)
2. Cochran, W.G.: Problems arising in the analysis of a series of similar experiments. Suppl. J. Roy. Stat. Soc. **4**(1), 102–118 (1937)
3. Deng, A., Knoblich, U., Lu, J.: Applying the delta method in metric analytics: a practical guide with novel ideas. In: Proceedings of the 24th ACM SIGKDD International Conference on Knowledge Discovery & Data Mining, pp. 233–242. ACM (2018)
4. Drutsa, A., Ufliand, A., Gusev, G.: Practical aspects of sensitivity in online experimentation with user engagement metrics. In: Proceedings of the 24th ACM International on Conference on Information and Knowledge Management, pp. 763–772. ACM (2015)
5. Freedman, D.A.: On regression adjustments to experimental data. Adv. Appl. Math. **40**(2), 180–193 (2008)
6. Kohavi, R., Tang, D., Xu, Y.: Trustworthy Online Controlled Experiments: A Practical Guide to A/B Testing. Cambridge University Press, Cambridge (2020)
7. Lin, W.: Agnostic notes on regression adjustments to experimental data: reexamining freedman's critique. Ann. Appl. Stat. **7**(1), 295–318 (2013)
8. Miratrix, L.W., Sekhon, J.S., Yu, B.: Adjusting treatment effect estimates by post stratification in randomized experiments. J. Roy. Stat. Soc. Ser. B Stat. Methodol. **75**(2), 369–396 (2013)
9. Sekhon, J.S., Shem-Tov, Y.: Inference on a new class of sample average treatment effects. J. Am. Stat. Assoc. 1–18 (2020)
10. Zhao, Z., Liu, M., Deb, A.: Safely and quickly deploying new features with a staged rollout framework using sequential test and adaptive experimental design. In: 2018 3rd International Conference on Computational Intelligence and Applications (ICCIA), pp. 59–70. IEEE (2018)

Risk Monitoring Services of Discharged SARS-CoV-2 Patients

Ada Bagozi, Devis Bianchini[✉], Valeria De Antonellis,
and Massimiliano Garda

Department of Information Engineering, University of Brescia,
Via Branze 38, 25123 Brescia, Italy
{a.bagozi,devis.bianchini,valeria.deantonellis,m.garda001}@unibs.it

Abstract. In the latest months, the outbreak of SARS-CoV-2 has forced worldwide healthcare systems to rethink their organisation. In this landscape, particular attention has been devoted to discharged patients. Remote monitoring on patients' health status is used, through dedicated web platforms and apps, to check home rehabilitation progress and, at the same time, promptly notify the arise of anomalies. Nevertheless, the variety of patients and the large volume of collected data call for models, tools and methods for data representation and exploration, in order to focus on relevant groups of patients only. Given our previous research efforts in the Big Data exploration field, we designed a Risk Monitoring Services ecosystem, devoted to support doctors (e.g., medical researchers, clinicians, analysts) in the analysis of data collected through app by: (i) identifying groups of SARS-CoV-2 discharged patients, built according to features such as sex, age, co-morbidities, prior therapies; (ii) monitoring the health status of patients, by extracting snapshots of patients' health parameters measurements, evolving over time, and comparing them with baseline or reference values within the same patients group; (iii) promptly notifying doctors when some measurements diverge from reference values for a group of patients.

Keywords: Multi-Dimensional Model · Patients monitoring ·
Anomaly detection services · SARS-CoV-2 outbreak

1 Introduction

The outbreak of SARS-CoV-2 has forced worldwide healthcare systems to rethink their organisation, to cope with growing hospitalisation workloads. In this landscape, particular attention has been devoted to those patients who present a clinical case permitting their discharge, albeit not negative to the virus yet. For discharged patients, remote monitoring, through dedicated web platforms and apps, is performed to supervise their health status, to check home rehabilitation progress and, at the same time, promptly notify the arise of anomalies [7]. This is part of an engagement process, urging patients in being

© Springer Nature Switzerland AG 2020
Z. Huang et al. (Eds.): WISE 2020, LNCS 12343, pp. 578–590, 2020.
https://doi.org/10.1007/978-3-030-62008-0_40

actively stimulated and involved in their personal health monitoring. Nevertheless, the variety of patients and the large volume of collected data call for models, tools and methods for data representation and exploration [2,10]. In particular, patients groups have to be identified by taking into account different features (e.g., habits, co-morbidities, physical connotations, prior therapies). Moreover, several categories of doctors (e.g., medical researchers, clinicians, analysts) should be able to focus on relevant groups only, i.e., those with SARS-CoV-2 or emerging paucisymptomatic conditions.

In this paper, we describe our contribution in the scope of an ongoing research project for remote monitoring of discharged SARS-CoV-2 patients. In particular, given our previous research efforts in the Big Data exploration field [1], we designed a Risk Monitoring Services ecosystem, to support doctors in the analysis of data collected through app by: (i) identifying patients groups based on characteristics like age, co-morbidities, BMI, prior therapies, thus organising measurements of patients' health parameters, such as temperature and heartbeat rate, according to a Multi-Dimensional Model (*Patients Groups Management Service*); (ii) monitoring the health status of patients, by extracting snapshots of parameters measurements, evolving over time, and comparing them with baseline or reference values within the same patients group (*Anomaly Detection Service*); (iii) promptly notifying doctors when some measurements diverge from reference values (*Patients Monitoring Service*). To this aim, incremental clustering techniques are applied to summarise collected measurements within patients groups and anomaly detection techniques are engaged to identify relevant measurements. Patients groups are exploited on the one hand to handle the variety of the domain and on the other hand to attract the attention of doctors on groups with high likelihood of persistent SARS-CoV-2 symptoms and paucisymptomatic conditions. In the context of the above mentioned research project, we will also describe the validation strategy of the presented approach.

The paper is organised as follows: in Sect. 2 the application context is explained, presenting the conceptual model for patients profiling and monitoring; Sect. 3 introduces the Patients Groups Management Service, relying on the Multi-Dimensional Model, while Anomaly Detection Service, based on incremental clustering and data relevance evaluation, is described in Sect. 4; Patients Monitoring Service, rooted on the Multi-Dimensional Model and anomaly detection techniques, is illustrated in Sect. 5, whereas Sect. 6 describes the implementation and the validation strategy of the approach; cutting-edge features of the proposed approach with respect to the literature are discussed in Sect. 7; finally, Sect. 8 closes the paper, sketching future research directions.

2   The Application Context

In this section, we describe the aim of a multidisciplinary research project we are involved in, in which the approach described in this paper has been applied. The project promotes the development of an app to ease the measure and subsequent evaluation of breath quality, analysing the dynamics of the respiratory act.

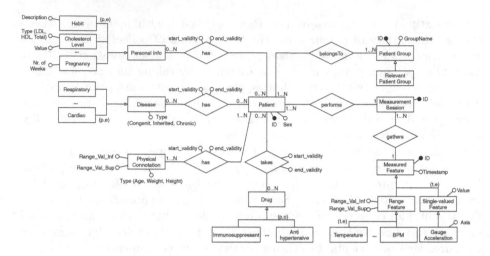

Fig. 1. E-R conceptual data model for patients' profiling and monitoring.

In particular, traces are recorded by the three axial accelerometers, embedded by a smartphone (positioned over the abdomen of the patient). Traces, plus other patient's health parameters (e.g., temperature, bpm), are exploited to evaluate the quality of breath, raising alerts in the presence of anomalies (e.g., evidences of shortness of breath). In this respect, the app could be employed as a self-evaluation instrument, albeit not conceived to substitute a medical device or the medical examination. Target users of the app are mainly discharged SARS-CoV-2 patients, for determining health improvements or emerging problems (e.g., uprising of breath difficulties). The app guides the user throughout the measurement session, with the support of predefined audio messages, to limit unwanted actions made by the user. A measurement session is articulated over two different phases: (a) an initial series of ten regulated respiratory acts (*controlled breath* phase); (b) free breathing followed by deep inspiration and forced expiration (*deep and short breath* phase). In this context, *Risk Monitoring Services* described in this paper have been developed to perform detection of breath anomalies.

2.1 Conceptual Model for Patients Profiling and Monitoring

In the following, we detail the data collected by the app, modelled through the Entity-Relationship (E-R) diagram (Fig. 1). The app collects data of both patients' profiling and monitoring data (collected during measurement sessions).

Profiling Data. When a patient registers to the app, the following information is collected for profiling: (i) *Physical Connotations*, concerning sex, age, weight and height; (ii) *Personal Information*, regarding habits of the patient (e.g., whether she is a smoker), possible ongoing pregnancy and cholesterol levels; (iii) *Diseases*, which are classified by their virtue (inherited, chronic, congenital) and by the part of the human body they affect; (iv) *Drugs*, assumed by patients and belonging to diverse classes, depending on their specific purpose (e.g., anti-hypertensive, immunosuppressant).

Profile data enables the creation of *Patients Groups*, uniquely identified by a combination of the aforementioned data (male patients, male patients over 65 years, etc.); amongst the (potentially vast) set of possible groups, *Relevant Patients Groups* (in brief, RPG) may be recognised, that is, groups which are under the lens of doctors' consideration, as they are more exposed to (a relapse of) the infection risk. RPG may be identified from well known clinical studies (e.g., male subjects are more likely of being infected by SARS-CoV-2 with respect to female ones), but also in a dynamic way, due to the emerging critical health status in several patients within the group.

Monitoring Data. Through the app, the patient can perform a *Measurement Session*, which consists in a two-phase exercise aimed at assessing the quality of her breath. Two types of features are measured: (a) *Range Features* such as the temperature and the bpm, sampled only once before the measurement session begins, and for which the patient has to select amongst predefined ranges (defined by clinicians and domain experts) the value falls in (e.g., if the bpm value is 77, it is included in the range $[75, 80]$); (b) *Single-valued Features*, regularly sampled multiple times within a measurement session (e.g., the gravity acceleration measured along the X axis within a session consists of $\approx 43k$ samples).

3 Patients Groups Management Service

Data collected from the app is explored according to different perspectives, to identify subsets of data upon which doctors' analysis must be focused, thus coping with the complexity and the variety of the domain (patients with different physical connotations, habits, diseases, etc.). Relevant patients groups might not be a-priori known, but they could be progressively detected through the ongoing anomaly detection procedure (performed by the Anomaly Detection Service described in the next section). Starting from these premises, we describe here a Multi-Dimensional Model grounded on four pivotal elements (dimensions, facets, features and measurements), descending from the conceptualisation provided in Sect. 2.1 and exploited by the Patients Groups Management Service to organise patients data, enabling an intuitive, structured and effective exploration of available information.

3.1 Dimensions and Exploration Facets

The proposed Multi-Dimensional Model (in brief, MDM) is exploited to guide the exploration over patients data, according to several exploration perspectives, and its formalisation is inspired by the literature on multi-dimensional data analysis [6]. In the following, we report the definition of the baseline constructs of the MDM for patients data organisation and exploration.

Grouping Levels

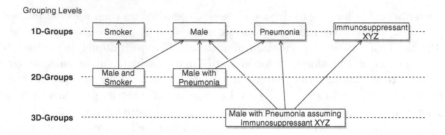

Fig. 2. Examples of facets (patients groups) with increasing grouping levels.

Definition 1 (Dimension). *A dimension d_i is an entity representing a single aspect of patient's profile, defined on categorical domain $Dom(d_i)$. We denote with $\mathcal{D} = \{d_1, \ldots, d_p\}$ the finite set of dimensions. An instance $v_{d_i}^j$ of $d_i \in \mathcal{D}$ is a categorical or range value, belonging to the domain of d_i, that is, $v_{d_i}^j \in Dom(d_i)$, $\forall i = 1, \ldots, p$ and $\forall j = 1, \ldots, |Dom(d_i)|$.*

Definition 2 (Exploration Facet). *An exploration facet ϕ_i (or, abbreviated, facet) is a combination of instances, belonging to different dimensions, apt to identify a specific patients group. For a facet ϕ_i, the grouping level $|\phi_i|$ denotes the number of dimension instances included in the facet. Let $\Phi = \{\phi_1, \ldots, \phi_m\}$ be the set of available facets.*

Given the set $\mathcal{D}$ of available dimensions, the extent of the space of patients groups (i.e., the cardinality of the set Φ, denoted as $|\Phi|$) can be very large (a sample excerpt is given in Fig. 2), as it spans all the possible combinations of dimension instances (by definition, $|\Phi| \leq 2^N - 1$, where $N = \sum_{i=1\ldots p} |Dom(d_i)|$, excluding the empty set combination and, generally, non combinable dimension instances). For this reason, doctors' focus should be on the *relevant patients groups*, meant to emphasise only specific groups of patients, whose composing dimension instances configure the clinical picture of a patient as harmful.

3.2 Features and Measurements

In the following, we provide a formalisation related to the physical quantities measured through the app, regarding a patient. Specifically, we will refer to *measurements* as the values associated with *features*, whose definitions are given as follows.

Definition 3 (Feature). *A feature f_i is a measurable quantity described as $\langle n_{f_i}, u_{f_i}, type_{f_i} \rangle$ where: (i) n_{f_i} is the feature name; (ii) u_{f_i} represents the unit of measure (if any) and (iii) $type_{f_i}$ is the type of feature (either range or single-valued). Let $\mathcal{F} = \{f_1, \ldots, f_d\}$ be the overall set of measurable features for a monitored patient.*

Definition 4 (Measurement). *A measurement $x_i(t)$ is a scalar value for the feature f_i, expressed in terms of the unit of measure u_{f_i} and the timestamp t, that represents the instant within the measurement session in which the measure has been collected. In the case of a range feature, the scalar value is obtained as the midpoint of the range and such value will be assimilated to a constant, for the entire duration of the measurement session.*

In the described context, measurements collected in each session are conceivable as a stream or a time series. We will adopt the vector notation $X_{\mathcal{F}}(t)$, to denote a record of measurements $\langle x_1(t), \ldots, x_d(t) \rangle$ for the set of features $\mathcal{F}$, synchronised with respect to the timestamp t.

4   Anomaly Detection Service

In this section, we introduce techniques to cope with the increasing volume, velocity and variety of patients' data, attracting the attention of doctors only on data deemed as relevant. Specifically, summarisation techniques have been applied in order to represent collected data characterising the health status of patients belonging to the same group, using a reduced amount of information. Additionally, anomaly detection techniques are applied on the summarised data instead of considering single measures, that can be affected by errors and false outliers due to measurement execution performed by non-expert patients. Both data summarisation and anomaly detection algorithms have been presented in our previous research effort [1] and are implemented here as sub-modules of the Anomaly Detection Service.

4.1   Clustering-Based Data Summarisation

Once focusing the attention on a specific patients group, the stream of records on a single patient can be used to ascertain whether the patient's health status is diverging from reference values associated with the group. To obtain an effective representation of the patient's health status, data summarisation is based on the incremental clustering algorithm we described in [1], relying on a lossless representation of a set of records close each others, denoted as *synthesis*. A synthesis corresponds to a cluster of records, composed of the measurements of the observed features. The clustering algorithm at a given time t produces a set of syntheses $\mathcal{S}(t)$, starting from records collected from timestamp $t - \Delta t$ to timestamp t and built on top of the previous set of syntheses $\mathcal{S}(t - \Delta t)$, for a given patients group. Roughly speaking, syntheses conceptually represent a specific state in a patient's health status. A set of syntheses is contained within a *snapshot*, a data structure defined as follows.

Definition 5 (Snapshot). *A snapshot $SN_i(t)$, stored at time t, is defined as the following tuple:*

$$SN_i(t) = \langle \mathcal{S}_i(t), F_{SN_i}, \phi_i, \rho_{SN_i} \rangle \tag{1}$$

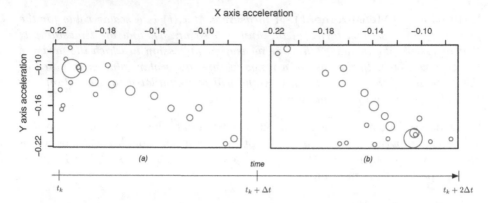

Fig. 3. Results of incremental clustering of a stream of records reporting X and Y axes acceleration values over time for a patient in the group "Males with age between 60 and 75 years" during controlled breath phase. Syntheses set changes from (a) to (b) denote weariness of the patient during expiration and inspiration activities.

where: (i) $\mathcal{S}_i(t)$ is a set of syntheses generated at time t, (ii) F_{SN_i} is the set of the monitored features; (iii) ϕ_i is the facet that identifies the patients group; (iv) ρ_{SN_i} is the breath phase (i.e., controlled breath, deep and short breath).

Figure 3 shows two snapshots taken at time $t_k + \Delta t$ and $t_k + 2\Delta t$, where measured features are accelerations over X and Y axes during controlled breath phase. Specifically, they are referred to a patient, belonging to a patients group gathering males, with age in the $[60, 75]$ range. Red circles represent identified syntheses within snapshots. A sequence of snapshots identifies a pattern, that is a behaviour related to the evolution of patient's health status. For example, in Fig. 3(b) a change in the syntheses set with respect to Fig. 3(a) is evident. Anomaly detection techniques described in the following are used to identify such changes.

4.2   Relevance-Based Anomaly Detection

We define as relevant those patients whose current health status is approaching an *anomalous status*, established for each group of patients (certified by relying on doctors' long-term expertise) and for each breath phase. The anomalous status is expressed through a set of thresholds for each observed feature within a specific patients group and a breath phase. Indeed, due to the fact that each measured feature is a physical quantity, it may present limits (bounds) that should not be violated. In particular, we distinguish between *warning* and *error* bounds: (i) a warning signals that values of a feature are getting closer to irreversible changes; (ii) an error identifies unacceptable values for a feature, determining health conditions in which a patient cannot withstand.

However, also the transition towards a warning or error condition is worth being detected. To this aim, in [1] we developed an anomaly detection mechanism based on the notion of *distance* between sets of syntheses within snapshots. Let's denote with *reference snapshot* $SN_i(t_0)$ the snapshot of a patient in healthy conditions after being discharged. The reference snapshot represents a baseline for all patients in normal health conditions given F_{SN_i}, ϕ_i and ρ_{SN_i}. Data relevance at time t is based on the computation of *distance* between the set of syntheses $\mathcal{S}_i(t)$, contained in the snapshot $SN_i(t)$ and $\mathcal{S}_i(t_0)$, in the reference snapshot $SN_i(t_0)$. Roughly speaking, relevance techniques allow to identify what are the syntheses that changed over time (namely, appeared, merged or removed) for a specific combination of F_{SN_i}, ϕ_i and ρ_{SN_i}. By detecting these changes, it is possible to focus in advance the attention of doctors on relevant syntheses that are approaching anomalous conditions, also enabling the prompt identification of unusual conditions on monitored patients, where warning or error bounds have not been defined yet, solely based on the notion of *data relevance*. We define *Relevant Patients Groups* the groups containing at least one relevant patient.

5   Patients Monitoring Service

Exploration of relevant patients groups is performed on top of the MDM presented in Sect. 3, in order for doctors to restrict the search space while monitoring patients' health status. Exploration is performed in two main steps: (i) firstly, relevance-based anomaly detection is used to identify relevant groups to start the exploration from; (ii) therefore, the data organisation imposed by the MDM is exploited to further guide the exploration.

How the Exploration Starts. In case the doctor is willing to focus her analysis on a specific patients group, she can directly choose a facet, drawn from the set of facets Φ. Conversely, in the case the doctor has explicit, albeit not completely defined, exploration demands, instead of indicating a single exploration facet as before, she may specify a set d^r of desired dimension instances for the dimensions she is interested in, where $d^r = \{v_{d_1^r}, \dots, v_{d_p^r}\}$ and $v_{d_i^r} \in Dom(d_i^r)$. Let's denote with $\Phi^r \subseteq \Phi$ the set of corresponding patients groups. In both cases, the doctor can be supported in the selection by proposing her the patients groups in Φ (resp., Φ^r) identified as relevant. Moreover, relevant patients groups are ranked considering the percentage of patients presenting anomalies inside each group; in this respect, doctor's attention will be attracted towards those groups in Φ (resp., Φ^r), ranked with higher percentage values.

How the Exploration Goes On. Let $\Phi_{rel} \subseteq \Phi$ (resp., $\Phi_{rel} \subseteq \Phi^r$) be the set of relevant patients groups. The doctor is guided by the MDM in order to explore the groups in Φ_{rel}. According to Fig. 2, relevant patients groups at highest grouping levels (e.g., "*males* with *pneumonia* assuming *immunosuppressant*") are proposed first. Starting from them, the doctor may split facets into composing dimensions, moving towards lower grouping levels. For example, starting from the patients group mentioned above, the doctor may inspect the percentage of relevant patients among *males*, among those affected by *pneumonia* and among those assuming *immunosuppressant*. This may help identifying facets and

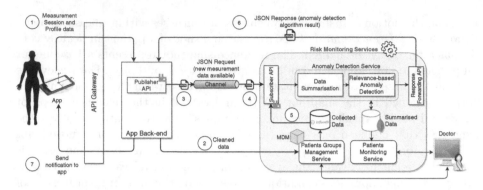

Fig. 4. Architecture of the proposed approach.

dimensions that are correlated the most with SARS-CoV-2 episodes, thus further increasing the knowledge on this pandemic phenomenon.

Once a relevant patients group has been selected in the set Φ_{rel}, the doctor may continue the exploration by adopting different strategies. On the one hand, she may decide to focus her attention on a specific patient, trying to diagnose the event that led to the anomalous situation (for instance, warning detected on features related to the acceleration over the three axes may be a symptom of shortness of breath). On the other hand, the doctor may carry out a comparative analysis over different patients of the same group, devoted to discover why anomalies detected in a single patient are somehow recurring in other patients of the group (for instance, due to a genetic defect shared by patients).

6   Implementation

Risk Monitoring Services are developed on top of a Glassfish Server 4 Open Source Edition and exchange messages with the App Back-end leveraging a publisher/subscriber mechanism based on the RabbitMQ middleware. Data is collected within an InfluxDB time series database, where measures are organised according to measurement sessions and labelled with the dimensions of the MDM. Summarised Data, obtained through incremental clustering algorithm, is stored within a MongoDB database. The numbers on the arrows in Fig. 4 highlight the order of the interaction flow between modules and services. Once the measurement session is completed, the related data is sent to the App Back-end (1), where undergoes a cleansing process (2) and is enriched with a reference to the dimensions of the MDM by the Patients Groups Management Service, before being stored within the InfluxDB store. Then, the Publisher API emits on the channel a JSON file containing the information about the aforementioned measurement session (3). As a result, the Subscriber API is notified of the presence of new available measurement data to process (4), thus retrieving it from the Collected Data store (5), which is subsequently elaborated by the Anomaly Detection Service. The output of the Anomaly Detection Service is stored within

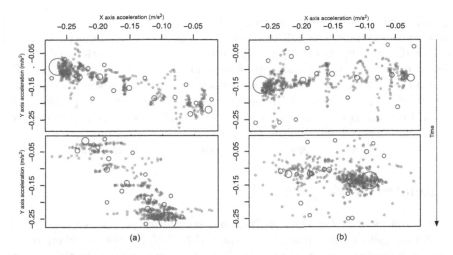

Fig. 5. Results of incremental clustering on discharged SARS-CoV-2 patients for controlled breath (a) and deep and short breath (b) phases.

the Summarised Data store and the Response Forwarding API sends back to the App Back-end possible alert messages, assembled as a JSON file containing the results (6), which in turn will be notified to the patient through the app (7).

Validation Strategy. Experiments are being conducted in order to assess: (i) the quality of data relevance evaluation techniques; (ii) processing time to verify whether summarisation and relevance evaluation techniques could face variable data acquisition rates and (iii) effectiveness of the data exploration techniques to properly attract the attention of experts on relevant patients groups. Preliminary results demonstrated that relevance-based anomaly detection techniques were capable of detecting injected variations (either gradual or sharp) in incoming data. Moreover, data summarisation and relevance-based anomaly detection could be efficiently carried out due to a thorough storage environment, ensuring processing time meeting high acquisition rates. Remarkably, relevance-based anomaly detection was particularly apt to ease data exploration, by identifying relevant snapshots, labelled with the facets of the MDM. Figure 5 shows the results of incremental clustering on a male patient data, with age in the [60, 75] range, on which worsening respiratory conditions have been detected. Visually, computed syntheses (red circles) are able to better detect a movement in acceleration values on the X axis, with respect to raw data (grey points), that are affected by variations and noise.

The validation of impact of the MDM on exploration will continue in the next months through an intense patients recruitment and testing. Recruitment will involve an initial sample of 50–100 discharged patients aged 18–75 years, 50% males with a SARS-CoV-2 diagnosis. App questionnaires for collecting profiling data from enrolled patients and patients inclusion/exclusion criteria are being defined. Exclusion criteria concern conditions affecting the capacity of the patient to provide informed consent and to use the app (e.g., physical and intellectual disability, dementia, current delusions or hallucinations). Selected patients will suffer of different kinds of respiratory and cardiac co-morbidities and assume different drugs such as immunosuppressant and anti-hypertensive therapies. For each group, a percentage of patients will be selected among smokers, as risk factor for SARS-CoV-2 episodes. GDPR procedures will be specified, in particular for data and contact tracing in remote medicine. Finally, usability experiments on the monitoring dashboard will be performed with the collaboration of different categories of doctors (e.g., general practitioners, medical researchers).

7   Related Work

In the recent months, Big Data analysis and exploration solutions in the health-care domain gained momentum, due to the SARS-CoV-2 outbreak, where remote monitoring services development is further being investigated. In this respect, we surveyed an excerpt of the literature focusing on how existing platforms and frameworks accomplish data exploration tasks properly addressing data volume and variety. The work presented in [5] suggest a web-based platform to interactively analyse patients flow data, to improve and optimise the quality of hospital services; herein, multi-perspective analysis of data is favoured by a thorough visualisation environment, employed to attract users' attention on data exceeding predefined thresholds. Conversely, the work proposed in [8] performs the integration of data by both hospital monitoring devices and mobile patients devices, as a telemedicine instrument to deliver more accurate diagnosis of patients. In the approach, data mining techniques are envisaged to perform risk analysis, disregarding data variety management techniques. Authors in [4] devise an OLAP-based system, through the creation of virtual analysis cubes, linked to multi-dimensional data exploration tools to support elderly care planning, but techniques to attract experts' attention during exploration are missing. The framework presented in [3] combines data summarisation (by means of clustering and classification), with a multi-dimensional model, to face the complexity of the showcased healthcare scenario. Lastly, the open-source framework proposed in [9] fosters an iterative clustering algorithm, paying attention on the feature engineering phase, coarsely describing exploratory aspects.

With respect to the surveyed literature, our proposal offers a comprehensive ecosystem of web services, employed in a real research project, to provide a solution for remote monitoring of the health status of discharged SARS-CoV-2 patients. To this aim, data relevance evaluation techniques allow, on the one hand, to detect anomaly triggers in patients' health status and, on the other hand, to attract doctors' attention on relevant groups only, thus reducing the search space while exploring data. In this respect, the MDM presented in this paper allows to set up proper exploration scenarios for supporting doctors.

8 Concluding Remarks

In this paper, we described our contribution in the scope of an ongoing, multi-disciplinary research project, designing an ecosystem of web services for remote monitoring of discharged SARS-CoV-2 patients. Services are meant to support doctors in the analysis of data collected through app by: (i) organising patients into groups according to features like sex, age, co-morbidities, prior therapies, to manage the variety of the domain; (ii) verifying patients' health status, relying on data summarisation and anomaly detection techniques and (iii) promptly notifying doctors when the values of monitored parameters diverge from reference values in the same group of patients. Future work will be devoted to perform an in-depth experimentation of Risk Monitoring Services, complying with the validation strategy, also considering other application contexts (e.g., sport, rehabilitation). Moreover, personalisation aspects in the exploration of data will be introduced, thus enabling doctors with different roles to monitor a subset of patients groups more directly related to their data analysis interests.

References

1. Bagozi, A., Bianchini, D., De Antonellis, V., Garda, M., Marini, A.: A relevance-based approach for big data exploration. Future Gener. Comput. Syst. **101**, 51–69 (2019)
2. Banerjee, A., Chakraborty, C., Kumar, A., Biswas, D.: Emerging trends in IoT and big data analytics for biomedical and health care technologies. In: Handbook of Data Science Approaches for Biomedical Engineering, pp. 121–152 (2020)
3. Bochicchio, M., Cuzzocrea, A., Vaira, L.: A big data analytics framework for supporting multidimensional mining over big healthcare data. In: Proceedings of 15th IEEE International Conference on Machine Learning and Applications (ICMLA 2016), Anaheim, California, USA, pp. 508–513 (2016)
4. da Cruz, H.F., Gebhardt, M., Becher, F., Schapranow, M.P.: Interactive data exploration supporting elderly care planning. In: Proceedings of 10th International Conference on eHealth, Telemedicine, and Social Medicine (eTELEMED 2018), Rome, Italy, p. 68 (2018)
5. Domova, V., Sander-Tavallaey, S.: Visualization for quality healthcare: patient flow exploration. In: 2019 IEEE International Conference on Big Data (Big Data 2019), Los Angeles, California, USA, pp. 1072–1079 (2019)
6. Golfarelli, M., Rizzi, S.: Data Warehouse Design: Modern Principles and Methodologies. McGraw-Hill, New York (2009)
7. Malasinghe, L.P., Ramzan, N., Dahal, K.: Remote patient monitoring: a comprehensive study. J. Ambient Intell. Humaniz. Comput. **10**(1), 57–76 (2017). https://doi.org/10.1007/s12652-017-0598-x

8. Opaliński, A., et al.: Medical data exploration based on the heterogeneous data sources aggregation system. In: Proceedings of 14th Federated Conference on Computer Science and Information Systems (FedCSIS 2019), Leipzig, Germany, pp. 591–597 (2019)

9. Rao, A.R., Clarke, D.: An open-source framework for the interactive exploration of Big Data: applications in understanding health care. In: 2017 International Joint Conference on Neural Networks (IJCNN 2017), Anchorage, Alaska, USA, pp. 1641–1648 (2017)

10. Shah, G., Shah, A., Shah, M.: Panacea of challenges in real-world application of big data analytics in healthcare sector. J. Data Inf. Manag. 1(3), 107–116 (2019). https://doi.org/10.1007/s42488-019-00010-1

UAVFog-Assisted Data-Driven Disaster Response: Architecture, Use Case, and Challenges

Xianglin Wei[1]([✉]), Li Li[2], Chaogang Tang[3], and Suresh Subramaniam[2]

[1] National University of Defense Technology, Nanjing 210007, China
wei_xianglin@163.com
[2] Department of Electrical and Computer Engineering,
George Washington University, Washington, DC 20052, USA
{lili1986,suresh}@gwu.edu
[3] School of Computer Science and Technology,
China University of Mining and Technology, Xuzhou 221116, China
cgtang@cumt.edu.cn

Abstract. It is a critical but difficult task to provide information transmission and computation services to first responders or rescue teams in disaster-hit areas as catastrophes may cause casualties and massive damage to human-made facilities. Unmanned aerial vehicles (UAVs) are a great choice to provide these services in such areas due to their inherent mobility and easy-to-deploy properties. This paper proposes a four-layer data-driven disaster response architecture, which can leverage UAV-carried Fog (UAVFog) nodes' communication and computation capabilities as well as deep learning's ability to extract mission-critical information from sensory data. In our proposal, UAVs are in charge of sensing and interconnecting disaster-hit areas. On the other hand, UAVFog nodes handle the data processing and decision-making tasks. We identify the functions and entities in each layer, and four key advantages of the proposed architecture are presented. The structure of the UAVFog node is detailed, and its technical requirements are summarized. Then, an injury diagnosis scenario is presented as a motivating use case with performance analysis. Finally, several open issues for future research are highlighted.

Keywords: Disaster relief · Fog computing · Information service · Unmanned aerial vehicle

1 Introduction

Natural disasters, such as earthquakes, tsunamis, forest fires, and floods, happen almost every day on earth, causing huge loss of lives and property. Many natural disasters would cause massive damage to human-made facilities, such as buildings, road networks, communications towers, power supply networks, etc.

© Springer Nature Switzerland AG 2020
Z. Huang et al. (Eds.): WISE 2020, LNCS 12343, pp. 591–606, 2020.
https://doi.org/10.1007/978-3-030-62008-0_41

To reduce the impact of disasters and save lives, timely and accurate disaster response and recovery are crucial. However, organizing effective and efficient disaster relief operations in the disaster-hit area is usually a challenging task, since the damaged or severely congested communications and computational infrastructures prevent first responders and rescue teams from sensing, sharing, and processing the data in a timely manner.

Post-disaster restoration of communications infrastructure, such as cellular networks, is time consuming and expensive, and one may resort to TETRA (Terrestrial Trunked Radio), walkie-talkie, satellite, or mobile ad hoc networks for disaster response. However, in general, the data rate of TETRA is low – walkie-talkie is mainly used for transmitting voice rather than data; the bandwidth of satellite links is relatively small and its transmission delay is too long to be useful; and the coverage of wireless ad hoc networks is very limited. In recent years, inspired by successes in a broad range of applications, Unmanned Aerial Vehicles (UAVs)-assisted aerial communication has been adopted to enhance public safety communications in disaster scenarios [1,2]. However, these designs only concentrated on the data transmission part and did not consider the computational needs, which play an equally important role in disaster relief operations; because, nowadays, many disaster response applications need to incorporate machine learning or deep learning algorithms that usually have high time and space complexity and are infeasible for end-user devices. Besides, how to extract critical information for disaster relief operations from the collected raw data has attracted little attention.

With this need as motivation, we propose a novel UAVFog-assisted data-driven disaster response architecture in this paper. In our architecture, UAVs are used for sensing, tracking, and connecting disaster-hit areas, and UAV-carried fog (UAVFog) nodes are in charge of handling data analysis tasks using machine learning algorithms to extract critical information in a smart manner. Our architecture includes four layers from the perspective of both the function and data domains, and the entities and data involved in each layer are identified. The structure of a UAVFog node is detailed with technical demands analysis. A use case, in which six different applications, i.e., image recognition, route plan, injury diagnosis, resources allocation, vehicle navigation, map construction, are involved, is introduced to demonstrate the performance of our architecture with different parameter settings. Finally, we discuss several open issues in the design and implementation of the proposed architecture.

2   Related Work

Recovering communication services in post-disaster areas has long been a challenging task due to the massive damages caused. A rapidly deployable, reliable, and mission-critical network is fundamental for the communication system during the recovery period. Many efforts have been made to set up such systems and prevent the occurrences of agnostic zones in disaster areas. IEEE USA MOVE project provides short-term communication, power supply, and network access to victims when natural disasters happen [3]. Arbia et al. discussed an Internet-of-Things (IoT)-based end-to-end emergency and disaster relief system [4]. Lu et al. proposed a data transmission network by connecting smart phones in disaster recovery [5].

UAVs equipped with communication devices, including 4G base stations, WiFi Access Points (APs), or ad hoc network equipments, are used to provide instant mobile/wireless network coverage. Zhou et al. suggested using UAVs to assist the interconnection of multiple ground vehicle networks in disaster rescue [6]. Erdelj et al. discussed the role of UAVs in different types of disasters [7]. In [8], Król et al. presented a UAV-assisted architecture that solves the communication outage issues caused by floods. UAVs or vehicles are adopted as communication relays that enable information sharing between clients and servers in [9]. Deruyck et al. have presented a UAV-assisted emergency network for large-area disasters [1]. Hayajneh et al. analyzed the performance of UAV-enabled disaster recovery cellular networks [2]. AT&T is developing a large, helicopter-like drone known as the "Flying COW", which can stay in the air 24-hours-a-day, and was deployed in Puerto Rico to provide emergency 4G coverage in the aftermath of Hurricane Maria. Zhao et al. have presented a multi-hop UAV relaying-base emergency networks framework in disaster scenarios [10]. Tang et al. have utilized the computation resources on the vehicles interconnected by UAVs for task processing in the disaster-hit [11]. A persistent monitoring architecture is presented by Noguchi et al. to conduct monitoring tasks in the disaster-hit areas [12].

These research works investigated disaster-relief operation-oriented service architectures from the viewpoint of communication. However, to our best knowledge, there is no unified service architecture that considers mobile devices' communication and the computational needs at the same time. Cheng et al. have considered an air-ground Integrated mobile edge computing (MEC) architecture [13]. Yu et al. have presented a UAV-aided MEC architecture with a mmWave backhaul [14]. However, these papers focused on a general scenario. This motivates us to design an architecture that effectively combines UAV communication, fog computing, and machine learning modules in this paper.

3 Data-Driven Disaster Response

3.1 Vision, Challenges, and Motivation

Natural disasters can cause different levels of life loss or property damages. In this paper, we focus on scenarios with the following settings: 1) a few zones are isolated from others because of floods, damaged bridges, or any obstacles; 2) access points or radio towers have malfunctioned due to power outage or building collapse; 3) there are a significant number of injuries in separated areas; 4) a few pre/post-deployed sensors and IoT gateways exist in the disaster area; 5) unmanned aerial vehicles (UAVs) are utilized by first responders or rescue teams for tracking and surveillance purpose; 6) satellite coverage is available but with limited bandwidth and long propagation latency; 7) a number of portable computational devices, such as laptops or hand-held computers, may be carried by the rescuers.

In this scenario, data-driven analytics holds tremendous potential for assisting many aspects of disaster response, such as injury pinpointing, damage assessing, resources allocation, etc. At first, raw data, including text, photos, videos, medical data, weather data, etc., are taken by sensors in the disaster area. Second, data are transmitted to the computation units through transmission links. Then, decisions are made and executed based on information extracted by the computation units using both traditional and machine learning algorithms.

There are challenges in each aforementioned step. First, data are hard to obtain in communication-isolated disaster areas. Second, there may not be a transmission link available for data gathering due to damaged communications infrastructure. Third, data processing may be difficult because of the lack of computational recourse. To deal with the first challenge, UAVs with multiple sensors, such as cameras, Synthetic Aperture Radars (SAR), Global Positioning System (GPS), could be deployed on-demand to monitor the disaster area. To overcome the second challenge, the UAVs could be equipped with wireless relaying radios that interconnect isolated units. The last challenge is mitigated by UAVFog nodes that carry computational units and are capable to process the data closer to where the data originates.

3.2 Architecture and Benefits

The data-driven disaster management architecture is illustrated in Fig. 1. On the right-hand of Fig. 1, four layers, i.e., Data Generation layer, Data Communication layer, Data Processing layer, and Data Output layer, are identified from the perspective of data domain. For each layer, its involved entities (humans, devices, software modules, or applications) are shown on the left from the viewpoint of function domain. In the function domain, the architecture is also constituted by four layers: Sensing layer, Transmission layer, Computation layer, and Application layer. We will introduce the architecture from the perspective of the data domain in the following description.

Data Generation Layer. Data generation layer is dedicated to capture data, which involves different types of entities that generate different kinds of data. Possible data sources are shown in the sensing layer in the function domain, including humans, UAVs, IoT sensors, cameras, smart phones, recorders etc. Generated data may include water level, temperature, constituent of atmosphere, deformation of the buildings, victims' physical conditions and locations, disaster images, video clips, SAR data, topographic data, voice call, text message, damage data, etc. Besides, the geospatial information of the data is needed to facilitate a location-based service.

Data Communication Layer. The functionality of this layer is setting up the communication links and transmitting the data generated in the bottom layer to the data processors or requesters. The wireless devices that may be involved

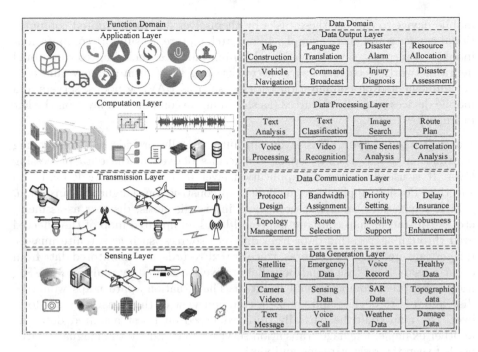

Fig. 1. The architecture of UAVFog-assisted Data-driven disaster response.

in this process include different types of sensors, IoT devices, wireless gateways or access points carried by UAVs, and wearable devices carried by victims or responders. Due to the heterogeneity of wireless devices, different types of communication links can be established. For example, Device-to-Device (D2D) links among wireless devices can be established for direct data sharing; Cellular/WiFi links could be set up between the smart phones and the cellular base stations/WiFi APs carried by UAVs; Sensors and IoT gateways can communicate through IoT links; UAVFog nodes communicate with each other via inter-vehicle links. To effectively utilize the available wireless links, attention should be paid to protocol design, bandwidth assignment, delay guarantees, topology management, route selection, mobility support, and robustness enhancement. A full realization of this vision may make the design of the communication module of the UAVFog nodes extremely challenging because of size, weight, cost, and energy supply limitations. Besides, for the purpose of providing priority-aware services for various types of network traffic, network middle boxes, such as load balancers, implemented by software or hardware may be adopted.

Data Processing Layer. This layer aims to derive information from the collected raw data in a timely manner, and plays the central role in the data-driven architecture. Text analysis and classification, route planning, image search, video recognition, audio processing, time series analysis and correlation analysis are typical functions of this layer. It is usually hard for an IoT or hand-held mobile device to execute these tasks by themselves due to its limited computational capacity and energy supply. In our architecture, to support computation-intensive functions, several machine learning modules, including convolutional neural networks, recurrent neural networks, long short-term memory, and etc., are implemented on computation entities, such as UAVFog nodes, servers and mobile devices with idle resources (as shown in the computation layer on the left in Fig. 1).

Data Output Layer. This layer can automatically provide decision-making information to the users based on the results derived in the data processing layer. It is usually called the application layer from the function domain perspective. Possible applications include map construction, language translation, vehicle navigation, command broadcast, injury diagnosis, disaster assessment, etc. For instance, map construction could build an updated map of the affected area through aggregating data from different sources, such as satellite images, images captured by UAV cameras, voice or text records, and monitored data from other sensors. This task could be collaboratively executed by multiple UAVFog nodes and other computational facilities. The situation-aware maps would be disseminated to devices and enable a common view of the disaster area. Moreover, the map could be partially updated based on latest acquired data. As another example, language translation service can eliminate the language barriers between the rescuer teams and victims.

We identify four key advantages of the proposed architecture:

1. From the sensing perspective, UAVs are deployed to enable a near-real-time and continuous surveillance of the disaster-hit area.
2. From the communication viewpoint, heterogeneous wireless devices can be interconnected or relayed by UAVs. This can connect the people and things together, and enable the flow of disaster-related data and information.
3. For computation and storage resources, UAVFog nodes are deployed near the data sources to enable near-source processing. They can process the collected data in disaster area to extract critical information. The short physical distances can greatly reduce the transmission latency.
4. Various disaster response applications, which can greatly enhance the efficiency of disaster relief operations in a smart manner, can be flexibly constructed or orchestrated based on data-driven analytics.

These advantages make the proposed architecture a necessary complement to many IoT application scenarios, such as smart cities, smart buildings, and intelligent transportation systems. It can provide information service in those areas where communication and computational facilities are damaged or in shortage.

4 UAVFog Node Design

UAVFog nodes play a key role in the implementation of the architecture shown in Fig. 1. This section presents the structure of a UAVFog node, and analyzes its technical requirements.

4.1 Structure

From the viewpoint of data generation layer, a UAVFog node carries heterogeneous sensors and actuators to serve a certain area. In data communication layer, a UAVFog node typically supports multiple mobile devices to collect and transmit data. To support processing functions listed in Fig. 1, in data processing layer, various machine learning algorithms should be implemented besides traditional analysis methods. As for data output layer, applications, application management, and user interface modules are designed in a UAVFog node. Besides, several management modules for security, storage, offloading task, heterogeneous connections are included.

The structure of a UAVFog node is illustrated in Fig. 2, which contains 10 modules.

- Communications Module. Wireless devices may implement different communication protocols. Interconnecting heterogeneous devices for data collection and sharing while fulfill the requirements including complexity, size, weight, and cost of the module, is very challenging. This module should support WiFi, LPWA (Low Power Wide Area), and cellular communication, which are the typical protocols used in end-user devices such as smart phones, laptops, cameras, and IoT sensors.
- Perception and Action Module. This module manages sensors, monitors, and actuators carried by the UAVs. Note that different types of sensors can be installed on a UAV on demand.
- Connection Management Module. This module is designed to manage communication components, as many wireless links may not be able to co-exist due to interference or may need to be shut down to save energy. Moreover, it could adaptively establish and select the most suitable data links for transmitting a particular message.
- Security Module. As a UAVFog node may store, transmit, or process sensitive identities and medical information, the security-related tasks, such as authentication, authority, and privacy, are processed by the security module.
- Data Storage Module. This module is in charge of data cleaning, fusion, and cache management. Outdated data are removed or transmitted to other specialized storage devices once the on-board storage space starts running out.
- Task Management Module. This module is responsible for processing incoming requests. It maintains a task queue and a scheduler, which schedules the offloaded applications or tasks based on their priorities, latencies and resource requirements.

- Data Processing Module. This module contains functional modules that are needed by various data output applications. Machine learning methods, such as decision-tree and Support Vector Machine (SVM), are deployed in this module to support diverse applications. Specifically, convolutional neural network and recurrent neural networks usually perform better than the conventional machine learning methods but with much higher complexity. They are included in a UAVFog node if the computational resource permits.
- Application Module. This module realizes disaster response-related applications, such as map construction, injury diagnosis, command broadcast, vehicle navigation, language translation, disaster assessment, resources allocation, etc.
- User Interface Module. This module helps users to make decisions by interacting with them and providing updated views of the disaster-hit area.
- Application Management Module. This module manages the applications on UAVFog nodes in an App Store-like manner; it can install, update, and uninstall applications as requested.

Note that a few modules or module components are optional and can be flexibly installed or removed on demand, such as satellite communication module and SAR.

4.2   Technical Requirements

The communication and computational requirements for a UAVFog node are determined by many factors, including the size of the served area, the expected number of served devices in the area, the required transmission bandwidth and the computation capability of each mobile device, and the latency constraint of offloaded tasks.

Denote n as the arrival rate of mobile devices in a UAVFog node's served area, and suppose it follows Poisson distribution. Let p be the offloading probability of a mobile device's task. The dataset sizes of the offloaded tasks are assumed to be exponentially distributed, and the average dataset size is s. Let c be the average amount of CPU cycles needed for processing per input bit, o be the average number of output bits for one input bit, f be the processing frequency of a virtual machine (VM) in the UAVFog node, m be the number of VMs, b_1/b_2 be the bandwidth of each mobile-to-UAVFog/UAVFog-to-mobile channel, and q be the number of channels. For simplicity, assume $b_2 = b$, and $b_1 = k \times b$, where $k \geq 1$ is a constant.

Based on the M/M/m queueing model, we can derive the average time each task experiences as follows [15]. First, the time spent in the transmission channel is $\frac{1}{\mu} + W$, where $\mu = \frac{k \times b}{s}$, and $W = \frac{\rho \times (q \times \rho)^q}{\lambda (1-\rho)^2 \times q! [\sum_{i=0}^{q-1} \frac{(q \times \rho)^i}{i!} + \frac{(q \times \rho)^q}{q!(1-\rho)}]}$. Here, $\lambda = n \times p$ and $\rho = \frac{\lambda}{\mu}$. Second, the processing time at the UAVFog node can be derived based on the same queueing model but with different service rate and number of servers. Figure 3 shows the average delay each task experiences with typical parameter settings. Here, n varies from 27 to 37; p varies from 0.6 to 0.8; b varies

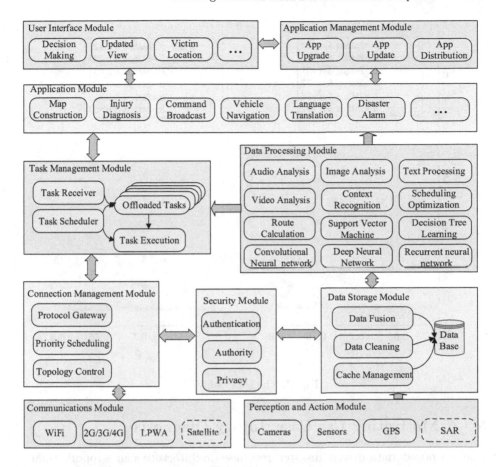

Fig. 2. UAVFog node structure.

from 2 Mbps to 5 Mbps; f, s, o, c, k, and q are set to be 2 GHz, 2 Mb, 0.1, 1000, 3, and 10 respectively. We can see that the delay of each task increases with increase of the offloading probability and device arrival rate. Moreover, by observing the two figures in Fig. 3, we conclude that large transmission bandwidth significantly reduces the task delay.

Based on this simple analysis, given the latency requirement of offloaded tasks, each UAVFog node can calculate the number of mobile devices that it can serve and the latency each task will experience based on other parameters obtained from historical data.

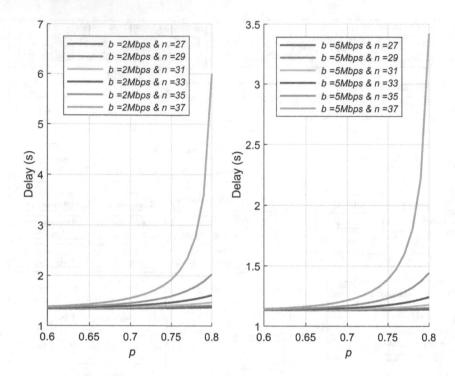

Fig. 3. The task delay.

5   A Motivating Use Case: Injury Diagnosis

The proposed data-driven disaster response architecture can support smart disaster-relief operations. This section takes injury diagnosis as a use case to show the workflow of the proposed architecture. The overall workflow of this use case is shown in Fig. 4.

In the studied case, we suppose a wounded person is found by a UAVFog node (U_1) which is equipped with a camera and can identify injury in captured photos using **image recognition** application (① and ② in Fig. 4). It could then notify the location of the wounded person, and transfer the photos to another UAVFog node (U_2) which connects to a nearby rescue team and could stay in the area for a while for further instruction (③ in Fig. 4). In step ④, U_2 sends the received information to the rescue team with the quickest path to the location calculated by the **route plan** application. After arriving at the location, the rescue team measures the medical data through wearable and portable devices in step ⑤. These data are transmitted to U_2 for a quick medical diagnosis in step ⑥. After receiving the data from the rescuer, U_2 starts the **Injury Diagnosis** application; the application calls the signal and image analysis functions in the Data Processing module, and gives out a preliminary diagnosis (⑦). These diagnosis results and corresponding treatment suggestions are then delivered back to the rescuer (⑧), and would help in the treatment for the wounded person (⑨).

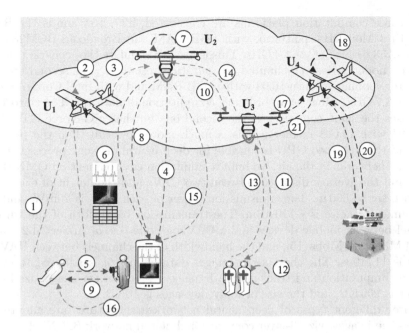

Fig. 4. A typical use case.

We can call this process a "quick response" because of its low latency compared to any kind of local processing. The wounded could get an immediate first-aid with the help of this process.

To get more professional diagnosis and treatment suggestions, U_2 sends the received data to another far-off UAVFog node U_3 (⑩), which maintains a connection with a remote medical team. Then, U_3 delivers the received data to the medical team (⑪), and gets professional treatment feedback (derived in step ⑫ by specialists) in step ⑬. This feedback is transmitted back via U_2 (⑭) to the rescue team (⑮) for potential further treatment according to the received guidance (⑯). This "remote diagnosis" process often incurs long delays but brings professional and accurate suggestions that are critical for treating the wounded.

The system could perform a series of "post-diagnosis" actions by the following procedures. U_3 collects rescue needs and sends related data to U_4 (⑰), which connects to a temporary hospital and ambulances. At the same time, U_4 derives the optimal **resources allocation** solutions, and the optimal path to the wounded using **vehicle navigation** and **map construction** applications in step (⑱). The calculated results are sent to the hospital in step (⑲), and the ambulance replies in step (⑳) to U_4. The reply and location of the ambulance are transmitted to U_3 (㉑) and finally to the rescue team to help them take further measures.

Prototype Evaluation. To evaluate the performance of the use case under different parameter settings, we have implemented a prototype. For each UAVFog node, three computation platforms are considered. The first one is the Raspberry Pi 3 Model B+ platform, with 1 GB RAM and Broadcom BCM2837B0, Cortex-A53 64-bit SoC @ 1.4 GHz. This is used to simulate the scenario that the UAVFog node carries very limited computation resources. The second one is a computer running Ubuntu 16.04 with 16 GB RAM and with Intel Core i7-8700K CPU @ 3.7 GHz, which is an ordinary computational platform. The third platform uses the same computer as the second one but has a GeForce GTX 1080 Ti GPU with 11 GB memory. This is much more powerful than the previous one, and the equipped GPU is suitable for deep neural network processing. For the UAV networks in the air, we built a simulation environment on OMNeT++ 4.4.1, and the average distance between two UAVs and the height of each UAV are both set to 500 m. The transmission power of each UAV is 20mW, and the environmental noise is −110 dBm. The transmission bandwidth of the wireless channel between mobile devices and a UAVFog node (i.e. b_1 in Sect. 4.2) varies from 1 Mbps to 2 Mbps [15], and the bandwidth of the channels between UAVFog nodes is 11 Mbps. The chest x-ray images dataset in [16] is adopted to evaluate the computation and transmission latencies. The average image size in the dataset is 200 KB, and the size of reply messages is 10KB.

Three different types of deep neural network architectures are adopted to process the images, viz., 3-layer convolutional neural network (CNN), Inception V3, and 18-layer deep convolutional neural network (ResNet-18). After refinement, the test accuracies of these three networks are 87.5%, 96.31%, and 97.28%, respectively.

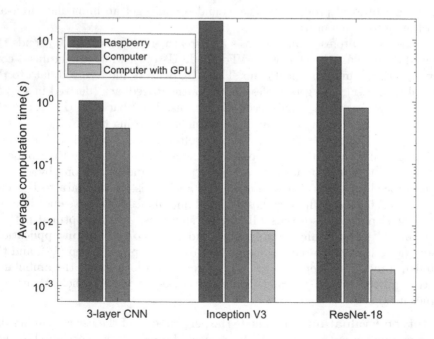

Fig. 5. The computation latencies of different cases on three computation platforms.

Figure 5 shows the computation time at the UAVFog node using three different computational platforms and three different neural network architectures. Note that a logarithmic scale is used in the y-axis in Fig. 5. As expected, the Raspberry Pi platform has the longest computation time, while the computation time of the GPU platform is negligible. The transmission latencies of three different cases, i.e. quick response, remote diagnosis, and post-diagnosis, are 3.93 (2.49), 19.18 (13.59), and 20.15 (16.41) seconds respectively when $b_1 = 1$ Mbps (2 Mbps). In short, the latency of quick response is significantly shorter than those of remote diagnosis and post-diagnosis process in all the cases.

6 Open Issues

Trajectory Planning for UAVFog Nodes. To cover a larger disaster-hit area collaboratively, the trajectories of all UAVFog nodes should be planned carefully. First, the entire area of interest should be covered or served by at least one UAVFog node when needed. Second, the energy consumption of a UAVFog node's planned trajectory should not exceed its power supply. Third, a negotiation scheme between neighbor UAVFog nodes is needed to process mobility, handover, and robustness issues. Reinforcement learning and deep reinforcement learning are promising solutions to tackle the trajectory optimization in disaster-response scenario. But how to achieve the tradeoff between complexity and performance needs further investigation.

Heterogeneous Data Fusion and Filtering. Heterogeneous data may be sent to a UAVFog node by diverse data sources. Moreover, collected raw data may be transmitted to other UAVFog nodes for further processing. In this circumstance, a lightweight heterogeneous data fusion and filtering method is necessary because of the limited energy supply and computational capacity of the UAVFog node.

Heterogeneous Network Interconnection. Various wireless devices are interconnected in a heterogeneous UAVFog node-relayed network, in which the gateway is carried by UAVFog nodes. This configuration implies that a UAVFog node must support different types of wireless standards. This complexity will significantly increase the deployment and management challenges and costs. First, it will increase the volume, weight, and cost of a UAVFog node. Second, simultaneously managing the wireless links with different standards is challenging because interference may occur among the wireless channels.

Energy Supply for the UAVFog Nodes. Sufficient power supply is critical for the continuous operation of UAVFog nodes. Helicopter-like large UAVFog node can rely on their self-carried fuels for electricity supply; for small drones that only carry cameras or smartphone-scale processing devices, a combination of rechargeable batteries and energy harvesting-based green energy may be a feasible solution.

Efficient Data Sharing. Searching and sharing the informative data among various disaster response units is a challenging task, because of the variety of data requests. For example, the medical specialists need the location and the physical condition of the wounded, while the rescue team wants to know the best route to its destination. A content-based request/reply matching scheme is suitable in this scenario. However, the implementation of such a comprehensive system raises many challenges such as content labeling, cache management, and semantic location-based service.

Resource Management at the UAVFog Nodes. Compared to powerful data centers, UAVFog nodes often have limited storage and computational capacities. To cope with many kinds of offloaded tasks from mobile devices, an efficient resource management and scheduling scheme is critical. The main design challenge is the trade-off between the latency constraint of offloaded tasks and energy consumption. Moreover, the radio resources should be carefully allocated to the mobile devices due to the limited bandwidth. A joint communication and computation optimization framework is established by Zhang et al. in [17]. But this area deserves much attention in the future.

Flexible Application Deployment. Disaster response-oriented applications play a key role in the data-driven architecture. Different disasters may require different types of applications. Therefore, it is very difficult to determine all possible applications on UAVFog nodes and deploy them on the mobile devices in advance. A middleware module that provides various functional modules and supports flexible application development is necessary for UAVFog nodes.

Data Security and Privacy Protection. In disaster response applications, sensitive medical data needs to be transmitted among UAVFog nodes. A lightweight data encryption and search scheme is needed in this circumstance to protect user's privacy.

Fault Tolerance. A UAVFog nodes may fail during the service process due to accidents such as power fault. Simultaneously maintaining redundant connections from an end-user device to multiple UAVFog nodes would alleviate this problem. But this solution requires a dense UAVFog node deployment which may not be feasible due to cost constraint. A fault-tolerant scheme is desired for enhancing the reliability of the proposed architecture.

7 Conclusion

In this paper, a four-layer UAVFog-assisted data-driven disaster response architecture is presented that can effectively combine the UAVs' flexibility and fog node's data processing capability. The functionalities and involved entities in each layer are identified and detailed. Then, a UAVFog node's structure is provided with technical demand analysis. Injury diagnosis is presented as a motivating use case with prototype experiments, in which three different disaster relief operations are considered. Finally, we highlighted a few open issues that may inspire future research.

References

1. Deruyck, M., Wyckmans, J., Joseph, W., Martens, L.: Designing UAV-aided emergency networks for large-scale disaster scenarios. EURASIP J. Wirel. Commun. Netw. **2018**(1), 1–12 (2018). https://doi.org/10.1186/s13638-018-1091-8
2. Hayajneh, A.M., Zaidi, S.A.R., McLernon, D.C., Di Renzo, M., Ghogho, M.: Performance analysis of UAV enabled disaster recovery networks: a stochastic geometric framework based on cluster processes. IEEE Access **6**, 26215–26230 (2018)
3. Conrad, J.M., et al.: The IEEE modular MOVE disaster relief project. In: 2017 IEEE Global Humanitarian Technology Conference (GHTC), pp. 1–6, October 2017
4. Arbia, D.B., Alam, M.M., Kadri, A., Hamida, E.B., Attia, R.: Enhanced IoT-based end-to-end emergency and disaster relief system. J. Sens. Actuator Netw. **6**(3), 19 (2017)
5. Lu, Z., Cao, G., La Porta, T.: TeamPhone: networking smartphones for disaster recovery. IEEE Trans. Mob. Comput. **16**(12), 3554–3567 (2017)
6. Zhou, Y., Cheng, N., Lu, N., Shen, X.S.: Multi-UAV-aided networks: aerial-ground cooperative vehicular networking architecture. IEEE Veh. Technol. Mag. **10**(4), 36–44 (2015)
7. Erdelj, M., Natalizio, E., Chowdhury, K.R., Akyildiz, I.F.: Help from the sky: leveraging UAVs for disaster management. IEEE Pervasive Comput. **16**(1), 24–32 (2017)
8. Król, M., Natalizio, E., Zema, N.R.: Tag-based data exchange in disaster relief scenarios. In: 2017 International Conference on Computing, Networking and Communications (ICNC), pp. 1068–1072, January 2017
9. Li, P., Miyazaki, T., Wang, K., Guo, S., Zhuang, W.: Vehicle-assist resilient information and network system for disaster management. IEEE Trans. Emerg. Top. Comput. **5**(3), 438–448 (2017)
10. Zhao, N., Lu, W., Sheng, M., Chen, Y., Tang, J., Yu, F.R., Wong, K.: UAV-assisted emergency networks in disasters. IEEE Wirel. Commun. **26**(1), 45–51 (2019)
11. Tang, C., Zhu, C., Wei, X., Peng, H., Wang, Y.: Integration of UAV and fog-enabled vehicle: application in post-disaster relief. In: 2019 IEEE 25th International Conference on Parallel and Distributed Systems (ICPADS), pp. 548–555. IEEE (2019)
12. Noguchi, T., Komiya, Y.: Persistent cooperative monitoring system of disaster areas using UAV networks. In: 2019 IEEE SmartWorld, Ubiquitous Intelligence Computing, Advanced Trusted Computing, Scalable Computing Communications, Cloud Big Data Computing, Internet of People and Smart City Innovation (SmartWorld/SCALCOM/UIC/ATC/CBDCom/IOP/SCI), pp. 1595–1600 (2019)
13. Cheng, N., et al.: Air-ground integrated mobile edge networks: architecture, challenges, and opportunities. IEEE Commun. Mag. **56**(8), 26–32 (2018)
14. Yu, Y., Bu, X., Yang, K., Yang, H., Han, Z.: UAV-aided low latency mobile edge computing with mmWave backhaul. In: ICC 2019–2019 IEEE International Conference on Communications (ICC), pp. 1–7, May 2019
15. Wei, X., Tang, C., Fan, J., Subramaniam, S.: Joint optimization of energy consumption and delay in cloud-to-thing continuum. IEEE Internet Things J. **6**(2), 2325–2337 (2019)

16. Kermany, D.S., et al.: Identifying medical diagnoses and treatable diseases by image-based deep learning. Cell **172**(5), 1122–1131 (2018)
17. Zhang, Q., Chen, J., Ji, L., Feng, Z., Han, Z., Chen, Z.: Response delay optimization in mobile edge computing enabled UAV swarm. IEEE Trans. Veh. Technol. **69**(3), 3280–3295 (2020)

Correction to: Fair Outlier Detection

Deepak P. and Savitha Sam Abraham

Correction to:
Chapter "Fair Outlier Detection" in: Z. Huang et al. (Eds.):
Web Information Systems Engineering – WISE 2020,
LNCS 12343, https://doi.org/10.1007/978-3-030-62008-0_31

The original version of this chapter was revised. The first author's name was corrected to Deepak P.

The updated version of this chapter can be found at
https://doi.org/10.1007/978-3-030-62008-0_31

Author Index

Printed in the United States
By Bookmasters